FRONTIER RESEARCH AND INNOVATION IN OPTOELECTRONICS TECHNOLOGY AND INDUSTRY

T0174028

PROCEEDINGS OF THE 11TH INTERNATIONAL SYMPOSIUM ON PHOTONICS AND OPTOELECTRONICS (SOPO 2018), KUNMING, CHINA, 18–20 AUGUST 2018

Frontier Research and Innovation in Optoelectronics Technology and Industry

Edited by

Khaled Habib
Kuwait Institute for Scientific Research (KISR), Kuwait

Elfed Lewis
University of Limerick, UK

CRC Press
Taylor & Francis Group
Boca Raton London New York

CRC Press is an imprint of the
Taylor & Francis Group, an **informa** business

A BALKEMA BOOK

Published by:
CRC Press/Balkema
P.O. Box 447, 2300 AK Leiden, The Netherlands
e-mail: Pub.NL@taylorandfrancis.com
www.crcpress.com – www.taylorandfrancis.com

First issued in paperback 2020

© 2019 by Taylor & Francis Group, LLC
CRC Press/Balkema is an imprint of the Taylor & Francis Group, an informa business

No claim to original U.S. Government works

ISBN 13: 978-0-367-73256-1 (pbk)
ISBN 13: 978-1-138-33178-5 (hbk)

This book contains information obtained from authentic and highly regarded sources. Reasonable efforts have been made to publish reliable data and information, but the author and publisher cannot assume responsibility for the validity of all materials or the consequences of their use. The authors and publishers have attempted to trace the copyright holders of all material reproduced in this publication and apologize to copyright holders if permission to publish in this form has not been obtained. If any copyright material has not been acknowledged please write and let us know so we may rectify in any future reprint.

Except as permitted under U.S. Copyright Law, no part of this book may be reprinted, reproduced, transmitted, or utilized in any form by any electronic, mechanical, or other means, now known or hereafter invented, including photocopying, microfilming, and recording, or in any information storage or retrieval system, without written permission from the publishers.

For permission to photocopy or use material electronically from this work, please access www.copyright.com (http://www.copyright.com/) or contact the Copyright Clearance Center, Inc. (CCC), 222 Rosewood Drive, Danvers, MA 01923, 978-750-8400. CCC is a not-for-profit organization that provides licenses and registration for a variety of users. For organizations that have been granted a photocopy license by the CCC, a separate system of payment has been arranged.

Trademark Notice: Product or corporate names may be trademarks or registered trademarks, and are used only for identification and explanation without intent to infringe.

Visit the Taylor & Francis Web site at
http://www.taylorandfrancis.com

and the CRC Press Web site at
http://www.crcpress.com

Typeset by V Publishing Solutions Pvt Ltd., Chennai, India

Library of Congress Cataloging-in-Publication Data

Names: Habib, Khaled, editor. | Lewis, E. (Elfed), editor.
Title: Frontier research and innovation in optoelectronics technology and industry/ Khaled Habib Kuwait Institute for Scientific Research (KISR), Kuwait, Elfed Lewis, University of Limerick, UK, editors.
Description: London, UK : CRC Press/Balkema, an imprint of the Taylor & Francis Group, [2019] | Includes bibliographical references and index.
Identifiers: LCCN 2018049107 (print) | LCCN 2018050842 (ebook) | ISBN 9780429447082 (ebook) | ISBN 9781138331785 (hardcover : alk. paper)
Subjects: LCSH: Optoelectronic devices—Research.
Classification: LCC TA1750 (ebook) | LCC TA1750 .F73 2019 (print) | DDC 621.3815/2—dc23
LC record available at https://lccn.loc.gov/2018049107

Frontier Research and Innovation in Optoelectronics Technology and Industry – Habib & Lewis (Eds)
© 2019 Taylor & Francis Group, London, ISBN 978-1-138-33178-5

Table of contents

Frontier Research and Innovation in Optoelectronics Technology and Industry – Habib & Lewis (Eds)
© 2019 Taylor & Francis Group, London, ISBN 978-1-138-33178-5

Preface

With the development of technology, a great variety of research results in the area of Photonics and Optoelectronics are emerging. By bringing together the high-quality technical content, influential speakers worldwide and latest technologies, SOPO2018 is to serve as a platform for exchanging information on recent advances and future trends for researchers and their peers. In the past nine years, more than thousand distinguished experts and scholars from all over the word shared their fresh and creative ideas with their colleagues via this platform. The proceedings consist of papers covering issues on Laser Technology and Applications, Optical Communications, Optoelectronic Devices and Integration, Energy Harvesting, Medical and Biological Applications and Image Processing. The proceedings provide details beyond what is possible to include in an oral presentation and constitute a concise but timely medium for the dissemination of recent research results.

Editors
2018

Image processing

Improving the accuracy of stripe center extraction in a structured light measurement system

B. Chen, Y.J. Zhang & J.R. Zhang
School of Mechanical Engineering, Xi'an Jiaotong University, Xi'an, Shaanxi, China

ABSTRACT: The center extraction of a stripe is an essential problem for the development of a structured light 3D measurement system. To obtain the center coordinates efficiently and accurately, a two-step method is proposed. Firstly, the grayscale degradation of a stripe image, which is caused by the curvature and reflectivity of the measured object, is eliminated by a grayscale adjustment method based on projecting another two uniform gray images. Then, a new self-adaptive threshold method is proposed based on the evaluation of the stripe asymmetry, and combined with the gray centroid method, the stripe centers are extracted. Experimental results demonstrate that the grayscale adjustment and center extraction method in this paper are effective and reliable.

1 INTRODUCTION

The structured light measurement method with stripe pattern projection has gained many application domains in industry for 3D shape reconstruction, precision inspection, and in other areas (D'Apuzzo, 2006a, 2006b; Song et al., 2005). The principle is to compute the measured object profile's world coordinates via the image coordinates of the stripe center. Therefore, the accuracy in extracting the stripe center from a captured image is a deciding factor of the measuring accuracy.

Up to now, many center extraction methods have been developed to obtain higher accuracy. The most common methods are those such as extreme value, gray centroid, curve fitting, Steger's algorithm, and their improved methods (Trucco et al., 1998a, 1998b; Steger, 1998). However, all of these methods to extract the center are based on the cross-section grayscale distribution of a captured stripe (Xue et al., 2014), which corresponds to a Gaussian profile. In fact, the grayscale distribution of the captured stripe is usually degraded by random noise, ambient light and the curvature or reflectivity of the measured object (Fisher & Naidu, 1996). Thus, errors are introduced into the center extraction results, which are bound to decrease the measuring accuracy.

Among those factors that degrade stripe grayscale distribution, the random noise can be removed by an appropriate filtering method (Ramachandra et al., 2014a, 2014b; Su et al., 2010), and the ambient light can be generally regarded as invariant for one scene. The measured object is the most uncertain part of the measurement system; its complicated surface and reflectivity are the main reasons for the degradation of the stripe grayscale distribution. Until now, few studies have been conducted on the elimination of the grayscale degradation that is caused by the measured object surface. Qi et al. (2017) proposed a stripe enhancement method to eliminate the influence of reflectivity, based on the phase-shift algorithm and the multi-exposure method, but in both cases there is a need to project and capture at least 12 images, which is too time-consuming. Hu et al. (2004) present a method to compensate the image bias that is caused by the surface curvature and surface normal, but it does not take the reflectivity into consideration, and it requires at least two cameras, which increases the system complexity.

In this paper, a method is put forward for improving the accuracy of stripe center extraction. The method is divided into two steps. The first step is to adjust the stripe gray distribution by projecting uniform grayscale images onto the measured object, which can eliminate the grayscale degradation of the captured stripe image. The second step is to calculate the coordinate of stripe center using an improved gray centroid method, based on a new self-adaptive threshold.

2 ADJUSTMENT OF STRIPE GRAYSCALE DISTRIBUTION

2.1 *Analysis of grayscale degradation*

The principle of structured light measurement is triangulation, as shown in Figure 1(a). In the measurement system, the surface of most measured objects is made up of flat or curved surfaces. Thus, in order to analyze the influence of a common measured surface on the grayscale distribution of captured stripes, a grayscale transfer model of structured light stripes, based on the theory of photometry, is simplified into a two-dimensional model, as shown in Figure 1(b). In this model, the projector is parallel to the X axis, and its projection direction is along the Y axis; the camera is parallel to the X' axis, and the capture direction of the camera is along the Y' axis. Therefore, α is the angle between the projector and the camera, and it is a constant during the measurement system. The curves S_1 and S_2 denote the grayscale profile of the projected stripe and captured stripe, respectively. Because the width of a single projected stripe is narrow, the local area of the measured object surface can be approximated by a circle with center at $O_c(x_o, y_o)$ and curvature radius r. The color of the object surface represents the reflectivity coefficient of the region, and P_k is the sudden change point of reflectivity.

Theoretically, the cross-section grayscale profile of the projected stripe has a Gaussian distribution. Its intensity I_p can be expressed as

$$I_p(x) = A \cdot \frac{1}{\sqrt{2\pi}\sigma} \cdot \exp\left[-\frac{(x-Xc)^2}{2\sigma^2}\right] \tag{1}$$

where A is the amplitude of the projected stripe grayscale, and σ is the standard deviation, which is related to the width of stripe. Xc denotes the X axis coordinate of the projected stripe center.

According to the Figure 1(b), for an arbitrary point X'_s on the X' axis, it intersects with the measured object surface along the captured direction, and the intersection point is $P_s(x_{ps}, y_{ps})$. At the P_s point, θ_s is the projection angle, and β_s is the capture angle. The cosine of θ_s and β_s can be expressed as

$$\begin{cases} \cos(\theta_s) = \dfrac{\sqrt{r^2 - (x_{Ps} - x_0)^2}}{r} \\ \cos(\beta_s) = \dfrac{|1 + k_1 k_2|}{\sqrt{(k_1 - k_2)^2 + (1 + k_1 k_2)^2}} \end{cases} \tag{2}$$

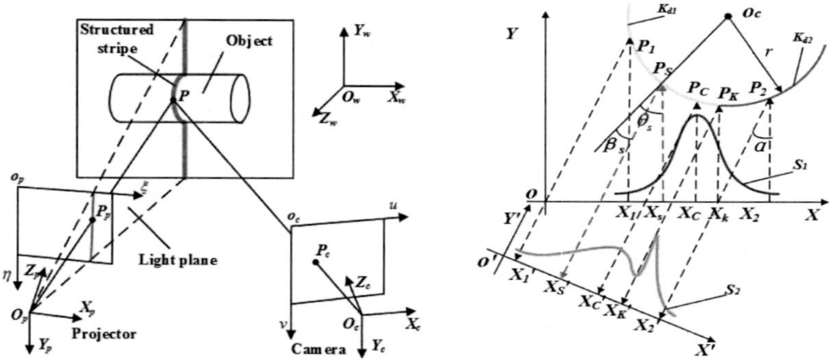

Figure 1.   Geometric model: (a) the measurement system; (b) grayscale transfer model of single stripe.

4

where k_1 is the slope of line O_cP_s, namely the slope of the normal at P_s, and $k_1 = (y_0 - y_{ps})/(x_o - x_{ps})$. k_2 is the slope of line $P_sX'_s$, and $k_2 = \cot(\alpha)$, which is a constant.

Because the width of the projected stripe is relatively narrow, the attenuation of the stripe grayscale on the cross-section can be ignored. According to the Lambert diffuse reflection model (Ikeuchi & Katsushi), the intensity of X'_s can be calculated as

$$\begin{cases} I_c(X'_s) = \dfrac{K_{d1} \cdot A}{\sqrt{2\pi}\sigma} \cdot \exp\left[-\dfrac{(x_{Ps} - Xc)^2}{2\sigma^2}\right] \cdot \dfrac{\cos(\theta_s)}{\cos(\beta_s)}, x_{Ps} < X_k \\ I_c(X'_s) = \dfrac{K_{d2} \cdot A}{\sqrt{2\pi}\sigma} \cdot \exp\left[-\dfrac{(x_{Ps} - Xc)^2}{2\sigma^2}\right] \cdot \dfrac{\cos(\theta_s)}{\cos(\beta_s)}, x_{Ps} > X_k \end{cases} \quad (3)$$

where I_c is the grayscale of the captured stripe, and K_{d1} and K_{d2} are reflectivity coefficients of different color areas on the measured object, as shown in Figure 1(b). X_k is the X axis coordinate of P_k.

According to Equation 3, the grayscale distribution of the captured stripe is modulated by the reflectivity coefficient of the object surface, the projection angle θ and the capture angle β, which are all related to the surface properties and the spatial location of the measured object.

2.2 Grayscale adjustment method

The objective of grayscale adjustment is to eliminate the degradation of the stripe gray distribution caused by the measured object surface. Based on Figure 1(a), P is an arbitrary point on the measured object surface, and $P_p(\xi_p, \eta_p)$ and $P_c(u_p, v_p)$ are the corresponding image coordinates of the projector and camera, respectively. Assume that K_{dp}, θ_P, β_P are the reflectivity coefficient, projection angle and capture angle of point P, respectively. According to Equation 3, we define the grayscale degradation coefficient as

$$f(K_{dp}, \theta_P, \beta_P) = K_{dp} \cdot \frac{\cos(\theta_P)}{\cos(\beta_P)} \quad (4)$$

Then, let $I(\xi_P, \eta_P)$ be the grayscale of P_p, $I(u_P, v_P)$ is the grayscale of P_c. In the process of the actual measurement, we take the influence of ambient light into consideration, and combine Equation 3 with Equation 4. The relationship between $I(u_P, v_P)$ and $I(\xi_P, \eta_P)$ can then be written as

$$I(u_P, v_P) = f(K_{dp}, \theta_P, \beta_P) \cdot I(\xi_P, \eta_P) + I_a(u_P, v_P) \quad (5)$$

where $I_a(u_P, v_P)$ is the intensity of ambient light at P_c, which is decided by the measuring environment. Generally, the ambient light of ambient light is constant, and can be easily eliminated by background subtraction.

Theoretically, with a known reflectivity coefficient, projection angle and capture angle at every point of the measured object, the grayscale degradation coefficient can be calculated directly using Equation 4. Then, the adjustment of the grayscale is carried out by eliminating the grayscale degradation coefficient, which can be expressed as

$$I'(u_P, v_P) = \frac{I(u_P, v_P) - I_a(u_P, v_P)}{f(K_{dp}, \theta_P, \beta_P)} \quad (6)$$

where $I'(u_P, v_P)$ is the new grayscale of P_c after grayscale adjustment.

However, according to the analysis in Section 2.1, it is impossible to obtain those parameters of the grayscale degradation coefficient directly. However, once the spatial location between the measurement system and the measured object is fixed, the intensity relationship of each pixel between the projected image and the captured image is invariant. Therefore, we proposed a method to obtain the grayscale degradation coefficient indirectly by projecting two uniform grayscale images onto the object. The uniform grayscale image means that the grayscales of each pixel are equal, namely $I(\xi, \eta) = C$, where C is constant and $C \in (0,255)$. The proposed grayscale adjustment method can be described as follows:

Step 1: Fix the measurement system and the environment. Then project two uniform grayscale images and one stripe image onto the measured object. The intensity of the two uniform grayscale images are C_1 and C_2, and $C_1 \neq C_2$. Assume that the captured images are I_1, I_2 and I_3, which can be expressed as

$$I_1(u_i,v_i)=f(K_{di},\theta_i,\beta_i)\cdot C_1 + I_a(u_i,v_i), \quad i=1,2,...N \tag{7}$$

$$I_2(u_i,v_i)=f(K_{di},\theta_i,\beta_i)\cdot C_2 + I_a(u_i,v_i), \quad i=1,2,...N \tag{8}$$

$$I_3(u_i,v_i)=f(K_{di},\theta_i,\beta_i)\cdot I(\xi_i,\eta_i) + I_a(u_i,v_i), \quad i=1,2,...N \tag{9}$$

where N denotes the number of measuring points, and (u_i, v_i) and (ξ_i,η_i) are the corresponding capture and projection image coordinates of the i'th measuring point, respectively. $f(K_{di},\theta_i,\beta_i)$ denotes the intensity degradation coefficient of the i'th measuring point.

Step 2: Calculate the grayscale degradation coefficient and ambient light term, point by point. According to Equation 8 and Equation 9, with known C_1 and C_2, we can get

$$f(K_{di},\theta_i,\beta_i)=\frac{I_1(u_i,v_i)-I_2(u_i,v_i)}{C_1-C_2}, \quad i=1,2,...N \tag{10}$$

After obtaining the grayscale degradation coefficient, according to Equation 8, the ambient light term can be expressed as

$$I_a(u_i,v_i)=I_1(u_i,v_i)-\frac{I_1(u_i,v_i)-I_2(u_i,v_i)}{C_1-C_2}\cdot C_1, \quad i=1,2,...N \tag{11}$$

Step 3: Adjust the grayscale distribution of the captured image I_3. Let the obtained intensity degradation coefficient and ambient light term into Equation 6, point by point. The adjustment process can be expressed as

$$I_3'(u_i,v_i)=\frac{I_3(u_i,v_i)-I_1(u_i,v_i)}{I_1(u_i,v_i)-I_2(u_i,v_i)}\cdot(C_1-C_2)+C_1, \quad i=1,2,...N \tag{12}$$

where I_3' denotes the new stripe image after grayscale adjustment. In order to avoid the quantization effect of the CCD camera, a Gauss filter is used to smooth the adjustment image I_3'.

3 EXTRACTION OF STRIPE CENTERS

With the adjusted stripe image, accurate center extraction can be conducted directly by using Steger's method (Steger, 1998). However, its computation efficiency is too low. In this section, an improved gray centroid method based on self-adaptive threshold is proposed. The algorithm details would be discussed as follows.

According to Figure 1(b), the grayscale distribution of captured stripes is not symmetric; namely, $|X_1'X'c| \neq |X'cX_2'|$. In order to take this kind of deterioration into consideration, an asymmetric degree D related to the threshold is defined. As shown in Figure 2, the computational procedure of D can be described as: (1) Let T denote the threshold, intersecting T with the grayscale profile; then get two intersection points P_L and P_R. (2) Find the peak point P_{MAX} of the grayscale profile, and calculate the midpoint P_M of P_L and P_R; then connect P_M and P_{MAX} to form a line l. (3) Calculate the angle α between l and the cross-section, and $\alpha \in (0, 90°)$. (4) Calculate the asymmetric degree $D = 90 - \alpha$. Therefore, the asymmetric degree describes the migration degree of the grayscale part which is larger than T.

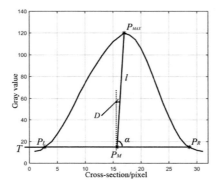

Figure 2. Calculation model of asymmetric degree.

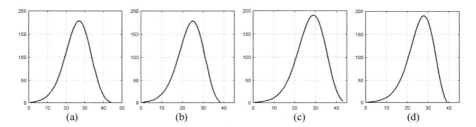

Figure 3. Simulation of asymmetrical stripe: (a) $r = 40$ mm, $\alpha = 30°$; (b) $r = 40$ mm, $\alpha = 40$; (c) $r = 30$ mm, $\alpha = 30°$; (d) $r = 30$ mm, $\alpha = 40°$.

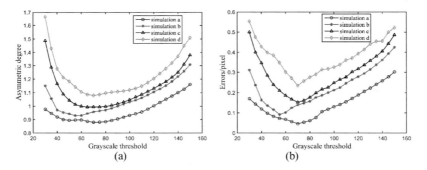

Figure 4. Analysis of simulation stripes: (a) asymmetric degree curves; (b) center extraction errors.

To illustrate the variation of asymmetrical stripe, a simulation base on the model in Figure 1(b) is carried out. During this simulation, the influence of reflectivity, cosine ratio of θ and β, is not considered. By changing the parameters r and α, four simulation stripes are obtained, as shown in Figure 3. Then, set the initial grayscale threshold to 25, and increase the grayscale threshold gradually with a step length of 5, until the grayscale threshold is 150, giving a threshold sequence $T_i(i = 1,2,3...)$. Subsequently, calculate the asymmetric degree of each simulation stripe at every grayscale threshold, and an asymmetric degree sequence $D_i(i = 1,2,3...)$ is obtained. Figure 4(a) shows the asymmetric degree curves along the direction of gray threshold.

For each threshold T_i in the threshold sequence, take out the grayscale part of the stripe which is larger than T_i. Then calculate the gray centroid of this grayscale part, which can be expressed as

$$C_i = \sum_{j=1}^{M_i} I_j \times x_j \Big/ \sum_{j=1}^{M_i} I_j \qquad (13)$$

where C_i is the stripe center when the threshold is T_i, x_j is the pixel coordinate of the cross-section, I_j is the gray value, and M_i is the number of the pixels whose gray value is larger than T_i.

The ideal stripe center is known during the simulation. Therefore, for every simulation stripe in Figure 3, the center extraction error of each threshold can be calculated. Figure 4(b) shows the error curves of the simulation stripes. Figure 4(a) and Figure 4(b) indicate that the optimal threshold for calculating the stripe center corresponds to the minimum asymmetric degree. Thus, for the adjusted stripe image I', the proposed gray centroid method based on self-adaptive threshold can be described as follows:

Step 1: Find the maximum grayscale G_{max} on a stripe cross-section. Initialize the threshold to $0.1G_{max}$; then gradually increase with a step length of 5, and calculate the asymmetric degree D, until the threshold is over $0.9G_{max}$.

Step 2: Choose the optimal threshold T_{opt} by finding the minimum asymmetric degree D_{min}; then keep the grayscale part which is larger than T_{opt}, and calculate the stripe center coordinate using the gray centroid method.

Step 3: Repeat steps 1 and 2, until all the stripe centers in the captured image are extracted.

4 EXPERIMENTAL RESULTS

In-laboratory experiments were completed to verify the validity of the proposed adjustment and center extraction method. The experimental system includes a DLP projector (model: Optoma DN322) with 1024×768 resolution, and a camera (model: Point Grey GX-FW-28-S5C/M-C) with 1600×1200 resolution. In this paper, two experiments are carried out, which are described as follows.

The first measured object is a black-and-white flat board. The projected images include a Gaussian stripe image, and two uniform grayscale images whose intensity are 160 and 200, respectively. Figure 5 shows the captured images corresponding to the projected images. As Figure 5(a) shows, the captured stripe image was seriously degraded by the reflectivity; the red lines in the image are some typical stripe cross-sections, which are low saturation stripe, distortion stripe and normal stripe in turn.

Combined with Figure 5(b) and Figure 5(c), the stripe image is processed by the proposed adjustment method. Figure 6 shows the grayscale distribution at the red lines before and after adjustment. Figure 6(a) shows that the grayscale features of the low saturation stripe are very vague, so that they are easy to be taken as background. After the adjustment, the contrast of the stripe is enhanced, which is beneficial to the center extraction. Figure 6(b) shows the distortion stripe caused by reflectivity mutation, and after the adjustment the distortion is eliminated. Figure 6(c) is the normal stripe with good quality, so its grayscale distribution before and after

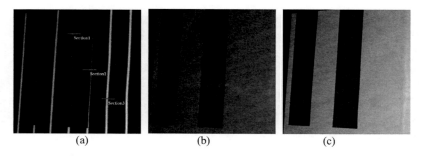

Figure 5. Captured images: (a) stripe image; (b) uniform gray image $[C_1 = 160]$; (c) uniform gray image $[C_2 = 200]$.

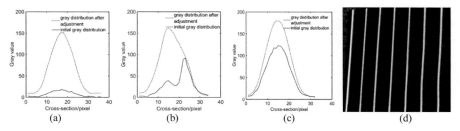

Figure 6. The grayscale distributions and the adjusted image: (a) low saturation stripe; (b) distortion stripe; (c) normal stripe; (d) adjusted stripe image.

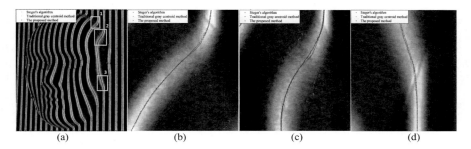

Figure 7. Results of center extraction: (a) full image; (b) to (d) are local enlarged images of rectangular regions 1 to 3 in (a).

Table 1. Comparison of the three methods.

Comparison parameter	Steger's algorithm	Traditional gray centroid method	The proposed method
Time costs/s	7.6912	0.2635	0.7311
Number of centers	21,188	18,169	17,806

adjustment shows no dramatic changes. Figure 6(d) is the adjusted stripe image, and compared with Figure 5(a), the overall quality of image is improved, and the grayscale of the stripes has no mutation along the direction of its extension. Therefore, the results indicate that the proposed adjustment method can solve the problems of low saturation and distortion effectively.

Subsequently, to verify the reliability of the proposed center extraction method, experiments on a complex face model are carried out. Figure 7(a) is the result of center extraction after grayscale adjustment. The red points, green points and blue points are stripe centers, extracted by Steger's algorithm, the traditional gray centroid method and the proposed method, respectively. The time costs of the three methods and the number of centers are shown in Table 1.

From the results of center extraction, Steger's algorithm can accurately extract the stripe center in the interior of the stripes. However, due to the Gauss convolution, there are a lot of interference points at the end of the stripes, and the operation is time-consuming. The traditional gray centroid method has the fastest extraction speed, but there are many noise points, because the threshold selection is not ideal. The proposed method has a relatively high accuracy without any noise points, and its processing efficiency can meet the requirement of real-time extraction.

Figures 7(b) to (d) are some local enlarged images from Figure 7(a). In order to quantitatively analyze the accuracy of the proposed method, for every local enlarged image, the center points of Steger's algorithm are fitted by cubic spline. Then, the distance between the center points are extracted by another two methods and the fitted curve is calculated, respectively. The error is then defined as the distance we have calculated. Table 2 shows the average (Ave.) error and the standard deviation (Std.) of errors. The results indicate that the accuracy of the proposed center extraction method is very close to that of Steger's algorithm.

Table 2. Statistical results of extraction error.

Extraction method	Rectangular region 1		Rectangular region 2		Rectangular region 3	
	Ave./pixel	Std./pixel	Ave./pixel	Std./pixel	Ave./pixel	Std./pixel
Gray centroid method	1.0595	0.7844	1.8464	1.0126	1.0312	0.6385
Proposed method	0.3275	0.2164	0.7964	0.4436	0.3103	0.1965

5 CONCLUSION

In this paper, the process of stripe grayscale degradation is analyzed by a photometric model, and a grayscale adjustment method is proposed. Subsequently, an asymmetric evaluation model of stripe cross-section is established, and an improved gray centroid method based on self-adaptive threshold is proposed. Experimental results show that the grayscale degradation of a captured image caused by the curvature and reflectivity of measured object is eliminated by the proposed grayscale adjustment method, and the accuracy of stripe center extraction is improved in the case of high extraction efficiency. In future works, for the application of more complex stripe images, we plan to consider the extension direction of stripes when extracting the center.

ACKNOWLEDGMENTS

This research is supported by Natural Foundation Research Project of Shaanxi Province (Grant No. 2017ZDJC-21) and National Science Foundation of China (Grant No. 51375377).

REFERENCES

D'Apuzzo, N. (2006), Overview of 3D surface digitization technologies in Europe. *Proceedings of SPIE—The International Society for Optical Engineering, 6056*, 5–13.

Fisher, R.B. & Naidu, D.K. (1996). A comparison of algorithms for subpixel peak detection. In *Image Technology* (pp. 385–404). Berlin, Germany: Springer, Berlin.

Hu, Q., Harding, K.G. & Hamilton, D. (2004). Image bias correction in structured light sensor. In *Two— and Three-Dimensional Vision Systems for Inspection, Control, and Metrology II* (pp. 117–123).

Ikeuchi, K. (2014). Lambertian reflectance. In *Encyclopedia of Computer Vision* (pp. 441–443). Springer.

Qi, Z., Wang, Z., Huang, J., Xue, Q. & Gao, J. (2017). Improving the quality of stripes in structured-light three-dimensional profile measurement. *Optical Engineering, 56*(3), 031208.

Ramachandra, V., Nash, J., Atanassov, K. & Goma, S. (2014). Structured light 3D depth map enhancement and gesture recognition using image content adaptive filtering. In *Computational Imaging XII*, 902005.

Song, L, Qu, X., Yang, Y., Chen, Y. & Ye, S. (2005). Application of structured lighting sensor for online measurement. *Optics and Lasers in Engineering, 43*(10), 1118–1126.

Steger, C. (1998). An unbiased detector of curvilinear structures. *IEEE Transactions on Pattern Analysis and Machine Intelligence, 20*(2), 113–125.

Su, W.H., Lee, C.K. & Lee, C.W. (2010). Noise reduction for fringe analysis using the empirical mode decomposition with the generalized analysis model. *Optics and Lasers in Engineering, 48*(2), 212–217.

Trucco, E., Fisher, R.B., Fitzgibbon, A.W. & Naidu, D.K. (1998). Calibration, data consistency and model acquisition with laser stripers. *International Journal of Computer Integrated Manufacturing, 11*(4), 293–310.

Xue, Q., Wang, Z., Huang, J. & Gao, J. (2014). Improving the measuring accuracy of structured light measurement system. *Optical Engineering, 53*(11), 112204.

Frontier Research and Innovation in Optoelectronics Technology and Industry – Habib & Lewis (Eds)
© 2019 Taylor & Francis Group, London, ISBN 978-1-138-33178-5

Deformation measurement of tooth-locked end closure flange of ultra high pressure vessel using digital image correlation method

Xiao-Yong Liu, Qing-Yu Kong, Yu-Guang Hou, Qi Guo & Rong-Li Li
School of Mechatronic Engineering, Changchun University of Technology, China

Chang-Jiang Xu
College of Biological and Agricultural Engineering, Jilin University, China;

ABSTRACT: In Ultra High Pressure Vessel (UHPV), Tooth-Locked Quick-Actuating (TLQA) seal device is often applied as top end closure to satisfy the requirements on ultra high pressure and frequent quick open-and-close operation. The structural parameters of TLQA end closure determined based on analysis or simulation are theoretical values, which are usually used in the previous works. However, in reality, these parameters may not satisfy the requirements of engineering application enough. In this paper, Digital Image Correlation (DIC) method was used to measure deformation field of tooth-locked end closure flange of UHPV for food processing. The basic principle of in-plane displacement measurement by DIC method was briefly given. The full-field displacement and strain distribution of tooth-locked end closure flange were computed under different internal pressure levels. Experimental results reveal that DIC method can be applied to practical full-field deformation measurement of TLQA seal device.

1 INTRODUCTION

Ultra high pressure (UHP) technology has been widely accepted and commonly used in the field of food science, chemical, petrochemical industry, nuclear engineering (Li 2013, Majid 2009 & Avrithi 2007). As the main equipment in UHP technology, ultra high pressure vessels have been ever-increasing demand. Tooth-locked quick-actuating seal device is a typical, commonly-used top end closure of UHPV, which consists of a tooth-locked flange and a top cover. A typical top end closure of UHPV usually weights 10–30% of the whole vessel, and costs about 15–40% (Chen 2007). An ideal top end closure should provide sufficient strength and have the lightest weight.

It should be noted that, previous work mainly focused on finite element analysis (FEA) and engineering design methods of TLQA (Avrithi 2007, Chen 2007 & Ma 2008). In the literature, structure strength, contact process of the teeth and cross-section configuration of the tooth-locked flange were investigated based on FEA. Then, the engineering design methods of the tooth-locked flange were proposed based on the analysis or simulation results. However, based on the author's knowledge, quantitative deformation measurement of the tooth-locked flange has yet to be reported. Therefore, it is useful to obtain the deformation characterization of the tooth-locked flange in reality for engineering design.

As a non-contact optical technique, DIC (Liu 2012a, Pan 2009a & Liu 2012b) has more attractive advantages: (1) full-field measurement (compared with contact strain gauges (Taylor 2012)); (2) simple experimental setup and preparation (compared with ultrasonic (Bray 2002), electronic speckle pattern interferometry (ESPI) (Pan 2009a) and moiré interferometry (MI) (Pan 2009a)); (3) low requirements in measurement environment (compared with ESPI and MI). Therefore, the digital image correlation method is selected for measuring the deformation distributions of the tooth-locked flange in this study. First, a series of digital speckle images were acquired under different pressure levels. In the following, the DIC method was used to compute the deformation field and strain field of the tooth-locked flange, and the experimental results were presented here.

2 DIGITAL IMAGE CORRELATION METHOD

Digital image correlation is a popular non-contact optical metrology tool for deformation and strain measurements, which has been increasingly used in experimental mechanics and other scientific fields. In DIC method, the test object surface must have random speckle patterns, which deform together with the object surface during loading. The underlying principle of DIC method is schematically shown in Figure 1. A reference subset centered at the interesting point from the reference image is chosen to determine its corresponding location in the searching subset chosen from the deformed image. For the best evaluation of the similarity between the reference and deformed subsets, the following correlation function is most commonly used (Liu 2012b):

$$C(X) = \frac{\sum_{i=1}^{n} \left[f(x_i, y_i) - f_m \right] \times \left[g(x_i', y_i') - g_m \right]}{\sqrt{\sum_{i=1}^{n} \left[f(x_i, y_i) - f_m \right]^2} \sqrt{\sum_{i=1}^{n} \left[g(x_i', y_i') - g_m \right]^2}}$$ (1)

where $f(x_i, y_i)$ and $g(x_i', y_i')$ are the intensity values at (x_i, y_i) in the reference subset and (x_i', y_i') in the deformed subset respectively; f_m and g_m are the mean intensity values of the reference and deformed subsets; n denotes the number of pixels contained in the reference subset; X is the desired deformation vector. The relationship of the coordinates (x_i, y_i) and (x_i', y_i') commonly was expressed by first-order displacement mapping function.

To optimize equation (1), the following classic Newton-Raphson iteration method (N-R method) was used (Bruck 1989). In the computations, a bicubic spline interpolation scheme with an interpolation area 4×4 pixels is implemented to obtain the gray value and first-order gray gradient at sub-pixel locations.

After the iterative computations, the displacements and displacement gradients can be obtained at the current point. The displacements and displacement gradients of a next point can be obtained by changing the reference point. Therefore, the full-field displacements can be obtained by repeating the above-mentioned process. In this paper, a pointwise local least squares fitting (PLS) algorithm (Pan 2009b) was employed for full-field strain determination.

3 EXPERIMENT INFORMATION

3.1 *Ultra high pressure vessel*

A two-layer shrink-fit ultra high pressure vessel (WHGR 600-10 × 64) for food processing was used in this experiment. The maximum internal pressure was 600 MPa. The UHPV was of 100 mm in inner diameter. The geometrical configuration of UHPV was shown in Figure 2(a).

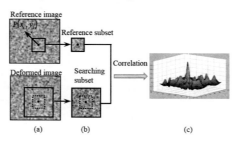

Figure 1. Principle of DIC; (a) two speckle images before and after deformation; (b) two subsets from the undeformed and deformed images; (c) distribution of the correlation coefficient.

Figure 2.   (a) Ultra high pressure vessel for food processing; (b) Top surface of end closure flange.

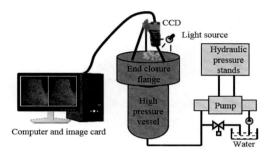

Figure 3.   Test system on ultra-high pressure vessel.

Tooth-locked quick-actuating structure is employed as the top end closure of UHPV. The tooth-locked flange connected with vessel by its thread, and connected with tooth cover by teeth. Equal teeth with same shape are distributed uniformly in circumferential direction at the end closure flange and the top cover, respectively. It can satisfy the requirements of quick open-and-close operation in frequent easily. The top view of tooth-locked flange is shown in Figure 2(b).

3.2   *Experimental setup*

Figure 3 illustrates the experimental setup for digital image correlation technology. A single CCD camera (MTV-1802CB) is placed on the top surface of end closure flange, and the optical axis is perpendicular to the surface. The CCD camera is used to record the surface images of the test specimen with a resolution of 795×596 pixels at 256 gray levels. The test surface was sprayed with a matt black paint, followed by a very fine dust of matt white paint. It is easy to match the image pairs of this pattern in digital image correlation method. The object surface is illuminated by a white-light source.

4   EXPERIMENTS

4.1   *Experimental procedure and experimental images*

In the whole test, the internal pressure of UHPV is gradually created by a pump feeding the water into the bottom of the vessel. The magnitude of internal pressure is recorded by a pressure gauge. The measurement zone is a single tooth surface of end closure flange. The procedure of deformation measurement of end closure flange involved in this paper is described below. First, the zoom lens of CCD camera is adjusted to let the region of interest fill the entire image. The magnification of the test system is 128.8 μm/pixel in this experiment.

Then, a reference speckle image at $P_0 = 0$ MPa was captured and recorded digitally as shown in Figure 4 (a). Subsequently, the internal pressure of UHPV was gradually increased by pump and a series of deformed speckle images are captured after the desired internal pressures are stable. In this experiment, 3 deformed images were consecutively obtained at

13

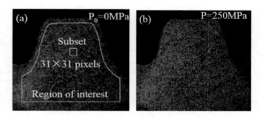

Figure 4. Speckle image of a tooth on end closure flange captured by imaging system at internal pressure of (a) $P_0 = 0$ MPa and (b) P = 250 MPa. The area enclosed by yellow lines in (a) is the region of interest where the deformation and strains are to be determined.

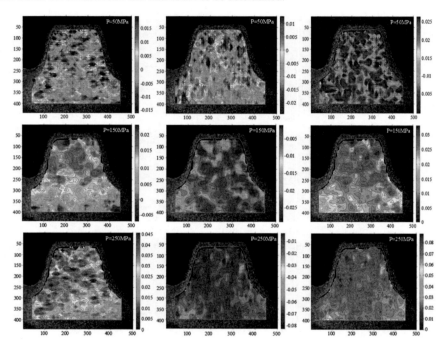

Figure 5. Displacement fields of test region under a pressure increment of 100 MPa (from 50 to 250 MPa), in millimeters, as indicated on the halftone bar. The first and second columns present the displacement fields in u (vertical) and v (horizontal), respectively. In the third column, the resultant displacements of test region are shown.

internal pressures from 50 MPa to 250 MPa with a pressure increment of 100 MPa at each step. Figure 4(b) show a typical speckle images for P = 250 MPa (P denotes the internal pressure of vessel). Finally, the speckle images are processed by the DIC technique to obtain the full-field displacements and strains under different internal pressure levels, which directly compares each deformed speckle image with the reference image.

In the analysis, the displacements were computed in the region of interest (shown in Figure 4(a)) with a reference subset size of 31×31 pixels and a grid step (i.e. distance between neighboring points) of 10 pixels. In total, the displacement vectors at 1554 ($= 37 \times 42$) discrete points were calculated. Afterwards, the strains with in region of interest were calculated by the PLS algorithm.

4.2   Experimental results and discussion

The full-field displacements of the test region of end closure flange under different pressure levels are computed using DIC software written with MATLAB language. Figure 5 shows the

14

u-field, v-field and resultant displacements under various internal pressures. From Figure 5, it can be seen that the distributions of displacement fields in u and v direction are non-homogenous. The main reason is that the nonuniform engaging status of teeth and the error of the image acquisition system result in a complex deformation state. Comparing of the distributions of displacement fields under different internal pressures, we can observe the displacements increasing with the internal pressure levels of UHPV. The maximum resultant displacement is less than 0.1 mm under the internal pressure of 250 MPa.

The correlation coefficient distribution of all measurement points in the region of interest under internal pressure 250 Mpa is given in Figure 6 due to the fact that these displacements are larger than others detected at other pressures. It can be seen form Figure 6 that the correlation coefficients are above 0.9 with a mean value of 0.9264. It indicates that the whole matching calculation is reliable and no evident decorrelation effect occurs. It means that the measured data are effective and credible.

Then, the strain fields were calculated by fitting local displacement fields using PLS algorithm. The computed normal strain in x and y direction as well as the shear strain under different pressure levels were presented in Figure 7. We can see that the strain fields are not fully harmoni-

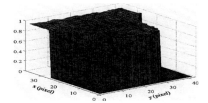

Figure 6.   Correlation coefficient map of test region at P = 250 MPa.

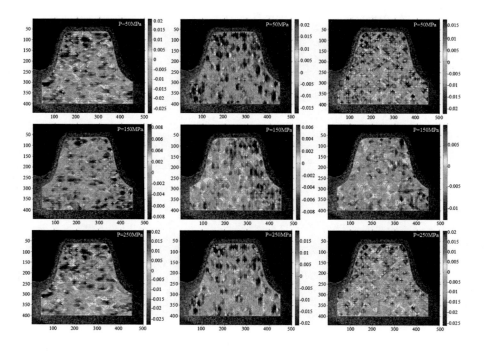

Figure 7.   Strain fields of test region at a pressure increment of 100 MPa (from 50 to 250 MPa). The three columns present the axial strain (εx) field, the circumferential strain (εy) field and the shear strain (γxy) fields, respectively.

15

ous due to the same reasons with the displacement measurement. We can also observe that the strain fields have not apparent strain concentrations appearing on the whole measuring surface. The maximum absolute value of ε_x, ε_y and γ_{xy} are less than 0.03, 0.025 and 0.03, respectively.

5 CONCLUSION

The measurement of tooth-locked flange deformation is a meaningful task in optimal design and materials selecting of TLQA seal device used in UHPV. In this work, a low-cost, non-contact full-field deformation measurement system based digital image correlation technique was successfully used to detect the displacement and strain fields of tooth-locked flange of TLQA structure. The full-field displacement and strain distributions of tooth-locked flange were acquired under various internal pressure levels. Results in this work reveal that the DIC method is capable of measuring full-field deformation of tooth-locked flange. In the future work, the DIC algorithm and measurement system will be investigated further to improve the measurement accuracy of TLQA. In addition, it is also worth noting that DIC method can be employed to measure other devices of UHPV, even more, to monitor the 3D full-field deformation of UHPV.

FUNDING

This work is funded by the Science and Technology Research Project in Education Department of Jilin Province during 13th Five-Year plan period (Grant No. JJKH20170558 KJ) and the National Natural Science Foundation of China (Grant Nos. 51605043 and 51505038).

REFERENCES

Avrithi, K., Bilal, M., Ayyub, A. 2007. Reliability-based approach for the design of nuclear piping for internal pressure. *Journal of Pressure Vessel Tech-ASME* 131(4): 041201–10.

Bray, D.E. 2002. Ultrasonic stress measurement and material characterization in pressure vessels, piping, and welds. *Journal of Pressure Vessel Tech-ASME* 124(3): 326–335.

Bruck, H.A., McNeil, S.R., Sutton, M.A., Peters, W.H. 1989. Digital image correlation using Newton-Raphson method of partial differential correlation. *Experimental Mechanics* 29(3): 261–267.

Chen, P., Qian, C.F., Zhang, Y.X. 2007. Design and Stress Analysis of a New Quick-opening Seal Device Connected by D-Shaped Shearing Bolts. *Journal of Pressure Vessel Tech-ASME* 129(3): 550–555.

Li, R.L., Liu, X.Y., Zhang, S.Q. 2013. The Design of Rigid Package Container with Telescopic Mechanism Used in Ultra High Pressure Food Processing. *Applied Mechanics and Materials* 303–306: 2744–2747.

Liu, X.Y., Tan, Q-C., Xiong, L., Liu, G.D. 2012a. A Subpixel Displacement Estimation Algorithm for Digital Image Correlation Based on a Nonlinear Intensity Change Model. *Lasers in Engineering* 23(1–2): 123–134.

Liu, X.Y., Tan, Q.C., Xiong, L., Liu, G.D., Liu, J.Y., Yang, X., Wang, C.Y. 2012b. Performance of iterative gradient-based algorithms with different intensity change models in digital image correlation. *Optics and Laser Technology* 44(4): 1060–1067.

Ma, X., Ning, Z.P., Chen, H.G., Zheng, J.Y. Modeling and optimization of ultra high pressure vessel with self-protective flat steel ribbons wound and tooth-locked quick-actuating end closure. In Artin A. Dermenjian (ed.), *High Pressure Technology; Nondestructive Evaluation Division; Student Paper Competition; Proc. ASME Pressure Vessel and Piping Division Conference, Chicago, 27–31 July 2008.*

Majid, M., Amir, H., Reza, K. 2009. Finite element analysis of deformation and fracture of an exploded gas cylinder. *Engineering Failure Analysis* 16(5): 1607–1615.

Pan, B., Asund, A., Xie, H.M., Gao, J.X. 2009b. Digital image correlation using iterative least squares and pointwise least squares for displacement field and strain field measurement. *Optics and Lasers in Engineering* 47(7–8): 865–874.

Pan, B., Qian, K., Xie, H.M., Asundi, A. 2009a. Two-dimensional digital image correlation for in-plane displacement and strain measurement: a review. *Measurement Science and Technology* 20(6): 1–17.

Taylor, D.J., Watkins, T.R., Hubbard, C.R., Hill, M.R., Meith, W.A. 2012. Residual stress measurements of explosively clad cylindrical pressure vessels. *Journal of Pressure Vessel Tech-ASME* 134(1): 011501–8.

Frontier Research and Innovation in Optoelectronics Technology and Industry – Habib & Lewis (Eds)
© 2019 Taylor & Francis Group, London, ISBN 978-1-138-33178-5

Single frame infrared image adaptive correction algorithm based on residual network

Xingang Mou, Junjie Lu, Xiao Zhou & Xuemin Wang
Wuhan University of Technology, Wuhan, China

ABSTRACT: The non-uniformity of the response between the detectors of the infrared detector is large, which seriously affects the imaging quality. And the detector response parameters will slowly drift, making it impossible to completely solve the non-uniformity problem through a single calibration method. Traditional adaptive correction algorithms require sufficient scene motion, otherwise there will be some problems such as degradation or non-convergence. This paper proposes a single-frame infrared image adaptive correction algorithm based on residual network. This method uses residual network to obtain the non-uniformity responsive model and brings in batch normalization to solve the problem of gradient disappearance. Compared with other several scene-based algorithms, this algorithm can achieve single-frame infrared image non-uniformity correction, solves the ghosting problem, and has a high degree of detail retention.

Keywords: non-uniformity; residual network; batch normalization

1 INTRODUCTION

The infrared imaging system's core device is the infrared focal plane array (IRFPA). IRFPA has a non-uniformity response which must be corrected during the application of infrared imaging systems. For the problem of non-uniformity of infrared images, there are mainly two major correction methods based on calibration and scene (Qian, C.H.E.N. 2013). The non-uniformity correction method based on calibration uses the uniform radiation blackbody images obtained at different temperatures. These images are calibrated to obtain the gain and bias of the detection unit response. Then, the expected value of the point to be calibrated is corrected by fitting calculation. The calibration method requires less calculation and the hardware is easy to implement, but it is not suitable for certain scenes which need to work continuously. To deal with the problem, people begin to study scene-based non-uniformity correction methods. The scene-based non-uniformity correction method means that the non-uniformity correction parameters of the infrared image are obtained through the scene. For example, the non-uniformity correction algorithm based on neural network and the non-uniformity correction algorithm based on image registration (Li, Q., Liu, S.Q., Wang, B.J., & Lai, R. 2007). Although those algorithms can satisfy the requirement of continuous operation, it is easy to produce ghost problems in scene and IRFPA motion process (Yu, Y., Li, L., Feng, H., Xu, Z., Li, Q., Chen, Y., & Hu, H. 2017).

In this paper, an adaptive algorithm of single frame infrared image based on residual network (He, K., Zhang, X., Ren, S., & Sun, J. 2016) is proposed for ghosting problems caused by scene-based non-uniformity correction. The method uses the original response of the detector in different working states and the corresponding scene thermal radiation estimation obtained by two-point correction as the training data. Then iterative training is carried out through convolution neural network and network parameters are adjusted continuously to obtain a detector non-uniformity correction model.

2 RESIDUAL NETWORK MODEL

The convolutional neural network has excellent performance in image processing. In this paper, our residual network is based on the adjustment of the deep residual network for Gaussian noise (Zhang, K., Zuo, W., Chen, Y., Meng, D., & Zhang, L. 2017). The network is mainly used for the removal of many types of noise, such as block-like effects due to Gaussian noise and JPEG compression, and image super-resolution tasks.

2.1 Sample data set production and packaging

After deeply studying the present situation of infrared detector correction technology and the engineering application of infrared detector non-uniformity correction method, this paper uses the original response data of infrared detector as sample data with non-uniformity response. Considering the imaging quality and the degree of approximation to the real scene thermal radiation, it is decided to adopt two-point correction results as the real scene thermal radiation in this chapter. The correction results are shown in Fig. 1.

This paper uses $10°C$ as the interval and uses two-point correction method to estimate the real heat radiation in five temperature regions to complete the preparation of the real scene thermal radiation training samples. In this paper, 800 frames of real thermal radiation and detector response data are selected. These data are labeled as real thermal radiation and detector response raw data.

We define the scale of scaling of the training data, which is 1.0,0.9,0.8,0.7. For each set of corresponding real scene thermal radiation and corresponding detector response, the tri-linear interpolation method is used to obtain their scaling results that enrich the data set. Finally, the data in the scaled data set is sampled and taken, and the feature block is prepared for the training of network learning. This chapter uses 40×40 scale, sampling step is 10 to produce the training data block. For the data with the input size of 320×240, the sampling step size is 10, and the sample size is 40×40. For the input size of 320×240, the sampled data block is 643712 blocks.

2.2 Residual network framework

The original response of the detector in this paper can be expressed as follows:

$$F(x) = x + R(x) \tag{1}$$

where $F(x)$ denotes the original response with non-uniformity, x denotes the corrected "true" response, and $R(x)$ denotes the non-uniformity response model. The previous non-uniformity correction algorithm based on neural network generally directly obtains the correction model. The proposed method uses the residual network to obtain $R(x)$, $R(x)$ will have fewer features than $F(x)$, also have better results. Fig. 2 shows the basic unit of residual network.

The network uses 17 convolution layers. Features are extracted in the first convolution layer, and then the activation output of the result is performed using the Rectified Linear Unit (ReLU). The later 15 convolution layers are connected to Batch Normalization. Batch Normalization is used to adopt the method of small sample processing in large samples. And

 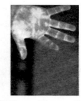

(a)original image   (b) Two-point correction result

Figure 1.   Results of two-point correction.

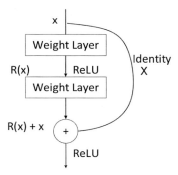

Figure 2.   Residual network basic unit.

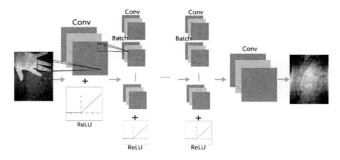

Figure 3.   Non-uniformity response predicts network structure.

it guarantees the convergence of learning and training, improves the training speed, and makes the convolution result nonlinear. The output result is reconstructed through the final convolution network. In this process, 64 convolution kernels of 3×3 scale are used to extract 64 characteristic planes. The network obtains the model of the distributed noise in the image by removing the standard image in the convolution process.

The predicted network structure of the nonuniform response model is shown in Fig. 3. The network is a feed-forward neural network.

2.3   *Loss function*

The input data of the proposed algorithm are the thermal radiation response estimation diagram and the original response diagram of infrared detector with non-uniformity. For the detector pixel, the degree of its non-uniformity response can be expressed as follows:

$$Diff_{ij} = \left| Nor_{ij} - Ori_{ij} \right| \tag{2}$$

where $Diff_{ij}$ denotes the difference between the pixel standard response and the non-uniformity response at the detector position (i, j), Nor_{ij} ndicates that the pixel at the detector position (i, j) receives the thermal radiation estimate and Ori_{ij} indicates the pixel non-uniformity response at the detector position (i, j). If $Diff_{ij}$ is larger, the non-uniformity of the pixel will be larger. And the greater the discomfort that the representation of the position will eventually bring to the observer in the infrared imaging system display device, the greater the difficulty for the subsequent target recognition and target tracking tasks. Non-uniformity response prediction network needs to complete the estimation task of infrared detector non-uniformity response model. In order to estimate the non-uniformity response model effectively, the real non-uniformity response of infrared detector is as close as possible. The mean square error between the estimated non-uniformity response and the real non-uniformity response is used to measure the learning quality of the network, and this criterion is used as the loss function of the deep neural network. Its specific expression is as follows:

$$l(\theta) = \frac{1}{2MN} \sum_{i=1}^{M} \sum_{j=1}^{N} R(\theta)_{ij} - Diff_{ij}^{2} \tag{3}$$

where θ denotes the residual error of non-uniformity response of depth neural network and $R(\theta)_{ij}$ indicates the non-uniformity residual response estimation of infrared detector position (i, j) under the trainable parameter.

3 TEST RESULTS

For the self-adaptive correction method of single-frame infrared image based on the convolutional neural network presented in this paper, the correction effect of the infrared detector non-uniformity response is demonstrated by the real infrared detector video sequence.

3.1 *The overall correction effect of an adaptive correction method for a single frame infrared image based on a convolutional neural network*

The overall correction result of the algorithm proposed in this paper for detector non-uniformity response is shown in Fig. 4. It can be seen from the figure that although there are some small non-uniformity responses of the scattered point distribution in the correction effect, the overall non-uniformity response of the infrared detector has been corrected. In this paper, the non-uniformity response model of predictive neural network is able to get by extracting a variety of features from the input data and implementing the nonlinearity of the network through the ReLU in the hidden layer, thereby obtaining the non-uniformity response of the detector.

3.2 *Single-frame correction capability based on adaptive correction of single-frame infrared image based on convolutional neural network*

The result that the algorithm proposed by this paper an algorithm compares with time-domain high-pass filtering non-uniformity correction (THPF-NUC) (Cao, X., Zhu, B., & Guo, L. 2013), airspace low-pass time-domain high-pass filtering non-uniformity correction (SLPF-NUC) (Qian, W., Chen, Q., & Gu, G. 2010), and bilateral filter time-domain high-

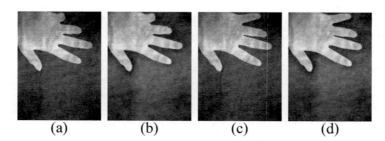

(a)   (b)   (c)   (d)

Figure 4. Correction results of the algorithm's response to the non-uniformity of the cooling detector. (a) The original response of the 1965 frame in the video sequence; (b) The correction effect of 1965 frame; (c) The original response of the 1967 frame in the video sequence; (d) The correction effect of 1967 frame.

pass filtering non-uniformity (BHPF-NUC) (Rossi, A., Diani, M., & Corsini, G. 2010) is shown in Figure 5. It can be seen from the figure that traditional scene-based non-uniformity correction methods take some time to reach convergence, and ghosts maybe appear during the entire non-uniformity correction process. By adjusting the parameters, the influence of ghosting can be properly reduced, but still can't fundamentally solve this problem. The algorithm in the paper can complete the non-uniformity correction of the infrared detector within a single frame, so there is no need to consider the convergence time and whether it can converge. It is fast and convenient. Secondly, since the non-uniformity response correction can be completed within a single frame, the problem of ghosting can be solved fundamentally.

3.3 *Local detail comparison of single-frame infrared image adaptive correction method based on convolutional neural network*

In this paper, the proposed algorithm compares with image block a priori single-frame infrared image adaptive calibration algorithm (SIAC-NUC) (Wang, J., Zhou, Z., Leng, H., & Wu, Q. 2015) and mid-infrared histogram equalization non-uniformity correction (MIHE-NUC) (Kang, C., Zhang, Q., & Zheng, Y. 2013) local amplification comparisons. The result is shown in Fig. 6.

From Fig. 6, (a1), (a2) and (a3) are the partial enlargement results of the same position in the respective correction results. (a1) shows that there are some obvious vertical stripe blocks. (a2) is not like the bar in (a1), but the non-uniformity of scatter distribution is serious. There is no bar in (a3) that exists in (a1), and the non-uniformity of the distribution of scattered dots is smaller than that in (a2). It is proved that the proposed algorithm is superior to the single-frame infrared image adaptive correction algorithm and intermediate infrared histogram equalization method for infrared detector non-uniformity response correction performance.

Besides, (b1) and (b2) and (b3) are also local zoom results at the same location in the correction results for different methods. From (b1) and (b2) in Fig. 6, it can be seen that there are some dark stripes in the correction results of various methods. This is the fixed pattern noise of infrared detectors. The fixed pattern noise is mainly caused by optical lens processing.

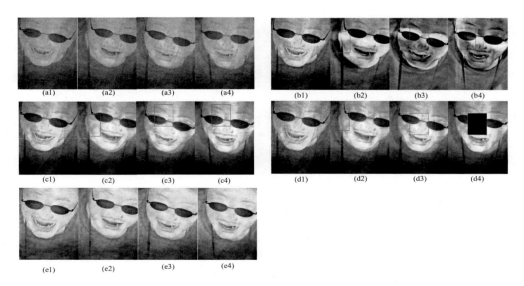

Figure 5. Comparison between the proposed algorithm and the traditional scene-based non-uniformity correction method. (a1)–(a4) are the original responses of the 10th, 50th, 150th, and 350th frames of the cooling detector video sequence respectively; (b1)–(b4) the correction results of THPF-NUC; (c1)–(c4) the correction results of SLPF-NUC; (d1)–(d4) the correction results of BHPF-NUC; (e1)–(e4) the correction results of the algorithm presented in this paper.

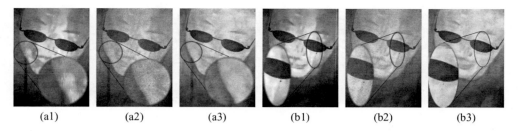

| (a1) | (a2) | (a3) | (b1) | (b2) | (b3) |

Figure 6. Comparison of the local correction effect of this algorithm and othersproposed in this paper. (a1), (b1) Partially magnified graphs of SIAC-NUC algorithm results (a2), (b2) Locally magnified graphs of MIHE-NUC algorithm results (a3), (b3) Partial magnified graphs of non-uniformity correction results of this algorithm.

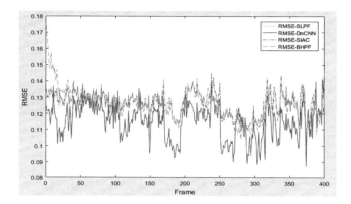

Figure 7. Root mean square error.

(b1) and (b2) both do not perform well in this fixed-pattern noise correction. It can be seen from (b3) that this stripe-fixed pattern non-uniformity is basically completed. The correction of the response shows that the correction ability of the algorithm for the fixed pattern non-uniformity response is superior to other algorithms.

3.4 *RMSE evaluation*

The Root Mean Square Error (RMSE) is used to evaluate the performance of the algorithm. The root mean square error formula is as follows:

$$\text{RMSE} = \sqrt{\frac{1}{MN} \sum_{i=1}^{M} \sum_{j=1}^{N} \left(\overline{Y}_{i,j} - Y_{i,j} \right)^2} \qquad (4)$$

In the formula, $\overline{Y}_{i,j}$ and $Y_{i,j}$ represent the correction result and the original noise map, respectively. If the value of RMSE is smaller, the correction effect will be better.

Fig. 7 compares the RMSE of SLPF-NUC, BHPF-NUC SIAC-NUC algorithm and the proposed algorithm of this paper (RMSE-DnCNNin Fig. 7 and the data is normalized by the results).

From Fig. 7, it can be seen that SLPF-NUC and BHPF-NUC algorithms initially have a decreasing RMSE during the entire video sequence. Because SLPF-NUC and BHPF-NUC algorithms require a certain number of frames of data for statistical iterative update, which obtains the appropriate non-uniformity correction parameters and gradually reaches the convergence level. Thought the entire process of SIAC-NUC has ups and downs, the overall trend is relatively stable. Single-frame infrared image self-adaptive correction method based on residual network has certain advantages in the expression of RMSE error compared with SLPF-NUC, BHPF-NUC, and SIAC-NUC. This illustrated that the correction result of this image adaptive correction method is closer to the "true" thermal radiation response of the detector.

4   CONCLUSION

At present, the single-frame infrared image adaptive correction algorithm based on residual network utilizes the "true" thermal radiation response estimation obtained by two-point correction, and the residual non-uniformity response model is obtained through residual network training to complete the calibration task. However, the network training sample set does not contain rich scenes due to limitations of experiment conditions and time. The next step is to enrich the training sample set and improve the ability to correct the response to the non-uniformity of infrared detectors. Secondly, network training efficiency is not high currently. We will try to modify the network, improve training efficiency, and reduce training time in the future.

ACKNOWLEDGMENT

This work is supported by the Fundamental Research Funds for the Central Universities (183204007), and by National Natural Science Foundation of China (61701357).

REFERENCES

Cao, X., Zhu, B., & Guo, L. 2013. Temporal High-pass Filtering Nonuniformity Correction with Adaptive Time Constant. *Opto-Electronic Engineering*, 40(7), 89–94.

He, K., Zhang, X., Ren, S., & Sun, J. 2016. Deep residual learning for image recognition. In *Proceedings of the IEEE conference on computer vision and pattern recognition* (pp. 770–778).

Kang, C., Zhang, Q., & Zheng, Y. 2013. Infrared image nonuniformity correction based on intermediate equilibrium histogram. *Laser and Infrared*, 43(11):1240–1242.

Li, Q., Liu, S.Q., Wang, B.J., & Lai, R. 2007. New nonuniformity correction algorithm for IRFPA based on neural network. *Infrared and Laser Engineering*, 36(3), 342.

Qian, C.H.E.N. 2013. The status and development trend of infrared image processing technology. *Infrared Technology*, 35(6), 311–318.

Qian, W., Chen, Q., & Gu, G. 2010. Space low-pass and temporal high-pass nonuniformity correction algorithm. *Optical review*, 17(1), 24–29.

Rossi, A., Diani, M., & Corsini, G. 2010. Bilateral filter-based adaptive nonuniformity correction for infrared focal-plane array systems. *Optical Engineering*, 49(5), 057003.

Wang, J., Zhou, Z., Leng, H., & Wu, Q. 2015. A non-uniformity correction algorithm based on single infrared image. *Journal of Information &Computational Science*, 12(1), 101–110.

Yu, Y., Li, L., Feng, H., Xu, Z., Li, Q., Chen, Y., & Hu, H. 2017. Adaptive neural network non-uniformity correction algorithm for infrared focal plane array based on bi-exponential edge-preserving smoother. In *AOPC 2017: Optical Sensing and Imaging Technology and Applications* (Vol. 10462, p. 1046215). International Society for Optics and Photonics.

Zhang, K., Zuo, W., Chen, Y., Meng, D., & Zhang, L. 2017. Beyond a gaussian denoiser: Residual learning of deep cnn for image denoising. *IEEE Transactions on Image Processing*, 26(7), 3142–3155.

Zuo, C., Chen, Q., Gu, G., & Qian, W. 2011. New temporal high-pass filter nonuniformity correction based on bilateral filter. *Optical Review*, 18(2), 197–202.

Frontier Research and Innovation in Optoelectronics Technology and Industry – Habib & Lewis (Eds)
© 2019 Taylor & Francis Group, London, ISBN 978-1-138-33178-5

Development of an efficient computer generated synthetic hologram method

X. Yang
Institute of Information Optics, Zhejiang Normal University, Jinhua, Zhejiang, China
Institute of Information Optics Engineering, Soochow University, Suzhou, Jiangsu, China
Hangzhou Lightin Inc., Hangzhou, Zhejiang, China

Y. Li, Q.H. Yan, H. Wang, F.Y. Xu & J.H. Zhang
Institute of Information Optics, Zhejiang Normal University, Jinhua, Zhejiang, China

H.B. Zhang
Department of Computer and Information Sciences, Virginia Military Institute, VA, USA

ABSTRACT: We present a simple method for large scale and high resolution computer generated synthetic hologram calculation. The frequencies of projected images of 3D object are known and mosaic in frequency domain to form the frequency of whole synthetic hologram without zero and conjugate order information. The synthetic hologram is the off-set value added to the real part of a two dimensional inverse Fourier transformation of sliced frequency. The results are evidences by one full-parallax synthetic hologram and synthetic color rainbow hologram at the resolution of 94340×94340 and size of 30 mm $\times$ 30 mm.

Keywords: Synthetic hologram, color rainbow hologram, computer generated hologram, Fourier transform, inverse Fourier transform

OCIS Codes: 090.0090; 090.1705; 090.1760; 090.5640

1 INTRODUCTION

A hologram can fully reproduce a wave-front of a 3D object. Embracing computing technology, it is possible to simulate the recording process of a hologram for digital hologram generation. Such a process is coined Computer Generated Holography (CGH) (Yoshikawa & Yamaguchi, 2009). Dynamic holographic 3D display (Qu et al., 2016) and large scale high resolution static holographic display have been extensively studied (Matsushima & Nakahara, 2010; Matsushima & Nakashrm, 2009). While progress has been made, challenges still exist. Due to the large pixel size and display panel of a Spatial Light Modulator (SLM) used for dynamic 3D display, the generated hologram using SLM has a limited viewing angle and field of view for practical applications. On the other hand, while the large size and high resolution holograms such as a Fresnel hologram or color rainbow hologram can produce very vivid 3D effects, the computation of a large scale high resolution hologram is computationally inefficient (Tsuchiyama & Matsushima, 2016; Yile et al., 2009). This inefficiency can be evidenced with an example: a fully computed holographic stereogram with the size of 20 mm $\times$ 20 mm and resolution of 20000×20000 requires about 32.9 hours (Hao & Yan, 2015).

We propose a simple method based on the frequency mosaic of frequencies of projected images without zero and conjugate order information to improve the computational efficiency of a synthetic hologram. The synthetic hologram relies on a Fourier transform in the frequency domain and only a fast Fourier transform is used for computation. Within the method, the calculation of the Fourier transform is only needed for the effective data areas, thus, being able to greatly reduce the computational time. We demonstrate that a synthetic hologram and color rainbow hologram with a size of 30 mm × 30 mm, at a resolution of 94340 × 94340 only needs 1.2 hours and 25 minutes respectively. The reconstructions are vivid in 3D and have a strong stereo effect.

2   METHODS

2.1   *The model of synthetic holography*

Figure 1(a) shows the reconstruction of the synthetic hologram, where P is one object point in space and A, B and C are the projected points of the object point P in the hologram plane.

In Figure 1(a), a small angle light wave is emitted from the points A, B and C. The light wave emitted by the points A and C can be perceived by the left and right eyes of the viewer, which can generate a stereo effect. The larger the light wave angle emitted by an object point, the greater the disparity of information it provides. Assume that the light emission angle θ of object point is divided into N_x parts along different viewing angle directions, and one viewing angle interval is $\Delta\theta_x = \theta_x/N$. The three-dimensional object is composed of a large number of object points, and each point has the same object light decomposition method. Figure 1(b) shows a projection diagram of points P_1 and P_2, and within it, their projection points are point A and point B, where the direction of the emitted light waves is the same as viewing interval. From the same perspective, the projections of all objects will form a projected image of a 3D object at this viewing angle. The bandwidth of the 3D object is determined by the spacing of the light emission angles $\Delta\theta_x$. The frequency center is determined by the angle bisector of the small angle.

Inspired by the decomposition principle, suppose that the two points l_1 and l_2 are on the light emitted from the point A as shown in Figure 2. The angle between l_1A and axis z is θ_{xi1} and the angle between l_2A and axis z is θ_{xi2}. $\Delta\theta = \theta_{xi2} - \theta_{xi1}$, when $\Delta\theta_x$ and θ_{xi} are smaller, the width of the frequency band determined by this parallax angle can be expressed as:

$$\Delta f_{xi} = \frac{\sin(\theta_{xi2}) - \sin(\theta_{xi1})}{\lambda} \approx \frac{\sin(\theta_{xi2} - \theta_{xi1})}{\lambda} = \frac{\sin(\Delta\theta_x)}{\lambda} \tag{1}$$

The location of the frequency center in x direction is:

$$f_{cxi} = \frac{\sin(\theta_{xi1} + \Delta\theta_x/2)}{\lambda} = \frac{\sin(\theta_{xi})}{\lambda} \tag{2}$$

In y direction, we can still get the same conclusion. The bandwidth corresponding to the j-th parallax angle in the y direction is:

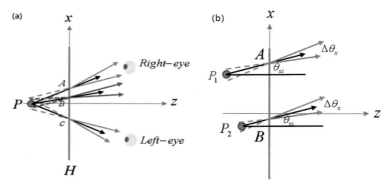

Figure 1.   Reconstruction of a synthetic hologram and parallax decomposition: (a) Reconstruction of a synthetic hologram; (b) Parallax decomposition.

26

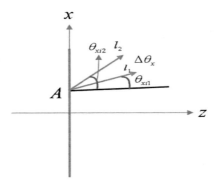

Figure 2. Diagram of a parallax angle.

$$\Delta f_{yj} = \frac{\sin(\theta_{yj2}) - \sin(\theta_{yj1})}{\lambda} \approx \frac{\sin(\theta_{yj2} - \theta_{yj1})}{\lambda} = \frac{\sin(\Delta\theta_y)}{\lambda} \tag{3}$$

The location of the frequency center in y direction is:

$$f_{cyj} = \frac{\sin(\theta_{yj1} + \Delta\theta_y / 2)}{\lambda} = \frac{\sin(\theta_{yj})}{\lambda} \tag{4}$$

The size of the hologram $w_h \times h_h$ is the same as the size of projected image $w_o \times h_o$ of the three-dimensional object, which can be expressed as:

$$w_o = w_h, h_o = h_h \tag{5}$$

The frequency sampling interval thus becomes:

$$\Delta f_{xo} = \frac{1}{h_o}, \Delta f_{yo} = \frac{1}{w_o} \tag{6}$$

According to the frequency bandwidth and frequency sampling interval, the resolution of the projected image is:

$$M = \frac{\Delta f_{xi}}{\Delta f_{xo}} = \frac{w_o \sin(\Delta\theta_x)}{\lambda}; N = \frac{\Delta f_{yi}}{\Delta f_{yo}} = \frac{h_o \sin(\Delta\theta_y)}{\lambda} \tag{7}$$

Assume the sampling interval of the hologram is d_h. The resolution of the hologram is:

$$M_h = \frac{w_o}{d_h}; N_h = \frac{h_o}{d_h} \tag{8}$$

The frequency domain range of the hologram can be expressed as:

$$-\frac{M_h}{2}\frac{1}{w_o} < f_{hx} < \frac{M_h}{2}\frac{1}{w_o}; -\frac{N_h}{2}\frac{1}{h_o} < f_{hy} < \frac{N_h}{2}\frac{1}{h_o} \tag{9}$$

The calculation of projected images is simple, which is shown in Figure 3(a). For the sake of simplicity, a one-dimensional directional projection is used to illustrate the calculation method of a projection image. The direction of the bisector of a small angle θ_{xm} in x direction is the same as the projection direction of all the points for a certain angle of view. The coordinates of the projection point x_{pk} of the k-th point $p(x_k, y_k, z_k)$ in Figure 3 can be expressed as:

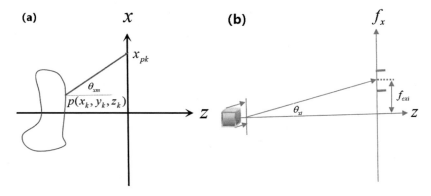

Figure 3. Calculation of the projected image and frequency of one projected image: (a) Model for projected image calculation; (b) The frequency calculation model.

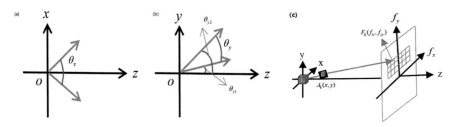

Figure 4. The design of the parallax angle and frequency mosaic calculation: (a) Parallax angle in x direction; (b) Parallax angle in y direction; (c) The mosaic of frequency.

$$x_{pk} = x_k + z_k \tan(\theta_{xm}) \tag{10}$$

Similarly, the coordinates of the projection point in the y direction can be calculated by $y_{pk} = y_k + z_k\tan(\theta_{ym})$, where θ_{ym} is the bisector of a small angle in y direction. After calculating the coordinates of the projection point, the amplitude of the corresponding point in the projection plane can be expressed as:

$$P(x_{pk}, y_{pk}) = A_k \tag{11}$$

where A_k is the amplitude of the k-th point. For a color object, A_k corresponds to r_k, g_k, b_k, which mean the amplitudes of three color components.

Figure 3(b) shows the frequency domain calculation diagram. For a certain point of view, we can calculate the projected image multiplied with rand phase and then perform the Fourier transform:

$$F(f_x, f_y) = \iint I_{i,j}(x,y)\exp[i\phi_{rand}(x,y)]\exp[-j2\pi(xf_x + yf_y)dxdy \tag{12}$$

The frequency of the image is shifted to form the frequency data corresponding to the angle of view in the holographic coordinate system, which can be expressed as:

$$F_{i,j}(f_x, f_y) = F(f_x + f_{xci}, f_y + f_{ycj}) \tag{13}$$

The full frequency corresponding to different views can be calculated sequentially.

Following the determination of frequency, we show the frequency mosaic of a full-parallax synthetic hologram in Figure 4. Within it, the parallax angles are θ_x and θ_y (Figures 4(a) and

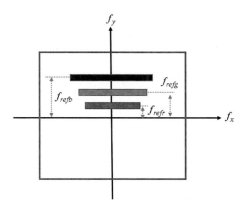

Figure 5.   Frequency of the color rainbow hologram.

(b)). In Figure 4(c), we show the frequency of the object wave of a full-parallax synthetic hologram in a rectangular region. It consists of the decomposition of a plurality of small rectangular regions; each rectangular region corresponds to the frequency spectrum of the projected image at a specific viewing angle. The light emitted by the object has a certain inclination angle in the y direction to avoid zero and conjugate order interference.

Figure 4(c) shows the object light splicing in the frequency domain. The projected image is calculated, linearly interpolated into an image with the resolution shown in Equation7, and Fourier transformed to obtain the local frequency of hologram. After all the frequencies are considered, the synthetic hologram can be calculated by:

$$H(x,y) = real\{ifft2[F(f_x, f_y)]\} + c \tag{14}$$

where $real\{*\}$ means taking real part of *, $ifft2[*]$ means 2D inverse Fourier transform of * and c is a constant, which keeps the intensity of hologram is greater or equal to zero.

For generation of a rainbow hologram, when the plane wave is used as the reference wave for calculation of a rainbow hologram, the first order frequency of the rainbow hologram containing three color frequency bands will separate under certain conditions as illustrated in Figure 5. The frequency shift caused by the reference wave can be expressed as:

$$f_{refr} = \frac{\sin(\theta_{refy})}{\lambda_r}; f_{refg} = \frac{\sin(\theta_{refy})}{\lambda_g}; f_{refb} = \frac{\sin(\theta_{refy})}{\lambda_b} \tag{15}$$

The projected images are calculated and interpolated according to Equation 7 with different wavelengths and transformed to the frequency domain for mosaic computation. After the frequency is obtained, Equation 14 can also be used for synthetic color rainbow hologram calculation.

The calculation of synthetic holograms with Equation (14) can be simplified by firstly applying a one-dimensional inverse Fourier transform in x direction only for the effective data areas (the areas with no zero data) and then a one-dimensional inverse Fourier transform in y direction. With this simplification, the calculation time can be reduced.

3   RESULTS

3.1   *The full-parallax synthetic hologram*

In our study, a 3D model with the size of $X \times Y \times Z = 14.97$ mm $\times 24.53$ mm $\times 12.78$ mm was used for full-parallax synthetic hologram calculation. The 882 projected images with a resolution of 1128×1128 were used. The sampling interval was $d_h = 0.318$ um. The size of generated hologram was 30 mm × 30 mm and the resolution was 94340 × 94340. The wavelength was 464 nm, θ_x was in range of −24° to 24° and θ_y is in the range of 7° to 24°. The horizontal and vertical angular intervals were 1°. The entire holographic calculation took 1.2 hours. Figure 6(a) is one perspective image of the 3D model. Figure 6(b–d) are three reconstructed images of the synthetic holograms with white LED for reconstruction, which are randomly captured from three positions.

3.2 *The synthetic color rainbow hologram*

The 3D model is the same size as the one used in the full-parallax synthetic hologram experiment. The three wavelengths used were $\lambda_r = 632\ nm$, $\lambda_g = 547\ nm$ and $\lambda_b = 464\ nm$. The parallax angle of the object light was set to 48° × 1° and the angle interval $\Delta\theta_x$ and $\Delta\theta_y$ was 1° in x and y direction, respectively. Forty eight images were used for calculation.

The three color channels used had resolutions of $M_r \times N_r = 829 \times 829$, $M_g \times N_g = 958 \times 958$ and $M_b \times N_b = 1130 \times 1130$. The angle of reference light was $\theta_{refy} = 18°$. The computation time for this color rainbow hologram was about 25 minutes.

The view of the 3D color model and three perspective reconstruction views are shown in Figure 7(a) and (b–d), respectively. The computer generated synthesized color rainbow hologram is accurate in color reproduction and also has a strong stereoscopic view effect.

The computing platform for the two hologram's calculation was a laptop with i7–7700H CPU, 16 G memory and 128 G SSD. Matlab2015b (Mathworks) was used for implementing the proposed algorithm. The calculated holograms were printed using a home-made hologram printing system at the Institute of Information Optics, ZheJiang Normal University (Yile et al., 2009). The hologram printing time was only related to the size of the hologram and 3.5 hours were needed for one hologram output. After developing, fixing and bleaching the holograms, a white LED was used to illuminate the holograms for optical reconstruction. The reconstructed images were clear and had a strong and smooth 3D effect for the human eye. The color of the rainbow hologram was vivid.

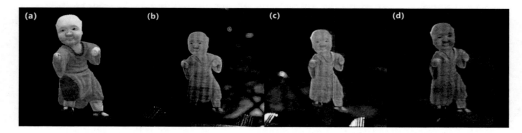

Figure 6.   Perspective view of the 3D model and optical reconstructed images: (a) 3D model; (b–d) Optical reconstructed images.

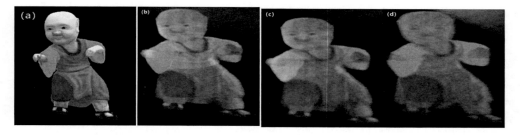

Figure 7.   3D model and reconstructed images: (a) View of 3D model; (b–d) Reconstructed images of the color rainbow hologram.

4 CONCLUSION

A simple method for computing a large scale high resolution synthetic hologram based on a frequency mosaic was proposed. A full-parallax synthetic hologram and synthetic color rainbow hologram at the resolution of 94340 × 94340 and size of 30 mm × 30 mm were calculated with a short computation time and optically reconstructed. The proposed method is effective and efficient, which is of great significance for digital holographic 3D applications.

REFERENCES

Hao, Z. & Yan, Z. (2015). Fully computed holographic stereogram based algorithm for computer-generated holograms with accurate depth cues. *Optics Express, 23*(4), 3901–3913.
Matsushima, K. & Nakahara, S. (2010). High-definition full parallax CGHs created by using the polygon-based method and the shifted angular spectrum method. *Proceedings of SPIE, 7619,* 761913.
Matsushima, K. & Nakashrm, S. (2009). Extremely high-definition full-parallax computer-generated hologram created by the polygon-based method. *Applied Optics, 48*(34), H54–H62.
Qu, W., Gu, H. & Tan, Q. (2016). Holographic projection with higher image quality. *Optics Express, 24*(17), 19179–19184.
Tsuchiyama, Y. & Matsushima, K. (2016). Full-color high-definition CGH using color filter and filter design based on simulation. *In Imaging and Applied Optics, DW5I.4.*
Yile, S. Hui, W. & Yong, L. (2009). Practical method for color computer-generated rainbow holograms of real existing objects. *Applied Optics, 48*(21), 4218–4226.
Yoshikawa, H. & Yamaguchi, T. (2009). Computer-generated holograms for 3D display. *Chinese Optics Letters, 7*(12), 1079–1082.

Frontier Research and Innovation in Optoelectronics Technology and Industry – Habib & Lewis (Eds)
© 2019 Taylor & Francis Group, London, ISBN 978-1-138-33178-5

The layered 3D non-local means for image de-noising

Changsheng Ying
School of Computer Science and Technology, Jilin Normal University, Siping, Jilin, China
School of Science, Changchun University of Science and Technology, Changchun, Jilin, China

Kun Hou
School of Computer Science and Technology, Jilin Normal University, Siping, Jilin, China

Ye Li
School of Science, Changchun University of Science and Technology, Changchun, Jilin, China

ABSTRACT: The Non-Local Means (NLM) image de-noising algorithm uses the similar structures of the whole image to improve the signal-to-noise ratio of the output image. There are some shifting and rotating changes in the time-domain images captured by the same camera. In this paper, a layered three dimensional non-local means de-noising algorithm is proposed by utilizing these rotating and shifting similarities to further improve the quality of the output image. In the first layer, all images are de-noised by the NLM algorithm respectively. These de-noised images are used to form a three dimensional tower in the second layer, and the similarities between the reference block and the matched blocks within the search window are calculated based on the Euclidean distance. The adaptive filter parameter is determined by the Otsu's method, and those matched blocks with small similarity values are excluded from the estimation of the output pixel. The experimental results show that our method can obtain better de-noising results as well as preserving more details.

1 INTRODUCTION

In the process of image acquisition, users can only obtain the degraded images due to the influences of the light source, the device dependent noises such as readout noise, the dark current of the charge coupled device (CCD) sensor, and so on. The higher level applications such as image classification, segmentation, recognition and target tracking can hardly be implemented with the noisy and dirty images. Image de-noising is the first step of image processing. The goal of image de-noising is to eliminate or reduce the noise so that the processed image should approximate the original image as much as possible, and the de-noised images should be more conducive to image classification, segmentation, recognition, fusion and human observation.

The classical de-noising methods commonly process the degraded images based on their characteristics in the spatial domain, the time domain or the frequency domain. The non-local means (NLM) (A. Buades et al., 2005) algorithm de-noises the image by utilizing all the pixels that have similar structures in the whole image. A very important issue in practice is that of the algorithm's computational efficiency. The pre-classification technique, the symmetry of the weight values and the lookup tables were used to speed up the NLM algorithm (A. Dauwe et al., 2008). Sun et al. (W.F. Sun & Y.H. Peng, 2010) got better results by means of using the similarities in the seven types of rotating blocks. Another improved version of the NLM method was proposed in the time domain (M. Wang et al., 2013). All images were de-noised by the NLM algorithm respectively, and the de-noised images were averaged to get the output image. Considering there are shifting and rotating changes between the sequential images captured by the same camera due to the moving objects or camera shaking, we proposed a layered three dimensional non-local means (NLM3D) de-noising method. All images will be processed by the NLM algorithm in the first layer by using a small noise factor, and then used to form a three dimensional tower. In the second layer, the output pixel value is the weighted sum of all the pixels whose similarities are larger enough in the tower.

This paper is organized as follows. The principles of the NLM algorithm are described in section 2. In section 3, the proposed NLM3D method is explained in detail. Section 4 shows the experimental results and section 5 concludes the paper.

2 NLM ALGORITHM

The mean filter and the median filter are classical de-noising methods in the spatial domain, and the correlations between the pixel being processed and the pixels in its neighborhood are used to obtain the output pixel. In the NLM algorithm, all pixels in the whole image have contributions to the output pixel according to their similarities with the pixel being processed.

Considering a noisy image V in a two or three dimensional grid I, the one dimensional form is,

$$V = \{v_i | i \in I\} \tag{1}$$

where v_i is the intensity value of the ith pixel in V. Let N_i be the vector that represents the intensity values of pixels in a $B \times B$ block (commonly 5×5 or 7×7) where the central position is i. The similarity between the reference block N_i and the matched block N_j is determined by the Euclidean distance,

$$ss_j = \left\| N_i - N_j \right\|_2^2 \tag{2}$$

The weight of v_j in the contribution list of v_i is,

$$\omega_{i,j} = \exp\left(-\frac{ss_j}{h^2} \right) \tag{3}$$

where the filter parameter h is a constant associated with σ^2, and gets the value of $10 \times \sigma$ usually. The weight $\omega_{i,j}$ decreases exponentially. The smaller ss_j, the higher $\omega_{i,j}$. The de-noised value $\hat{v}_i$ of v_i is a linear combination of all the pixels in the whole image,

$$\hat{v}_i = \frac{\sum_{j \in V} \omega_{i,j} v_j}{\sum_{j \in V} \omega_{i,j}} \tag{4}$$

3 LAYERED THREE DIMENTIONAL NON-LOCAL MEANS ALGORITHM

3.1 *Principles of the NLM3D*

The shifting and rotating changes in the sequential images captured by the same camera in the time domain are shown in Figure 1.

Besides the similar structures in the same image, there are plenty of similar structures in the shifting and rotating images. Considering the high correlations between these similar structures in the spatial domain and the time domain, we proposed the NLM3D method. The NLM3D algorithm consists of two layers. In the first layer, the sequential images will are de-noised separately by the NLM method as shown in Figure 2(a). As can be seen from Figure 2(b), these de-noised images are used to form a three dimensional tower. The output pixels in the de-noised image are obtained by weighting the matched pixels in the tower that have the similarities high enough.

(a)   (b)   (c)

Figure 1. The shifting and rotating changes in the sequential images. The original image (a), the left rotating image (b) and the right rotating image (c).

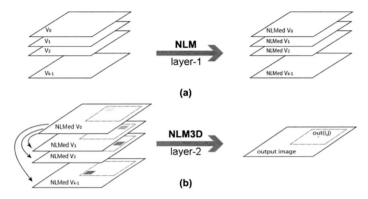

Figure 2. The schematic diagram of NLM3D.

The first layer is the basis of the second layer, and it aims to filter the noise partially and promote the accuracy of the second layer. The sizes of the neighborhood window and the searching window affect the results and the speed of the de-noising process. When de-noising the sequential images, the key of the first layer processing is using a smaller noise factor which ensures that part of the noise is filtered as well as preserving the details. The second layer of NLM3D de-noising is based on the fact that the image being processed has similar structures in the spatial domain and strong correlations in the neighboring regions of the time-domain images. Assuming that the total number of the shifting and rotating images is k, and any of these images can be selected as the reference image V_0. Let v_i be the ith pixel in V_0, N_i the neighborhood window of size $(2w+1) \times (2w+1)$ with i as the center, S_j the searching window of size $(2sw+1) \times (2sw+1)$ which is several times larger than N_i, and the total number of the matched blocks in the whole tower corresponding to v_i is $(2sw+1) \times (2sw+1) \times k-1$. The matched block N_j corresponding to the reference block N_i is determined by searching all the $S_j(j=1,2,...,k)$ in the tower. We get n blocks by rotating N_i or N_j in a small range, and the total number of the matched blocks in the searching windows of the whole tower is $(2sw+1) \times (2sw+1) \times k \times n-1$. Once the similarity ss_j between N_i and N_j has been calculated, ss_j and v_j will be stored in a $n \times 2$ vector M_i for later processing.

3.2 The adaptive selection of the filter parameter

When all the similarities between N_i and N_j have been obtained, the filter parameter h in the second layer will be determined by the Otsu's method (N. Otsu, 1979). Let T_h be the similarity threshold, n_0 the number of pixels whose values larger than T_h, n_1 the number of pixels whose values lower than T_h, μ_0 the mean of n_0, and μ_1 the mean of n_1. The mean of all the similarities can be expressed as,

$$\mu = \eta_0\mu_0 + \eta_1\mu_1 \qquad (5)$$

35

where $\eta_0 = n_0 / (n_0 + n_1)$ and $\eta_1 = n_1 / (n_0 + n_1)$. The variance between the two kinds of similarities can be expressed as,

$$g = \eta_0 \left(\mu_0 - \mu \right)^2 + \eta_1 \left(\mu_1 - \mu \right)^2 \tag{6}$$

Finally, we substitute $\mu = \eta_0 \mu_0 + \eta_1 \mu_1$ to express g in terms of μ_0, μ_1, η_0 and η_1,

$$g = \eta_0 \eta_1 (\mu_0 - \mu_1)^2 \tag{7}$$

When g gets the maximum value, the variance between the two classes of similarities also is the largest, and T_h is the best threshold value. The filter parameter h can be obtained by applying the Otsu's method on all the similarities stored in M_i twice,

$$h = \begin{cases} h_1 & , & ss_j \geq T_{h1} \\ h_2 & , & T_{h2} \leq ss_j < T_{h1} \end{cases} \tag{8}$$

The pixel v_j whose similarity satisfies the relationship of $ss_j < T_{h2}$ will be excluded from Equation (4). These pixels have almost no contribution to $\hat{v}_i$, and even worse blur the details of the output image. So, the final expression of $\hat{v}_i$ is,

$$\hat{v}_i = \sum_j \frac{\exp(-\frac{ss_j}{h^2})}{\sum_j \exp(-\frac{ss_j}{h^2})} v_j \ \ s.t. \ ss_j \geq T_{h2} \tag{9}$$

4   EXPERIMENTAL RESULTS

In the experiment, the standard images such as Barbara, Boat, Cameraman, House, Lena and Peppers were used to test our proposed method. A Gaussian white noise with mean 0 and standard deviation 30 was added to all images, and then all images were rotated with a random angle less than one degree. To evaluate the processed images such as de-noising, restoration or SR reconstruction, the peak signal-to-noise ratio (PSNR), which is associated with mean-square error (MSE), is a commonly used objective metric. The PSNR is defined as,

$$PSNR = 10 \times \log_{10} \left(\frac{L^2}{MSE} \right) \tag{10}$$

where L is the maximum possible intensity value of the image. PSNR is an objective image quality evaluation method based on the difference of the pixels between the processed image and the original image. The de-noising performance of our proposed method is compared to Median filter, the NLM, the fast NLM (FNLM), the K-SVD (M. Aharon et al., 2006), and WANG's method. A 5×5 neighborhood window, 21×21 searching window were used for all the NLM methods. The number of the sequential images was 3, and the noise factor in the first layer was $\sigma_1 = 1.25$. The noise factors used for the NLM, FNLM, K-SVD, WANG and the second layer of the proposed method were $\sigma_2 = 20, 30, 40$. The results are shown in Table 1, Table 2 and Table 3.

As can be seen from Table 1 to Table 3, the median filter performed poor due to the large noise added to the images, and different versions of the NLM method got better results on the images that having rich structures and textures such as Barbara, Boat, House and Lena. The proposed method outperformed all the other de-noising methods no matter the estimated noise factors were lower or larger than the actual noise added, and obtained the highest values in all the cases. The de-noised images of Barbara, Lena and Peppers are shown in Figure 3, Figure 4 and Figure 5, and the noise factor used in the de-noising process was $\sigma_2 = 30$.

Table 1. The PSNRs of different de-noising methods ($\sigma_2 = 20$).

Images	Median	NLM	FNLM	K-SVD	WANG	NLM3D
Barbara	19.07	23.18	21.21	23.15	24.51	25.64
Boat	19.05	24.06	22.82	23.02	24.06	25.28
Cameraman	19.07	24.34	24.25	23.48	24.16	25.53
House	19.05	25.02	23.94	23.48	25.09	26.42
Lena	19.09	23.53	21.07	22.07	24.16	25.34
Peppers	19.11	22.50	19.11	21.42	23.81	23.87

Table 2. The PSNRs of different de-noising methods ($\sigma_2 = 30$).

Images	Median	NLM	FNLM	K-SVD	WANG	NLM3D
Barbara	19.58	24.17	21.21	23.10	24.80	26.30
Boat	19.62	23.81	22.82	25.10	22.90	25.89
Cameraman	20.01	23.11	21.22	24.16	23.69	25.70
House	19.62	24.18	23.94	25.32	24.81	26.92
Lena	19.68	24.89	24.25	25.72	23.97	26.72
Peppers	19.69	21.70	23.17	18.40	21.04	23.45

Table 3. The PSNRs of different de-noising methods ($\sigma_2 = 40$).

Images	Median	NLM	FNLM	K-SVD	WANG	NLM3D
Barbara	20.32	22.20	21.21	21.71	22.20	23.65
Boat	20.44	21.59	22.82	22.79	21.23	23.29
Cameraman	20.87	19.10	21.22	22.92	18.80	23.22
House	20.52	22.07	23.94	24.16	21.08	25.93
Lena	20.59	22.45	24.25	24.22	22.39	25.05
Peppers	20.51	18.89	21.59	17.64	18.84	23.37

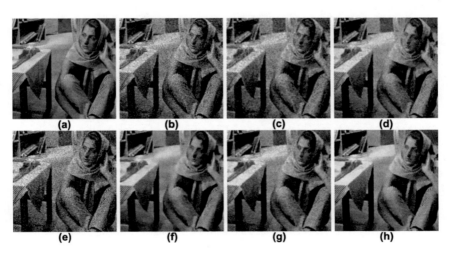

Figure 3. De-noised Images of Barbara. (a) original; (b) noisy image; (c) median; (d) NLM; (e) FNLM; (f) K-SVD; (g) WANG's method; (h) proposed.

As shown in Figure 3 and Figure 4, when the images are seriously degraded by the noise, the de-noising effect of the median filter is poor, the de-noised images still have a lot of spots after processing, and the details are submerged by the remaining noise. The de-noised images of the FNLM method are still noisy, just like the result images of the median filter. A possible

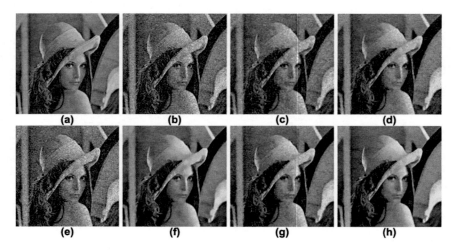

Figure 4. De-noised Images of Lena. (a) original; (b) noisy image; (c) median; (d) NLM; (e) FNLM; (f) K-SVD; (g) WANG's method; (h) proposed.

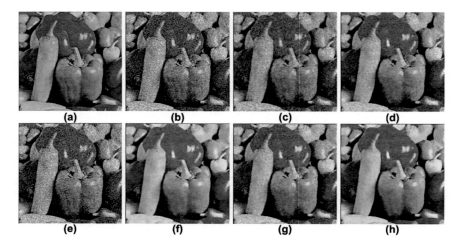

Figure 5. De-noised Images of Peppers. (a) original; (b) noisy image; (c) median; (d) NLM; (e) FNLM; (f) K-SVD; (g) WANG's method; (h) proposed.

explanation of the unexpected results is that the noise is so great that many of the similar structures are lost due to the speed up techniques. The K-SVD processed images are much softer, but the details such as the textures and details of the desktop, the palm of Barbara, the hat and its decoration of Lena are lost. The result images of NLM and WANG's methods are much better, but still have a certain degree of noise, especially in the uniform area such as the background, the skins of Barbara and Lena. The result images of our proposed method have the best visual qualities, and the textures and details such as the desktop, the scarf and the palm of Barbara, the hat and the decoration of Lena are almost preserved.

It can be seen from Figure 5 that the results of all the different versions of the NLM methods do not perform as better as the formers since the textures and sharp edges in Peppers are fewer than those in Barbara and Lena. We can also get this trend from Table 1 to Table 3, since that the PSNRs of the images lacking of the edges and textures are lower than those ones containing rich textures. But, it should be noted that our proposed method outperform all the methods above not only in the PSNR values but also in the visual qualities and the preserved details.

When the number of the sequential images is 3, we have found in the experiment that there exists a balance between the de-noising results and the computational efficiency. Based on the existing acceleration techniques, the proposed NLM3D algorithm still needs to be further optimized to improve the computational efficiency.

5 CONCLUSION

This paper presents a NLM3D de-noising method for the consideration that there exist similar structures between the shifting and rotating images captured by the same camera. Images are de-noised by the NLM algorithm using a small noise factor in the first layer in order to preserve as much details as possible. After calculating the similarities between the image blocks, the algorithm adaptively adjusts the filtering parameters in the second layer by using the Otsu method. The pixels with low similarities are excluded from the contribution list in order to ensure better de-noising effect while retaining more details of the image. Experimental results show that our algorithm is better than the similar algorithms no matter what values the estimated noise factors are.

REFERENCES

Aharon, M., M. Elad and A. Bruckstein, 2006, K-SVD: An algorithm for designing overcomplete dictionaries for sparse representation, *IEEE Transactions on signal processing,* 54(11): 4311–4322.

Buades, A., B. Coll and J. Ml Morel, 2005, A non-local algorithm for image denoising, *IEEE Computer Society Conference on Computer Vision and Pattern Recognition,* 2(7): 60–65.

Dauwe, A., B. Goossens and H. Luong, 2008, A fast non-local image denoising algorithm,*Image Processing: Algorithms and Systems,* 6812: 6812101–6812108.

Otsu, N. 1979, A threshold selection method from gray-level histograms, *IEEE transactions on systems, man, and cybernetics,* 9(1): 62–66.

Sun, W.F., and Y.H. Peng, 2010, An improved non-local means de-noising approach, *Acta Electronica Sinica,* 38(4): 923–928.

Wang, M., H.J. Wang and G.Y. Sun, 2013, Three-dimensional image denoising algorithm based on non-local means, *Infrared Technology,* 35(4): 238–241.

Frontier Research and Innovation in Optoelectronics Technology and Industry – Habib & Lewis (Eds)
© 2019 Taylor & Francis Group, London, ISBN 978-1-138-33178-5

The measurement of depth of field in microscope system based on image clarity evaluation

Lu Zhang, Wen-Jian Chen, Wu-Sen Li & Kun-Feng Zhan
Department of Electronic Engineering and Photoelectric Technology, Nanjing University of Science and Technology, Nanjing, China

ABSTRACT: In order to realize the digital measurement of the depth of field in microscope systems, a method of measuring the depth of field in microscope systems based on image clarity evaluation is proposed. The basic principle of the method is: pay attention to the image clarity evaluation based on wavelet transform when changing the distance between the microscope lens and the sample. When the measurement is reduced to the threshold value, the image returns blurred at once. Then the distance of the movement between the microscope and the sample is the result. The method is tested in many different microscope systems. The results show that the method is feasible, uses less time and has better robustness. The accuracy can reach more than 95.34%, proving that image processing technology can be applied in the area of the measurement of the depth of field.

Keywords: Microscope; Depth of field; Image clarity evaluation; Wavelet transform; Robustness

1 INTRODUCTION

There are visual microscopes and digital microscopes because of the difference of image acquisition. The depth of field is one of the important parameters of the microscope system. When the microscope focuses on the object plane, which is called the reference plane, the axial distance between the two farthest planes that can be clearly observed is the imaging depth of the microscope, namely "depth of field" (C. Maurer et al. 2013). Based on the domestic literature, there has been no reports except that some scholars have used MTF to simulate the depth of field of the microscopic objective lens (Mei-Mei et al. 2008). While the value is usually calculated by the traditional formula. The traditional calculation of the depth of field in visual microscopes is the sum of geometric depth of field, physical depth of field and the adjustment of the depth of field of the eyes, where geometric depth of field is calculated in the finite optical path, the physical depth of field is generated by the diffraction effect, and the other is calculated based on the observer's ability of the adjustment. As for a digital microscope, the depth of field is mainly affected by the diffraction effect, so the depth of field is only effected by the physical depth of field. Some researchers have studied the traditional calculation of the depth of field in microscopes and there are some proposed improvements (Li-Cun Sun et al. 2013; Min-Jun Wang et al. 2010, Zhi-Jian Wang & Peng Wang. 2000), however the method is complicated.

In order to get the value more efficiently, we propose a method using image processing. Through the image clarity evaluation based on the wavelet transform, when the image clarity evaluation decreases to the definition value from the highest when changing the distance between the microscope and the sample, the depth of field can be measured. The method is tested in many different microscope systems. The results show that the method is feasible and more efficient.

2   ANALYSIS OF TRADITIONAL CALCULATION

Because of the diffraction effect, the depth of field in the digital microscope is affected by the physical depth of field as mentioned:

$$Al = \frac{n\lambda}{NA^2} \qquad (1)$$

where n is the refractive index of object space, λ is the wavelength of the light, NA is the numerical aperture of objective lens.

3   THE PRINCIPLE OF MEASURING SYSTEM

3.1   *The measuring system*

The measuring system includes a microscope, a detector, a computer and a motion mechanism, as shown in Figure 1. (a) samples are sampled at different magnification rates in the microscope; (b) detector collects the images of different positions; (c) computer is responsible for the image processing and the image output; (d) motion mechanism changes the distance between the microscope and the samples. The measurement process is shown in Figure 2, where the core part of the system is the image clarity evaluation based on the wavelet transform.

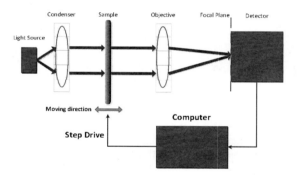

Figure 1.   The measuring system.

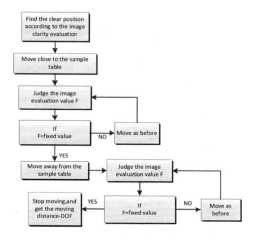

Figure 2.   The processing flow of detection method.

3.2 *Image clarity evaluation based on wavelet transform*

Wavelet transform is a kind of mathematical transformation. It is a set of functions obtained by stretching and translating a mother wavelet function (Min-Jun Wang et al. 2010; Daubechies I. & Sweldens W. 1998). To be used in the image processing is to transform the data in horizontal and vertical directions, as shown below:

$$W_f(a, b_x, b_y) = \iint f(x, y) \overline{\varphi_{a, b_x, b_y}(x, y)} \, dx \, dy \qquad (2)$$

where b_x, b_y represent the translation of two dimensions.

We can devide the image into four parts by it: the main information part of the image (LL1), the horizontal direction information part of the image (HL1), the vertical direction information part of the image (LH1), the diagonal direction information part of the image (HH1). The wavelet transform is used to decompose the main information (LL1) and the second level of wavelet can be obtained (as shown in Figure 3).

The signal and noise have different characteristics in different levels. We get the function of the clarity evaluation by multiplying wavelet coefficients of different levels with texture coefficients as shown:

$$F_j = P(h) \left| f_{2^j}^h \times f_{2^{j+1}}^h \right| + P(v) \left| f_{2^j}^v \times f_{2^{j+1}}^v \right| + \left| P(d) \left| f_{2^j}^d \times f_{2^{j+1}}^d \right| \right| \qquad (3)$$

where h, v, d are the three directions of horizontal, vertical and diagonal, $P(h), P(v), P(d)$ are the characteristic coefficients for three directions.

The microscope system $(20 \times (NA = 0.5))$ is used to test the lymphatic cancer cells in observation positions of different focuses. The classical laplace function, variance function and basic wavelet function are selected for comparative experiments. From Figure 4(a), we can see that without noise, each function satisfies the unimodal property and can identify the clearest position. However, the algorithm used in this system has higher sensitivity and better performance. As you can see from Figure 4(b), Laplace function, variance function and wavelet function are very sensitive to noise. After comparison, our system can still judge clarity correctly.

Figure 3.   Wavelet decomposition.

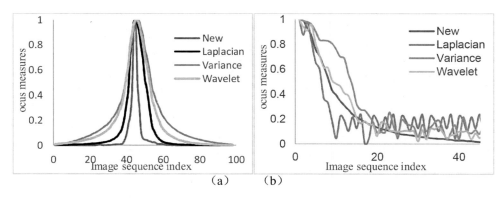

(a)　　　(b)

Figure 4.   (a) The normalized computed accuracy with none noise; (b) The normalized computed accuracy with noise.

It shows that the evaluation algorithms in our system has a better sensitivity compared with other algorithms under noise affects, which ensures the robustness of the system.

4 EXPERIMENT

The microscope takes 1.25 μm as a step to move to collect images. It is tested in three specifications of the microscope system: $10\times |NA = 0.25|$ $20\times |NA = 0.5|$ $40\times |NA = 0.65|$. The testing equipment of the lens specification of $20\times |NA = 0.5|$ is shown in Figure 5(a). Our system chooses test the lymphatic cancer cells samples as shown in Figure 5(b). We found that the samples of different observation positions perform the distinct differences transformed by one wavelet transform and two level wavelet transform as shown in Figure 6.

4.1 *Definition value*

Before the start of the measurement, the definition value needs to be confirmed. It is found that the image with the image clarity value less than 80% has obvious blurred state. So that we select different appropriate threshold values to test the best property for the system, as shown in Table 1.

Figure 5.   (a) Testing equipment; (b) Test sample.

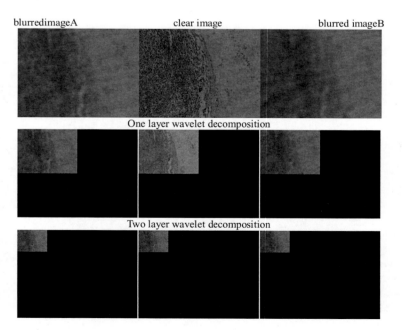

Figure 6.   Wavelet decomposition of the sample.

We list the results of testing threshold values in the system of $20 \times |NA = 0.5|$ and $40 \times NA = 0.65$ as shown in the table below.

It can be seen that we can get satisfactory results in different systems when the definition value is selected as 85%. And the system of the bigger depth of field has smaller errors, which means the system of $20 \times (NA = 0.5)$ has better results than the system of $40 \times (NA = 0.65)$. And the test in the system of $40 \times (NA = 0.65)$ has shown the definition value of 85% has already met the optical testing requirements. Therefore, 85% is chosen as the definition value of the measuring system.

4.2 Measuring results

85% is chosen as the definition value of the measuring system as mentioned. The motion mechanism chooses 1.25 μm as a step to drive the microscope to move to collect different pictures. The value of depth of field is obtained by the system:

$$\Delta = 2.1734 \ \mu m$$

The absolute value of relative error is calculated as:

$$\delta = \frac{\Delta}{L} \times 100\% = \frac{2.2 \ \mu m - 2.1734 \ \mu m}{2.2 \ \mu m} \times 100\% = 1.21\%$$

That result is within the extent permitted proving that the method is feasible for measuring the depth of field for the microscope system. The same method is used to test three other microscope systems as shown in Table 2.

Table 1. DOF results in microscopes with different definition values.

		0.80	0.85	0.90	0.95
	Definition value	0.80	0.85	0.90	0.95
	Real value of depth of field (μm)		2.2		
$20 \times (NA = 0.5)$	Measurement value of depth of field (μm)	2.7924	2.1934	1.8297	1.4316
	Absolute value of relative error (%)	26.93	1.91	16.83	34.93
	Critical value of fixed focus	0.80	0.85	0.90	0.95
	Real value of depth of field (μm)		1.3		
$40 \times (NA = 0.65)$	Measurement value of depth of field (μm)	3.3224	1.2394	0.8092	0.5224
	Absolute value of relative error (%)	>100	4.66	37.75	59.81

Table 2. DOF results for different microscopes.

Objective magnification	Numerical aperture	Depth of field (μm)	Measured value (μm)	Relative error
$10 \times$	0.3	8.5	8.6176	1.38%
$20 \times$	0.5	2.2	2.1934	1.91%
$40 \times$	0.65	1.3	1.2394	4.66%

The testing absolute values of relative error are all permitted, so that measurement accuracy meets the requirements of optical testing. In addition, from the traditional formula of the depth of field, with the increase of magnification and numerical aperture, the depth of field will gradually decrease. The testing data proves the measuring data is consistent with the theoretical reality. All results show that the measurement accuracy can reach over 95.34%, so that it is feasible to be used in the field of optical detection and some other fields related.

5 CONCLUSIONS

In this study, a method for measuring the depth of field for the microscope systems is designed by studying the image clarity function based on wavelet transform. It is tested in different microscope systems, where the measurement accuracy is more than 95.34%. As for the core of the system, a new image clarity evaluation function has been proposed. The method is compared with other classical methods, in the environments of noises and no noise. The results prove that our method has higher sensitivity and better performance, so that the robustness of the system is guaranteed. This system applies image processing technology to the measurement of depth of field in the microscope system, which has good market prospects.

REFERENCES

Chen Xiaodong et al. 2011. New method for determining the depth of field of microscope systems. Applied Optics, 50(10):5524–5533.

Daubechies I. & Sweldens W. 1998. Factoring wavelet transforms into lifting steps. J Fourier Anal. 4(3):245–267.

Li-Cun Sun et al. 2013, Determination and measurement of depth of field electronical eyepiece microscope. Optical precision engineering, 21(5):1152–1159.

Maurer C. et al. 2010. Depth of field multiplexing in microscopy. Opt. Express 18, 3023–3034.

Mei-Mei Han & Ze-Xin Han. 2008. Simulation measurement of depth of field of the objective lens in microscope system based on MTF. Infrared and laser engineering, 37:312–315.

Min-Jun Wang et al. 2010. Image clarity evaluation based on the wavelet transform. Computer Systems and Applications. 19(8):164–167.

Ze-Ying Chi & Wen-Jian Chen. 2013. Basis of applied optics and optical design. Beijing: Advanced education press.

Zhi-Jian Wang & Peng Wang. 2000. Rayleigh judgment and stauliere criterion (with focus accuracy of depth of field in optical system and resolution method. Journal of Electronics, 29(7):621–625.

Frontier Research and Innovation in Optoelectronics Technology and Industry – Habib & Lewis (Eds)
© *2019 Taylor & Francis Group, London, ISBN 978-1-138-33178-5*

An occlusion consistency processing method based on virtual-real fusion

C. Zhang, R.C. Xu, C. Han & H.Y. Zhai
School of Computer Science and Technology, Changchun University of Science and Technology, Changchun, Jilin, China

ABSTRACT: In the process of the virtual and the real occlusion coherence processing of digital film and television, the spatial coherence relationship between the real scene and the virtual scene will directly affect the realism-rendering effect of the virtual-real fusion scene. So, the occlusion integration process of the virtual-real fusion plays an important role in the seamless synthesis of a digital movie screen. In order to reduce the complexity of the virtual and the real occlusion consistency process and improve the universality of the virtual and the real occlusion consistency method, a virtual and real occlusion consistency processing method based on hierarchical segmentation is proposed. This method uses three steps to realize the dynamic occlusion rendering of the virtual and real scenes through the experimental analysis. They are hierarchical clustering of depth data, optimization of the layered texture maps, and the consistency rendering of the virtual and real occlusion. Through the experimental results and mathematical analysis, we can see that the virtual-real fusion effect obtained by this method is better for dealing with the virtual and real occlusion consistency. Also, the ideal virtual and real occlusion effect has been achieved, which is totally consistent with the human visual perception mode.

Keywords: augmented reality; virtual and real occlusion; hierarchical segmentation; layered texture mapping

1 INTRODUCTION

With the rapid development of augmented reality technology in recent years, the accuracy requirements of the virtual and real occlusion processing have become more and more (Xu et al., 2013). In the scene of the virtual-real fusion, one of the problems is that the augmented reality technology needs to solve the issue that makes the virtual objects and the real objects completely conform to the human visual pattern in the perspective. Judging whether there is an occlusion relationship between the virtual scene in the real scene and the synthesis of the real scene, it is necessary to compare both at the same time with the same spatial coordinate system. The correctness of the occlusion relationship between the virtual scenes and real scenes in three-dimensional space directly affects the effect of virtual-real fusion from the view of geometric consistency of virtual-real fusion (Yokoya, 1999). Now the commonly used methods to solve the problem of the virtual reality occlusion include: the virtual and real occlusion method for 3D model reconstructing of the real scene, and the virtual and real occlusion method by using the stereoscopic vision to generate a real scene depth map.

The 3D reconstruction of the real scene model realizes the consistency of the virtual and real occlusion based on the virtual reality occlusion mode of scene modeling. In the process of virtual-real fusion, only the modeled real object model needs to be processed to achieve the effect that the virtual object is shielded by the real object if the virtual object is placed behind the real object. In the literature (Newcombe et al., 2011), a depth map of the entire scene is first generated by Kinect in order to perform the real-time fast three-dimensional reconstruction of the real objects in the scene. Then the real-time reconstruction of the scene is combined with the GPU and the scene occlusion edges are accurately extracted to achieve the virtual and the real occlusions. The scene depth information is first acquired through binocular stereo vision based on the depth information, and then it can be judged whether there is any occlusion phenomenon in the virtual scene rendering position (Zheng, 2014). Kim et al. (2003) used binocular vision to construct scene depth information and improved the stereo matching method to avoid real-time computation. The acquired scene depth information accuracy was relatively low. Based on the virtual reality occlusion mode of a weak perspective model, Zhang et al. (2015) proposed a set of implementations of an augmented reality system with a combination of simple combination markers. One kind of 3D pose reduction method using redundant features to implement marking objects is studied in the literature. This method can be adapted to relatively complex enhanced scenes and can effectively reduce the registration module error, but its processing cannot guarantee real-time performance. Then Lepetit and Berger (2000) proposed a virtual and real occlusion recovery strategy based on foreground object contour tracking. The strategy realizes the object contour tracking in the current frame, according to the artificially defined foreground contour on the key frame. The system uses the contour to extract foreground pixels and overlays it on the virtual scene to complete the virtual and real occlusion. However, this method stipulates that the inner pixels of the foreground object outline have the same depth and are always superimposed on the virtual scene. Moreover, this method does not apply to foreground objects whose shape can change (non-rigid). Liu (2011) designed an imaginary scene occlusion recovery strategy based on the virtual reality occlusion mode of stereo vision and foreground contour depth recovery. It can effectively solve the problem of the virtual and real scene occlusion of desktop augmented reality systems. That greatly improves the reality of the system.

This paper proposes a virtual and real occlusion processing method based on the hierarchical segmentation in the virtual and real occlusion consistency process, in order to reduce complexity and improve its universality. This method will construct a layered texture from the perspective of performing clustering of the depth information on a real scene, so as to achieve the dynamic occlusion rendering of the real scenes and the virtual scenes. As can be seen from the experiment above, a virtual and real occlusion consistency processing method based on the hierarchical segmentation is used to ensure that the depth relationship between the real scenes conforms, and that the visual perception characteristics and the preprocessing time is greatly decreased under the premise of the limited resolution scale. Moreover, the process of the occlusion relationship is transformed from the point processing to the surface processing. At the same time, the proposed method makes it possible that the process of the virtual-real integration is not limited by the movement of the viewpoint and the change of the dynamic scene under the condition of the satisfying real-time without the prior conditions.

2 VIRTUAL AND REAL OCCLUSION CONSISTENCY PROCESSING METHOD BASED ON HIERARCHICAL SEGMENTATION

Due to the complexity and poor real-time performance of 3D modeling of real scenes in the virtual and real occlusal consistency process, acquiring the depth map of the scene is relatively easy and can meet real-time requirements. Therefore, a virtual and real occlusion consistency processing method based on hierarchical segmentation is proposed. This method realizes the hierarchical overlay of virtual and real fusion through real-time rendering of hierarchical texture images.

The process of virtual and real occlusion consistency based on hierarchical segmentation is shown in Figure 1. Firstly, depth data clustering is performed according to the depth information corresponding to the original image, and the data clustering results at multiple levels are obtained, wherein: the parameters obtained by the binocular vision calibration can restore the depth information image to the true scene depth; then, combined with the use of depth data clustering results and the original color texture image to obtain a hierarchical texture image corresponding level; then the hierarchical texture image can be rendered to a

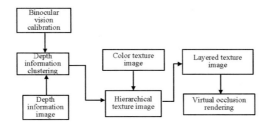

Figure 1.   Virtual occlusion conformance processing based on hierarchical segmentation.

three-dimensional scene in a texture map; finally, you can achieve the current virtual reality under the viewpoint occlusion rendering.

2.1   *Hierarchical clustering of depth data*

Data clustering is a process that divides the collection of sample objects into similar objects, and the sample objects can be physical or abstract. Clustering analysis can be widely applied to many areas, such as market analysis, pattern recognition, and image processing. It solves the same problem: forming different categories of groups, and then assigning each individual to the appropriate group.

In order to realize the hierarchical clustering of deep data, K samples are first selected randomly as the initial cluster center; then according to the Euclidean formula, the distances of the remaining samples to the center of each cluster are calculated, and the sample that is closest to which cluster center, and the samples are divided into clusters with the nearest cluster center; if an error packet is readjusted, after all the sample adjustment after the recalculation of the cluster center; finally, it is checked whether each data sample is assigned to the correct cluster class, that is, whether the objective function has reached a minimum or remains unchanged, and if so, then the depth data clustering ends.

2.2   *Optimization of layered texture mapping*

For simple occlusion of virtual reality, virtual scenes are always placed before real scenes. Virtual scenes will never be obscured by real scenes. In the normal situation, not only can the virtual scene block the real scene, but also the real scene can block the virtual scene.

After deep-data hierarchical clustering, we get the depth information clustering results at different levels before and after. Then we can get the hierarchical texture map of the 3D virtual environment rendering, based on the clustering results of the raw texture and depth information.

2.3   *Consistency rendering of virtual and real occlusion*

The state of view can be described by a one-dimensional vector of seven elements, which depicts the displacement and rotation changes of the point of view. In the process of texture rendering of a 3D virtual scene, the texture rendering plane will react appropriately with the movement of the viewpoint. That is, there is a directional relationship between the texture rendering plane and the viewpoint.

3   EXPERIMENTAL RESULTS AND MATHEMATICAL ANALYSIS

In order to solve the problem of occlusion conformance processing in virtual reality fusion, this paper proposes a hierarchical segmentation-based method of virtual and real occlusion conformance processing. In order to verify the feasibility, effectiveness and robustness of the algorithm proposed in this paper, qualitative and quantitative analysis is carried out using multiple sets of test images in the experimental results and mathematical analysis process. In order to verify the feasibility and validity of the method for processing the virtual and real occlusions based on hierarchical segmentation, the specific processing effects of each stage of the method will be explained in detail from the aspects of hierarchical clustering of texture mapping and the coherence of virtual and real occlusions. Also, the verification image used during the experiment comes from a standard test data set or a special request image.

Figure 2 shows hierarchical texture map clustering (experiment effect one). Figure 2(a) is the real scene texture image from the standard test data set. Figure 2(f) is the visual histogram representation of depth information corresponding to Figure 2(a). According to histogram statistics of disparity depth corresponding to Figure 2(a), the depth histogram statistics data is classified according to the depth data hierarchical clustering method. From Figure 2(f) real scene depth information can be seen in the histogram, clustering obviously there are six sets of depth information, but due to the proximity between group difference in depth may appear too small. Therefore, we use four cluster centers to cluster the depth histograms and divide the real scene texture image into four parts: table lamp texture map, plaster image texture map, desk body texture map, and bookcase wall texture map. Thus, the texture map clustering results shown in Figure 2(b) to Figure 2(e) are obtained.

From the division results of the texture map (layer 4) of Figure 2(b) texture mapping (layer 1) to Figure 2(e), it can be seen that the depth histogram statistical clustering of the real scene texture image can be consistent with the real depth-of-field layered texture map. Therefore, it provides a strong support for the process of virtual reality shielding consistency (experimental effect 1) in Figure 3.

The result of occlusion and superposition of virtual reality fusion is shown in Figure 3(a). The local amplification effect of virtual reality occlusion and superposition is shown in Figure 3(b). The stereo synthesis effect of the left eye image and the right eye image after virtual and real occlusion is shown in Figure 3(c). From Figure 3(b), we can see that the virtual

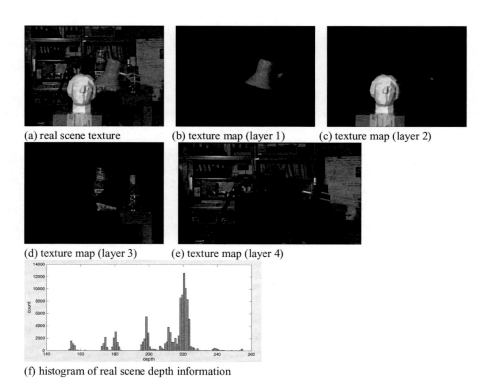

(a) real scene texture   (b) texture map (layer 1)   (c) texture map (layer 2)

(d) texture map (layer 3)   (e) texture map (layer 4)

(f) histogram of real scene depth information

Figure 2.   Hierarchical texture mapping clustering (experimental results 1).

50

character model is covered by the plaster image, and the virtual character model also shields the bookcase wall of the real scene. Further, the occlusion transition between the virtual character and the real scene is also relatively smooth, and the ideal virtual and real occlusion coherence processing effect is obtained.

Figure 4 shows hierarchical texture map clustering (experiment effect two). Figure 4(a) is the real scene texture image from the standard test data set. Figure 4(f) is the histogram

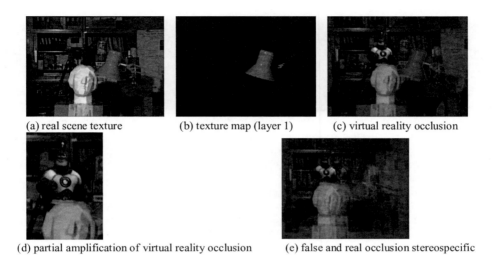

(a) real scene texture   (b) texture map (layer 1)   (c) virtual reality occlusion

(d) partial amplification of virtual reality occlusion   (e) false and real occlusion stereospecific

Figure 3.   Virtual reality occlusion conformance processing (experimental results 1).

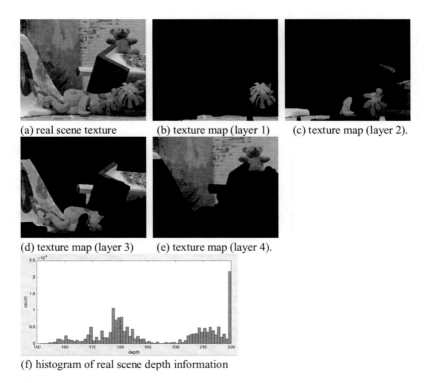

(a) real scene texture   (b) texture map (layer 1)   (c) texture map (layer 2).

(d) texture map (layer 3)   (e) texture map (layer 4).

(f) histogram of real scene depth information

Figure 4.   Hierarchical texture mapping clustering (experiment effect two).

51

representation of the depth information corresponding to Figure 4(a). According to histogram statistics of disparity depth corresponding to Figure 4(a), the depth histogram statistics data is classified according to the depth data hierarchical clustering method. From the histogram of the real scene depth information in Figure 4(f), it can be seen that there are obviously five clusters of depth information. However, because the depth difference between adjacent groups may be too small, we chose to use four cluster centers to cluster the depth histograms. Thus, the texture map clustering results shown in Figure 4(b) to Figure 4(e) are obtained.

From the division results of the texture map (layer 4) of Figure 4(b) texture mapping (layer 1) to Figure 4(e), it can be seen that the depth histogram statistical clustering of the real scene texture image can be consistent with the real depth-of-field layered texture map. Therefore, it provides a strong support for the process of virtual reality shielding consistency (experimental effect 2) in Figure 5.

The result of the occlusion and superposition of virtual reality fusion is shown in Figure 5(a). The local amplification effect of virtual reality occlusion and superposition is shown in Figure 5(b). The stereo synthesis effect of the left eye image and the right eye image after virtual and real occlusion is shown in Figure 5(c). From Figure 5(b), we can see that the virtual character model is covered by the green doll and white soft cloth, and the virtual character model also shields the blue palette of the real scene. In addition, the occlusion transition between the virtual character and the real scene is also relatively smooth, and the ideal virtual and real occlusion coherence processing effect is obtained.

Figure 6 shows hierarchical texture map clustering (experiment effect three). Figure 6(a) is the real scene texture image from the standard test data set. Figure 6(f) is the histogram representation of depth information corresponding to Figure 6(a). According to histogram statistics of disparity depth corresponding to Figure 6(a), and using the depth data hierarchical clustering method, we divide the depth histogram data hierarchically, so that we get the texture map clustering results shown in Figure 6(b) to Figure 6(e).

From the division results of the texture map (layer 4) of Figure 6(b) texture mapping (layer 1) to Figure 6(e), it can be seen that the depth histogram statistical clustering of the real scene texture image can be consistent with the real depth-of-field layered texture map. Therefore, it provides a strong support for the process of virtual reality shielding consistency (experimental effect 3) in Figure 7.

As shown in Figure 7(a), the result of occlusion and superposition is shown as virtual reality fusion. Figure 7(b) shows the local amplification effect of virtual reality occlusion. Figure 7(c) shows the stereo synthesis effect of the left eye image and the right eye image after virtual and real occlusion. From Figure 7(b), we can see that the virtual character model is covered by the right foot of carton and doll. Further, the occlusion transition between the virtual character and the real scene is also relatively smooth, and the ideal virtual and real occlusion coherence processing effect is obtained.

Figure 8 shows hierarchical texture map clustering (experiment effect four). Figure 8(a) is the real scene texture image from the standard test data set. Figure 8(f) is the histogram representation of depth information corresponding to Figure 8(a). According to histogram statistics of disparity depth corresponding to Figure 8(a), and the hierarchical clustering

(a) virtual reality occlusion

(b) partial amplification of virtual reality occlusion

(c) virtual reality occlusion stereospecific

Figure 5.   Virtual reality occlusion conformance processing (experiment effect two).

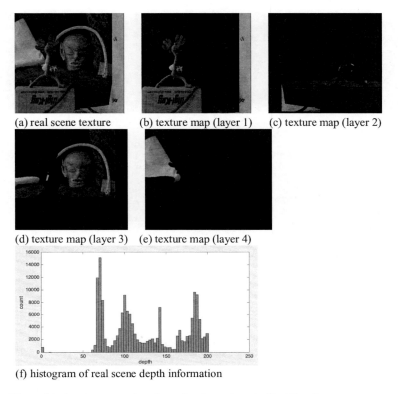

(a) real scene texture   (b) texture map (layer 1)   (c) texture map (layer 2)

(d) texture map (layer 3)   (e) texture map (layer 4)

(f) histogram of real scene depth information

Figure 6.   Hierarchical texture mapping clustering (experiment effect three).

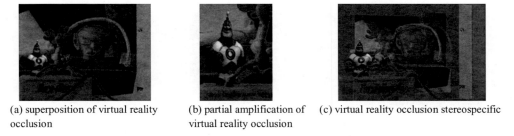

(a) superposition of virtual reality occlusion   (b) partial amplification of virtual reality occlusion   (c) virtual reality occlusion stereospecific

Figure 7.   Virtual reality occlusion conformance processing (experiment effect three).

method of deep data, we divide the depth histogram data into layers, and get the texture mapping results shown in Figure 8(b) to Figure 8(e).

From the division results of the texture map (layer 4) of Figure 8(b) texture mapping (layer 1) to Figure 8(e), it can be seen that the depth histogram statistical clustering of the real scene texture image can be consistent with the real depth-of-field layered texture map. Therefore, it provides a strong support for the process of virtual reality shielding consistency (experimental effect 4) in Figure 9.

The result of occlusion and superposition of virtual reality fusion is shown in Figure 9(a). The local amplification effect of virtual reality occlusion and superposition is shown in Figure 9(b). The stereoscopic synthesis effect diagram of left eye image and right eye image is shown in Figure 9(c). From Figure 9(b), the occlusion local amplification effect can be seen in the virtual character model and brush shielding, virtual characters of occlusion of the right hand brush, and brush on the upper body of the virtual characters and occlusion. Further,

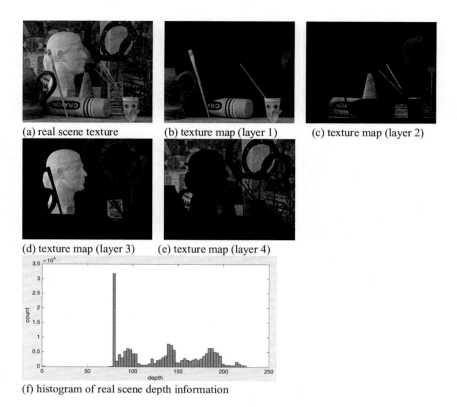

(a) real scene texture   (b) texture map (layer 1)   (c) texture map (layer 2)

(d) texture map (layer 3)   (e) texture map (layer 4)

(f) histogram of real scene depth information

Figure 8.   Hierarchical texture mapping clustering (experiment effect four).

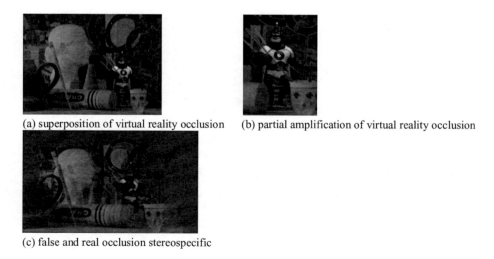

(a) superposition of virtual reality occlusion   (b) partial amplification of virtual reality occlusion

(c) false and real occlusion stereospecific

Figure 9.   Virtual reality occlusion conformance processing (experiment effect four).

the occlusion transition between the virtual character and the real scene is also relatively smooth, and the ideal virtual and real occlusion coherence processing effect is obtained.

Through the above-mentioned four groups of experiments, we can see that the virtual and real consistency processing method based on hierarchical segmentation can effectively achieve the occlusion processing between virtual scenes and real scenes. The method achieves

Table 1. Comparison with other methods.

Occlusion method	Scene change	View movement	Priori conditions	Real-time
literature (Breen et al., 1996)	static state	limited	fixed scene	non-real-time
literature (Berger, 1997)	static state	limited	nothing	real-time
literature (Ong et al., 1998)	dynamic	limited	real scene outline	real-time
literature (Lepetit et al., 2000)	dynamic	limited	real scene outline	real-time
literature (Schmidt et al., 2002)	static state	limited	nothing	real-time
literature (Hayashi et al., 2005)	static state	limited	specific identification	real-time
literature (Fortin et al., 2006)	dynamic	arbitrarily	specific identification	real-time
this method	dynamic	arbitrarily	nothing	real-time

the expected virtual and real occlusion effect and verifies the feasibility and effectiveness of the method.

In order to further illustrate that this method has more advantages than other existing methods, the following four aspects will be compared: whether to support scene change, whether the motion of the viewpoint is limited, whether it needs prior condition, and the real-time property. From the comparisons shown in Table 1 and other existing methods, we can see that the real-time occlusion processing method based on hierarchical segmentation not only supports virtual reality occlusion without prior conditions, but also supports real-time virtual reality occlusion processing under an arbitrary viewpoint.

This paper proposes a virtual and real occlusion processing method based on the hierarchical segmentation in the virtual and real occlusion consistency process in order to reduce complexity and improve its universality. This method will construct a layered texture from the perspective of performing clustering of the depth information on a real scene, so as to achieve the dynamic occlusion rendering of the real scenes and the virtual scenes. As can be seen from the experiment above, a virtual and real occlusion consistency processing method based on the hierarchical segmentation is used to ensure the depth relationship between the real scenes conforms, and the visual perception characteristics and the preprocessing time are greatly decreased under the premise of the limited resolution scale. Moreover, the process of the occlusion relationship is transformed from the point processing to the surface processing. At the same time, the proposed method makes it possible that the process of the virtual-real integration is not limited by the movement of the viewpoint and the change of the dynamic scene under the condition of the satisfying real-time without the prior conditions.

4 CONCLUSION

Aiming at the problems of high modeling complexity, strong restrictive scene change, and poor real-time performance in the virtual and real occlusion coherence processing methods (Guo et al., 2017), this paper proposes a virtual and real occlusion coherence processing method based on hierarchical segmentation. The method consists of three steps: hierarchical clustering of deep data, optimization of hierarchical texture maps, and rendering of virtual and real occlusion consistency. The virtual and real occlusion coherence processing method based on hierarchical segmentation realizes dynamic stratification of a real scene reconstruction model. Therefore, it is more self-adjusting when judging the shielding relationship between a virtual scene and a real scene, and it also reduces the time complexity of the processing process. Through the experimental results and mathematical analysis, we can see that the virtual-real fusion effect obtained by this method is better for dealing with the consistency between the virtual and the real occlusion, and achieves the ideal virtual reality occlusion effect, which is totally consistent with the human visual perception mode.

ACKNOWLEDGMENTS

This work was supported in part by the Jilin Province Science and Technology Project under Grant 20170203003GX and in part by the Jilin Province Science and Technology Project under Grant 20170203004GX. Thanks to the reviewers for their comments and suggestions that led to an improved manuscript.

REFERENCES

Berger, M.O. (1997). Resolving occlusion in augmented reality: A contour based approach without 3D reconstruction. *Conference on Computer Vision and Pattern Recognition* (p. 91). IEEE Computer Society.

Breen, D.E., Whitaker, R.T., Rose, E. & Tuceryan, M. (1996). Interactive occlusion and automatic object placement for augmented reality. In *Computer Graphics Forum* (*15*(3), pp. 11–22). Edinburgh, UK: Blackwell Science Ltd.

Fortin, P.A. & Hebert, P. (2006). Handling occlusions in real-time augmented reality: Dealing with movable real and virtual objects. *The Canadian Conference on Computer and Robot Vision* (p. 54). IEEE Computer Society.

Guo, F. et al. (2017). Research on the problem of seamless integration of virtual reality in augmented reality system. *International Conference on Education, Management*. Information and Mechanical Engineering.

Hayashi, K., Kato, H. & Nishida, S. (2005). Occlusion detection of real objects using contour based stereo matching. *International Conference on Augmented Tele-existence* (pp. 180–186). ACM.

Kim, H., Yang, S.J. & Sohn, K. (2003). 3D reconstruction of stereo images for interaction between real and virtual worlds. *Proceedings of the 2nd IEEE and ACM International Symposium on Mixed and Augmented Reality* (pp. 169–176). Los Alamitos: IEEE Computer Society Press.

Lepetit, V. & Berger, M.O. (2000a). A semi-automatic method for resolving occlusion in augmented reality. In *Proceedings IEEE Conference on Computer Vision and Pattern Recognition* (*2*, pp. 225–230). IEEE.

Lepetit, V. & Berger, M.O. (2000b). Handling occlusion in augmented reality systems: A semi-automatic method. *IEEE and ACM International Symposium on Augmented Reality* (pp. 137–146). IEEE.

Liu, L. (2011). Research on virtual reality occlusion method based on contour depth recovery in augmented reality. *Computer Application and Software*, *28*(1), 220–222.

Newcombe, R.A., Izadi, S., Hilliges, O., Molyneaux, D., Kim, D., Davison, A.J., … Fitzgibbon, A. (2011). KinectFusion: Real-time dense surface mapping and tracking. *Proceedings of the 10th IEEE International Symposium on Mixed and Augmented Reality* (pp. 127–136). Los Alamitos: IEEE Computer Society Press.

Ong, K.C., Teh, H.C. & Tan, T.S. (1998). Resolving occlusion in image sequence made easy. *The Visual Computer*, *14*(4), 153–165.

Schmidt, J., Niemann, H. & Vogt, S. (2002). Dense disparity maps in real-time with an application to augmented reality. *IEEE Workshop on Applications of Computer Vision. IEEE Computer Society* (p. 225). IEEE.

Xu, W.P. et al. (2013). An overview of virtual reality occlusion in augmented reality. *Computer Aided Design and Graphics*, *25*(11), 1635–1642.

Yokoya, N. (1999). Stereo vision based video see-through mixed reality. *Proceedings of the 1st International Symposia on Mixed Reality* (pp. 85–94).

Zhang, G.L. et al. (2015). Research on modeling and registration error of augmented reality system based on visual markup. *Computer Science*, *42*(6), 299–302.

Zheng, Y. (2014). Review and prospect of augmented reality occlusion method. *System Simulation*, *26*(1), 1–10.

Frontier Research and Innovation in Optoelectronics Technology and Industry – Habib & Lewis (Eds)
© 2019 Taylor & Francis Group, London, ISBN 978-1-138-33178-5

Improving the measuring accuracy and reliability of a binocular structured light measurement system

J.R. Zhang, Y.J. Zhang & B. Chen
School of Mechanical Engineering, Xi'an Jiaotong University, Xi'an, Shaanxi, China

ABSTRACT: An approach is proposed to improve the measuring accuracy and reliability of a binocular structured light measurement system, based on multiple measurements. In this study, the traditional binocular structured light measurement system is rebuilt into two monocular structured light measurement systems and one traditional binocular structured light measurement system. Then, the object can be measured independently three times simultaneously, and three pieces of point cloud can be obtained. Finally, a piece of point cloud with higher precision and reliability can be achieved by the multi-view point-data Iterative Closest Points (ICP) registration based on the Moving Least-Squares (MLS) surface. Experiment results show the effectiveness of the proposed method.

Keywords: binocular structured light measurement system, monocular structured light measurement system, multiple measuring, multi-view point-data registration

1   INTRODUCTION

The optical Three-Dimensional (3D) shape measurement technique has been extensively applied in such areas as industrial inspections, reverse engineering and heritage conservation (Chen et al., 2000). However, the accuracy of the optical 3D shape measurement techniques is always influenced by the environment and the reflectivity of the object. Therefore, error is easily introduced into the measurement result. As a traditional optical measurement method, the binocular structured light measurement technique is also affected by these unfavorable factors (Barone et al., 2003).

In order to improve the accuracy and reliability of the structured light measurement technique, much research has been done on this topic. Zhang proposed the accurate calibration method to reduce error in the calibration process for structured light measurement (Zhang, 2014). Gupta introduced the micro phase shifting technique to overcome a variety of global illumination and illumination-defocus effects in measurement results (Gupta et al., 2012). Although the accuracy and reliability of the structured light measurement system is improved by the previous methods, some data with large error still exists in the measuring result. In this study, a method is put forward to enhance the measuring accuracy and reliability of the binocular structured light measurement system. As shown in Figure 1, a typical binocular structured light measurement system is composed of two cameras and a projector. In the actual measuring procedure, a projector projects a series of patterns onto the object, and two cameras capture the images. Then, the interested points are matched in a pair of stereo images (Salvi et al., 2004). Finally, the 3D profile of the object can be calculated with the principle of measurement (Lanman & Taubin, 2009). Although the binocular structured light system inherits the advantage of stereo vision and digital fringe projection techniques, it does not realize all of the potential abilities of these two methods. In the current binocular structured light system, the projector only projects the patterns onto the object to achieve exact matching between two fields of view, but it fails to play a role in computing the 3D profile of the object, which is different from the function of the projector in the monocular structured light technique.

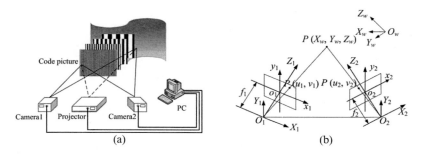

Figure 1.   The principle of a binocular structured light system: (a) system components; (b) coordinate systems of the measurement system.

In the proposed method, the traditional binocular structured light measurement system can be reconstructed into one binocular structured light measurement system and two monocular structured light measurement systems. In other words, the object can be measured three times simultaneously, and three pieces of independent point cloud of the object can be received. Finally, a piece of point cloud with higher precision and reliability can be achieved by the multi-view point-data Iterative Closest Points (ICP) registration based on the Moving Least-Squares (MLS) surface.

The paper is organized as follows. Section 2 introduces the mathematical model of the method. The proposed method is analyzed in Section 3. The experimental results are presented in Section 4. The summary is shown in Section 5.

2   MATHEMATICAL MODEL

2.1   *Feature points matching*

In general, the absolute phase is the best choice to ensure interested points matching (Burton et al., 1995). In this paper, the absolute phase is obtained by the combined encoding method of gray-code and four-step phase shifting. The four-step phase shifting algorithm can be described as follows.

$$I_i(x, y) = I'(x, y) + I''(x, y)\cos[\phi(x, y) + 2i\pi / N] \tag{1}$$

where $I'(x, y)$ is the average intensity, $I''(x, y)$ the intensity modulation, $i = 1, 2, 3, \cdots, N$ and $\phi(x, y)$ can be solved as follows.

$$\phi(x, y) = \arctan\left[\frac{I_4(x, y) - I_2(x, y)}{I_1(x, y) - I_3(x, y)}\right] \tag{2}$$

$\phi(x, y)$ is a wrapped phase, which lies between 0 and 2π rad. However, the absolute phase is required, which can be calculated by adding or subtracting multiples of 2π, as

$$\theta(x, y) = \phi(x, y) + 2k\pi \tag{3}$$

where $\theta(x, y)$ represents the absolute phase, k is the number of the stripe period. In this paper, a temporal phase unwrapping method based on gray-coded fringe is used to obtain the stripe period k (Wang et al., 2013).

2.2   *System calibration and 3D reconstruction*

In order to calculate the 3D point-data of an object, besides interested points matching, parameters of all devices are also very important. Although many camera calibration methods have been proposed in past decades, Zhang's method is widely used in most community-developed tools (Zhang, 1999). In this paper, Zhang's method is used to calibrate the camera and projector. Finally, with the interested points matching results and the device's model, and according to the Line-Plane intersection or Line-Line intersection principle, the 3D profile of the measuring object can be reconstructed (Lanman & Taubin, 2009).

3 THE PROPOSED MEASURING METHOD

3.1 System analysis

In the proposed method, the traditional binocular structured light system is reconstructed into one binocular structured light system and two monocular structured light systems without adding any hardware. The systems are shown in Figure 2.

3.2 Multi-view point-data registration

The ICP algorithm and its variants (Yu, 2013) are classical approaches for registration. However, these kinds of methods always need a perfect initial estimate of transformation between the source point set and the target point set to ensure a global convergence. To improve the accuracy of registration and simplify building the initial estimate of the point-data for the ICP algorithm, a new multi-view point-data registration method is adopted. This method includes three steps. Firstly, the overlapping area of the multi-view point-data is determined. Secondly, the moving least-squares surface is fitted in the overlapping zone of the point-data (Levin, 1998). Finally, an ICP approach is used to register the points onto the MLS surface. The specific processing is described as follows.

Step 1: Determining overlapping area of the multi-view point-data
Given three pieces of input point cloud P_1, P_2, P_3, and $p_i \in P_1$, $p_j \in P_2$, $p_k \in P_3$, and then build up a spherical neighborhood. The spherical center is represented by p_i, r is the radius of the sphere, so this spherical neighborhood contains points of P_2 and P_3 that can be defined as $\bigcup_{p_i \in P_1} (B_1(P_2, P_3, p_i, r))$. In the same way, $\bigcup_{p_j \in P_2} (B_2(P_1, P_3, p_j, r))$ and $\bigcup_{p_k \in P_3} (B_3(P_1, P_2, p_k, r))$ represent the same spherical neighborhood in point-data P_2 and P_3. Finally, the overlapping area among three pieces of point cloud is shown as follows.

$$P_0 = \left[\bigcup_{p_i \in P_1} (B_1(P_2, P_3, p_i, r)) \right] \cup \left[\bigcup_{p_j \in P_2} (B_2(P_1, P_3, p_j, r)) \right] \cup \left[\bigcup_{p_k \in P_3} (B_3(P_1, P_2, p_k, r)) \right] \quad (4)$$

Generally, r is decided by the density of points, which is always as much as about 3 to 5 times the point density.
Step 2: MLS surface definition
Input point cloud $P_0 = \{p_r\}$ and construct set $V_0 = \{v_r\}$, v_r is unit normal vector of p_r. The directional function $n(x)$: $R^3 \rightarrow R^3$ and energy function $e(x,a)$: $R^3 \cdot R^3 \rightarrow R^3$ are defined as follows.

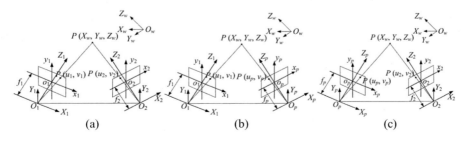

Figure 2. The proposed measurement systems: (a) binocular structured light system; (b) monocular structured light system 1; (c) monocular structured light system 2.

$$n(x) = \frac{\sum_{r=1}^{k} v_r \theta(\| x - p_r \|)}{\| \sum_{r=1}^{k} v_r \theta(\| x - p_r \|) \|}$$ (5)

$$e(x, a) = \sum_{r=1}^{k} \langle x - p_r, a \rangle^2 \theta(\| x - p_r \|)$$ (6)

where p_r represents k points in point cloud P_0, these points are close to the space point x, $\theta(x) = e^{-x^2/h^2}$ is a Gaussian weight function, h is a scale parameter.

By analyzing, the points on the MLS surface are the extreme points. They should make the energy function reach the minimum value along the direction function. So the MLS surface S_{MLS} can be defined as follows.

$$S_{MLS}(x) = \{x \mid x = y, \min_{y \in L(x, n(x))} e(y, n(x))\}$$ (7)

According to a necessary condition for the existence of extreme points for a function, the derivation calculus is shown in Equation 7, and the result is represented as follows.

$$S_{MLS}(x) = (n(x))^T \cdot \frac{\partial}{\partial y}(e(y, n(x)))|_{y=x} = 0$$ (8)

Notice that the MLS surface is an implicit surface, which has smoothing continuity and is suitable for approximating the original sampling surface.
Step 3: Multi-view point-data ICP registration based on MLS surface

The concrete process using the ICP and MLS surface to solve the multi-view point-data registration is explained as follows.

- Input point cloud $p_1' = \{p_i'\}$, and then calculate extreme projection point q_i of point p_i' on the MLS surface, which is defined by P_0. (p_i', q_i) which presents a pair of corresponding points.
- Select the objective function, which is defined by the minimum square sum of the Euclidean distance between those corresponding points: $\min d(R_i, t_i) = \sum_{i=1}^{k} \| q_i - R_i \cdot p_i - t_i \|^2$. Using ICP algorithm for every point in p_1', when $d(R_{k-1}, t_{k-1}) - d(R_k, t_k) \le \varepsilon$, the iteration processes stop, and the rotation matrix R_k and the translation vector t_k have been obtained. P_1' has been transformed into P_1''.
- Based on the definition of overlapping area P_0, the overlapping zone among P_1'', P_2', P_3' can be expressed as follows. $P_0' = [\bigcup_{p_i \in P_1''} (B_1(P_2', P_3', p_i, r))] \cup [\bigcup_{p_j \in P_2'} (B_2(P_1'', P_3', p_j, r))] \cup [\bigcup_{p_k \in P_3'} (B_3(P_1'', P_2', p_k, r))]$ and the MLS surface is rebuilt by P_0'.
- Repeat the three steps above, until all pieces of point cloud match onto the MLS surface.

4 EXPERIMENTS

The experimental system is composed of a Digital Light Processing (DLP) projector and two cameras with two 12 mm focal length. The system is calibrated by Zhang's method. In the experiments, a flat board (40 mm × 30 mm) and a ball (radius: 20 mm) have been measured, respectively. Figure 3 shows the measurement results of the flat board. Figure 3(a) shows the results of the binocular structured light system. There are some points with larger error which must be eliminated for a higher accuracy measurement. Figures 3(b) and 3(c) present the results of two monocular structured light systems respectively. The measurement results of the proposed method are shown in Figure 3(d). For further comparison, Table 1 shows the statistics results of the measurement results of the flat board. The error of measuring results is defined as the distance between measuring point and the fitted plane. Figure 3 and Table 1 show that the proposed method is effective.

The measurement results of the ball are shown in Figure 4. Table 2 shows the statistics of the measuring error. The measurement error of the ball is defined as the distance from the measuring point-data to the fitted spherical surface. The radius of the ball is 20.050 mm, which is measured by a Coordinate Measuring Machine (CMM). The measurement results of the binocular structured light measurement system are: the fitted radius of the ball is 20.252 mm, and the max error and standard deviation of error are 1.792 mm and 0.218 mm, respectively. The results of the measuring data processed by the proposed method are: the fitted radius of the ball reaches to 19.923 mm, which is the closest one to the actual radius, the max error reduces to 0.565 mm, and the standard deviation of error is 0.134 mm. These errors are smaller than the results of binocular structured light system.

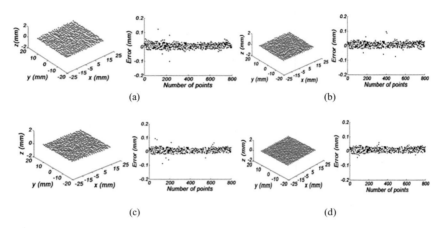

Figure 3. Measurement results of the flat board: (a) binocular structured light system; (b) monocular structured light system 1; (c) monocular structured light system 2; (d) the proposed method.

Table 1. Measurement results of the flat board (mm).

Statistics	Binocular structured light system	Monocular structured light system 1	Monocular structured light system 2	Proposed method
Max error	0.168	0.124	0.117	0.086
Std.	0.023	0.026	0.025	0.021

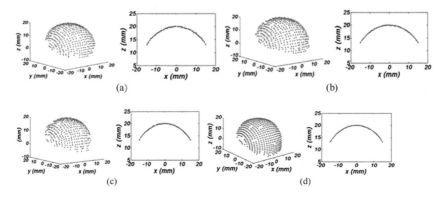

Figure 4. Measurement results of the ball: (a) binocular structured light system; (b) monocular structured light system 1; (c) monocular structured light system 2; (d) the proposed method.

61

Table 2. Measurement results of the ball (mm).

Statistics	Binocular structured light system	Monocular structured light system 1	Monocular structured light system 1	Proposed method
Fitted radius	20.252	19.821	19.836	19.923
Max error	1.792	1.834	1.801	0.565
Std.	0.218	0.185	0.179	0.134

5 CONCLUSION

In this study, the conventional binocular structured light measurement system is reconstructed into one binocular structured light measurement system and two monocular structured light measurement systems. Therefore, three pieces of independent point cloud of the same region of the object can be received. Finally, the data processing is achieved by ICP registration based on MLS surface, and the large error of measurement can be removed. The experiment results verified that a higher measuring accuracy and reliability can be achieved for the binocular structured light measurement system by our method. Of course, this method is not trouble-free. Comparing with the traditional binocular structured light measuring method, the consumed time for computing is a little longer, because of the complex data processing procedure.

ACKNOWLEDGMENTS

The work is supported by National Natural Science Foundation of China (NSFC 51375377) and Natural Foundation Research Project of Shaanxi Province (2017ZDJC-21).

REFERENCES

Barone, S., Curcio, A. & Razionale, A.V. (2003). A structured light stereo system for reverse engineering applications. In *IV Seminario Italo-Espanol, Reverse Engineering Techniques and Applications*. Cassino, Italy.

Burton, D.R., Goodall, A.J., Atkinson, J.T. & Lalor, M.J. (1995). The use of carrier frequency shifting for the elimination of phase discontinuities in Fourier transform profilometry. *Optics and Lasers in Engineering, 23*(4), 245–257.

Chen, F., Brown, G.M. & Song, M. (2000). Overview of 3-D shape measurement using optical methods. *Optical Engineering, 39*(1), 10–22.

Gupta, M. & Nayar, S.K. (2012). Micro phase shifting. In *Computer Vision and Pattern Recognition (CVPR); Proc. IEEE., RI. USA, 16–21 June 2012* (pp. 813–820). IEEE.

Lanman, D. & Taubin, G. (2009). Build your own 3D scanner: 3D photography for beginners. In *ACM SIGGRAPH 2009 Courses* (p. 8). New Orleans, Louisiana: ACM.

Levin, D. (1998). The approximation power of moving least-squares. *Mathematics of Computation, 67*(224), 1517–1531.

Salvi, J., Pagès, J. & Batlle, J. (2004). Pattern codification strategies in structured light systems. *Pattern Recognition, 37*(4), 827–849.

Wang, Z.G., Fu, Y., Yang, J., Xia, G. & Wang, J. (2013). Three-dimensional shape measurement based on a combination of gray-code and phase-shift light projection. In *Sixth International Symposium on Precision Mechanical Measurements; Proc. SPIE., Guiyang, China, 10 October 2013* (pp. 89163Y-1–7). SPIE.

Yu, M.R., Zhang, Y., Li, Y. & Zhang, D. (2013). Adaptive sampling method for inspection planning on CMM for free-form surfaces. *The International Journal of Advanced Manufacturing Technology, 67*(9–12), 1967–1975.

Zhang, Y. (2014). Fast approach to checkerboard corner detection for calibration. *Optical Engineering, 53*(11), 112203-1-112203-6.

Zhang, Z.Y. (1999). Flexible camera calibration by viewing a plane from unknown orientations. *The 7th International Conference on Computer Vision; Proc. IEEE, Kerkyra, Greece, 20–27 September 1999.*

Frontier Research and Innovation in Optoelectronics Technology and Industry – Habib & Lewis (Eds)
© 2019 Taylor & Francis Group, London, ISBN 978-1-138-33178-5

Design of an autofocus method based on an off-axis reflective zoom optical system

J.J. Zhu, W.W. Zhu, K. Zhang, X.X. Gao, M.J. Sun & Y. Liu
China North Vehicle Research Institute, Beijing, China

ABSTRACT: An automatic focus technique has been developed on an off-axis reflective zoom optical system for the first time, to the best of our knowledge. The off-axis three-mirror reflective zoom optical system is based on three mirrors. The focus measurement function and the hill-climb search algorithm have been developed and optimized in the system. The results pave the way for a novel object detection technology based on an off-axis reflective zoom optical system.

Keywords: Autofocus; reflective zoom; off axis; hill climb search algorithm

1   INTRODUCTION

Nowadays, intelligent visual detection technology has become increasingly important. The requirement for precise and fast functioning instruments has been great in recent years, particularly for visual detection devices and image identification technologies. Therefore, the automatic focus technique has been developed to replace the manual focus system.

In 1973, Komine developed an autozoom device for a motion-picture camera that can make a changeover from an electric-motor-driven autozoom to a manual zoom (Komine, 1974). And now, the autozoom system is widely used in the optical system of digital cameras (Nozaki, 2001), holography (Ferrante, 1988), microscopy (Hsu et al., 2009) and mobile terminals (Jung & Jung, 2008). However, most of the traditional automatic focus systems are based on transmission zoom optical systems. Compared with transmission systems, reflective systems have unique advantages because there is no central obstruction and no consequent restriction on field of view (Wetherell & Womble, 1978). The off-axis reflective zoom optical system has a wide range of application, because it is free of chromatic aberration and possesses high thermal stability. After the all-reflective zoom optical system was proposed, many kinds of reflective zoom systems have been developed to give a widened field of view and larger relative aperture. In comparison with the traditional coaxial reflective zoom system, the off-axis system has no central obscuration, especially in a large field of view (Korsch, 1980). However, the alignment of an off-axis reflective system is a complex task as it does not have axial symmetry, and thus may take more time to change focus.

In this paper, we report for the first time an autofocus technique on the off-axis three-mirror reflective zoom system, which significantly simplifies the alignment procedure. This smart autozoom system has been analyzed, and the focus measure function has been adjusted and can evaluate the image definition exactly. In the meantime, the hill-climb search algorithm was optimized to match the two-mirror focusing automatically. The procedure obtains high-quality images when the system is in focus. As the design alignment and focusing become far more efficient, the off-axis three-mirror reflective zoom system can be widely used in object detection and other fields.

2   ZOOM METHOD OF THE OFF-AXIS REFLECTIVE SYSTEM

The reflective system in this device contains three mirrors: M_1, M_2 and M_3. They refer to the primary, secondary and tertiary mirrors, respectively. In this threefold zoom system, the primary mirror and Charge-Coupled Diode (CCD) are designed as fixed, and the distance between the secondary mirror and the tertiary mirror is changed with the zoom. This means that M_2 and M_3 need to be adjusted together when focusing. As the optic system was a discontinuous zoom system and has two foci, this indicates that a linear guided stage could be used to move M_2 and M_3 in this system.

The automatic system was then set up. The system realizes the zoom by moving the mirrors. Mirror M_1 and the CCD were fixed and mirrors M_2 and M_3 were set up on a linear guided stage and controlled by stepper motors, respectively. All the mirrors were made of aluminum with silver plating. The CCD (510×492) was fixed at the focus plane and connected with a Data Acquisition board (DAQ), which can translate the analog signals into digital signals. When the computer receives the digital signal from the DAQ, the images can be shown. The stepper motors were controlled by a motor controller. Thus the autozoom was realized. The structure of the autozoom system is shown in Figure 1 and the autozoom operating process is shown in Figure 2.

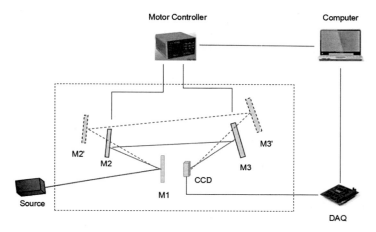

Figure 1. Autozoom system structure.

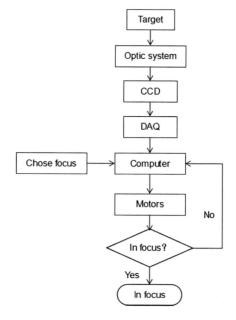

Figure 2. Whole autozoom operating process.

3 ANALYSIS OF FOCUS MEASUREMENT FUNCTION

A Focus Measurement Function (FMF) is calculated to characterize the sharpness of an image. Traditional algorithms include the Tenengrad algorithm, the energy gradient algorithm, the Brenner algorithm, the variance algorithm, and so on (Wang et al., 2008). In order to realize an autozoom, an FMF was developed in this system. The image quality of the off-axis reflective zoom optical system is more sensitive than the traditional coaxial transmission optical system. The image pre-processing was done before the focus was measured. In the focus measurement, we first used parallel light to check the focus center of the image, then calculated the FMF of the bright part.

Six traditional FMF algorithms were compared in the examination: the standard deviation algorithm, the entropy function algorithm, the Sobel algorithm, the Laplace algorithm, the gradient square algorithm and the Prewitt algorithm. In the experiment, mirrors M_2 and M_3 were controlled by the stepper motors and moved step by step; then the FMF values were calculated. We computed 20 images taken by this optical system (as seen in Figure 3a) from defocus to focus to defocus. A resolution testing board and parallel light source were used for imaging. All the FMF values were normalized (as seen in Figure 3b). In Figure 3b, the gradient square algorithm using Equation 1 shows higher precision, stronger unimodal and unbiasedness distribution than other algorithms in the FMF calculation. It also can be seen from Figure 3c that this focus measurement involves less computing time. This means that the gradient square algorithm is the most suitable for our system.

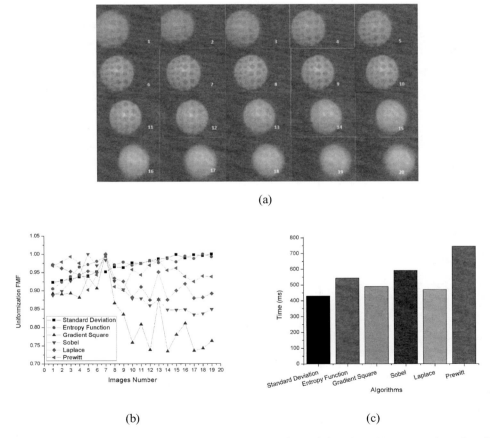

(a)

(b) (c)

Figure 3. Calculated results: (a) images from defocusing–focus–defocusing; (b) computed results of different algorithms; (c) computing time of different algorithms.

$$V = \sum {}_x \sum {}_y \left\{ \left[f(x,y) - f(x+1,y) \right]^2 + \left[f(x,y) - f(x,y+1) \right]^2 \right\} \qquad (1)$$

4 RESEARCH ON THE AUTO-ZOOM CONTROL ALGORITHM

The autozoom control algorithm is used to control the motors moving and searching the focus positions. The hill-climb algorithm has been used widely to control traditional autozoom systems (Zandifar et al., 2002). However the autozoom system is a discontinuous focusing system, so M_2 and M_3 should be controlled independently for focusing. It is very different from traditional coaxial optical systems. The traditional hill-climb algorithm is unsuitable in our system and a new control algorithm for our proposed autozoom system should be developed.

This discontinuous zoom optical system has only two focus positions. This means that if M_2 and M_3 are out of the focus range, there will no image on the image plane, so no matter how M_2 or M_3 move, we cannot get any image from the CCD. Because of this, all the positions of the two mirrors have to be checked. Therefore, the control directions of the mirrors should be known. When we focus manually, M_3 has to be adjusted first as it is more sensitive to the quality of the image. In addition, the focus range of M_3 is shorter than that of M_2. The step lengths of the two mirrors are also different according to the design of the optical system, with the step of M_2 being about twice as large as that of M_3.

In our system, the hill-climb algorithm has been improved to adapt our two-mirror system. The two mirrors controlling the hill-climb algorithm contain six steps (as seen in Figure 4): 1) checking the position of M_3 and M_2 and the move into the focus range; 2) controlling M_3 by coverage navigation and the calculation of the FMF value; 3) moving M_3 to the position where the FMF value is highest; 4) controlling M_2 by coverage navigation and calculation of the FMF value; 5) moving M_2 to the position where the FMF value is highest; 6) in focus. The computer can control the motors with different moving steps in short and long focus, and the step sizes are also different between M_2 and M_3.

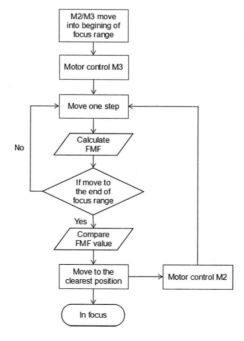

Figure 4.   Controlling algorithm flowchart.

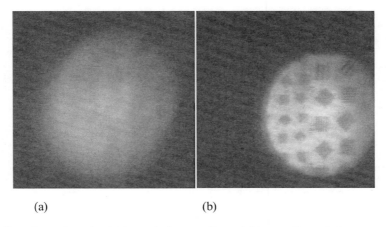

| (a) | (b) |

Figure 5.   Experimental results: (a) image before autofocus; (b) image after autofocus.

5   EXPERIMENTAL RESULTS

Display control operation software was written to realize the automatic off-axis reflective system in which the gradient square algorithm and the improved hill-climb algorithm were used. In the software interface, the real-time image and the virtual control buttons were shown in the window. Through use of the buttons, the system can realize the focus and set the motor step size.

The system showed excellent autofocus action in the experiment. The experimental results using the virtual control panel are shown in Figure 5. In the experiment, mirrors M_2 and M_3 were located in random positions, firstly where the optic system was unable to image. After the automatic control software judged the position and controlled the motors, the mirrors were moved to the beginning of the focus range and a fuzzy image can be seen (as shown in Figure 5a). Then the automatic control software calculated the FMF value and controlled the motors step by step, and the off-axis reflective zoom optical system came into focus, as shown in Figure 5b. Comparing Figure 5b with the images in Figure 3a, it can be seen clearly that the optimal clarity of the manual and auto focus systems are almost the same. The system has been autofocused successfully.

However, we can see that the images in focus are not quite as clear as the traditional transmission optical system. The main reason is that the mirrors of a traditional transmission optical system are usually made of glass, but in this system the mirrors are made of aluminum. The smoothness of the surface is different between the two. Nevertheless, the results prove the feasibility of the smart autofocus method.

6   CONCLUSION

In this paper, a new kind of automatic focus method based on an off-axis reflective zoom optical system has been proposed and demonstrated. In addition, to our knowledge this is the first time an autozoom has been achieved in a reflective optical system. The focus measurement function and the control algorithm were developed to match the smart system. Experimental results support the gradient square algorithm and the hill-climb search algorithm applied in this research. With this autofocus technique, the alignment of an off-axis reflective system is greatly simplified in operation.

REFERENCES

Ferrante, R.A. (1988). *Multidirectional holographic scanner*. U.S. Patent No. US4794237 A. Washington, DC: U.S. Patent and Trademark Office.

Hsu, W.-Y., Lee, C.-S., Chen, P.-J., Chen, N.-T., Chen, F.-Z., Yu, Z.-R., ... Hwang, C.-H. (2009). Development of the fast astigmatic auto-focus microscope system. *Measurement Science & Technology*, *20*(4), 045902.

Jung, J.Y. & Jung, J.I. (2008). *Apparatus and method for controlling an auto-zooming operation of a mobile terminal*. U.S. Patent No. US7330607B2. Washington, DC: U.S. Patent and Trademark Office.

Komine, Y. (1974). *Auto zoom device for a motion picture camera*. U.S. Patent No. US3834796 A. Washington, DC: U.S. Patent and Trademark Office.

Korsch, D. (1980). Design and optimization technique for three-mirror telescopes. *Applied Optics*, *19*(21), 3640–3645.

Nozaki, H. (2001). *Multi-point auto-focus digital camera including electronic zoom*. U.S. Patent No. US20010022626 A1. Washington, DC: U.S. Patent and Trademark Office.

Wang, X., et al. (2008). The research of CCD camera auto-focusing technology based on image definition criterion. *Journal of Changchun University of Science & Technology (Natural Science Edition)*, *31*(1), 11–14.

Wetherell, W.B. & Womble, D.A. (1978). *All-reflective three element objective*. U.S. Patent No. US05967535. Washington, DC: U.S. Patent and Trademark Office.

Zandifar, A., Duraiswami, R., Chahine, A. & Davis, L.S. (2002). A video based interface to textual information for the visually impaired. In *Proceedings of Fourth IEEE International Conference on Multimodal Interfaces, Pittsburgh, PA, USA* (pp. 325–330). doi:10.1109/ICMI.2002.1167016.

Laser technology and applications

Frontier Research and Innovation in Optoelectronics Technology and Industry – Habib & Lewis (Eds)
© 2019 Taylor & Francis Group, London, ISBN 978-1-138-33178-5

Extinction characteristics of biological aggregated particles with different porosity in the far infrared band

X. Chen, Y.H. Hu, Y.L. Gu, X.Y. Zhao & X.Y. Wang
*State Key Laboratory of Pulsed Power Laser Technology (National University of Defense Technology),
Hefei, China*
Key Laboratory of Electronic Restriction Technology, Hefei, China

ABSTRACT: Taking the squash method to measure the reflection spectra of BB0919 spores in the range of 2.5~15 μm, and the complex refractive index of biological material in the range of 8~14 μm was calculated based on the Kramers-Kroning (K-K) relation. Then, we simulated five biological aggregated particles with the same original particle number and different porosity by a cluster-cluster model, the discrete dipole approximation was used to calculate the extinction characteristics of the aggregated particles with different spatial structures in the far infrared band. The results indicated that the larger the porosity of the aggregated particles, the better the extinction performance in the far infrared band. When the value of the original particle number, particle radius, and mass density is 50, 1.5 μm, 1120 kg/m^3 respectively, the mass extinction coefficient of the aggregated particles with the porosity of 0.9033 takes a maximum of 2.262 m^2/g and a minimum of 1.041 m^2/g in the far infrared band.

Keywords: Biological material porosity mass extinction coefficient

1 INTRODUCTION

Biological aerosol is an aerosol including bacteria, viruses, sensitized pollen, mold spores, fern spores, and parasite eggs, which is an important part of atmospheric aerosols. It also has an important impact on environmental monitoring, atmospheric remote sensing, target detection and development. Studies have shown that biological extinction materials have the characteristics of short production cycle, safety and environmental protection. These materials have the potential to be developed into new type of extinction materials.

At present, researches on the optical properties of biological materials have achieved some results: Maria Velazco (Velazco, M. and Dzhongova, E. 2008) calculated the optical constants of Bacillus subtilis within the wavelength of 400~1200 nm based on the T-matrix method and the RGD approximation. Wang Peng (Wang, P. and Liu, H.X. 2016) used Fourier transform infrared spectroscopy to measure the light reflection and absorption of seven microbial materials in the 2.5~25 μm band and calculated the mass extinction coefficient based on the specular reflection method. Gu Youlin (Gu, Y.L. and Wang, C. 2015) calculated the mass extinction coefficient of aspergillus niger spores before and after in activation in the infrared waveband. Li Le (Li, L. and Hu, Y.H. 2015) studied the effect of light incidence angle on the complex refractive index of pear pollen at the range of 2.5~15 μm. These studies are mainly focused on the extinction characteristics of biological particles with a single structure, and the research on the extinction characteristics of biological aggregated particles with different porosity remains to be carried out. Based on the previous study on the biological extinction materials, the reflection spectra of BB0919 spores in the 2.5~15 μm band was measured by the squash method. Based on the measured dates, the complex refractive index of BB0919 spores in the far infrared band was calculated by Kramers-Kroning (K-K) relation. Taking a cluster-cluster model (CCA) to simulate some biological aggregated particles with different porosity.

The mass extinction coefficient and the extinction efficiency factor of the aggregated particles in the 8~14 μm band were calculated by discrete dipole approximation (DDA).

2 THEORETICAL BASIS

2.1 *Material preparation*

The biological extinction material BB0919 spores used in this paper were provided by the Key Laboratory of Ion Beam Biological Engineering, Chinese Academy of Sciences. The sample of BB0919 spores was prepared through the process of "bacteria activation → cultivating in flasks → cleaning with pure water → drying with vacuum freeze-dryer → crushing with Chinese medicine mills". The sample was stored in a drying dish at room temperature. Hitachi Electronics' JSM-6700 scanning electron microscope was used to obtain the scanning electron micrograph of BB0919 spores. Fig. 1 is an scanning electron micrograph of BB0919 spores. It can be seen from Fig. 1 that the biological particles of BB0919 spores are spherical with wrinkles, hollows, and protrusions on the surface, the particles size are distributed in 2~4 μm. The original particles collide with each other to form aggregated particles.

2.2 *Calculation model*

Porosity is a parameter that characterizes the tightness of an aggregated particles' structure, which can be represented by the radius of gyration Rg and the number of original particle N (Kozasa, T. and Blum, T. 1992):

$$P = 1 - N \left[r / \left(\sqrt{5/3} R_g \right) \right]^3 \qquad (1)$$

where r is particle radius and P is porosity. The expression of Rg (Xie Y.X. and Luo W.F. 2004) is:

$$R_g^2 = \frac{1}{2N^2} \sum_{i=1}^{N} \sum_{j=1}^{N} \left| r_i - r_j \right|^2 \qquad (2)$$

where r_i and r_j are the coordinates of the i-th and j-th original particle in the space. In order to study the effect of porosity on the extinction characteristics of biological aggregated particles

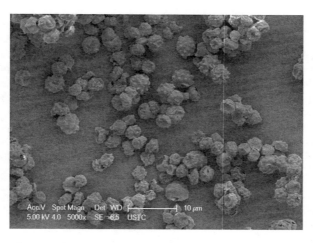

Figure 1.   Scanning electron micrograph of BB0919 spores.

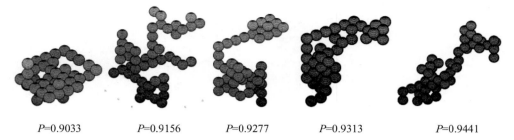

| P=0.9033 | P=0.9156 | P=0.9277 | P=0.9313 | P=0.9441 |

Figure 2.　Schematic of biological aggregated particles.

in the far infrared band, a cluster-cluster (CCA) model was used to simulate five biological aggregated particles with different porosity, the number of original particle $N = 50$, the radius $r = 1.5$ μm. The result is shown in Fig. 2. From the results we can see that when the number of original particle of biological aggregated particles is the same, The smaller the porosity, the tighter the spatial structure of the particles.

Discrete dipole approximation (DDA) is an important theoretical method for studying the extinction characteristics of aggregated particles. This method has the advantage of high iterative efficiency and it can be free from the object structure limitation. Therefore, this paper used the DDA method to calculate the extinction characteristics of the BB0919 biological aggregated particles in the far infrared band. When the spatial structure, radius, and the number of original particle are determined, the DDA method can be used to calculate the scattering parameters such as the extinction efficiency factor Q_{ext}, absorption efficiency factor Q_{abs}, scattering efficiency factor Q_{sca} and asymmetry factor g. The mass extinction coefficient of biological aggregated particles α_e can be calculated by the extinction efficiency factor Q_{ext}:

$$C_{ext} = Q_{ext} \cdot \pi \cdot r_{eff} \tag{3}$$

$$\alpha_e = \frac{C_{ext}}{N \cdot \frac{4}{3}\pi r^3 \rho} \tag{4}$$

where r_{eff} is the equivalent radius of the aggregated particles, ρ is the mass density of BB0919 spores.

3　EXPERIMENT AND SIMULATION

3.1　Complex refractive index

A certain amount of biological material BB0919 spores was weighed, and a powder press machine model number 769YP-15 A was used to make the material into tablets. The reflectance of the tablets in the band of 2.5 to 15 μm was measured by an microscopic infrared spectrometer. We select three points on the tablets and use the average of measurements as the reflection spectrum. The spot size of the microscopic infrared spectrometer is 100 μm × 100 μm, and the speckle within the spot is more smooth and flat. The reflection spectrum measured by microscopic is closer to the ideal specular reflection spectrum. The result is shown in Fig. 3. As the figure shows, in the 2.5~15 μm band, the reflectance of bb0919 spores does not exceed 6% and no less than 1%. In the range of 2.5~10 μm, the change of the reflectance with wavelength is more obvious than in the range of 10~15 μm.

To obtain the complex refractive index of the biological material BB0919 spores, it is necessary to measure the specular reflection spectrum of its entire wave band, but in an actual experimental measurement, it is difficult to measure the reflection spectrum of the entire wave band due to the angle limitation of the experiment. We used a constant extrapolation method to extrapolate the reflectivity $R(\lambda)$ of BB0919 spores in the 2.5~15 μm band to the entire frequency band, where $R(\lambda)$ in the 0~2.5 μm band equals $R(2.5$ μm$)$ and in the 15~100 μm band equals $R(15$ μm$)$. Since the reflectance of the material at wavelengths above 100 μm has negligible effect on the complex refractive index of materials in the 2.5~15 μm band, set 100 μm as the upper integral limit, then we got the $R(\lambda)$ over the entire integration range (0~100 μm) of the sample. Based on the Kramers-Kroning (K-K) relation, the complex refractive index of the microbial sample in the 8~14 μm band was calculated. The real part of complex refractive index n shows in Fig. 4(a) and the imaginary part of complex refractive index k shows in Fig. 4(b). As it can be seen from the figure that the complex refractive index of BB0919 spores increases with wavelength in the 8~10 μm band and decreases with wavelength in the 10~14 μm band, there are some fluctuates.

3.3 *Mass extinction coefficient*

In order to study the effect of porosity on the extinction characteristics in the far infrared band, according to the five different aggregated particles simulated in Fig. 2, we calculated the change of the extinction efficiency factor with wavelength in the 8~14 μm band. The result is shown in Fig. 5. It can be seen from the figure that when the wavelength is over 9.5 μm, the extinction efficiency factor decreases with the wavelength. The mass extinction

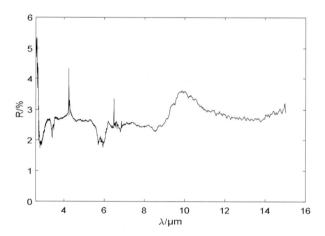

Figure 3.   Reflection spectrum of BB0919 spores within waveband of 2.5~15 μm.

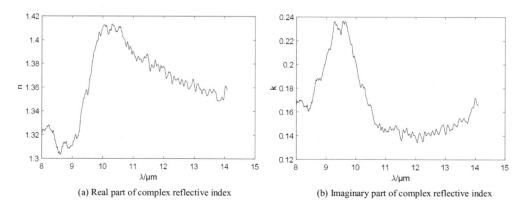

(a) Real part of complex reflective index   (b) Imaginary part of complex reflective index

Figure 4.   Complex reflective index of BB0919 spores in the 8~14 μm band.

74

coefficient of the aggregated particles takes the maximum at 9.5 μm and the minimum at 13 μm. The trend of the extinction efficiency factor changed with the wavelength is basically the same. The smaller the porosity of the aggregated particles, the larger the extinction efficiency factor at the same wavelength. The reason is that when the original particle number of the aggregated particles is the same, the larger the porosity, the more loose the structure of the aggregated particles, thus reducing the absorption and scattering of incident light, and the extinction efficiency factor is the sum of the scattering efficiency factor and the absorption efficiency factor. Therefore, at the same incident wavelength, the larger the porosity, the smaller the extinction efficiency factor.

The mass extinction coefficient of the aggregated particles in the 8~14 μm band can be obtained by the extinction efficiency factor calculated in Fig. 5, set the particle radius $r = 1.5$ μm and the particle mass density $\rho = 1120$ kg/m². The result is shown in Fig 6. The figure shows that the trend of the mass extinction coefficient changed with wavelength is consistent with the extinction efficiency factor, and the aggregated particles with larger porosity have smaller mass extinction coefficient. According to Formula 3 and 4, when the original particle number, particle radius, and particle mass density are determined, the mass extinction coefficient of the aggregated particles depends only on the extinction efficiency factor. The result shows that the mass extinction coefficient of the aggregated particles with the porosity of 0.9033 reaches a maximum of 2.262 m²/g at a wavelength of 9.5 μm, and a minimum of 1.040 m²/g at a wavelength of 13 μm; For the aggregated particles with the porosity of 0.9441, the mass extinction coefficient has a maximum value of 1.288 m²/g and the minimum value of 0.584 m²/g.

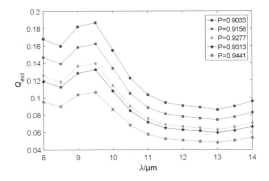

Figure 5.   Change of extinction efficiency factor with wavelength.

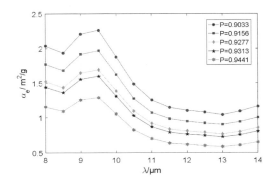

Figure 6.   Change of mass extinction coefficient with wavelength.

4 CONCLUSIONS

The reflectance spectra of BB0919 spores was measured by the squash method, the complex refractive index of the biological material in the 8~14 μm band was calculated based on the Kramers-Kroning (K-K) relation. A cluster-cluster model was used to simulate the structure of biological aggregated particles with different porosity. The extinction efficiency factor and mass extinction coefficient of the biological aggregated particles in the 8~14 μm band were calculated by discrete dipole approximation. The results show that:

1. The extinction characteristics of biological aggregated particles depend on the number of original particle, the spatial structure. In the 8~14 μm band, the aggregated particles with smaller porosity have better extinction characteristics.
2. When the original particle number of biological aggregated particles is 50, the particle radius is 1.5 μm, and the particle mass density is 1120 kg/m³, in the 8~14 μm band, the mass extinction coefficient of the aggregated particles with the porosity of 0.9033 takes the maximum value of 2.262 m²/g and the minimum of 1.041 m²/g.

REFERENCES

Gu, Y.L. and Wang, C. 2015. Infrared extinction before and after aspergillus niger spores inactivation. Infrared and Laser Engineering 44(1):36–41.

Kozasa, T. and Blum, T. 1992. Optical properties of dust aggregates. Astronomy and Astrophysics. 263(1–2): 423–432.

Li, L. and Hu, Y.H. 2015. Measurement and analysis on complex refraction indices of pear pollen in infrared band. Spectroscopy and Spectral Analysis 35(1):89–92.

Velazco, M and Dzhongova, E. 2008. Complex refractive index of nonspherical particles in the visible near infrared region—application to Bacillus subtilis spores. Applied Optics 47(33):6183–6189.

Wang, P. and Liu, H.X. 2016. Electromagnetic attenuation characteristics of microbial materials in the infrared band. Applied Spectroscopy 70(9):1456–1463.

Xie Y.X. and Luo W.F. 2004. Fractal characteristic of atmospheric particulate matters. World Sci-Tech R & D 26(6):24–29.

Frontier Research and Innovation in Optoelectronics Technology and Industry – Habib & Lewis (Eds)
© 2019 Taylor & Francis Group, London, ISBN 978-1-138-33178-5

Research into the frequency-modulation efficiency of laser diodes

Xiya Chen, Qiong Yao & Shuidong Xiong
College of Meteorology and Oceanology, National University of Defense Technology, Changsha, Hunan, China

ABSTRACT: Frequency modulation of laser diodes is a key component in Phase-Generated Carrier (PGC) systems which use Michelson interferometers. The value of C is an important parameter in measuring the frequency-modulation properties of laser diodes and plays an important role in PGC demodulation expression. The original expression of C depends on the largest frequency shift and arm difference between the two arms of the Michelson interferometer. However, according to our analysis, the value of C also depends on modulation frequency. With an increase in modulation frequency, the value of C will decrease, indicating that the modulation efficiency also decreases.

Keywords: Frequency modulation, PGC, value of C, laser diodes

1 INTRODUCTION

A laser diode is a key component of an optical fiber sensing system, directly influencing its properties. For a Phase-Generated Carrier (PGC) system in practical use, the frequency-modulation property of the laser diode plays an important role in PGC demodulation expression (Shi et al., 2010; Wu et al., 2012).

In most studies, the value of C depends on the largest frequency shift and arm difference between the two arms of the Michelson interferometer, and this has been clearly demonstrated in a number of papers. It is clear that this is the case in most studies of PGC systems, whether they use Polarization-Maintaining Fiber (PMF) Michelson interferometers, Mach–Zehnder interferometers (Li et al., 1999), or other types. This cannot, however, always be the case when the modulation frequency in the laser diode changes; in such situations, there might be unnoticed problems that could influence the effectiveness of the sensor system.

According to our analysis, the value of C is related to modulation frequency, and when modulation frequency is large enough, the value of C will drop drastically, leading to decreased frequency-modulation efficiency.

2 ORIGINAL EXPRESSION OF C IN A PGC SYSTEM USING A MICHELSON INTERFEROMETER

The optical system uses an unbalanced polarization-maintaining structure in which the two sensor arms of the interferometer are not equal; its structure is depicted in Figure 1.

A Frequency-Modulation Laser (FML) emits a laser beam to the PMF that arrives at a coupler with a split ratio of 3 dB, and this is transmitted into a reference arm and a sensor arm. These beams are then reflected at the ends of the two arms and travel back to create interference that is identified by a detector. The signal of sound pressure and acceleration is transferred into the phase signal of the laser, denoted as ϕ_s.

If the laser is not modulated, the interference signal is defined as follows (Dandridge et al., 1982):

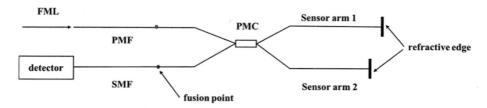

Figure 1. Schematic of polarization-preserving Michelson interferometer light path based on laser source modulation.

$$I = A + B\cos(\phi_0 + \phi_s) \tag{1}$$

where A is a constant. It is related to the input light intensity of the interferometer, polarizer, and coupler insertion loss. The amplitude of the interferometer signal is B and is expressed as $B = kA, k < 1$ (k being interference fringe coherence) and is related to input light intensity, coupling of the coupler, extinction ratio, and reflective end-face loss. The initial phase is ϕ_0. In the practical system, ϕ_0 drifts in the range of 0–2π. This is the reason why the output signal is unstable, also known as signal fading; it has nothing to do with the value of C when there is no frequency modulation in the laser diode.

By modulating the frequency of the laser source represented in Figure 1, a large amplitude of phase modulation is produced so that the working point of this system sweeps dynamically between 0 and 2π. From this, it can be seen that the working point will not stay in the incentive zone. Finally, the system will achieve stable output by using the PGC method and this sufficiently resolves the signal-fading phenomenon.

The path difference between the sensor arm and the reference arm is l, the coherence length of the laser source is L, and the refractive index of the fiber core is n. Coherence can be achieved when $2nl < L$. The phase difference of the interferometer is as follows (Liang, 2008):

$$\phi = \frac{2\pi \times 2nl}{c} v \tag{2}$$

where v is laser frequency and c is vacuum speed. When the laser frequency changes (Δv), the phase change of the interferometer is:

$$\Delta\phi = \frac{2\pi \times (2nl)}{c} \Delta v \tag{3}$$

If laser frequency is a sinusoidal modulation, the phase change can be denoted as $\Delta v = v_0 \cos \omega_0 t$ (where ω_0 is the modulation angular frequency of the laser source and v_0 is the modulation amplitude of the laser frequency). The phase modulation $\Delta\phi$ is:

$$\Delta\phi = \frac{2\pi \times (2nl)}{c} v_0 \cos \omega_0 t = C \cos \omega_0 t \tag{4}$$

and the expression of C is:

$$C = \frac{2\pi \times (2nl)}{c} v_0 \tag{5}$$

This is also known as the phase carrier modulation amplitude. It is related to the modulation amplitude of laser frequency v_0, arm difference, light speed and fiber refractive index.

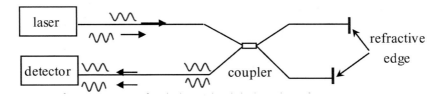

Figure 2.   Interference wave of unbalanced Michelson interferometer.

With laser frequency modulation, the detected signal expression of the fiber interferometer as given in Equation 1 can be altered to:

$$I = A + B\cos[C\cos\omega_0 t + \varphi(t)] \tag{6}$$

Here,

$$\varphi(t) = \sum_i D_i \cos\omega_{si} t + \varphi_0(t), \quad i = 1,2,3,\cdots,n \tag{7}$$

includes low-frequency signal of sound or acceleration signal and random fluctuation of phase $\varphi_0(t)$, which is caused by environmental disturbance. After PGC demodulation processes, we can acquire the signal that the system detected.

PGC demodulation is an intuitive process. In Figure 2 we can see that both laser beams are set off at different times from the laser source, with the timing difference being the source of interference phase modulation caused by laser frequency modulation.

According to our analysis, there might be some drawbacks to previous statements about the value of C, which cannot fully explain the efficiency properties of frequency modulation. To address this, we make a further calculation of the value of C and attempt to explain it more accurately.

3   IMPROVED EXPRESSION OF C AND PROPERTY ANALYSIS

Both laser beams are produced by the same laser source. They are transmitted along an optical fiber axially and have the same polarization direction. This means we can ignore the symbol of the vector and initial phase during the process of derivation.

At distance 0 from the interferometer, expression of the electrical field component of the light wave can be denoted as:

$$E = A\exp[-j(\omega t)] \tag{8}$$

After a distance of z, it becomes:

$$E = A\exp[-j(\omega t - kz)] \tag{9}$$

It can also be expressed as:

$$E = A\exp\{-j[\omega(t - \tau)\} \tag{10}$$

where $\tau = nz/c$, the time that the laser beam takes to travel a distance of z, the refractive index is n. By directly current-modulating the semiconductor laser, frequency modulation is produced. If we ignore all forms of relaxation processes, the light wave at a distance of 0 can be expressed as follows (Rao, 2012):

$$E = A\exp\{-j[\omega t + \frac{\Delta\omega}{\omega_m}\sin(\omega_m t)]\} \tag{11}$$

The derivative of phase of time is the modulation of cosine laser frequency $\omega + \Delta\omega\cos(\omega_m t)$ and ω_m is the frequency of laser source modulation. The greatest frequency shift is $\Delta\omega$. After time τ_0, the expression becomes:

$$E = A\exp\{-j[\omega(t-\tau) + \frac{\Delta\omega}{\omega_m}\sin(\omega_m(t-\tau))]\} \tag{12}$$

The laser from the source is transmitted to the Michelson interferometer and both laser beams will be reflected back to the detector, causing interference. These can be expressed separately as:

$$\begin{cases} E_1 = A_1\exp\{-j[\omega(t-\tau_1) + \frac{\Delta\omega}{\omega_m}\sin\omega_m(t-\tau_1)]\} \\ E_2 = A_2\exp\{-j[\omega(t-\tau_2) + \frac{\Delta\omega}{\omega_m}\sin\omega_m(t-\tau_2)]\} \end{cases} \tag{13}$$

Here, $\tau_1 = 2nl_1/c$ and $\tau_2 = 2nl_2/c$ are the total times of the two laser beams-to transfer from source to detector. The following expression can be deduced from coherence formula $I = <(E_1 + E_2)(E_1 + E_2)^* >$:

$$I = A_1^2 + A_2^2 + 2A_1 A_2\cos\{\omega(\tau_2 - \tau_1) + 2\frac{\Delta\omega}{\omega_m}\sin(\omega_m(\frac{\tau_2 - \tau_1}{2}))\cos[\omega_m(t - \frac{\tau_2 + \tau_1}{2})]\} \tag{14}$$

This can also be expressed as:

$$I = A + B\cos[\omega\tau + 2\frac{\Delta\omega}{\omega_m}\sin(\omega_m\tau/2)\cos\omega_m(t-T)] \tag{15}$$

We set $\tau = \tau_2 - \tau_1$ (time difference between two laser beams transferring from source to detector), $T = (\tau_2 + \tau_1)/2$, $A = A_1^2 + A_2^2$ and $B = 2A_1 A_2$.
The expression above can be written as:

$$I = A + B\cos[\omega\tau + 2\frac{\Delta\omega}{\omega_m}\sin(\omega_m\tau/2)\cos\omega_m(t-T)] \tag{16}$$

We assume that $l = l_2 - l_1$. This is the arm difference between sensor arms 1 and 2. The first item of phase in this formula is $\omega\tau = \frac{2}{c}\omega\cdot n\cdot l$; this is irrelevant to modulation. It is the product of the constant center frequency of the laser source, the refractive index of the optical fiber, and the arm difference of the interferometer. We can see that the variation of this item comes from the shift of laser source center frequency, fluctuation of optical fiber refractive index, and length of sensor arm caused by external environmental influence.

In the formula, the second item of the cosine function is related to modulation. Compared with the former expression, the value of C here should be written as:

$$C = 2\frac{\Delta\omega}{\omega_m}\sin(\frac{\omega_m\tau}{2}) \tag{17}$$

In most cases, $\omega_m\tau/2 << 1$. High-order items of Taylor expansion can be ignored, and the interference formula can be denoted as:

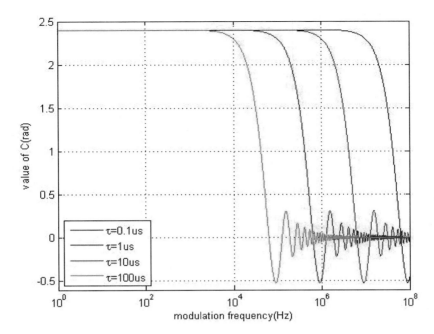

Figure 3.   Value of C changing with modulation frequency and time difference.

$$I = A + B\cos[C \cdot \cos(\omega_m t - \omega_m T) + \varphi] \qquad (18)$$

Here, $C = \Delta\omega \cdot \tau = \dfrac{2nl}{c}\Delta\omega$ is modulation depth, which is consistent with other published research.

In addition,

$$\tau = \tau_2 - \tau_1 = 2n(l_1 - l_2)/c \qquad (19)$$

This means that time difference depends on arm difference between sensor arms 1 and 2. We can see that the value of C relies on the largest frequency shift of the laser diode and arm difference.

However, when $\omega_m \tau/2$ is too large to be ignored it is obvious that the value of C will not be a constant and is also related to modulation frequency: when modulation frequency changes, C will also change. It is necessary to simulate the relationship between C and other parameters to achieve an intuitive expression.

From Figure 3 we can see that when time difference remains unchanged (which means arm difference does not change) within a certain range, and when modulation frequency increases, the value of C will fall drastically, meaning that frequency-modulation efficiency also falls. We can also see that with the increase of time difference, the value of C falls earlier. This means that modulation frequency should be rather small to ensure that modulation frequency efficiency remains stable.

In practical fiber optics, the refractive index is 1.456. According to Equation 19, we know that when $\tau = 0.1$ μs and laser speed is 3×10^8 m/s, arm difference is approximately 10.3 m and the value of C can remain stable when modulation frequency reaches at least 1 MHz. This is quite large for practical PGC systems and demonstrates that the value of C can be quite robust and useful in practical application over a large range of modulation frequencies.

4 CONCLUSIONS

The former expression of modulation depth C is an approximation under the circumstance of $\omega_m \tau \ll 1$. When $\omega_m \tau / 2$ is sufficiently large it cannot be ignored, and we have to consider the influence of frequency modulation. When modulation frequency increases, the frequency-modulation efficiency of the laser diode will decrease. When time difference increases, which also means arm difference increases, the frequency-modulation efficiency falls earlier. The value of C remains quite stable in practical systems for a large range of modulation frequencies.

REFERENCES

Dandridge, A., Tvetan, A. & Gaillorenzi, T. (1982). Homodyne demodulation scheme for fiber optic sensors using phase generated carrier. *IEEE Journal of Quantum Electronics, 18*(10), 1647–1653.

Li, X.-Y., Zhang, L.-K. & Zhang, X.-Z. (1999). Analysis and simulation of PGC detection of interferometric fiber-optic sensor. *Journal of Harbin Engineering University, 1999*(2), 58–64.

Liang, X. (2008). Investigation of Noise Analysis and Suppression Technologies in Fiber Optic Hydrophone System. [D]. *National University of Defense Technology.*

Rao, W. (2012). Study on key techniques of fiber optic vector hydrophone for high resolution seafloor strata detection [D]. *National University of Defense Technology.*

Shi, Q., Tian, Q., Wang, L., Tian, C., Zhang, H., Zhang, M., ... Huang, L. (2010). Performance improvement of phase-generated carrier method by eliminating laser-intensity modulation for optical seismometer. *Optical Engineering, 49*(2), 193–213.

Wu, X., Tao, R., Zhang, Q., Zhang, G., Li, L., Peng, J. & Yu, B. (2012). Eliminating additional laser intensity modulation with an analog divider for fiber-optic interferometers. *Optics Communications, 285*(5), 738–741.

Zhou, X., Tang, W. & Zhou, W. (1998). Realization of PGC scheme interferometric fiber-optic sensor. *Chinese Journal of Lasers, 25,* 411–414.

Frontier Research and Innovation in Optoelectronics Technology and Industry – Habib & Lewis (Eds)
© 2019 Taylor & Francis Group, London, ISBN 978-1-138-33178-5

Experimental study on interference threshold of HgCdTe refrigeration gaze thermal imager in medium wave by pulse laser

X.Z. Cheng, M. Shao, X.K. Miao, Z.F. Hou, B. Bai & L.L. Zhang
Luoyang Electronic Equipment Test Center of China, Luoyang, Henan Province, China
Key Laboratory of Electro-Optical Countermeasures Test and Evaluation Technology, Luoyang, Henan Province, China

ABSTRACT: The research on the interference effect of laser on infrared imaging system is a hot topic recently. At present, there are few researches on the interference of mercury cadmium telluride (be short for HgCdTe) refrigeration gaze thermal imager in medium wave by pulse laser. In this paper, a 3 kHz pulsed laser was used to irradiate the thermal imager. The interference threshold was obtained and the mathematical relationship between the number of saturated pixels and the interference laser power density was obtained by fitting.

Keywords: pulsed laser, HgCdTe refrigeration gaze thermal imager in medium wave, interference threshold, fitting

1 INTRODUCTION

With the rapid development of optoelectronic technology, various types of photo detectors and photoelectric targeting devices have emerged in an endless stream, which have greatly expanded the scope of human vision, and have been widely used in military surveillance such as reconnaissance and early warning, satellite remote sensing, imaging observation, precision guidance, and range measurement (Shi Wenyuan and Dong Liang, 2015; Zhao Xinyu, Qiao Yanfeng, Guo Xiaohai, 2012). Infrared imaging technology has long been a breakthrough in military applications as a night vision equipment, and has been used for search, tracking, guidance and other fields (Che Jinxi and Wang Dong, 2011; Yang Aifen, Zhang Jia and Li Gang, 2015). As an effective means of photoelectric countermeasures, the research on the interference and damage mechanism of laser to infrared imaging system has increasingly attracted the attention of the military of all countries (Chen Zhaobing, Cao Lihua and Wang Bing, 2013). At present, there are few studies on the interference of HgCdTe refrigeration gaze thermal imager in medium wave by pulse laser and it is necessary to carry out the research. In this paper, a laser with a wavelength of 3 μm–5 μm, a repeat frequency of 3 kHz, and a pulse width of 82 ns was used as the interference source to irradiate the HgCdTe refrigeration gaze thermal imager in medium wave. The interference threshold and the mathematical relationship between the number of saturated pixels and the laser power density are obtained.

2 EXPERIMENTAL LAYOUT AND STEPS

The experimental pulse laser is used to interfere with the thermal imager. The output wavelength of the pulsed laser is 3 μm-5 μm. Two of the peak wavelengths are 3.7 μm and 4.1 μm respectively. The average power is about 1.2 W and adjustable. The repetition frequency is 3 kHz and the pulse width is 82 ns. The thermal imager response wavelength is 3.7 μm–4.8 μm, using the 320×256 pixel gaze refrigeration HgCdTe focal plane detector, and the NETD is about 16 mK. The layout of the experimental system is shown as Figure 1. At the exit of the laser, an attenuator is added or the drive current is adjusted to reduce

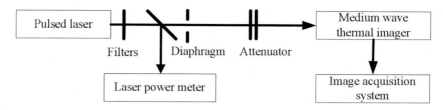

Figure 1. Experimental layout diagram.

the output power appropriately. After the laser being splited, the laser beam splitter will transmit light into the laser power meter, and the other way will directly interfere with the thermal imager. The image acquisition system collects and stores the output image of the thermal imager in real time.

The experimental steps are as follows: Adjusting the azimuth pitching of the laser, so that the laser is vertically irradiated on the lens of the thermal imager. By changing the attenuation multiple of the attenuator or adjusting the driving current, the laser power density received by the thermal imager varies from small to large. The output image of the thermal imager is observed in real time, and the experiment is stopped when the thermal imager is seriously saturated, and the interference images of the thermal imager and corresponding interference laser power are recorded and stored by the computer.

3 EXPERIMENTAL RESULTS

3.1 Detector linear working area

According to the measurements, in the condition where the camera gain is 0 and the bias is 5, when the laser power incident on the optical system is not more than 1.1×10^{-3} W, the thermal imaging detector is in a linear working state, and the corresponding power density should be less than 4.4×10^{-4} W/cm². The laser spot and the three-dimensional energy distribution are shown as Fig. 2a and Fig. 2b respectively. The white spot in the center of the picture is the image point in the stronger part of the laser beam.

3.2 Detector point saturation

According to the measurements, in the condition where the camera gain is 0 and the bias is 5, when the laser power incident on the optical system is 1.1×10^{-3} W, the detector just appears saturation state, the irradiation point is a small bright spot when the laser is irradiated, the gray value of the center area reaches 255, the bright spot is immediately restored after the laser irradiation, and the radiation point can be recovered normal imaging. The calculated power density of laser radiation is 4.4×10^{-4} W/cm², and the point saturation statistical threshold detector is shown in Table 1.

3.3 Detector moderate saturation

According to the measurements, in the condition where the camera gain is 0 and the bias is 5, to further increase the laser power, when the laser power incident to the optical lens is between 1.1×10^{-3} W–2.9×10^{-1} W, it can be seen from the image that the saturation bright spot begins to diffuse into a bright spot when the laser irradiation is irradiated. With the increase of the incident laser power, the surface product of the bright spot increases correspondingly. After the laser irradiation, the laser irradiation is over. The spots then resume and the radiation points can be normal. At this time, the laser radiation corresponding to the power density of the optical lens is 4.4×10^{-4} W/cm²–1.1×10^{-1} W/cm². Laser spot imaging and three-dimensional energy distribution are shown as Fig. 3a and Fig. 3b respectively.

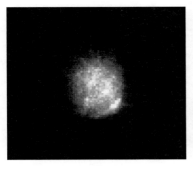

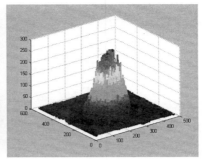

Fig. 2a) Laser spot   Fig. 2b) Three-dimensional energy distribution

Figure 2.   Laser spot imaging image and three-dimensional energy distribution when the detector operates in the linear working area.

Table 1.   Calculation results of detector point saturation threshold.

No.	Spot power at the lens (W)	Saturation threshold (W/cm²)	Average value (W/cm²)	Standard deviation (W/cm²)	Percentage standard deviation	Maximum deviation	Confidence interval (W/cm²)
1	1.08×10^{-3}	4.3×10^{-4}					
2	1.10×10^{-3}	4.4×10^{-4}					
3	1.08×10^{-3}	4.3×10^{-4}					
4	1.23×10^{-3}	4.9×10^{-4}					
5	1.13×10^{-3}	4.5×10^{-4}	4.4×10^{-4}	0.2×10^{-4}	4.55%	15.9%	4.3×10^{-4} -4.5×10^{-4}
6	1.05×10^{-3}	4.2×10^{-4}					
7	1.05×10^{-3}	4.2×10^{-4}					
8	1.10×10^{-3}	4.4×10^{-4}					
9	1.08×10^{-3}	4.3×10^{-4}					
10	1.05×10^{-3}	4.2×10^{-4}					

Note: Percent standard deviation = standard deviation/average; maximum deviation = (data maximum – data minimum)/average; confidence interval confidence is 95%.

Fig. 3a) Laser spot 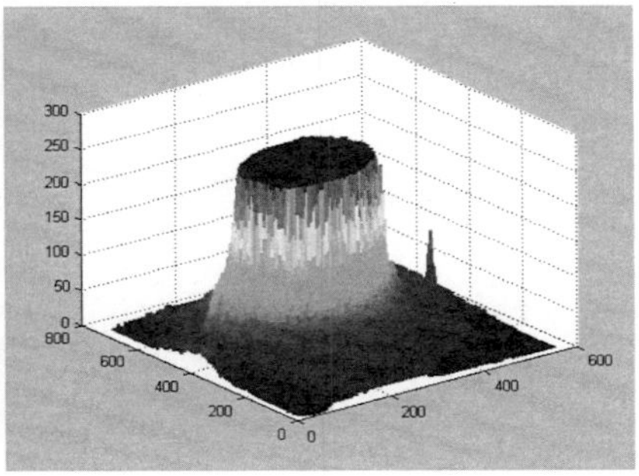   Fig. 3b) Three-dimensional energy distribution

Figure 3.   Laser spot imaging image and three-dimensional energy distribution when the detector is moderately saturated.

3.4   *Detector severe saturation*

According to the measurements, in the condition where the camera gain is 0 and the bias is 5, to further increase the laser power, when the laser power incident to the optical lens is 3.9×10^{-1} W, it can be seen from the image that the saturation bright spot begins to diffuse

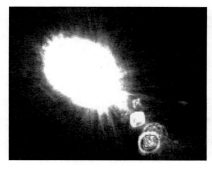
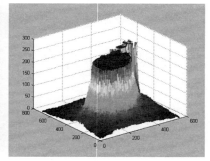

Fig. 4a) Laser spot   Fig. 4b) Three-dimensional energy distribution

Figure 4.   Laser spot imaging image and three-dimensional energy distribution when the detector is severely saturated.

into a bright spot when the laser irradiation is irradiated. When the irradiation spot is irradiated with laser light, the laser spot becomes further larger, and a secondary spot is generated on the laser light path. After the laser irradiation is finished, the bright spot irradiated by the laser cannot be recovered immediately, resulting in the radiation spot not working normally. After a certain period of time, the bright spot can return to normal, and the recovery time increases with the increase of the interference laser power. At the same time, the bright spot can also be completely eliminated by correcting the thermal imager. It is calculated that the power density of the laser radiation corresponding to the optical lens at this time is 1.5×10^{-1} W/cm^2. The laser spot imaging image and three-dimensional energy distribution are shown as Fig. 4a and Fig. 4b respectively.

4   RESULTS ANALYSIS

The pulsed laser was used to interfere with the HgCdTe refrigeration gaze thermal imager in medium wave. With the increase of the laser power density, the detector appeared pixel saturation, moderate saturation, severe saturation and point damage (the damage experiment was not done). The thresholds were 4.4×10^{-4} W/cm^2, 4.4×10^{-4} W/cm^2–1×10^{-1} W/cm^2, 1.5×10^{-1} W/cm^2 when the saturation point was just reached point saturation, moderate saturation and severe saturation.

Because of the short time of action, the damage to the detector mainly depends on the peak power of the pulse laser, and the damage of the continuous laser is mainly dependent on the long time due to the low power. At present, the peak power of the pulse laser is several thousand times higher than that of the continuous laser. Because the detection system is often high-speed, the incident laser has a short standing time in the photosensitive surface of the detector, so the short pulse high power laser is more vulnerable to damage the detector in the short standing time.

According to the experiments, when the laser is irradiated, the spot area increases with the increase of the incident laser power, and the power distribution of the spot also develops from a small spike to a large cylinder. Since the infrared detector is mainly used for weak signal detection at a long distance, the sensitivity of the detector is high. Even if the laser power does not cause damage to the detector, it can interfere with it. When the detector is in point saturation, the saturated pixels have only 1 or 2 pixels, but in the moderate saturation, the number of saturated pixels has increased to about 800. In severe saturation, the number of saturated pixels reaches more than 3000, which has caused great interference with the thermal imager. If interference occurs at a long distance, the number of saturated pixels will increase further due to diffraction of the optical system (Zhang Yuansheng, Xu Liang and

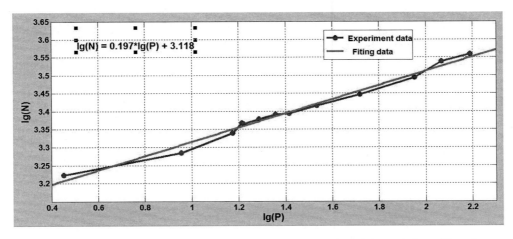

Figure 5. The relationship between the number of saturated pixels and the variation of interference laser power density.

Chen Fang. 2017). By data fitting, the logarithmic relationship between the logarithm of the number of saturated pixels and the incident laser power density is shown in Figure 5. The relationship between the two is $\lg(N) = 0.197 \times \lg(P) + 3.118$.

5 CONCLUSION

Using a pulsed laser as the laser interference source, interference experiments were performed on the HgCdTe refrigeration gaze thermal imager in medium wave. The saturation threshold and logarithmic relationship between the logarithm of the number of saturated pixels and the incident laser power density were obtained, providing a reference for the design, development and finalization of medium-wave laser interference equipment.

REFERENCES

Che Jinxi and Wang Dong. 2011. Experimental study on the effect of laser parameters on the interference effect of infrared imaging systems. *Applied Optics*, 32(5): 992–997.

Chen Zhaobing, Cao Lihua and Wang Bing. 2013. Experimental research on the long-distance interference infrared detector of medium-wave infrared laser. *Infrared and Laser Engineering*, 42(7): 1700–1707.

Shi Wenyuan and Dong Liang. 2015. Wave-infrared laser far-field power measurement system in imaging method. *Optoelectronics Technology*, 30(2): 11–13.

Yang Aifen, Zhang Jia and Li Gang. 2015. Medium wave infrared laser technology for directional infrared countermeasure, *Applied Optics*, 36 (1): 119–125.

Zhang Yuansheng, Xu Liang and Chen Fang. 2017. Medium wave infrared laser and key technology of airborne directional infrared countermeasure system. *Electro-optic and Control*, (5): 56–59.

Zhao Xinyu, Qiao Yanfeng and Guo Xiaohai. 2012, Near-field far-field test method and application of medium-wave infrared laser. *Infrared and Laser Engineering*, 41(1):49–52.

Frontier Research and Innovation in Optoelectronics Technology and Industry – Habib & Lewis (Eds)
© *2019 Taylor & Francis Group, London, ISBN 978-1-138-33178-5*

Designing and technical analysis of the use of combination of PhCs based hydrogel with an enzyme hydrogel as biosensors

Kamal Nain Chopra
MAIT, GGSIP University, Rohini, New Delhi, India

ABSTRACT: This paper presents the Designing and Technical Analysis of the use of combination of PhCs based Hydrogel with an Enzyme Hydrogel as Biosensors; and discusses the qualitative results of various types of such Biosensors. Some numerical results have also been presented along with some explanations. The paper is a very useful attempt to introduce the intricacies of the subject to the budding researchers, scientists, and designers and engineers working in Photonic Crystals based Hydrogel Sensors as Biosensors.

Keywords: Photonic Crystals based Hydrogel Sensors as Biosensors, Sensors based on a combination of an enzyme hydrogel with a PhC hydrogel layer

1 INTRODUCTION

Photonic Crystals (PhCs) based Hydrogel Sensors have recently been used extensively as biosensors; and in fact, have been found to be replacing the other types of biosensors. This is because of the ease of making them, and also controlling their characteristics. In addition, the designing of this type of biosensors depends on two parameters: designing and feedback from the experimentalists and the users. This feedback is useful in finalizing the design of the Photonic Crystals based Hydrogel Sensors. Another advantage is that such biosensors have all the other inherent qualities of Photonic Crystals. Out of all the various types of systems being employed, sensing biomolecules by using a combination of an enzyme hydrogel with a PhC hydrogel layer have been observed to be the best, and is discussed in detail in this paper.

Tavakoli and Tang (2017) have presented an updated review of the hydrogel based sensors for biomedical applications. They have discussed that the biosensors for detecting and converting biological reactions to a measurable signal have gained much attention in recent years. They have considered bioreceptors to be immobilized on hydrogel based biosensors, their advantages and disadvantages, and immobilization techniques. Inan et al (2017) have discussed (i) the recent applications of PC-based biosensors incorporated with emerging technologies, including telemedicine, flexible and wearable sensing, smart materials, and metamaterials; and (ii) the current challenges associated with existing biosensors. Also, an outlook for PC-based biosensors and their promise at the point-of-care (POC) has been provided.

Chen et al (2017) have discussed that (i) the photonic crystal (PC) materials exhibit unique structural colors that originate from their intrinsic photonic band gap, (ii) because of their highly ordered structure and distinct optical characteristics, PC-based biomaterials have advantages in the multiplex detection, biomolecular screening, and real-time monitoring of biomolecules, and (iii) the PCs provide good platforms for drug loading and biomolecule modification, which could be applied to biosensors and biological carriers. At present, various methods are available to fabricate PC materials with variable structure colors, which could be applied in biomedicine. Emphasis is given to the description of various applications of PC materials in biomedicine, including drug delivery, biodetection and tumor screening. Chen et al (2017) have reproduced the results of the pNIPAM hydrogel inverse opal and the schematic of the self-reporting feature of the pNIPAM hydrogel IO particles during drug release. It has been emphasized that their paper will promote greater communication among researchers in the fields of chemistry, material science, biology, medicine, and pharmacy. It is now well established that the Lectin proteins, such as the highly toxic lectin protein, ricin, and the immunochemically important lectin, jacalin, play significant roles in many biological functions, and therefore, it is highly desirable to develop a simple and efficient method to selectively detect lectin proteins. Cai et al (2017) have reported the development of carbohydrate containing responsive hydrogel sensing materials for the selective detection of lectin proteins. It has been explained that the copolymerization of a vinyl linked carbohydrate monomer with acrylamide and acrylic acid forms a carbohydrate hydrogel, which shows specific "multivalent" binding to lectin proteins', and the resulting carbohydrate hydrogels are attached to 2-D photonic crystals (PCs), which brightly diffract visible light. It has been stated that this diffraction provides an optical readout, which sensitively monitors the hydrogel volume. Cai et al (2017) have utilized lactose, galactose, and mannose containing hydrogels to fabricate a series of 2-D PC sensors, which show strong selective binding to the lectin proteins ricin, jacalin, and concanavalin A (Con A), and have discussed that this binding causes a carbohydrate hydrogel shrinkage, which significantly shifts the diffraction wavelength. It has been emphasized that (i) the resulting 2-D PC sensors can selectively detect the lectin proteins ricin, jacalin, and Con A, and (ii) these unoptimized 2-D PC hydrogel sensors show a limit of detection (LoD) of 7.5×10^{-8} M for ricin, 2.3×10^{-7} M for jacalin, and 3.8×10^{-8} M for Con A. It has been pointed out that this sensor fabrication approach may enable numerous sensors for the selective detection of numerous lectin proteins.

Xu et al (2017) have developed a novel three-dimensional photonic crystal hydrogel, which was hydrolyzed by sodium hydroxide (NaOH), and immobilized with butyrylcholinesterase (BuChE) by 1-(3-Dimethylaminopropyl)-3-ethylcarbodiimide hydrochloride (EDC), which have been demonstrated to be excellent in response to sarin (a gas) and a limit of detection (LOD) of 1×10^{-9} mg mL^{-1} has been achieved.

Fenzl et al (2015) have presented a new scheme for sensing biomolecules by combining an enzyme hydrogel with a photonic crystal hydrogel layer, which responds to ionic strength and pH changes, and have demonstrated this unique combination by detecting acetylcholine (ACh) and acetyl cholinesterase (AChE) inhibitors. In this study, the sandwich assembly is composed of layers of photonic crystals, and a polyacrylamide hydrogel functionalized with AChE. It has been pointed out that the photonic crystal film has a red color, and turns dark purple within 2–6 minutes of the enzymatic reaction upon analyze addition, and more importantly that this 3D photonic crystal sensor responds to acetylcholine in the 1 nM to 10 µM concentration range, which in fact includes the relevant range of ACh concentrations in human body fluids. It is really interesting to note that the presence of the acetyl cholinesterase inhibitor neostigmine at concentrations as low as 1 fM has been demonstrated, which is lower than the necessary detection limit for clinical diagnostics. It has been emphasized that this novel concept can be useful in its application in clinical diagnostics, for pesticide and nerve agent detection.

It is now well established that Acetylcholine (ACh) is an important neurotransmitters, which plays an important role in the regulation of body processes like an activator of skeletal muscles. In human cerebrospinal fluid, the concentration of ACh is approximately 5 pmol mL^{-1}; and the measurement of ACh concentrations in human fluids is important for diagnostic analysis and evaluation treatment effects, since a lack of ACh exists in certain diseases e.g. Alzheimer disease. The working is quite simple: The enzyme acetylcholinesterase (AChE) is present at neuromuscular junctions, and hydrolyses ACh, which gets inactivated and the concentration of the transmitter is regulated at the synapses. Interestingly, this mechanism can be inhibited by acetylcholinesterase inhibitors (AChEIs), which results in the hindering of ACh breakdown, which leads to an increase in the concentration of ACh, and thereby affecting the parasympathetic nervous system like reducing heart rate and muscle contractions. However, the AChEI detection schemes have a number of drawbacks e.g. the preparation, characterization, modification and readout of a porous silicon wafer are time-consuming, and require specially trained personnel, and the sensitivity of this approach is only ~ 1/100 of the relevant concentrations of neostigmine in body fluids, which limits the use in point-of-care diagnostics. Further, photonic crystal (PhC) chemical sensors have already been developed for various analytes, and in fact consist of periodically arranged structures of a dielectric material, often stabilized by a polymer. The working principle is simple, in that they reflect light of a certain wavelength that is dependent on the angle of the incident light, the distances of particles within the structure, and the refractive indices of the structure and the surrounding medium; and is based on Bragg's law of diffraction combined with Snell's law of reflection, described by the following equation (Fudouzi 2004, Lee & Braun 2003, Aguirre et al 2010):

$$\sqrt{\frac{8}{3}} D (n^2_{eff} - \sin^2 \theta)^{\frac{1}{2}} = m\lambda$$

where D is the center-to-center distance between particles, n_{eff} is the mean effective refractive index (RI), θ is the angle of incident light, m is the order of diffraction, and λ is the wavelength of the reflected light. The mean effective refractive index n_{eff} is given by the following equation:

$$n^2_{eff} = n^2_p V_p + n^2_m V_m,$$

where n_p and n_m denote respectively the refractive indices of the particles and surrounding medium, and V_p and V_m denote the respective volume fractions.

Preparation of sensor films is quite cumbersome. The hydrogel-based photonic crystal layer is prepared by a technique analogous to the protocol (Fenzl et al 2013, Fenzl et al 2014). Acrylamide (50 mg) and N,N′-methylenebisacrylamide (2.5 mg) are dissolved in a suspension of 1 mL of poly (styrene-co-sodium styrenesulfonate nanoparticles (60 g L^{-1}) in ultrapure water, and subsequently, a solution of 10 µL of 2,2-diethoxyacetophenone in 10 µL of DMSO and 160 mg of ion-exchange resin are added. This is followed by intense vortexing and sonication, and oxygen is removed by bubbling nitrogen through the vial. Then this suspension is injected into a polymerization cell consisting of two microscope slides, and sidewalls consisting of a Parafilm™ spacer with a thickness of 125 µm, and the gel is photo polymerized by UV irradiation at 366 nm (6 W) for 5 h, after which it is fully polymerized before completely drying. Then the film is strongly washed with water, by which the amino groups are hydrolyzed to form carboxy groups by applying a mixture of 900 µL 1 M sodium hydroxide and 100 µL TEMED, and after ~ 6 min, the film is again thoroughly washed with ultrapure water. The enzyme gel is prepared in a comparatively simpler way; a polyacrylamide film is prepared as described above, but the difference is that instead of the photonic crystal dispersion only ultrapure water is used. The transparent gel is allowed to swell in ultrapure water for ~ 24 h; followed by placing the gel in a solution of 0.5 mg AChE (422 U) in 4 mL ultrapure water for 48 h, where the diffusion of the enzyme into the matrix is achieved. AChE is then covalently attached to the hydrogel with 1-ethyl-3-(3-dimethylaminopropyl) carboiimide (EDC) (25 mg mL^{-1}) for 1.5 h. Finally, the gel is washed several times to remove unbounded enzyme. It is important to note that the enzyme films are stored at 4°C.

For studying the Time-resolved effect of solutions of acetylcholine at varying concentrations on the reflected wavelength on the PhC–enzyme film assembly, the reflected light is recorded over time for about 15–16 min for determining the sensitivity of the sensor arrangement towards acetylcholine, and for examining the overall kinetics of the PhC–enzyme gel assembly. Fenzl et al (2014) have studied this in detail, and their results have verified that an increase in the ACh concentration, results in the blue shifting of the reflected wavelength, which correlates with an increasing formation of acetic acid. The most interesting and important observation is, that at a concentration of 10^{-9} M, the formation of the resulting acetic acid is too low to affect the PhC film; and as a result, only a slight shift of the reflected wavelength takes place. In addition, it has been observed that reflected wavelength is maximum in this case, and remains constant for a period as long as 15 minutes even. Hence, the designer has to see that some optimum performance around this value is chosen after taking feedback from the experimentalists and the users.

2 PHOTONIC CRYSTAL–BASED SENSING AND IMAGING OF POTASSIUM IONS

Fenzl et al (2014) have reported on a method for selective optical sensing and imaging of potassium ions using a sandwich assembly composed of layers of photonic crystals and an ion-selective membrane, which in fact represents a new scheme for sensing ions in that an ionic strength-sensitive photonic crystal hydrogel layer is combined with a K^+-selective membrane, the latter being consisting of plasticized poly(vinyl chloride) doped with the K^+-selective ion carrier, valinomycin. It has been observed that the film has a red color, if immersed into plain water, but is green in 5 mM KCl, and purple at KCl concentrations of 100 mM or higher. It has been emphasized that this 3D photonic crystal sensor responds to K^+ ions in the 1 to 50 mM concentration range, which includes the K^+ concentration range encountered in blood, and shows high selectivity over ammonium and sodium ions. Sensor films have also been imaged with a digital camera for studying the effect of the salt concentration on the wavelength of the reflected light of colloidal PhCs in a hydrolyzed PAM hydrogel covered with a K^+-selective membrane. The film is immersed into solutions of KCl, KNO_3, NaCl and NH_4Cl of varying concentrations. From the results of Fenzl et al (2014), it has been verified that for the Potasium compounds, the reflected wavelength/nm falls quite rapidly at salt concentration higher than 10^{-3} mol-L^{-1}. However, for sodium and Ammonium compounds, it remains nearly constant.

3 CELLULOSE-BASED HYDROGELS

Navarra et al (2015) have made an interesting study of the Cellulose-based hydrogels, obtained by tuned, low-cost synthetic routes, are proposed as convenient gel electrolyte membranes. Interestingly, the hydrogels may be prepared from different types of cellulose by optimizing the solubilization and cross linking steps, and the gel membranes can be characterized by various characterizing techniques like infrared spectroscopy, scanning electron microscopy, thermo gravimetric analysis, and mechanical tests in order to investigate respectively the cross linking occurrence and modifications of cellulose resulting from the synthetic process, morphology of the hydrogels, their thermal stability, and viscoelastic-extensional properties. The successful applicability of the proposed membranes as gel electrolytes for electrochemical devices, can be judged by evaluating the Hydrogels liquid uptake capability and ionic conductivity, based on the absorption of aqueous electrolytic solutions. This can be done by studying the redox behavior of electroactive species entrapped into the hydrogels by cyclic voltammetry tests, which in fact show very high reversibility and ion diffusivity.

It is good to note that the hydrogels derived from natural polymers like polysaccharides, are very useful and suitable materials for their applicability in many fields including agriculture, tissue engineering, drug delivery, and biosensors (Chang and Zhang 2011), because of their advantage of being prepared from environmentally-friendly, renewable, and low cost raw materials. It has been established that cellulose combines hydrophilicity with good mechanical properties, which are the competitive characteristics because of the numerous hydroxyl groups, which interact by hydrogen bonds preferentially with water or with hydroxyl groups of adjacent polymer chains (Klemm et al 1998) From the results reported in the literature (Navarra et al 2015), it can be established that the band at 1630 cm^{-1} (adsorbed water) is more intense in hA than in hC. In addition, clear modification in the region 1500–899 cm^{-1} is seen, which can be attributed to an alteration of cellulose crystalline organization because of the alkali treatment. Also, the decreased intensity of the band at 1430 cm^{-1}, known as the crystallinity band, reflects a decrease of the crystallinity degree of cellulose. It has now been established that the etherification reaction leads to the formation of the local hydrophilic domains for water sorption, and also stops the adjacent cellulose chains from establishing intermolecular hydrogen bonds.

Time dependence of the swelling ratio of hC and hA immersed in 0.5 M Na_2SO_4 aqueous solution at RT, is important to study, as it gives an idea about the ability of the membrane to reversibly absorb the swelling solution. Swelling Measurements are done by taking the samples from washed hydrogels, and then vacuum-drying for 24 h, and finally by immersing them in 0.5 M Na_2SO_4 aqueous solution, within a corked Falcon tube. The swelling ratios (SR%) of the hydrogels can be calculated by using the expression given below:

$$SR(\%) = \left\{ \frac{(W_s - W_d)}{W} \right\} \times 100$$

where W_d and W_s represent respectively the weights of dry and swollen hydrogels,. It is customary to take the swollen weight on hydrogel wiped with moistened filter paper for removing any excess liquid. The results of Navarra et al (2015) on the Time dependence of the swelling ratio of hC and hA immersed in 0.5 M Na_2SO_4 aqueous solution at RT as reported in the literature (Navarra et al 2015) make it clear that the hydrogels reach a saturation state within three to five days, and thereafter they are able to hold the absorbed solution without leaking nor drying for a long time ~ six months, when kept in a covered container. It has been noticed that, even after drying, the membrane is able to reversibly absorb the swelling solution. The higher SR of hA may be considered to be related to a wider extension of the chemical cross linking with respect to hC, which implies a higher number of oxygen atoms resulting from the etherification reaction, that are available to interact with water through hydrogen bond. It has also been observed and reported that the variability of the SR is higher in hA than in hC because of the difficulty to manipulate samples of hA, which is softer and stickier than those of hC. Hence, the designer has to study the characteristics of the system in which the PhC hydrogel sensor is to be used, and then design accordingly for the suitable value of the swelling ratio.

Wang et al (2013) have suggested that the hydrogel photonic crystal microparticles (HPCMs) with inverse opal structure can be prepared by a combination of microfluidic and templating technique, the Schematic of whose preparation has been given below:

Recently, Chopra (2018) has studied in detail the Modeling and Designing of Hydrogel Sensors based on Photonic Crystals in Medicine and Biosensors. In this paper, Chopra (2018) has explained explicitly the Functionalization and Designing of photonic crystal hydrogels by using the relevant mathematical equations based on the Flory–Rehner equation for hydrogel swelling and the Bragg-Snell equation for diffraction, and has outlined the design considerations for the photonic crystal hydrogels. In addition, Chopra (2018) has done a detailed overview and technical analysis of the Photonic Crystals and their characterization with emphasis on computation and designing of Photonic Band structure.

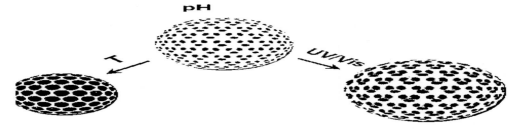

Figure 1.   Schematic of the preparation of Hydrogel Photonic-Crystal microparticles (HPCMs) with inverse opal structure, for example, N-isopropylacrylamide (NIPAm) and metylacrylic acid (MAA) (modified from Wang et al. 2013).

4 CONCLUDING REMARKS

Recently, the importance of hydrogel sensors has been realized in other fields, and hence has resulted in multidisciplinary field research of such sensors, One such marked work is by Qin et al (2017), who have made an interesting study, and have fabricated a hydrogel sensing material by using three dimensional, polymerized colloidal photonic crystals (PCPCs) for highly sensitive and selective detection of mercury ions (Hg^{2+}) in sea water. It has been argued that due to a periodically ordered lattice of the embedded photonic crystals, the hydrogel diffracts light in visible spectral range on the basis of the Bragg's law. Qin et al (2017) have functionalized the hydrogel by spatially distributed –SH groups through cleaving –S–S– bonds in grafted N,N'-cystaminebisacrylamide molecules, and selectively binding with Hg^{2+} ions to form –S–Hg–S– bridge bonds in sea water. It has been reported that the Hg^{2+} binding causes the volume of the hydrogel to shrink, corresponding to a wavelength shift of the light diffracted by the hydrogel, and the shifted wavelength is found to be proportional to the amount of bound Hg^{2+} ions, which enables the quantitative evaluation of Hg^{2+} ions with a limit of detection at 10^{-9} M level in seawater, as measured by a portable spectrometer. It has been emphasized that with this intelligent hydrogel sensors, a simple method for the rapid in situ monitoring and detection of Hg^{2+} ions at low concentrations in sea water, has been demonstrated. Thus, it can be concluded that the topic of Hydrogel Sensors based on Photonic Crystals in Medicine and Biosensors is evolving quite fast with a number of novel applications.

ACKNOWLEDGEMENTS

The author is grateful to the Dr. Nand Kishore Garg, Chairman, Maharaja Agrasen Institute of Technology, GGSIP University, Delhi for providing the facilities for carrying out this research work, and also for his moral support. The author is thankful to Dr. M.L. Goyal, Director General for encouragement. Thanks are also due to Dr. V.K. Jain, Deputy Director for his support during the course of the work. The author is thankful to Prof. V.K. Tripathi, Lasers and Photonics Group, Department of Physics, Indian Institute of Technology, Delhi for useful discussions and various suggestions which have significantly improved the contents of the paper. Finally thanks are due to Dr Rolin and Dr Mary of the Organizing Committee of SOPO 2018, for encouragement and guidance in the preparation and submission of the manuscript.

REFERENCES

Aguirre C. I., Reguera E. & Stein A., 2010. Tunable Colors in Opals and Inverse Opal Photonic Crystals, Adv. Funct. Mater., 20: 2565–2578.

Cai Zhongyu, Sasmal Aniruddha, Liu Xinyu & Asher Sanford A., 2017. Responsive Photonic Crystal Carbohydrate Hydrogel Sensor Materials for Selective and Sensitive Lectin Protein Detection, ACS Sens., 2 (10): 1474–1481.

Chang, C. & Zhang, L. 2011. Cellulose-based hydrogels: Present status and application prospects. Carbohydr. Polym., 84: 40–53.

Chen Huadong, Lou Rong, Chen Yanxiao, Chen Lili, Lu Jingya, & Dong Qianqian, 2017. Photonic crystal materials and their application in biomedicine, Drug Deliv 24: 775–780.

Chopra Kamal Nain, 2018. A detailed overview and technical analysis of the Photonic Crystals and their characterization with emphasis on computation and designing of Photonic Band structure, Atti Fond G. Ronchi 73: 177–215.

Chopra Kamal Nain, 2018. Modeling and Designing of Hydrogel Sensors based on Photonic Crystals in Medicine and Biosensors, Atti Fond G. Ronchi 73: In Press, PP 1–11.

Fenzl C., Kirchinger M., Hirsch T., & Wolfbeis O. S., 2014. Photonic Crystal-Based Sensing and Imaging of Potassium Ions,Chemosensors, 2: 207–218.

Fenzl C., Wilhelm S., Hirsch T., & Wolfbeis O.S., 2013. Optical Sensing of the Ionic Strength Using Photonic Crystals in a Hydrogel Matrix, ACS Appl. Mater. Interfaces, 5: 173–178.

Fenzl Christoph, Genslein Christa, Zöpfl Alexander, Baeumner Antje J., & Hirsch Thomas, 2015. A photonic crystal based sensing scheme for acetylcholine and acetylcholinesterase inhibitors, J. Mater. Chem. B, 3: 2089–2095.

Fudouzi Hiroshi, 2004. Fabricating high-quality opal films with uniform structure over a large area, Journal of Colloid and Interface Science, 275: 277–283.

Inan Hakan, Poyraz, Muhammet, Inci Fatih, Lifson Mark A., Baday Murat, Cunningham Brian T., & Demirci Utkan, 2017. Photonic crystals: emerging biosensors and their promise for point-of-care applications, Chem Soc Rev. 46:366–388.

Klemm, D., Philipp, B., Heinze, T., Heinze, U., & Wagenknecht, W. 1998. Comprehensive Cellulose Chemistry; Wiley-VCH: Weinheim, Germany; Volume 1.

Lee Y.-J. & Braun P. V., 2003. Tunable Inverse Opal Hydrogel pH Sensors, Adv. Mater., 15: 563–566.

Navarra Maria Assunta , Bosco Chiara Dal , Moreno Judith Serra , Vitucci Francesco Maria , Paolone Annalisa, & Panero Stefania, 2015. Synthesis and Characterization of Cellulose-Based Hydrogels to Be Used as Gel Electrolytes, Membranes, 5: 810–823.

Qin Junjie, Dong Bohua, Li Xue, Han Jiwei, Gao Rongjie, Su Ge, Cao Lixin, & Wang Wei, 2017. Fabrication of intelligent photonic crystal hydrogel sensors for selective detection of trace mercury ions in seawater, J. Mater. Chem. C, 5: 8482–8488.

Tavakoli Javad & Tang Youhong, 2017. Hydrogel Based Sensors for Biomedical Applications: An Updated Review, Polymers 9:364.

Wang Jianying, Hu Yuandu, Deng Renhua, Liang Ruijing, Li Weikun, Liu Shanqin, & Zhu Jintao, 2013. Multiresponsive Hydrogel Photonic Crystal Microparticles with Inverse-Opal Structure, Langmuir 29: 8825–8834.

Xu Jiayu, Yan Chunxiao, Liu Chao, Zhou Chaohua, Hu Xiaochun, & Qi Fenglian, 2017. Photonic crystal hydrogel sensor for detection of nerve agent, IOP Conf. Ser. Mater. Sci. Eng. 167: 012024.

Frontier Research and Innovation in Optoelectronics Technology and Industry – Habib & Lewis (Eds)
© 2019 Taylor & Francis Group, London, ISBN 978-1-138-33178-5

RLG coatings – characterization and optimization for improving laser damage threshold, and losses

Kamal Nain Chopra

MAIT, GGSIP University, Rohini, New Delhi, India

ABSTRACT: Ring Laser Gyroscope (RLG) is a very useful electro-optical device used for navigation purposes in Missiles. It is considered much more suitable than the mechanical Gyroscope. The system is based on the interference between two opposite co-rotating laser beams, and the fringes' variation provides very accurate information about the motion of the missile, which is used for the mid course correction. The device being dependent on the electro-optic effect provides very precise information. The key component of the RLG is the mirror, which is required to have scattering loss –3 ppm or better, since scattering results in the lock in problem in the RLG. And so there is no movement of the fringes, because of which the RLG stops working. This paper presents the qualitative results of various RLG Mirror coatings developed with different coating materials, different rates of coating, and different coatings designs. Some numerical results have also been presented along with the possible explanations. The paper seems to be a very useful attempt to introduce the intricacies of the subject to the budding researchers, scientists, and coating designers and engineers.

Keywords: Ring Laser Coatings, Dual Ion Beam Sputtering, Differential Interference Contrast Microscope, Atomic Force Microscope

1 INTRODUCTION

The RLG Mirrors, because of their special characteristics of minimum scattering losses, can be fabricated by using only sophisticated coating units based on Electron Beam (EB) Deposition or Dual Ion Beam Sputtering (DIBS), after optimizing the various Process control parameters like rates of deposition, flow rate of O2 (only in case of EB evaporation for depositing oxide coatings), degree of vacuum inside the coating chamber, and temperature inside the coating chamber while the deposition is going on; and other Process parameters like the effect of overcoats, rate of rotation of the substrate holder for achieving the uniformity of the coatings, choice of glass for substrates, the coating materials, and the Coating designs (Chopra & Maini, 2010, Chopra a, 2018). It is important to note that the optimization is done by performing a large number of experiments with various combinations of the parameters; and that these are applicable only for the unit in use. Of course, this set of parameters can be applied to any other coating system, with minor modifications of the parameters done by the thin films engineers. It is interesting to note that each unit requires a different set of optimized parameters. These optimized parameters are obtained by characterization of coatings and application of optical testing techniques Ring Laser Gyro (RLG) Mirrors are the Laser Mirrors with very low scattering loss (5–10 ppm) and also negligible absorption loss; which in fact, are the requirements for avoiding the lock-in problem encountered in the RLG used for navigation purposes in the missiles (Chopra & Maini, 2010, Chopra a, 2018). These requirements result in their fabrication becoming a very difficult and tedious proposition, along with greatly enhancing their production costs, due to the requirements of the very costly state of art dual ion beam sputtering (DIBS) deposition system, very pure coating materials, zerodur glass for making the RLG substrates with zero expansion, extra clean air room facilities, advanced Techniques for optical testing (like Differential Interference Contrast Microscope), and Thin Films Characterization (like IR/VIS Spectrophotometer and FTIR) for Reflectivity/Transmission measurements, Scatter meter, Cavity loss meter for total loss measurements, and Atomic Force Microscope (AFM) for studying the structure of the coatings. In addition, different aspects like coating materials with absorption edge away from the operational wavelengths and their, different coatings designs including optimum pair design for reducing the losses, and the effect of overcoats for increasing the damage threshold of the RLG Mirror Coatings, have to be thoroughly studied and understood (Chopra b 2015, Chopra c 2015, Zhang et al 2014, Husu et al 2014, Conference Taiwan 2013, Conference USA 2014, and Conference Czhec Republic 2014).

2 OPTIMIZED COATING PARAMETERS

A number of experiments have been performed for optimizing each parameter, and the results are as given below:

2.1 *Choice of glass material for the substrates*

The glass to be used for the RĹG Mirrors should have (i) Nearly zero expansion, (ii) polish able, (iii) high conductivity, (iv) high resistance to scratch, and (v) high thermal runaway (refers to the situations where an increase in temperature changes the conditions in a way that causes a further increase in temperature, often leading to a destructive result. It is a kind of uncontrolled positive feedback). Zerodur and fused silica have been found to be suitable for this purpose. The substrates should have flatness of the order of $\lambda/10$ (λ being the wavelength in use); scratch/dig values equal to 3/5, and microroughness of the order of 3 Angstroms (A°), because the coatings can provide at the most the existing micro roughness, which in fact is required for making the RLG Mirrors with scattering loss of around 3 ppm. It is important to note that microroughness of this value is obtained by float polishing technique. Based on keeping the substrate rotating and floating on suitable fluid, and is obviously a very slow and time consuming process carried out by expert optical engineers.

Scratch/Dig parameter measures the allowable defects in a coating or on the surface of an optical element, and is specified as a numerical value.

Scratch: Scratch numbers are the apparent widths of hairline scratches allowed in units of 0.001 mm. A scratch number of 80 is really 0.08 mm wide.

Dig: Digs represent the apparent diameters of allowable defects such as bubbles, pinholes and inclusions on the surface of lens or coating. Digs are specified in units of 0.01 mm, so a dig value of 50 is actually 0.5 mm diameter allowable inclusion. The allowable number of maximum size digs within the useful area of the lens is one, and the sum of the diameters of all digs cannot exceed twice the diameter of the minimum size dig number specified.

An optical flat is an optical-grade piece of glass lapped and polished to be extremely flat on one or both sides, usually within a few millionths of an inch (about 25 nanometres). They are used with a monochromatic light to determine the flatness of other optical surfaces by interference.

When an optical flat is placed on another surface and illuminated, the light waves reflect off both the bottom surface of the flat and the surface it is resting on. This causes a phenomenon similar to thin-film interference. The reflected waves interfere, creating a pattern of interference fringes visible as light and dark bands. The spacing between the fringes is smaller where the gap is changing more rapidly, indicating a departure from flatness in one of the two surfaces, in a similar way to the contour lines on a map.

A flat surface is indicated by a pattern of straight, parallel fringes with equal spacing, while other patterns indicate uneven surfaces

As can be seen from the results, commonly reported in the literature, and also verified experimentally, the rms value of the micro roughness of the RLG Mirror is very close to that of the uncoated substrate, but not less than that.

2.2 *Effect of overcoat of half wave of SiO2 layer on top layer of RLG Mirror*

It has been found that an overcoat of half wave of SiO2 layer on top layer of RLG Mirror produces the following effects:

i. Increases the mechanical strength of the mirror;
ii. Reduces the residual stress of the multilayer stack of the mirror;
iii. Greatly reduces the interaction with humidity and chemicals of the atmosphere on exposure to the environment; and
iv. Increases the damage threshold of the mirrors by about 30%.

It may be noted that the half wave SiO2 layer is also termed as Absentee layer, since it does not have any effect on the R and T values of the mirrors; and the effect of each half of this layer is cancelled by the effect of the other.

2.3 *Choice of coating materials and coating design*

The coating materials should be pure – 99.999% purity, having absorption edge away from the operating wavelength, along with good adhesion and hardness; and also least residual stress.

For E.B. deposition, the best combination has been found to be TiO2/SiO2. In this case, the coatings have to be carried out in pure O2 atmosphere. Proper rates of flow of O2 have to be experimentally found out, especially for TiO2, since O2 deficiency does not give pure TiO2 deposition, but in fact a mixture of Ti and TiO2; and excess of it forms some TiO3 along with TiO2. So, to get correct stoichiometry, flow rate of O2 has to be optimized, which by our repeated experimentation comes out to be 4 A°/s and 7 A°/s respectively for TiO2 and SiO2.

However, in case of DIBS deposition, SiO2 and Ta2O5 materials are found more useful. Actually, the deposition of Ta2O5 by EB deposition is not easy and does not give good results.

It is well known that the normal mirror design is based on having a multilayer stack of alternate layers of quarter wave thickness of high and low refractive indices, with high refractive index layer as the top layer and bottom layer. It is represented as:

$$(HLHLH \text{ ------ } LHLHLH)n,$$

where H and L denote the quarter wave layers of high refractive index and low refractive index respectively; and n denotes the total number of layers. This design is based on constructive interference of light; and gives very high reflectivity (about 99.9%) for n = 23 or so. In this case, the maximum of the static electric field in the mirror lies on the boundary between the layers; and thus results in high total loss.

However, if we use optimum pair design, in which either (i) the first and last layers are 0.5 H, or (ii) each layer thickness bears a constant ratio (r) to the previous one (say around 0.9) on each side of the middle layer. In each of these cases, the maximum of the static electric field inside the multilayer stack lies away from the boundary, and hence the losses are reduced. These designs are represented by:

$$(0.5HLHLH. \text{ ------ } LHLHL0.5H)n; \text{ and}$$
$$(H \text{ } rL \text{ } r2H \text{ ------ } r2H \text{ } rLH)n$$

In the 1st case the total loss is reduced by about 20%, and in 2nd case by about 30%. However, in the second case, the controlling of layer thicknesses is quite difficult.

In addition, (i) the substrate holder must have planetary motion; (ii)the temperature inside the coating chamber during deposition should be about 3 rotations per minute, so that uniform coatings are obtained.

2.4 *Some other important process parameters*

To achieve high quality coatings, cleaning of substrates and the coating chamber are absolutely necessary. The substrates have to be manually cleaned by ionized water, acetone, and alcohol, followed by gentle rubbing with clean tissue papers. Ultrasonic cleaning is also done, again followed by gentle rubbing with clean tissue papers by giving strokes in same directions diagonally. The substrate after cleaning and also after deposition is observed in DIC Microscope to ascertain that the digs and scratches are seen within the specified range, The image is like as given below.

The chamber has also to be cleaned quite regularly by cleansing agents and thorough drying with pure nitrogen. For best results, the inside walls of the chamber and shutters should be covered by thin Al foils for protection and thereby reducing the effort of cleaning.

2.5   Important test for the smoothness of the RLG Mirror (Measurement of the change in the shape of the reflected wave front from the incident one)

2.5.1   Change in the reflected wave front from the RLG Mirror

It has been observed from the literature, and the experimentation that with the DIBS coated RLG Mirror, the change in the reflected wave front is only 0.005, as compared to 015 in case of EB deposition, and 0.10 in case of IAD (Ion assisted deposition based on using one ion beam for assisting the deposition of layers) So, for the final approval of the successful coating is done by undertaking this test. It has been empirically observed that the RLG Mirror with this much change of wave front meets the scattering and microroughness specifications. It is confirmed by taking observations at various points on the mirror by surface Profiler and the AFM. An idea about the AFM images of SiO2 coatings can be obtained from the following figure:

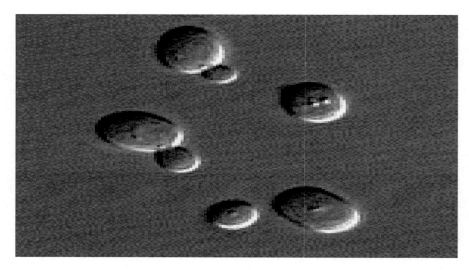

Figure 1.   Differential Interference Contrast (DIC) microscopy—cultured cells imaged by DIC microscopy.

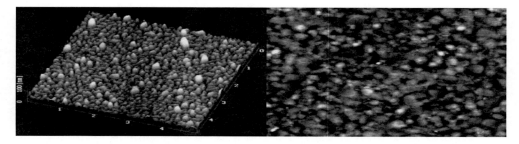

Figure 2.   AFM images of SiO2 coatings under different conditions (Left: Bigger grain Size, Right: Smaller grain Size).

2.5.2 *Optimization of total loss*

The optical cavity is composed of two or more highly reflective mirrors. As the light bounces back and forth inside the cavity, a small portion is transmitted through each mirror. This transmitted light is monitored at one of the mirrors as a function of time. The resulting exponential signal decay (I) can be expressed as:

$$I = Ioexp[-(1-R)tc/2\,L],$$

where (Io) is the initial signal magnitude, (1–R) is the cumulative transmission through the mirrors assuming no absorption or scatter, (t) is time, (c) is the speed of light, and (L) is the cavity length.

The loss is measured by an apparatus given below:

2.5.3 *Environmental testing*

Though the DIBS coated RLG Mirrors in general always meet Environmental tests, yet some tests are performed to verify them. These tests are as given below:

- Water Immersion Test
- A standard test typically performed over 24 hours.
- Salt Solution Test
- A standard test typically performed over 24 hours. By customer request this can be extended up to 7 days.
- Salt Spray (Fog) Test
- This test is typically performed over 24 hours at 47°C. By customer request this can be extended up to 90 days.
- Temperature Test
- A standard test which cycles between hot and cold temperatures, typically between –62°C and + 71°C for 5 hours each. The temperature and duration can be varied to customer's specific requirements. We also have the capability to perform thermal shock tests upon request.
- Abrasion Test

A standard test varying from moderate abrasion with cheesecloth to severe abrasion with an eraser at a known pressure. Other abrasion tests that can be carried out include sand abrasion and windscreen wiper abrasion.

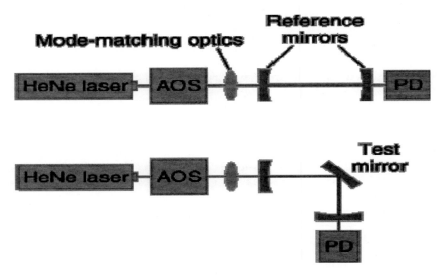

Figure 3. Block diagram of experimental setup for measurement of loss of RLG Mirrors; AOS denotes the Acousto-optical frequency shifter, and PD denotes the Photodiode.

- Adhesion Test
  A standard test performed using adhesive tape to a known specification
- Solubility and Clean ability Test
  Typical chemicals used would be acetone, IPA, or AG101.
- Chemical Degradation Test
  A coated witness sample would be subjected to chemical attack using customer specified chemicals.
- Shock and Vibration Test

This testing is used to simulate the worst care handling and shipping environments of mounted components and complex optical systems to the limits of lifetime exposure. This testing protocol assures products delivered are robust.

3 RESULTS AND DISCUSSION

Ring Laser Gyro (RLG) Mirrors are the Laser Mirrors with very low scattering loss (5–10 ppm) and also negligible absorption loss; which in fact, are the requirements for avoiding the lock-in problem encountered in the RLG used for navigation purposes in the missiles. These requirements result in their fabrication becoming a very difficult and tedious proposition, along with greatly enhancing their production costs, due to the requirements of the very costly state of art dual ion beam sputtering (DIBS) deposition system, very pure coating materials, zerodur glass for making the RLG substrates with zero expansion, extra clean air room facilities, advanced Techniques for optical testing (like Differential Interference Contrast Microscope), and Thin Films Characterization (like IR/VIS Spectrophotometer and FTIR) for Reflectivity/Transmission measurements, Scatter meter, Cavity loss meter for total loss measurements, and Atomic Force Microscope (AFM) for studying the structure of the coatings. In addition, different aspects like coating materials with absorption edge away from the operational wavelengths and their, different coatings designs including optimum pair design for reducing the losses, and the effect of overcoats for increasing the damage threshold of the RLG Mirror Coatings, have to be thoroughly studied and understood.

From the description of the complexity of the experimentation, and very sophisticated instrumentation regarding the Thin films characterization, and Optical testing techniques, it becomes evident that the development of RLG Mirrors is a very difficult and tedious work. Therefore, many establishments engaged in developing RLGs are finding it convenient to purchase the RLG Mirrors from the firm's making and supplying them. This explains as to why only a couple of Thin Films Companies (including M/s OCLI, USA) are able to develop them successfully. The making of such mirrors being difficult, some successful efforts have also been made to develop RLGs by using optical prisms based on total internal reflection, especially by those having great expertise in fabricating very high quality optical components.

4 CONCLUSION

Optimized process control parameters have been presented for the designing and developing of the RLG Mirrors. Practically all the aspects of the process- choice of glass required for making the substrates, their polishing and microroughness requirements, choice of the coating materials, different coating techniques with emphasis on using the state of art DIBS deposition system, various thin films coating designs, and their merits, characterization techniques for optics and the coatings, have been discussed quite at length. These results should be of tremendous use not only for the researchers starting their careers in this fascinating and complicated field, but also for the Optical engineers and designers well entrenched in the field. Though, as discussed in this paper, the fabrication of the RLG Mirrors is very difficult indeed, yet it is felt that the results presented in this paper will definitely motivate the new researchers to undertake this work, and take the field to higher levels.

ACKNOWLEDGEMENTS

The author is grateful to the Dr. Nand Kishore Garg, Chairman, Maharaja Agrasen Institute of Technology, GGSIP University, Delhi for providing the facilities for carrying out this research work, and also for his moral support. The author is thankful to Dr. M.L. Goyal, Director General for encouragement. Thanks are also due to Dr. V.K. Jain, Deputy Director for his support during the course of the work. The author is thankful to Prof. V.K. Tripathi, Department of Physics, Indian Institute of Technology, Delhi for useful discussions and various suggestions which have significantly improved the contents of the paper. The author is also grateful to Ms Mary for her constant support and guidance resulting in considerable improvement in the presentation and readability of the paper.

REFERENCES

Chopra Kamal Nain & Maini Anil Kumar (2010), Book on "Thin Films Coatings and their Applications in Military and Civil Sectors"; Defence Scientific Information and Documentation Centre (DESIDOC), Defence Research and Development Organization (DRDO), Ministry of Defence, Metcalfe House, Delhi, INDIA.

Chopra Kamal Nain, Invited Talk on "Ring Laser Gyro Coatings and their Characterization" (2018) in the Inaugural Session of the National Conference on Optics, Photonics and Synchrotron Radiation for Technological Applications (OPSR-2018) held at Raja Ramanna Centre for Advanced Technology (RRCAT), Indore, INDIA during 29th April – 2nd May, 2018.

Chopra Kamal Nain, Optical Testing of Optical Elements – Technical Analysis and Overview, Atti Fond G. Ronchi, ITALY,70 (2015) 77–101.

Chopra Kamal Nain, Scientific Analysis of Characterization Techniques for Optical Thin Films – An Overview, Atti Fond G. Ronchi, ITALY, 70 (2015) 465–478.

Conference on Optics and Measurement, (Liberec, Czech Republic, 07–10 Oct 2014).

Conference on Optics and Photonics, (San Diego, USA, 17–21 Aug 2014).

Husu H., Saastamoinen T., Laukkanen J., Siitonen S., Turunen J., & Lassila A., Scatterometer for characterization of diffractive optical elements, Meas. Sci. Technol., 25 (2014) 044019.

Intern. Conference on Optical Design and Testing, OPTIC 2013, Optics and Photonics Taiwan, (Taiwan, Dec 2013).

Zhang Wang, Chen Shouqian, Hao Chenglong, Wang Honghao, Zuo Baojun, & Fan Zhigang, Conformal dome aberration correction with gradient index optical elements, Opt. Express, 22, (2014) 3514–3525.

Frontier Research and Innovation in Optoelectronics Technology and Industry – Habib & Lewis (Eds)
© *2019 Taylor & Francis Group, London, ISBN 978-1-138-33178-5*

Investigation of sub-diffraction mode characteristics in a semiconductor plasmonic nanolaser at telecom wavelength

M. Ferdosian & H. Kaatuzian

Department of Electrical Engineering, Amirkabir University of Technology, Tehran, Iran

ABSTRACT: Nanolasers are the key component in photonic integration and in this work we investigate the mode characteristics of an optically pumped Metal-Insulator-Semiconductor (MIS) plasmonic nanolaser structure at the sub-diffraction limit. The nanolaser structure consists of Ag/oxide/InGaAsP materials and operates at the telecom wavelength of 1330 nm. For two different oxides (Al_2O_3, MgF_2) and waveguide sizes, mode-characteristic parameters such as effective refractive index, propagation loss, mode area and confinement factor are calculated. In addition, the threshold gain for each mode is investigated and the appropriate waveguide size according to the required threshold gain is estimated. Based on our calculations, the hybrid mode can propagate with low loss and sub-diffraction size in waveguides of 650 to 800 nm width.

Keywords: plasmonic nanolasers, quantum well, eigenmodes, effective refractive index, propagation loss, mode area, confinement factor, threshold gain

1 INTRODUCTION

Active photonic components such as lasers and light-emitting diodes have important roles in modern optical devices. But like any other photonic component, they have a minimum size limit of half the wavelength, called the diffraction limit. This results in a minimum attainable volume of $(\lambda_0/2n)^3$ for photonic devices and, given the typical telecom wavelength, means a considerable size of hundreds of nanometers while electronic components are capable of integrating with sizes as small as tens of nanometers (Sun et al., 2014). To allow better integration capabilities and competition with their electronic counterparts, researchers have started to investigate new ways and possibilities of breaking the diffraction size limit in these photonic devices, leading to the emergence of the nanophotonic field.

One of the most promising fields of research in realizing these nanophotonic devices is the use of Surface Plasmon Polaritons (SPPs). SPPs can propagate on the surface of a metal waveguide that is only a couple of nanometers thick and could be used to realize nanoscale photonic components (Keshavarz & Kaatuzian, 2015; Livani & Kaatuzian, 2015; Sun et al., 2014; Taheri & Kaatuzian, 2014). However, the most important challenge in realizing these plasmonic devices is the high energy absorption loss in the metal, especially at high frequencies that considerably limit the propagation length (L_{prop}) of plasmonic modes to hundreds of nanometers. To overcome this challenge, researchers have studied active plasmonic devices with a gain medium to miniaturize the plasmonic losses. This has led to the introduction of plasmonic amplifiers and nanolasers (Berini & De Leon, 2012; Livani & Kaatuzian, 2015).

Nanolasers using surface plasmons (SPASERs) have been demonstrated with different structures, from nanoparticles to nanowires and slab waveguides, and a frequency range of near ultra-violet to mid-infrared and even terahertz (Hill et al., 2007; Oulton, 2008). One of the most promising of the structures reported to realize photonic integration is the plasmonic nanolaser using hybrid semiconductor structures, plane waveguides and quantum wells in active mediums (Costantini et al., 2013; Lee et al., 2017). These nanolasers demonstrated a high modulation bandwidth with a three-dimensional confinement of the optical mode below the diffraction limit and a very small footprint.

In this paper, we investigate the modal behaviors in an optically pumped hybrid Metal-Insulator(oxide)-Semiconductor (MIS) plasmonic nanolaser structure based on propagating long-range SPPs. The modal characteristics of the nanolaser around and below the diffraction limit for different propagating eigenmodes are investigated and parameters such as the effective refractive indices, propagation length, mode area of plasmonic and hybrid modes, and mode confinement factors are calculated. Using these parameters, the threshold optical gain for the lossless propagation of each eigenmode is determined and compared with the other modes. This analysis will lead to an estimation of the applicable size of a laser waveguide for operating at the desired telecom wavelength of 1330 nm.

2   NANOLASER STRUCTURE AND MODAL CHARACTERISTICS

In recent years, different structures with different performances have been proposed to realize the concept of plasmonic nanolasers but not all of these structures are suitable for photonic integration applications and, among them, structures based on nanowires and planar waveguides are the most promising candidates. According to our survey of previous work (Hill et al., 2007; Oulton, 2008), it was obvious that structures with a laser diode configuration and using active gain media consisting of quantum wells have superior performance and a lower lasing threshold in comparison to structures using directly pumped dielectric nanowires or planar layers of dielectric as gain media (Oulton, 2008, 2010). In addition, it was notable that the plasmonic contact area of a wider structure has a lower lasing threshold.

Accordingly, we chose a plasmonic nanolaser with a planar waveguide structure (Lee et al., 2017) as the base structure for our study. The nanolaser has an MIS structure and can provide low-threshold lasing owing to its wide plasmonic contact area and Multiple Quantum Well (MQW) gain medium.

The active region is formed by an InGaAsP material system, a well-known and common four-element semiconductor material for active photonic devices operating in the telecom wavelength of 1.3–1.55 um. The radiation wavelength can be adjusted to the desired frequency by changing the component composition and stress levels of the Quantum Wells (QWs) and barrier layers.

Silver was used for the metallic part of the plasmonic waveguide, which has a small permittivity imaginary part and can lower the plasmonic mode attenuation. For the guiding part of the plasmonic waveguide, two different oxide materials, Al_2O_3 and MgF_2, were studied to investigate their refractive index in relation to the nanolaser gain. Al_2O_3 has a relatively high refractive index of 1.75 (at 1330 nm) and MgF_2 is transparent in the visible and infrared bands of the spectrum and has a lower refractive index of 1.4 at 1330 nm.

Figure 1 shows the X–Y plane cross section of the plasmonic nanolaser structure, consisting of 200 nm silver, 5 nm of oxide and a 140-nm-thick dielectric layer. The dielectric

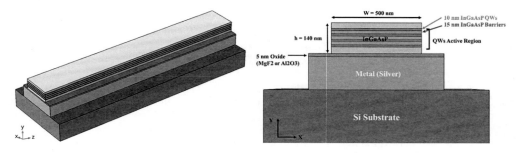

Figure 1.   Schematic of the plasmonic nanolaser: (a) 3D view; (b) X–Y plane with details of dimensions and materials. The waveguide in the Z direction is 5 μm in length.

material is an InGaAsP semiconductor including four QW layers of 10 nm thickness and 15 nm barrier layers with different compositions, sandwiched between two InGaAsP spacer layers.

The two spacer layers confine the optical mode inside the semiconductor waveguide region. Moreover, the bottom layer prevents the depletion of photo-generated electrons and holes in the metallic layer. It should be mentioned that a thick bottom spacer layer will prevent efficient coupling between the plasmonic modes and the quantum-well active region and there is a trade-off between nonradioactive losses and mode coupling in designing the nanolaser waveguide.

As defined in the literature (Maier, 2007), plasmonic structures with oxides of higher refractive index have a larger effective mode index, a higher propagation loss and a smaller propagation length. It means that the plasmonic mode needs a higher marital gain to overcome the losses induced by the large imaginary part of the metal permittivity. On the other hand, there is a well-known trade-off between increasing the propagation length and decreasing the confinement factor that directly leads to shrinkage in the interaction of the plasmonic mode with the gain medium.

Here we compare the plasmonic nanolaser performance for oxides of different refractive index, with MgF_2 and Al_2O_3 being the low and high refractive index oxides, respectively, to determine the material most suitable for the nanolaser structure. First, the mode characteristics of the nanolasers with the two different oxides are calculated and compared. Theoretically (Maier, 2007), only the Transverse Magnetic (TM) mode can propagate on the surface of a pure plasmonic waveguide (metal-insulator). On the other hand, the semiconductor waveguide supports the propagation of Transverse Electric (TE) modes with the electric field parallel to the quantum-well axis. But in the hybrid plasmonic waveguide, a hybrid or quasi-TM mode (EH) can also propagate (Figure 2). Therefore, to investigate the modal behaviors of the structure, we choose three different modes, the first fundamental plasmonic mode (SPP0), the first hybrid mode (EH00) and the fundamental photonic mode.

Figure 2 shows $|E(x,y)|$ and the electric field distribution direction (white arrows) for plasmonic, hybrid and photonic eigenmodes in the nanolaser with a 5 nm MgF_2 oxide layer and a semiconductor waveguide width of 600 nm (in the case of the EH mode, the waveguide width was 900 nm). The fundamental photonic mode was created by placing the semiconductor waveguide on a quartz (n = 1.53) substrate layer.

Figure 2 clearly shows the confinement of the electrical field inside the oxide region for the plasmonic mode (Figure 2a). For the SPP0 and photonic eigenmodes, polarization (direction of white arrows) is mostly perpendicular or parallel, respectively, to the metal surface, and shows the propagation of the TM and TE modes. In the case of the hybrid EH00 eigenmode, polarization is a combination of these two pure modes, with perpendicular polarization near the metal surface and polarization parallel to the metal and quantum wells in the semiconductor waveguide core. This phenomenon will take the energy of the optical mode created by photons generated inside the quantum-well cavities and transfer it to the highly confined plasmonic mode, allowing the hybrid mode to travel below diffraction limits. The calculated electric field intensities for these modes inside the waveguide are illustrated in Figure 3.

Figure 2.   Calculated electric field distribution in the X–Y plane ($|E(x,y)|$) and direction for different eigenmodes for $\lambda = 1330$ nm, a 5 nm layer of MgF_2 and $w = 600$ nm: (a) SPP0 mode; (b) EH00 mode; (c) photonic mode.

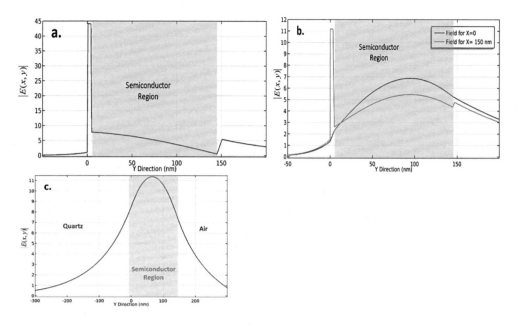

Figure 3.   Confinement of electric field intensity inside different parts of the nanolaser structure along the *Y* and mid-*X* axes: (a) SPP0 mode; (b) EH00 mode; (c) photonic mode.

3   RESULTS AND DISCUSSIONS

Figure 3 shows the amount of electrical energy confined by the different eigenmodes in different parts of the waveguide. As can be seen, for plasmonic eigenmodes the majority of the electrical energy is concentrated in the oxide, with little intensity inside the metallic layer, and the remaining energy distributed in the semiconductor. For the photonic mode, the electrical energy is mostly inside the semiconductor. In the case of the hybrid modes, the electrical energy is present in both the oxide and the semiconductor. As discussed further below, this will increase both the confinement factor and the propagation length of the mode.

Figure 4 shows the calculated effective mode index, propagation loss and the propagation length/diffraction limit of the various eigenmodes for different waveguide widths for the 5 nm oxide layers of Al_2O_3 and MgF_2. The field distribution and eigenmode refractive indices were calculated using the Finite Element Method in FEMLAB modeling software (COMSOL Inc., Stockholm, Sweden).

The photonic-mode loss was much smaller compared with the plasmonic modes and has been excluded from the plots in Figures 4b and 4c. Figure 4a shows that the parameters of the two plasmonic (SPP0) modes and the two hybrid modes are behaving similarly. As expected, and as apparent in Figure 4a, the first fundamental plasmonic mode (SPP0) doesn't have a cutoff width. The hybrid modes are cut off below the 450 nm waveguide width, which indicates the minimum possible width for these modes to propagate inside the nanolaser. However, the photonic mode can propagate inside waveguides as small as 300 nm. Inside the Al_2O_3 layer, the plasmonic mode experiences higher loss compared to the MgF_2 layer and has a smaller propagation length. It can be understood that the propagation loss for the EH00 hybrid modes decreases rapidly with wider waveguides and indicates more power in traveling as photonic mode than plasmonic. For each eigenmode, the effective mode index n_{eff} was evaluated as the real part of the complex mode refractive index in the waveguide, and the propagation loss and distance were obtained from its imaginary part as follows:

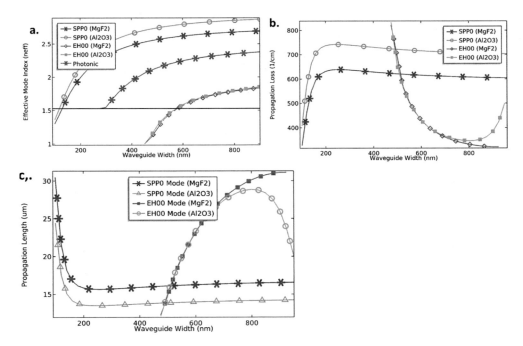

Figure 4. Mode characteristics vs waveguide width for the plasmonic and hybrid eigenmodes in 5 nm of Al_2O_3 and MgF_2, and the photonic eigenmode: (a) effective mode index; (b) propagation loss; (c) propagation length. The black solid line in (a) represents the quartz refractive index and indicates the photonic-mode cutoff width. The hybrid modes are cut off at widths smaller than 450 nm.

$$Loss_P = 2 \times \mathrm{Im}(\beta) = \frac{4 \times \pi}{\lambda}\mathrm{Im}(N) \qquad (1)$$

$$L_P = \frac{1}{Loss_P} \qquad (2)$$

Figure 5 shows the calculated mode areas and confinement factors in 5 nm of Al_2O_3 and MgF_2. The mode area A_m is evaluated as the ratio of the total energy and the maximum energy density for each mode. The confinement factor is defined as the ratio of the electrical density $|E(x,y)|^2$ in the active area of the nanolaser and the whole-structure cross-sectional area in the $X–Y$ plane.

To give a better understanding of mode area value, Figure 5a is shown as the ratio of mode area and diffraction limit, which is defined as $(\lambda/2n_{eff})^2$.

According to our calculations of the results shown in Figures 4 and 5, modes with higher propagation length also have a higher confinement factor, which was as expected from their trade-off. To determine which oxide can provide better performance for the plasmonic nanolaser, the threshold gain (g_{th}) required of the material to overcome the propagation loss is calculated and shown in Figure 6. Note that g_{th} is defined as $Loss_P/\Gamma$ where the amount of material gain reached in each mode equals the waveguide loss.

According to Figure 6, for waveguides wider than 600 nm, the plasmonic (SPP0) modes, as expected, need higher gain levels for lossless propagation as a result of higher propagation loss and a poor confinement factor. Propagation loss in small waveguides increments because more energy propagates inside the metal.

In addition, the nanolaser structure using Al_2O_3 needs less gain to overcome the plasmonic mode losses than that using MgF_2. For the hybrid modes, the threshold gain is nearly one

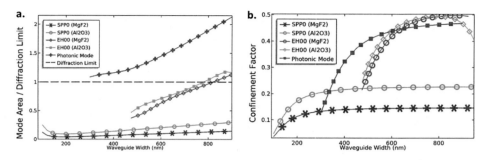

Figure 5. SPP0 and EH00 modes in 5 nm of Al_2O_3 and MgF_2, and photonic mode: (a) mode area to diffraction limit ratios; (b) confinement factors. The gray dashed line in (a) marks where the mode area equals the diffraction limit.

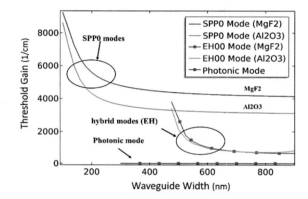

Figure 6. Threshold gain for lossless propagation of SPP and hybrid modes in 5-nm-thick Al_2O_3 and MgF_2, and photonic mode. Photonic-mode gain has been depicted without scaling to show the small amount of gain needed for this mode.

order of magnitude (700 1/cm) smaller than that of the plasmonic modes and the choice of oxide material does not make a significant difference.

From Figure 6 it is apparent that the hybrid modes' threshold gain increases rapidly at smaller waveguides and for widths near 450 nm (hybrid-mode cutoff width), and the SPP and EH modes need similar gain levels. Figure 6 also shows the required threshold gain for the photonic mode is much smaller than the other two modes.

It should be mentioned that our results were derived assuming a loss-free semiconductor waveguide to match the approach of previous work. The inclusion of waveguide losses in our calculations makes a noticeable change, particularly in propagation length and Γ factor. Moreover, other parameters such as the quality and the Purcell factor will also contribute to the nanolaser performance and output and should be considered when trying to predict or simulate a plasmonic nanolaser structure output.

4 CONCLUSIONS

In summary, we have investigated the mode-characteristic parameters and gain threshold of a semiconductor plasmonic nanolaser with multiple InGaAsP quantum wells active layer at a wavelength of 1330 nm. Effective mode index, propagation loss, mode area and confinement factor for two different oxides (Al_2O_3, MgF_2), three different propagating modes (SPP0, EH00 and photonic) and waveguide-width ranges of 100 nm to 900 nm were obtained and used to calculate the threshold gain. The results show that the SPP0 mode has no cutoff and propagates in waveguides as narrow as 100 nm or less, but the propagation loss increases rapidly. The plasmonic mode always propagates with a sub-diffraction mode area. Hybrid modes have a cutoff width near 450 nm and at waveguides broader than 650 nm can propagate with losses one order of magnitude smaller than the plasmonic mode and, in addition, below the diffraction limit. For waveguides wider than 800 nm, this mode reaches the diffraction limit. The photonic mode needs a considerably lower threshold gain for lossless propagation but can only propagate above the diffraction limit.

REFERENCES

Berini, P. & De Leon, I. (2012). Surface plasmon–polariton amplifiers and lasers. *Nature Photonics*, *6*, 16–24.

Costantini, D., Greusard, L., Bousseksou, A., De Wilde, Y., Habert, B., Marquier, F., … Colombelli, R. (2013). A hybrid plasmonic semiconductor laser. *Applied Physics Letters*, *102*(10), 101106. doi:10.1063/1.4794175.

Hill, M.T., Oei, Y.-S., Smalbrugge, B., Zhu, Y., de Vries, T., van Veldhoven, P.J., … Smit, M.K. (2007). Lasing in metallic-coated nanocavities. *Nature Photonics*, *1*, 589–594.

Keshavarz Moazzam, M. & Kaatuzian, H. (2015). Design and investigation of N-type metal/insulator/semiconductor/metal structure two-port electro-plasmonic addressed routing switch. *Applied Optics*, *54*, 6199–6207.

Lee, C.-J., Yeh, H., Cheng, F., Su, P.-H., Her, T.-H., Chen, Y.-C., … Chang, W.-H. (2017). Low-threshold plasmonic lasers on a single-crystalline epitaxial silver platform at telecom wavelength. *ACS Photonics*, *4*(6), 1431–1439.

Liu, L., Han, Z. & He, S. (2005). Novel surface plasmon waveguide for high integration. *Optics Express*, *13*(17), 6645–6650.

Livani, A.M. & Kaatuzian, H. (2015). Analysis and simulation of nonlinearity and effects of spontaneous emission in Schottky–junction–based plasmonic amplifiers. *Applied Optics*, *54*, 6103–6110.

Maier, S.A. (2007). *Plasmonics: Fundamentals and applications*. New York, NY; Springer.

Oulton, R.F. (2008). A hybrid plasmonic waveguide for sub-wavelength confinement and long-range propagation. *Nature Photonics*, *2*, 495–500.

Sun, Y., Majumdar, A., Cheng, C.-W., Martin, R.M., Bruce, R.L., Yau, J.-B., … Leobandung, E. (2014). High-performance CMOS-compatible self-aligned In0.53Ga0.47 As MOSFETs with GMSAT over 2200 µS/µm at VDD = 0.5 V. In *2014 IEEE International Electron Devices Meeting: Technical digest*. Piscataway, NJ: Institute of Electrical and Electronics Engineers. doi:10.1109/IEDM.2014.7047106.

Taheri, A.N. & Kaatuzian, H. (2014). Numerical investigation of a nano-scale electro-plasmonic switch based on metal-insulator-metal stub filter. *Optical and Quantum Electronics*, *47*(2), 159–168.

Frontier Research and Innovation in Optoelectronics Technology and Industry – Habib & Lewis (Eds)
© 2019 Taylor & Francis Group, London, ISBN 978-1-138-33178-5

Theoretical study on spot forming of stainless steel surface by pulse laser

Jiang Huang, Wenqing Shi, Yuping Xie, Yongqiang Li & Huixian Wu
College of Electronics and Information Engineering, Guangdong Ocean University,
Zhanjiang, Guangdong, China

Fenju An
School of Mechanical and Power Engineering, Guangdong Ocean University,
Zhanjiang, Guangdong, China

ABSTRACT: XL – 800 WF fiber transmission multi-function laser processing system is used to irradiate stainless steel plate with pulse, which does not form melting spots after the steel plate is solidified. In this paper, the diameter of the laser spot is calculated based on the size of the circular channel of the spot. The forming mechanism of the shape of the fused spot spherical cap was explained from the condition of non-spontaneous nucleation. On the basis of experiments, relevant data are sorted out, the wetting angle of the melting spot is calculated, and the conclusion that the work of non-spontaneous nucleation is far less than that of spontaneous nucleation is obtained.

1   INTRODUCTION

Laser is one of the most important inventions in the 20th century. Since the advent of laser, laser technology has been widely used in industry, agriculture, national defense, etc. Such as laser ranging (Hahnel, A and Burgard, W, 2003), laser scanning (Denk. W, Strickler. J.H and Webb. W.W, 1990), laser radar (Heiselberg. H and Busck. J. 2004), optical fiber communication (Rao. Y. J, Zhu. T, Ran. Z. L, et al. 2004), etc. In the field of laser processing, laser is favored because it has a series of unique advantages such as high power density, high local temperature, fast processing speed and small heat affected area. Laser equipment is even more indispensable in cutting (Schuettler. M, Stiess. S, King. B. V, et al. 2005), marking (Novotny. V, Alexandru. L. 2010), welding (Shi. W, Wang. W, Huang. Y. 2016) and cladding (Shepeleva. L, Medres. B, Kaplan. W. D, et al. 2000) of metal materials.

Although laser processing has been widely used, the interaction mechanism between laser and metal materials has not been fully explained. The interaction between laser and metal materials leads to the destruction of metal materials. the interaction process mainly depends on two parts: first, laser parameters, including laser wavelength, energy, power, pulse width and frequency of action, etc. these factors will have an important influence on the process of action. Secondly, the characteristics of metal materials, including specific heat capacity of solids, electrical conductivity, absorption rate, melting point, latent heat of fusion, crystal structure and so on, also affect the interaction effect. Thus, the interaction between laser and metal materials is complicated, involving metal melting physics (Jiang. G. C, Wu. Y. Q. 2012), phase transition theory (David. A. P, Kenneth. E. E, Mohamed. Y. S, 2013), quantum theory (Zhou. S. X, 2002) and solidification principle (Hu. H. C, 2015), etc. At present, there are many studies on the interaction between laser and metal materials. for example, Zhao Yijun (Zhao. Y. J. 1989) has studied the movement law of ablation vapor in the interaction between laser and metal, and calculated the radiation absorption coefficient and transport coefficient from the atomic molecular data theory. Huang guoxiu (Huang. G. X. 2008) used finite element method to study the temperature field on the upper and lower surfaces of metal under the interaction between laser and metal. Yang Nan et al (Yang. N, Yang. X. C, 2008) studied the interaction between laser and metal powder particles during laser cladding, and established the motion model and thermal model of metal powder particles by adjusting process parameters.

In this experiment, the stainless steel plate was irradiated by laser pulse. the laser melted the irradiated part of the stainless steel plate and formed a laser spot after solidification. The microscopic pattern was obtained by white light interferometer and step meter. Based on the condition of non-spontaneous nucleation, the work and shape factor of non-spontaneous nucleation are derived. The image fitting method was used to calculate the specific value of the shape factor, and the corresponding explanation was given for its physical reason.

2 EXPERIMENT

Xl – 800 WF optical fiber transmission multifunctional laser processing system. The laser device has a maximum output power of 800 w, a laser wavelength of 1064 nm, an effective spot size of 0.8 mm, a single-pulse maximum laser energy of 90 j, a pulse width of 0.5 ms to 5 ms, and a pulse frequency of 1 to 40 Hz. the focal length of the focusing lens used in operation is 160 mm. the device is equipped with a CCD monitoring system using an optical fiber input method.

In the experiment, the laser parameters are: current 150a, pulse duration 2.5 ma, laser output power about 500 w, and single pulse laser output energy about 20 j. Use 06 Cr 19 ni9 s 30408 stainless steel plate. The surface is smooth and smooth, and the surface area is large enough not to affect heat diffusion. The steel plate is relatively thick and will not melt through during laser irradiation. The fixed point of stainless steel plate was irradiated by a single laser pulse in the experiment, and the spot formed after cooling and solidification of the irradiated point. The laser spot was microscopically imaged under a microscope. Fig. 1a is a micrograph of a stainless steel plate, and Fig. 1b is a white light measurement diagram of the stainless steel plate. The experimental results in Fig. 1 show that the periphery of the laser spot presents a circular channel, and the inside is a spherical crown-shaped protrusion as a whole. The formation mechanism of the circular channel and the spherical cap-shaped protrusion will be explained below.

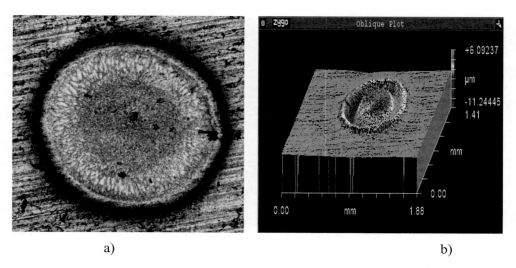

a)   b)

Figure 1.   A single laser pulse irradiates the stainless steel plate. a) microscopic image b) the white light measuring image.

3 FORMATION MECHANISM OF FUSED SPHERICAL CAP

This section uses the non-spontaneous nucleation condition to explain the formation mechanism of the spherical cap-type melting spot, and calculates the wetting angle through numerical simulation. The laser with high energy density acts on the surface of the solid metal and melts it. the liquid metal solidifies after cooling, showing its original solid state. Metal solidification can be divided into two ways: spontaneous nucleation and non-spontaneous nucleation. Since the material in this experiment is 06 Cr 19 ni9 s 30408 stainless steel plate, it is obvious that the solidification of metal in this experiment belongs to non-spontaneous nucleation solidification. Therefore, this paper calculates the change of Gibbs free energy from the interface energy and volume free energy of the theory of non-spontaneous nucleation and solidification.

3.1 *Modeling*

The following model is used to model the spherical crown-type melting spot. First of all, it should be noted that this experiment meets the following prerequisites: (1) stainless steel plate is planar; (2) the substrate area of inclusion is much larger than the contact area of crystal blank; (3) the wetting angle has nothing to do with the initial temperature of welding and is a constant.

The spherical cap type melting spot can be represented by Fig. 2, in which σ_{AS}, σ_{AC} and σ_{CS} are the interfacial tensions formed between air and the crystal blank, between air and the inner surface, and between the inner surface and the crystal blank.

According to the same tension in the horizontal direction, get

$$\sigma_{AC} = \sigma_{CS} + \sigma_{AS}\cos\theta \qquad (1)$$

That is

$$\cos\theta = \frac{\sigma_{AC} - \sigma_{CS}}{\sigma_{AS}} \qquad (2)$$

where θ is the wetting angle. The volume of the melting spot of the spherical cap is V, the contact area with the inner surface is A_1, and the contact area with the air is A_2. Calculation of V, A_1, A_2

$$A_1 = \pi(r\sin\theta)^2 \qquad (3)$$

$$A_2 = 2\pi r^2(1-\sin\theta) \qquad (4)$$

$$V = \pi r^3\left(\frac{2-3\cos\theta+\cos^3\theta}{3}\right) \qquad (5)$$

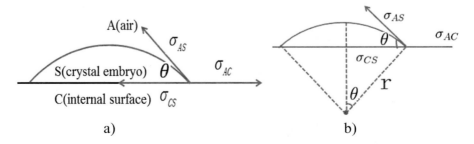

Figure 2. a) The schematic diagram of the spherical shape non-spontaneous nucleation. b) The calculation diagram of the spherical shape non-spontaneous nucleation.

Before nucleation, the gas is in contact with the inner surface and its interface can be

$$\sigma_{AC} A_1 = \sigma_{AC} \pi (r \sin \theta)^2 \tag{6}$$

After the crystal nucleus is formed, its interface can consist of two interfaces of A_1 and A_2

$$\sigma_{AS} A_2 + \sigma_{CS} A_1 = \sigma_{AS} 2\pi r^2 (1 - \sin \theta) + \sigma_{CS} \pi (r \sin \theta)^2 \tag{7}$$

Therefore, the interfacial energy change ΔG_i before and after nucleation is

$$\Delta G_i = \sigma_{AS} A_2 + \sigma_{CS} A_1 - \sigma_{AC} A_1 = \sigma_{AS} 2\pi r^2 (1 - \sin \theta) + \pi (r \sin \theta)^2 (\sigma_{CS} - \sigma_{AS}) \tag{8}$$

Substituting equation (2) into equation (8) to obtain

$$\Delta G_i = \pi r^2 \sigma_{AS} (2 - 3\cos \theta + \cos^3 \theta) \tag{9}$$

The volume Gibbs free energy after nucleation is

$$V \Delta G_m = \pi r^3 \left(\frac{2 - 3\cos \theta + \cos^3 \theta}{3} \right) \Delta G_m \tag{10}$$

Therefore, the total Gibbs free energy change of the system during nucleation is

$$\Delta G_{he} = V \Delta G_m + \Delta G_i = \left[\frac{4}{3} \pi r^3 \Delta G_m + 4\pi r^2 \sigma_{AS} \right] \left(\frac{2 - 3\cos \theta + \cos^3 \theta}{4} \right) \tag{11}$$

3.2 Calculation

Now, the work of non-spontaneous nucleation is calculated under the critical radius condition $\frac{d \Delta G_{he}}{dr} = 0$ so that

$$r = -\frac{2\sigma_{AS}}{\Delta G_m} \tag{12}$$

Therefore, the work of non-spontaneous nucleation under critical conditions is

$$\Delta G_{he} = \frac{16 \pi \sigma_{AS}^2}{3 \Delta G_m^2} \left(\frac{2 - 3\cos \theta + \cos^3 \theta}{4} \right) = \Delta G_{ho} f(\theta) \tag{13}$$

where $\Delta G_{ho} = \frac{16 \pi \sigma_{AS}^2}{3 \Delta G_m^2}$ is spontaneous nucleation work.

$$f(\theta) = \left(\frac{2 - 3\cos \theta + \cos^3 \theta}{4} \right) \tag{14}$$

where equation (14) is a shape factor, which is the ratio of non-spontaneous nucleation work to spontaneous nucleation work. It is a parameter that measures the difficulty degree of non-spontaneous nucleation work and is directly proportional to the wetting angle.

3.3 Analysis

Fig. 3 shows the data measured by white light interferometer and step meter, in which Fig. 3a is a white light meter and Fig. 3b is a step meter. According to the results of the bench meter, the radius and height of the melting spot of the spherical cap were fitted.

116

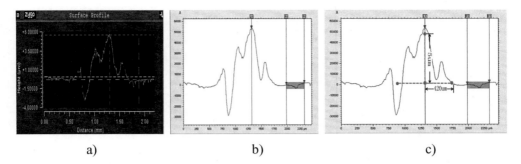

a) b) c)

Figure 3. a) The white light interferometer measurement diagram of the single pulse laser interacting with stainless steel plate, b) The step profiler measurement diagram, c) Fitted the radius and height.

$$r \approx 420 \, um \tag{15}$$

$$h \approx 1 \, um \tag{16}$$

With reference to Fig. 2b, the cosine value of the wetting angle is easily calculated as

$$\cos\theta = \frac{r-h}{r} \approx \frac{420-5}{420} = \frac{83}{84} \tag{17}$$

Substituting equation (17) into equation (14), the shape factor $f(\theta)$ can be calculated as

$$f(\theta) = \left(\frac{2-3\cos\theta+\cos^3\theta}{4}\right) \approx 10^{-4} \tag{18}$$

Equation (17) shows that in this experiment, the work of non-spontaneous nucleation is much smaller than that of spontaneous nucleation. Therefore, the nucleation of laser spot is mainly in the form of non-spontaneous nucleation. The nucleus of non-spontaneous nucleation mainly comes from two parts: impurities in the liquid metal of laser spot and the inner surface of the spot. Thus, this conclusion is in good agreement with the parameters given by the sheet metal.

4   UNIVERSALITY ANALYSIS OF EXPERIMENTAL AND THEORETICAL RESULTS

In order to explain the universality of this experiment and the theoretical results, we irradiated the same type of 06 Cr 19 ni9 s 30408 stainless steel plate with multiple laser pulses. Figure 4 shows micrographs of three, five and ten exposures to the same location of the stainless steel plate respectively. Fig. 5 is a white light interferometer picture of three, five and ten exposures to the same position of the stainless steel plate respectively. Fig. 6 is an outline picture of three, five and ten exposures to the same position of the stainless steel plate respectively. The results show that the results of multiple pulse laser irradiation on stainless steel plate are very similar to those of single pulse laser irradiation. The micrographs of Fig. 4 and Fig. 5 and the white light interferometer both show that the stainless steel plate irradiated by multiple pulses of laser forms circular melting spots and assumes a spherical crown shape appearance. The outline diagram of Fig. 6 further highlights the characteristics of the spherical crown shape appearance of the melting spots.

117

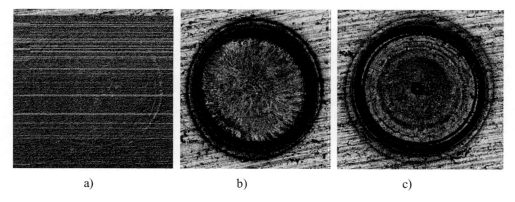

a)   b)   c)

Figure 4.   a), b) and c) are the micrograph of stainless steel plate irradiated by three, five, ten times laser pulse irradiates the stainless steel plate, respectively.

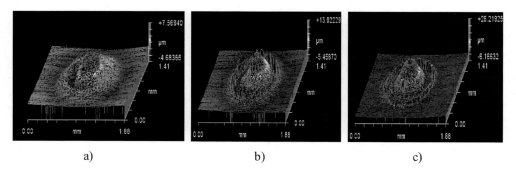

a)   b)   c)

Figure 5.   a), b) and c) are three, five, ten times laser pulse irradiates the stainless steel plate, respectively.

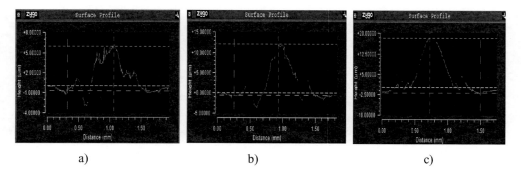

a)   b)   c)

Figure 6.   a), b) and c) are the outline picture of three, five, ten times pulse laser acting on stainless steel plate, respectively.

5   CONCLUSION

1. In this paper, XL – 800 WF fiber transmission multifunctional laser processing system is used to irradiate 06 Cr 19 ni9 s 30408 stainless steel plate with a single pulse. the stainless steel plate is melted and solidified into a spherical crown-shaped melting spot. The effective diameter of the laser is estimated through the circular channel of the fused spot. Based on the data of white light interferometer and step meter, the wetting angle is obtained from

the condition of non-spontaneous nucleation, which explains the forming principle of spherical crown-type fused spot.

2. In view of the fact that the microscopic mechanism of laser-metal interaction is not completely clear, this thesis has made a preliminary exploration on the microscopic mechanism, which will promote the research on the microscopic mechanism of laser-metal interaction to a certain extent.

ACKNOWLEDGMENT

Projection of enhancing school with innovation of Guangdong Ocean University (No. GDOU2017052504), Outstanding young backbone teachers (No. HDYQ2017005) and University students' innovation and entrepreneurship team (No. CCTD201823).

REFERENCES

David. A. P, Kenneth. E. E, Mohamed. Y. S, 2013, Phase Transformation in Metals and Alloys, Taylor & Francis Group LLC.

Denk. W, Strickler. J.H, Webb. W.W. Two-photon laser scanning fluorescence microscopy. Science, 248(4951):73–76.

Hahnel. D, Burgard. W, Fox. D, et al. 2003, An efficient fast SLAM algorithm for generating maps of large-scale cyclic environments from raw laser range measurements, International Conference on Intelligent Robots and Systems. IEEE:206–211, vol.1.

Heiselberg. H, Busck. J. 2004, Gated viewing and high-accuracy three-dimensional laser radar [J]. Applied Optics, 43(24):4705.

Hu. H. C, 2015, Metal Solidification Principle, Beijing: China Machine Press.

Huang. G.X. 2008, Analysis of temperature field of interaction between laser and metal. Changchun university of technology.

Jiang. G. C, Wu. Y.Q. 2012, Access to Molten Metal Physics, Beijing: Metallurgical industry press.

Novotny. V, Alexandru. L. 2010, Laser marking in dye-polymer systems. Journal of Applied Polymer Science, 24(5):1321–1328.

Rao. Y. J, Zhu. T, Ran. Z. L, et al. 2004, Novel long-period fiber gratings written by high-frequency CO 2, laser pulses and applications in optical fiber communication. Optics Communications, 229(1):209–221.

Schuettler. M, Stiess. S, King. B. V, et al. 2005, Fabrication of implantable microelectrode arrays by laser cutting of silicone rubber and platinum foil. Journal of Neural Engineering, 2(1): S121–S128.

Shepeleva. L, Medres. B, Kaplan. W. D, et al. 2000, Laser cladding of turbine blades. Surface & Coatings Technology, 125(1):45–48.

Shi. W, Wang. W, Huang. Y. 2016, Laser micro-welding of Cu-Al dissimilar metals. International Journal of Advanced Manufacturing Technology, 85(1–4):185–189.

Yang. N, Yang. X. C, 2008, Model of interaction between metal powder particle and laser beam in laser cladding, 29(9):1745–1750.

Zhao. Y.J. 1989, Several problems of atomic and molecular physics related to the interactions of laser beam with metals, Progress in Physics: 429–450.

Zhou. S. X, 2002, Quantum Dynamics, Beijing: Higher Education Press.

Experimental study into ultraviolet degradation of abamectin based on absorption spectrum

R.D. Ji & X.Y. Wang
Jiangsu Laboratory of Lake Environment Remote Sensing Technologies, Huaiyin Institute of Technology, Huaian, Jiangsu Province, China
Huaiyin Institute of Technology, Huaian, Jiangsu Province, China

J.Y. He
Huaiyin Institute of Technology, Huaian, Jiangsu Province, China

T.Z. Zhu, Y.S. Yu, X. Yang & Y.L. Zhang
Jiangsu Laboratory of Lake Environment Remote Sensing Technologies, Huaiyin Institute of Technology, Huaian, Jiangsu Province, China
Huaiyin Institute of Technology, Huaian, Jiangsu Province, China

ABSTRACT: The absorption spectrum of abamectin was studied using a spectrophotometer. The experimental results show that the most intensive characteristic peak (219 nm) was found in the abamectin standard solution spectrum. The function relationship between abamectin concentration and absorbance was obtained with a good linear relationship (R > 0.99). The hardware platforms for pesticide residue degradation are designed and built based on ultraviolet light technology, and abamectin degradation experiments were conducted using ultraviolet light. It is put forward to describe the degradation effect according to the change in the pesticide's absorbance, and the prediction model of pesticide residues degradation was obtained according to the relationship between degradation time and absorbance. Combining with the abamectin pesticide residue detection model function based on the absorption spectrum, the degradation rate is calculated corresponding to different degradation times.

1 INTRODUCTION

Abamectin belongs to the family of avermectins, which are macrocyclic lactones produced by the actinomycete Streptomyces avemitilis (Zhang, 1998; Celestina et al., 2010). Abamectin has wide applications as it acts on insects by interfering with their neural and neuromuscular transmission (Xie et al., 2005). For example, abamectin is used to control insect and mite pests of a range of agronomic, fruit, vegetable and ornamental crops. However the standard of Maximum Residue Level (MRL) for avermectins in vegetables and fruits is very strict all over the world (Zhao et al., 2012). There are many kinds of analysis methods relating to abamectin pesticide residue detection, for example, high-performance liquid chromatography-ultraviolet detection (Xu et al., 2010), high performance liquid chromatography-mass spectrometry (He et al., 2012), and fluorescence spectroscopy (Ji et al., 2015).

The natural decomposition of pesticides alone is obviously insufficient to meet the security needs of human beings. We must apply advanced technology to promote the degradation of pesticides, in order to improve the safety of agricultural products. So, the degradation of pesticide residues is an important research field to reduce the side effects of environment pesticides. At present, the degradation of pesticide residues mainly includes physical method, chemical method and biological method. The physical method mainly contains ultrasonic wave and ionization irradiation (Shriwas & Gogate, 2011; Chen et al., 2003). Chemical method is one of the most studied pesticide residue degradation technologies, including photochemical degradation, chemical oxidation and photocatalysis method (de Castro Peixoto & Teixeira, 2014; Ong et al., 1996; Baranda et al., 2014). Microbial method degrades pesticide residues by using bacteria, fungi, actinomycetes and other micro life (Mandal et al., 2013; Cheng et al., 1998). The principle of ultraviolet irradiation degradation of pesticide residues is consistent with that of photochemical degradation (Shawaqfeh & Al Momani, 2010). The pesticide will chemically react after ultraviolet irradiation (253.7 nm), which can break the double bond of the pesticide, and destroy the combination of organic carbon and other elements that constitute the pesticide. Subsequently, organic matter will be decomposed into smaller molecule substances.

The degradation of abamectin residue using ultraviolet light is proposed in this paper. The degradation effect is characterized through the change of the abamectin absorption spectra intensity, and the relationship between degradation time and degradation rate is then derived. Research results provide a reference value for the detection and degradation technology of abamectin pesticide residues and the characterization method of degradation effect.

2 EXPERIMENT

2.1 *Materials*

Abamectin was purchased from Shanghai Pesticide Research Institute (Shanghai, China).

2.2 *Apparatus*

The absorption spectrum was recorded by using a UV-3600PC spectrophotometer (Shimadzu, Japan) equipped with 1.0 cm quartz cell. The self-designed equipment for UV photodegradation of pesticides (Wang et al., 2018).

2.3 *Procedures*

Firstly, abamectin was diluted to different concentrations using deionized water. The absorption spectra of various solutions were obtained using the UV-3600PC. Secondly, 3 ml of each abamectin solution was placed in the sample pool of the ultraviolet degradation device using a cuvette. The absorption spectrum of the pesticide was recorded, by using the UV-3600PC after the solution was irradiated under ultraviolet light for different time periods. Finally, the relationship between the illumination time and characteristic peak absorbance of the pesticide was deduced.

3 RESULTS AND DISCUSSIONS

3.1 *Absorption spectrum of abamectin*

Figure 1 shows absorption spectrum standard curve of the abamectin solution. The abscissa indicates the light wavelength while the ordinate indicates absorbance. In addition, in Figure 1 the abamectin solution corresponds to 0.0781, 0.1563, 0.3125, 0.625, 1.25, 2.5 and 5 μg/ml for curves 1 to 7, respectively. It can be seen that absorbance increases with the concentration of abamectin, and that there is a maximal absorption shoulder peak at 219 nm. So, 219 nm can be used as the characteristic peak of abamectin residues.

In order to analyze the relationship between abamectin solution concentration and absorbance, a unitary linearity regression method was applied between absorbance and abamectin solution with different concentrations. The results are shown in Figure 2. It is found that the relationship between abamectin concentration and absorbance is linear when the wavelength is at 219 nm. The correlation coefficient is 0.9994, and the function equation of the concentration prediction model is:

$$Y = 0.1981 + 0.2875x \tag{1}$$

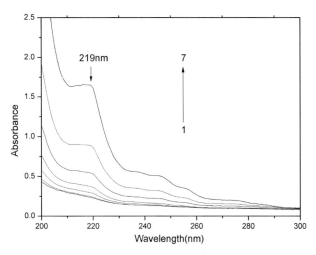

Figure 1. Absorption spectrum of abamectin at different concentrations (solution concentration was 0.0781, 0.1563, 0.3125, 0.625, 1.25, 2.5 and 5 μg/ml for curve 1 to curve 7, respectively).

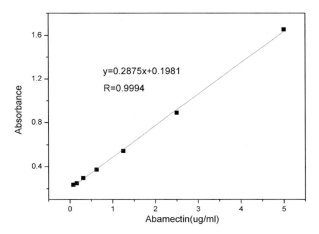

Figure 2. Linear relationship between abamectin concentration and absorbance at 219 nm.

3.2 *Abamectin degradation using ultraviolet irradiation*

Abamectin solution (5.5 μg/ml) was degraded through ultraviolet irradiation for different time periods. The absorption spectrum was detected using the UV-3600PC after each degradation. The absorption spectrum results are shown in Figure 3, in which the abscissa represents wavelength, and the ordinate represents absorbance. The degradation time of the abamectin solution was set to 0, 1, 2, 3, 5, 10, 15 and 20 minutes, respectively. As shown in Figure 3, the curves from 1 to 8 represent the corresponding absorption spectra. It is obvious that the peak value is correspondingly reduced from 1.79 to 1.29 (absorbance) with the increase of degradation time, but the peak position remains stable.

For further analysis of the abamectin degradation trend, the regression modeling method was applied between ultraviolet irradiation time and absorbance at 219 nm. The results are shown in Figure 4. The correlation coefficient is 0.9965, and the function equation is:

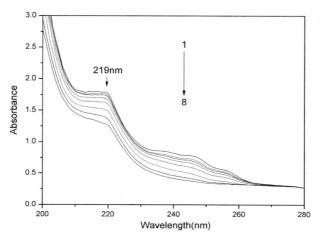

Figure 3. Absorption spectrum of abamectin solution degraded using ultraviolet irradiation for different time periods (irradiation time was set to 0, 1, 2, 3, 5, 10, 15 and 20 minutes from curve 1 to curve 8, respectively).

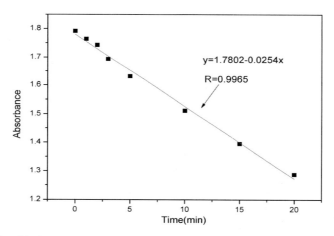

Figure 4. Relationship between ultraviolet irradiation time and abamectin absorbance at 219 nm.

Table 1. Abamectin concentration and degradation rate under different degradation times.

Degradation time (min)	1	2	3	5	10	15	20
Abamectin concentration (μg/ml)	5.4466	5.3736	5.2031	4.9840	4.5701	4.1631	3.7875
Degradation rate (%)	1.76	3.07	6.15	10.10	17.57	24.91	31.68

$$Y = 1.7802 - 0.0254x \qquad (2)$$

The equivalent concentration of abamectin solution after being degraded using ultraviolet irradiation for different time periods was obtained according to Equation 1, and the degradation rate of abamectin could be further calculated, based on the abamectin concentration before and after degradation. The results are shown in Table 1. The calculation formula of

124

degradation rate is as shown in Equation 3, in which C_0 is initial concentration of abamectin solution before degradation, C_t is concentration after degradation using ultraviolet irradiation for time t, and D represents degradation rate.

$$D = \frac{(C_0 - C_t)}{C_0} \times 100\% \tag{3}$$

The results shown in Table 1 indicate that abamectin concentration was decreased from 5.5 µg/ml to 5.4466 µg/ml after ultraviolet irradiation for 1 minute. When ultraviolet irradiation time was increased to 10 minutes, the abamectin concentration was 4.5701 µg/ml. When ultraviolet irradiation time was set to 20 minutes, the abamectin concentration was 3.7875 µg/ml.

4 CONCLUSIONS

The absorbance spectra of abamectin solutions with different concentrations were detected using a UV-3600PC in this paper, and the analysis results show that 219 nm is its characteristic peak. A unitary linearity regression method was applied between the absorbance and the abamectin solution with different concentrations, and the function equation of the abamectin concentration prediction model was deduced with a correlation coefficient of 0.9994.

The abamectin degradation experiment with ultraviolet irradiation was conducted, and then the degradation process was characterized by the change in the absorbance spectrum. The degradation model function and degradation rate were finally obtained through the change of abamectin concentration. The results show that the characteristic peak of abamectin spectrum decreased with the degradation time increasing, and the degradation rate was increased accordingly. For example, when the degradation time was set to 20 minutes, the degradation rate reached up to 31.68%.

In this paper the degradation effect of abamectin is characterized using ultraviolet light based on absorbance change. The results show that pesticide degradation can be realized by using ultraviolet irradiation, and this provides a reference value for the selection of pesticide degradation technology.

ACKNOWLEDGMENTS

This work was supported by the National Natural Science Foundation of China (61704063 and 61801188), the Science Project of the National Ministry of Housing and Urban-Rural Development (2016 K4069), the Natural Science Research Project of Higher Education Institutions in Jiangsu Province (17 KJA510001), Jiangsu Overseas Visiting Scholar Program for University Prominent Yong. Jiangsu University "Blue Project" Funding. The school-level achievement conversion special fund, and the Natural Science Foundation Project of Huaiyin Institute of Technology (16HGZ004).

REFERENCES

Baranda, A.B., Fundazuri, O. & de Marañón, I.M. (2014). Photodegradation of several triazidic and organophosphorus pesticides in water by pulsed light technology. *Journal of Photochemistry and Photobiology A: Chemistry*, *286*(15), 29–39.

Celestina, T.V., Kolar, L., Gobec, I., Kužner, J., Flajs, V.C., Pogačnik, M. & Eržen, N.K. (2010). Factors influencing dissipation of avermectins in sheep faeces. *Ecotoxicology and Environmental Safety*, *73*(1), 18–23.

Chen, M.H., Zhang, Y. & Cheng, S.H. (2003). Study on degradation of pesticide residues by ionizing radiation. *Ningxia Agricultural Science and Technology*, *25*(12), 27–30.

Cheng, G.F., Li, S.P., Shen, B. & Yang, J.H. (1998). Microbial degradation of pesticide residues in vegetables. *Journal of Applied and Environmental Biology*, *4*(1), 81–84.

de Castro Peixoto, A.L. & Teixeira, A.C.S.C. (2014). Degradation of amicarbazone herbicide by photochemical processes. *Journal of Photochemistry and Photobiology A: Chemistry*, *275*, 54–64.

He, H.M., Zhao, H., Zhao, C.R., Hu, X.Q.,Pin, L.F., Wu, M., Zhang, C.P., Cai, X.M., Zhu, Y.H. & Li, Z. (2012). Determination of abamectin residue in paddy rice by ultra performance liquid chromatography-tandem mass spectrometry. *Chinese Journal of Analytical Chemistry*, *40*(1), 140–144.

Ji, R.D., Zhao, Z.M., Zhu, X.Y. & Yu, R.S. (2015). Determination of abamectin residues in fruit juice by fluorescence spectrum. *Agro Food Industry Hi-Tech*, *26*(1), 8–11.

Mandal, K., Singh, B., Jariyal, M. & Gupta, V.K. (2013). Microbial degradation of fipronil by Bacillus thuringiensis. *Ecotoxicology and Environmental Safety*, *93*, 87–92.

Ong, K.C., Cash, J.N., Zabik, M.J., Siddiq, M. & Jones, A.L. (1996). Chlorine and ozone washes for pesticide removal from apples and processed apple sauce. *Food Chemistry*, *55*(2), 153–160.

Shawaqfeh, A.T. & Al Momani, F.A. (2010). Photocatalytic treatment of water soluble pesticide by advanced oxidation technologies using UV light and solar energy. *Solar Energy*, *84*(7), 1157–1165.

Shriwas, A.K. & Gogate, P.R. (2011). Ultrasonic degradation of methyl Parathion in aqueous solutions: Intensification using additives and scale up aspects. *Separation and Purification Technology*, *79*(1), 1–7.

Wang, X.Y., Ji, R.D., Zhang, Y.L., Yang, Y., Fu, C. & Yang, D. (2018). Research on characterization and modeling for ultraviolet degradation of imidacloprid based on absorbance change. *Optik-International Journal for Light and Electron Optics*, *154*, 315–319.

Xie, X.C., Zhang, S.H., Wang, D.S., Huangfu, W.G., Yang, T. & He, X.H. (2005). Determination of abamectin and its toxicological metabolite in vegetables and fruits by HPLC with pre-column fluorescent derivatization. *Scientia Agricultura Sinica*, *38*(11), 2254–2260.

Xu, H.R., Yang, R.B., Fu, Q. & Liao, H.Y. (2010). Abamectin residue in water, soil and rice. *Environmental Science and Management*, *35*(4), 35–37.

Zhang, M.H. (1998). New broad-spectrum insecticide to kill avermectin abamectin. *Pesticide*, *3*, 36–37.

Zhao, X.H., Cao, Z.Y., Mou, R.X., Xu, P. & Chen, M.X. (2012). Determination of 5 avermectin residues in fruits and vegetables by liquid chromatography-tandem mass spectrometry. *Journal of Instrumental Analysis*, *31*(10), 1266–1271.

Frontier Research and Innovation in Optoelectronics Technology and Industry – Habib & Lewis (Eds)
© 2019 Taylor & Francis Group, London, ISBN 978-1-138-33178-5

Generating an enhanced optical comb based on a gain switched laser by optimizing the modulation ratio

Y.D. Li
Communications Engineering College, PLA Army Engineering University, Nanjing, China
PLA Army Military Transportation University, Zhenjiang, China

T. Pu, H.T. Zhu, P. Xiang, J.L. Zheng, J. Li & X. Zhang
College of Communications Engineering, PLA Army Engineering University, Nanjing, China

M. Hu & Y. Liu
PLA Army Military Transportation University, Zhenjiang, China

ABSTRACT: A simple and robust technique to generate an optical comb is based on a Gain Switched Laser (GSL), which is usually realized by directly modulating a Distributed-Feedback (DFB) semiconductor laser. However, unfortunately the large temporal jitter, which reduces the phase correlation and the comb tone to the background Noise Ratio (TNR) of the optical comb, is also generated across the direct modulation technique. To acquire as large as possible TNR of the discernible optical comb at different modulation frequencies, the modulation ratio of the DFB laser, which is defined as the ratio between the operating current of the RF modulation signal and the bias current of the DFB laser, influences on the TNR and bandwidth of the discernible optical comb is studied in this paper. The experiment results show that there is a trade-off between the TNR and the bandwidth along with the modulation ratio changes, and to generate the discernible optical comb, the modulation ratio should be gradually increased or decreased as the modulation frequency gradually increases or decreases. By optimizing the modulation ratio, the discernible optical comb with a TNR larger than 30 dB is achieved at different modulation frequencies in experiment.

1 INTRODUCTION

The optical comb offers several attractive applications, such as remote sensing (Scotti et al., 2014), modern instrumentations (Coddington et al., 2009), optical arbitrary waveform generation (Jiang et al., 2007), radio-over-fiber systems (Shao et al., 2015), and wavelength-division-multiplexing systems (Roslund et al., 2013). The mode-locked semiconductor laser is a basic method for optical comb generation. However, the linewidth of the generated optical comb is large (Kim & Song, 2016). In contrast to mode-locked lasers, the external optical modulator based optical comb generation method can overcome the linewidth issue by using a narrow linewidth laser. The number of optical comb tones can be enhanced by cascading multiple external optical modulators and nonlinear fiber. However, the system configuration is complicated and the insertion loss is usually large (Wu et al., 2013). One of the simplest and most reliable methods for optical comb generation is based on the gain switched Distributed-Feedback (DFB) semiconductor laser (Anandarajah et al., 2009). Since the temporal jitter and frequency chirp are generated across the generation of the optical comb, thus there will no discernible optical comb. To overcome this issue, external injection is usually introduced and then the discernible optical comb can be generated (Zhou et al., 2011). Recently, without external injection, an optical comb generation method via a gain switching DFB laser greater than or equal to its relaxation oscillation frequency has been reported (Anandarajah et al.,

2015). However, the modulation frequency range is restricted. To acquire the discernible optical comb with as large as the TNR at any modulation frequency. In this paper, the modulation ratio, which is defined as the ratio between the operating current of the RF modulation signal and the bias current of the DFB laser, influences on the TNR and bandwidth of the discernible optical comb is studied. The experiment results show there is a trade-off between the TNR and the bandwidth of the discernible optical comb, and to generate the discernible optical comb with as large a TNR, the modulation ratio should be increased or decreased as the modulation frequency increases or decreases. By optimizing the modulation ratio, the discernible optical comb with a TNR larger than 30 dB can be achieved at different modulation frequencies.

2 EXPERIMENTAL SETUP AND RESULTS

2.1 *Experimental setup*

The experimental setup for the frequency response measurement of the DFB laser and the generated optical comb measurement are shown in Figure 1(a) and 1(b) respectively.

The DFB laser is a common commercially available device with the threshold current (I_{th}) of 17 mA at a temperature of 22.5°C. A Laser Diode Controller (LDC) (ILX 3724) is used to control the temperature and the bias current of the DFB laser. The sinusoidal RF signal comes from the RF source within the range of 50 MHz to 12.5 GHz. The amplifier with the RF gain of 25 dB and the 3 dB bandwidth of 18 GHz is used to amplify the RF signal. The variable attenuator with the 3 dB bandwidth of 18 GHz is used to attenuate the amplified RF signal. The RF signal output from the variable attenuator is used to directly modulate the DFB laser. A 50:50 optical coupler is connected to the DFB laser, and the output signals are sent to a scope, and to an Optical Spectrum Analyzer (OSA) with a resolution of 150 MHz.

2.2 *Simulations results*

According to the single mode rate equations of the DFB laser (Cartledge & Srinivasan, 1997), the impacts of I_m and I_b on the generated optical comb are studied and simulated at the modulation frequency of 10 GHz. The simulation result is shown in Figure 2.

As shown in Figure 2(a) and Figure 2(b), the TNR of the optical comb decreases as the I_m increases from 20 mA to 600 mA with a step of 20 mA, but the 3 dB bandwidth of the optical comb increases. This is because the temporal jitter increases as the I_m increases, and thus the TNR decreases. The frequency chirp increases as the I_m increases, and thus the 3 dB bandwidth

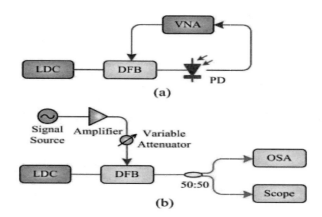

Figure 1. The experimental setup for: (a) the frequency response measurement; (b) the optical comb tests.

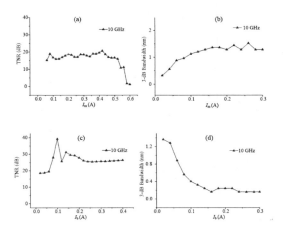

Figure 2. The variation of: (a) TNR and (b) 3 dB bandwidth, under different I_m; (c) TNR and (d) 3 dB bandwidth, under different I_b.

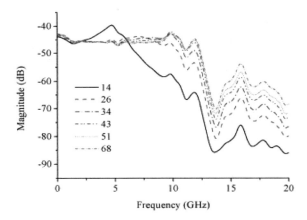

Figure 3. The frequency responses of the DFB laser under different bias currents.

of the optical comb increases. There is a trade-off between the TNR and the 3 dB bandwidth of the optical comb.

As shown in Figure 2(c) and Figure 2(d), the TNR of the optical comb increases as the I_b increases from 20 mA to 400 mA with a step of 20 mA, but the 3 dB bandwidth of the optical comb decreases. This is because the temporal jitter decreases as the I_b increases, and thus the TNR increases. The frequency chirp decreases as the I_b increases, and thus the 3 dB bandwidth of the optical comb decreases. There is a trade-off between the TNR and the 3 dB bandwidth of the optical comb.

2.3 *Experimental simulations results*

The frequency responses of the DFB laser under different I_b are shown in Figure 3. As can be seen, the relaxation oscillation frequency is 4.7 GHz and the 3 dB bandwidth is 5.4 GHz when I_b is 14 mA, which is less than the threshold current. As I_b increases, both the relaxation oscillation frequency and the 3 dB bandwidth increase. When I_b is more than 34 mA, the relaxation oscillation frequency is about 9.8 GHz.

The optical spectra of the gain switched DFB laser under different modulation frequencies, attenuators and bias currents are shown in Figure 4.

129

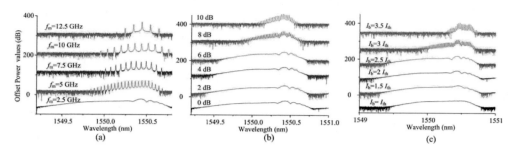

Figure 4. The measured optical spectra of DFB laser under: (a) different modulation frequencies; (b) different attenuators; (c) different bias currents.

In Figure 4(a), the bias current (I_b) of the DFB laser is fixed as two times the threshold ($2I_{th}$) and the attenuator is fixed as 0 dB. When the modulation frequency (f_m) is set as 2.5 GHz, there is no discernible optical comb, but when the modulation frequency is set as 5 GHz, 7.5 GHz, 10 GHz, or 12.5 GHz, respectively, there is a discernible optical comb. The reason is shown in Anandarajah et al. (2015), that when the modulation frequency is lower than the relaxation frequency of the gain switched DFB laser, the optical pulse decays to zero and the following optical pulse builds from spontaneous emission, resulting in the large temporal jitter. As the modulation frequency is increased to be equal to or larger than the relaxation frequency, then the optical pulses no longer decay to zero before the laser gain recovers and the following pulse is created. In short, the Root Mean Square (RMS) temporal jitter is gradually decreased as the modulation frequency increases. Thus the transition of the optical comb is gradually from there being no discernible optical comb to a discernible optical comb, and then the TNR is gradually increased. However, the bandwidth of the optical comb is gradually decreased as the modulation frequency increases, and thus there is a trade-off between the TNR and the bandwidth.

In Figure 4(b), the modulation frequency (f_m) is fixed as 2.5 GHz and the I_b is fixed as $2I_{th}$. When the attenuator is fixed as 2 dB, 4 dB, or 6 dB, respectively, there is no discernible optical comb. Once the attenuator is larger than 6 dB, such as 8 dB or 10 dB, the optical comb is discernible. The reason is that when the attenuator is increased, the modulation ratio is decreased, and then the variable density of carrier density in the active region of the DFB laser will be decreased; thus, the RMS temporal jitter will be gradually decreased. Once the temporal jitter is lower than a certain value, the optical comb will be discernible. Along with the attenuator increases from 0 dB to 10 dB, the 3 dB bandwidth of the optical comb is gradually decreased from 0.075 nm to 0.042 nm, but the TNR of the optical comb is gradually increased from 0 to 28 dB. There is a trade-off between the TNR and the bandwidth.

In Figure 4(c), the attenuator is fixed as 0 dB and the f_m is fixed as 2.5 GHz. When the bias current (I_b) of the DFB laser is set as $1.5I_{th}$, $2I_{th}$, or $2.5I_{th}$, respectively, there is no discernible optical comb. Once the I_b is larger than $2.5I_{th}$, such as $3I_{th}$, or $3.5I_{th}$, the optical comb will be discernible. The reason is that when the I_b is increased, the modulation ratio is decreased, and then the variable density of the carrier density in the active region of the DFB laser will be decreased; thus, the RMS temporal jitter will be gradually decreased. Once the temporal jitter is lower than a certain value, the optical comb will be discernible. Along with the I_b increases from I_{th} to $3.5I_{th}$, the 3 dB bandwidth is gradually decreased from 0.41 to 0.02 nm, but the TNR is gradually increased from 0 to 33 dB. There is a trade-off between the TNR and the bandwidth. In short, to generate the discernible optical comb, the modulation ratio should be gradually decreased or increased as the modulation frequency gradually decreases or increases, and there is a trade-off between the TNR and the bandwidth.

To test the above conclusion, the f_m is fixed as the different frequencies, such as 2.5 GHz, 5 GHz, 10 GHz, respectively, the attenuator is fixed as 0 dB, and the modulation ratio is

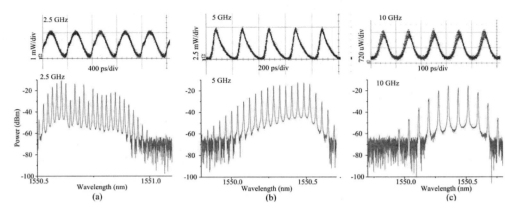

Figure 5. The measured waveform and optical spectrum of the DFB laser modulated by: (a) 2.5 GHz; (b) 5 GHz; (c) 10 GHz RF signal.

optimized by adjusting the I_b to generate the discernible optical comb to be as large as the TNR. The experiment results are shown in Figure 5.

When the I_b is $5.6I_{th}$, the discernible optical comb with the 34 dB TNR can be realized at 2.5 GHz, and the waveform is a sinusoidal-like shape. The optical spectrum and the waveform of the discernible optical comb are shown in Figure 5(a). When the I_b is $2I_{th}$, the discernible optical comb with the 33 dB TNR can be realized at 5 GHz, and the waveform is a sawtooth-like shape. The optical spectrum and the waveform of the discernible optical comb are shown in Figure 5(b). When the I_b is $2I_{th}$, the discernible optical comb with the 42 dB TNR can be realized at 10 GHz, and the waveform is a triangle-like shape. The optical spectrum and the waveform of the discernible optical comb are shown in Figure 5(c).

As we can see from Figure 5, firstly, by optimizing the modulation ratios, the discernible optical comb can be generated at different modulation frequencies. Secondly, compare Figure 4(a) and Figure 4(b) – to generate the discernible optical comb, the modulation ratio should be gradually increased as the modulation frequency gradually increases. Thirdly, compare Figure 4(b) and Figure 4(c) – the I_b does not change, but due to the fact that the RMS temporal jitter is gradually decreased as the modulation frequency increases[10], thus the TNR in Figure 4(b) is lower than the TNR in Figure 4(c).

3 DISCUSSIONS AND CONCLUSIONS

In this paper, the modulation ratio influences on the TNR and bandwidth of the optical comb are studied, which can be tuned by adjusting the I_m and I_b separately or together. As I_m increases, the TNR will decrease and the bandwidth will increase. However, with the enlargement of I_b, the TNR will increase and the bandwidth will decrease. Thus, there is a trade-off between the TNR and the bandwidth when the modulation is changed. To generate the discernible optical comb as large as the TNR, the modulation ratio should be gradually decreased or increased as the modulation frequency gradually increases or decreases. By optimizing the modulation ratio, the discernible optical comb with a TNR larger than 30 dB is achieved at different modulation frequencies.

ACKNOWLEDGMENTS

The work was supported in part by the National Natural Science Foundation of China under the Grant No. 61475193, No. 61174199, No. 61504170, No. 61671306, and the Jiangsu Province Natural Science Foundation Council under the Grant No. BK20140069.

REFERENCES

Anandarajah, P.M., Dúill, S.P.Ó., Zhou, R. & Barry, L.P. (2015). Enhanced optical comb generation by gain-switching a single-mode semiconductor laser close to its relaxation oscillation frequency. *IEEE Journal of Selected Topics in Quantum Electronics*, *21*(6), 592–600.

Anandarajah, P.M., Shi, K., O'Carroll, J., Kaszubowska, A., Phelan, R., Barry, L.P., ... O'Gorman, J. (2009). Phase shift keyed systems based on a gain switched laser transmitter. *Optics Express*, *17*(15), 12668.

Cartledge, J.C. & Srinivasan, R.C. (1997). Extraction of DFB laser rate equation parameters for system simulation purposes. *Journal of Lightwave Technology*, *15*(5), 852–860.

Coddington, I., Swann, W.C., Nenadovic, L. & Newbury, N.R. (2009). Rapid and precise absolute distance measurements at long range. *Nature Photonics*, *3*(6), 351.

Jiang, Z., Huang, C.B., Leaird, D.E. & Weiner, A.M. (2007). Optical arbitrary waveform processing of more than 100 spectral comb lines. *Nature Photonics*, *1*(8), 463–467.

Kim, J. & Song, Y. (2016). Ultralow-noise mode-locked fiber lasers and frequency combs: Principles, status, and applications. *Advances in Optics and Photonics*, *8*(3), 465.

Roslund, J., De Araujo, R. M., Jiang, S., Fabre, C. & Treps, N. (2013). Wavelength-multiplexed quantum networks with ultrafast frequency combs. *Nature Photonics*, *8*(2), 109–112.

Scotti, F., Laghezza, F., Serafino, G., Pinna, S., Onori, D., Ghelfi, P. & Bogoni, A. (2014). In-field experiments of the first photonics-based software-defined coherent radar. *Journal of Lightwave Technology*, *32*(20), 3365–3372.

Shao, T., Martin, E., Prince, A.M. & Barry, L.P. (2015). DM-DD OFDM-RoF system with adaptive modulation using a gain-switched laser. *IEEE Photonics Technology Letters*, *27*(8), 856–859.

Wu, R., Leaird, D.E. & Weiner, A.M. (2013). Supercontinuum-based 10-GHz flat-topped optical frequency comb generation. *Optics Express*, *21*(5), 6045.

Zhou, R., Latkowski, S., O'Carroll, J., Phelan, R., Barry, L.P. & Anandarajah, P. (2011). 40 nm wavelength tunable gain-switched optical comb source. *Optics Express*, *19*(26), B415.

Frontier Research and Innovation in Optoelectronics Technology and Industry – Habib & Lewis (Eds)
© 2019 Taylor & Francis Group, London, ISBN 978-1-138-33178-5

Second harmonic measurement of multi-beam laser heterodyne with ultra-precision for the glass thickness based on oscillating mirror sinusoidal modulation

Y.C. Li, Y. Liu & W.D. Zhong
Electronic Engineering College, Heilongjiang University, Harbin, Heilongjiang, China

ABSTRACT: In order to improve both the measurement accuracy of glass thickness and the signal processing speed of operation, this paper proposes a novel method of second harmonic measurement of multi-beam laser heterodyne to obtain the glass thickness, which is based on the combination of Doppler effect and heterodyne technology, loading the information of the glass thickness to the frequency difference of second harmonic of the multi-beam laser heterodyne signal by frequency modulation of the oscillating mirror, which is in the light path. Heterodyne signal frequency can be obtained by fast Fourier transform, and values of the glass thickness can be accurately obtained after the multi-beam laser heterodyne signal demodulation. This novel method is used to simulate measurement for thickness of glass by MATLAB. The obtained result shows that the relative measurement error of this method is just 0.08%.

Keywords: Laser heterodyne, glass thickness, Doppler effect, Fourier analysis, sinusoidal modulation

1 INTRODUCTION

Precision glass thickness measurement is always a problem in the engineering field. With the development of science and technology, methods of thickness measurement continuously innovate, including the optical measurement method (Ulrich & Torge, 1973; Nemoto, 1992), the fractional fringe order method (Tsai, Huang et al., 1999; Tsai, Tian et al., 1999), interferometry (Novikov et al., 2004; Protopopov et al., 2006), and the astigmatism method (Liu & Li, 2008). Using these methods does not generally reach the requirement of high-precision measurement.

Since the optical thickness measurement using a non-contact, high-precision, simple structure receives much concern, the optical thickness measurement method has been more widely used. In the optical thickness measurement method, the laser heterodyne measurement techniques inherits the many advantages of laser heterodyne technology (Yan-Chao et al., 2011; Yan-Chao & Chun-Hui, 2012; Li et al., 2012a; Li et al., 2009; Li et al., 2012b), and is one of the ultra-precision measurement methods.

To be able to collect a better laser difference frequency signal and increase the signal processing speed of operation, this paper proposes a novel method of a second harmonic of multi-beam laser heterodyne measurement for the glass thickness, which is based on the combination of Doppler effect and heterodyne technology, loaded the information of the glass thickness to the frequency difference of the multi-beam laser heterodyne signal by frequency modulation of the oscillating mirror, which is in the light path. Heterodyne signal frequency can be obtained by fast Fourier transform, and can obtain values of the glass thickness accurately after the multi-beam laser heterodyne signal demodulation. This paper describes in detail the principle of second harmonic of multi-beam laser heterodyne measurement for the glass thickness and the simulation result.

2 THE OPTICAL PATH DESIGN AND MEASURING PRINCIPLE OF THE GLASS THICKNESS

2.1 *Optical path design*

The scheme of second harmonic measurement of multi-beam laser heterodynes for the glass thickness based on oscillating mirror modulation is shown in Figure 1. The scheme mainly consists of a solid state laser, mirror, quarter-wave plate, oscillating mirror, Polarizing Beam Splitter (PBS), converging lens, plate glass, photodetector, and signal processing system which consists of a preamplifier, filter, A/D converter and Digital Signal Processor (DSP). The oscillating mirror moves as simple harmonic motion under the action of the signal source, and the advantage of the oscillating mirror is in the fact that the laser incidence on the oscillating mirror in different moment can be modulated in frequency.

Firstly, turning on the solid state laser, we make the linearly polarized beams go through the PBS and quarter-wave plate, then reaching the front surface of the oscillating mirror. Therefore, the reflected light in different moment, modulated by the oscillating mirror, propagates through the quarter-wave plate and PBS oblique incidence at the mirror which agglutinates on the surface of the standard beam. The light transmission through the former surface of the glass is reflected by the former surface of the glass, and the light reflected by the last surface of the former surface of the glass is converged together onto the sensitive surface of the photodetector by the converging lens. Finally, the electrical signal converted by the photodetector propagates through the preamplifier, A/D converter and DSP, and we obtain the thickness of the glass under test.

2.2 *Principle of second harmonic of multi-beam laser heterodyne measurement for the glass thickness*

As is shown in Figure 2, due to the beam reflecting and refracting continuously between the former and rear surfaces of the plate glass, the reflection and refraction have a contribution to the interference between the reflected and transmitted light at the infinite distance or the focal plate of the lens. Therefore, when discussing interference, we must consider multiple reflection and refraction effects; hence, we should discuss the multi-beam laser interference.

However, the magnitude of the two difference frequency signals have a difference of two to three orders of magnitude, generated by mixing the reflected beam former surface of

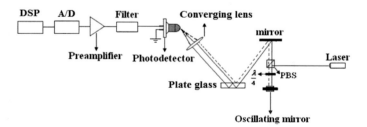

Figure 1. Optical path design.

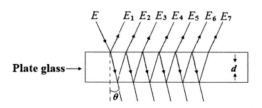

Figure 2. Multi-beam laser interferometer schematic.

134

glass and the transmitted beam generated by k and $k+2$ times the reflected beam of the rear surface of plate glass. Therefore, in order to collect a better signal of the laser difference frequency and to improve the signal processing computing speed after Fourier transform, here we only consider the second harmonic frequency difference generated by mixing the k times reflected beam E_k and $k+2$ times reflected beam E_{k+2} of the rear surface of the plate glass.

Suppose that the incident glass thickness of the laser is θ_0, and that the incident field is:

$$E(t) = E_0 \exp(i\omega_0 t) \tag{1}$$

The simple harmonic motion equation is:

$$x(t) = x_0 \cos(\omega_c t) \tag{2}$$

The rate equation of the oscillating mirror is:

$$v(t) = -\omega_c x_0 \sin(\omega_c t) \tag{3}$$

Because of the movement of the oscillating mirror, based on Doppler effect, the frequency of reflected light changes to:

$$\omega = \omega_0 (1 - 2\omega_c x_0 \sin(\omega_c t)/c) \tag{4}$$

where ω_0 is the angular frequency of the laser, x_0 is amplitude of the oscillating mirror, ω_c is angular frequency of the oscillating mirror, and c is light speed. So, optical field arrived at former surface of plate glass at $t - l/c$ time is:

$$\begin{aligned}E_1(t) = \alpha E_I \exp\{i[\omega_0(1 - 2\omega_c x_0 \sin(\omega_c(t - l/c))/c) \\ (t - l/c) + \omega_0 x_0 \cos(\omega_c(t - l/c))/c]\}\end{aligned} \tag{5}$$

The beam is reflected in different time by the rear surface of the plate glass, and the reflected light can be described as follows:

$$\begin{aligned}E_2(t) &= \alpha\alpha_1^2 E_I \exp\{i[\omega_0(1 - 2\omega_c x_0 \sin(\omega_c(t - (l + 2nd\cos\theta)/c))/c) \\ &\quad (t - (l + 2nd\cos\theta)/c) + \omega_0 x_0 \cos(\omega_c(t - (l + 2nd\cos\theta)/c))/c]\} \\ E_3(t) &= \alpha^3\alpha_1^2 E_I \exp\{i[\omega_0(1 - 2\omega_c x_0 \sin(\omega_c(t - (l + 4nd\cos\theta)/c))/c) \\ &\quad (t - (l + 4nd\cos\theta)/c) + \omega_0 x_0 \cos(\omega_c(t - (l + 4nd\cos\theta)/c))/c]\} \\ E_4(t) &= \alpha^5\alpha_1^2 E_I \exp\{i[\omega_0(1 - 2\omega_c x_0 \sin(\omega_c(t - (l + 6nd\cos\theta)/c))/c) \\ &\quad (t - (l + 6nd\cos\theta)/c) + \omega_0 x_0 \cos(\omega_c(t - (l + 6nd\cos\theta)/c))/c]\}\end{aligned} \tag{6}$$

$$\begin{aligned}\bullet \\ \bullet \\ \bullet\end{aligned}$$

$$\begin{aligned}E_m(t) &= \alpha^{2m-1}\alpha_1^2 E_I \exp\{i[\omega_0(1 - 2\omega_c x_0 \sin(\omega_c(t - (l + 2mnd\cos\theta)/c))/c) \\ &\quad (t - (l + 2mnd\cos\theta)/c) + \omega_0 x_0 \cos(\omega_c(t - (l + 2mnd\cos\theta)/c))/c]\}\end{aligned}$$

where m is a positive integer, n is the refractive index, $\alpha_1 = r$, $\alpha_2 = \beta^2 r$, ..., $\alpha_m = \beta^2 r^{(2m-3)}$, r stands for reflectivity of light from the surrounding media into plate glass, β stands for transitivity, d is the thickness of the plate glass, θ is refraction angle, and l is distance to the surface of the glass from the mirror.

So, the total optical field received by the photodetector is:

$$E(t) = E_1(t) + E_2(t) + \cdots + E_m(t) \tag{7}$$

Then, the output photocurrent of the photodetector is:

135

$$I = \frac{\eta e}{h\nu} \frac{1}{Z} \iint_D \frac{1}{2} [E_1(t) + E_2(t) + \cdots + E_m(t) + \cdots][E_1(t) + E_2(t) + \cdots + E_m(t) + \cdots]^* \, ds \qquad (8)$$

where e is electronic charge, Z is intrinsic impedance of the medium in the surface of the photodetector, η is quantum efficiency, S is sensitive surface area of photodetector, h is Planck constant and v is laser frequency.

We only consider the second harmonic frequency difference generated by mixing the k times reflected beam E_k and $k+2$ times reflected beam E_{k+2} of the rear surface of the plate glass. The DC component can be filtered after passing through the low-pass filter. Therefore, considering the AC component only, and call this AC component to be an intermediate frequency current, we can coordinate and obtain the intermediate frequency current as:

$$I_{IF} = \frac{\eta e}{2h\nu} \frac{1}{Z} \iint_s \sum_{p=0}^{\infty} \sum_{j=p+2}^{\infty} (E_p(t)E_j^*(t) + E_p^*(t)E_j(t)) \, ds \qquad (9)$$

Substitute Equation 5 and Equation 6 into Equation 9, and calculate the integral by software and obtain the results as:

$$I_{IF} = \frac{\eta e}{h\nu} \frac{\pi}{Z} E_0^2 \sum_{p=0}^{m-1} \sum_{j=0}^{m-p} \alpha_{j+p} \alpha_j \cos\left[\frac{8nd\cos\theta\omega_0\omega_c^2 x_0}{c^2}t + \frac{2\omega_0 x_0}{c} - \frac{4nd\omega_0\cos\theta}{c} \right. $$
$$\left. - \frac{8nd\cos\theta\omega_0\omega_c^2 x_0(l+2pnd\cos\theta)}{c^3} \right] \qquad (10)$$

After ignoring the minor term of $1/c^3$, Equation 10 can be simplified as:

$$I_{IF} = \frac{\eta e}{h\nu} \frac{\pi}{Z} E_0^2 \sum_{p=0}^{m-1} \sum_{j=0}^{m-p} \alpha_{j+p} \alpha_j \cos\left(\frac{8nd\cos\theta\omega_0\omega_c^2 x_0}{c^2}t + \frac{2\omega_0 x_0 - 4nd\omega_0\cos\theta}{c} \right) \qquad (11)$$

where p is a positive integer.

From Equation 11, it can be seen that the information of the thickness d of the plate glass for the frequency difference of the Intermediate Frequency (IF) component can be obtained from the multi-beam heterodyne measurement. Analyze frequency difference of IF component primarily, since the use of Fast Fourier Transform (FFT) makes it easy to implement frequency measurement. According to Equation 11, the frequency of the interference signal can be recorded as:

$$f = 8nd\cos\theta\omega_0\omega_c^2 x_0 /(2\pi c^2) = 4n\omega_0\omega_c^2 x_0 \cos\theta/(\pi c^2) = Kd \qquad (12)$$

Equation 12 shows that the frequency of the interference signal is in proportion to the glass thickness, and the scale factor can be shown as:

$$K = 4n\omega_0\omega_c^2 x_0 \cos\theta/(\pi c^2) \qquad (13)$$

which depends on the angular frequency of light source ω_0, refractive index n and refraction angle θ, vibration acceleration a and angular frequency of oscillating mirror ω_c.

It should be noted that in the formula derivation process, the phase shift of 180° has already been represented by symbol β, based on the Fresnel formula.

3   NUMERICAL INVESTIGATION AND RESULT ANALYSIS

According to the theoretical analysis, H_o solid laser, for example its wavelength is 2,050 nm, and the refractive index of planar glass is 1.493983. The photosurface diameter of the detector is 1 mm with the sensitivity of 1 A/W, and the amplitude of the oscillating mirror is 0.0001 m.

Figure 3 shows the Fourier transform spectrum of normal and oblique incidence hetero-dyne signals in the process of measuring the thickness of glass d. Among them, the solid line is relative to the oblique incidence heterodyne signal, and the dotted line is relative to normal incidence heterodyne signal.

The purpose of Figure 3 is to obtain the refractive angle when it is unknown, by using the ratio. Equation 8 shows that the ratio of the two center frequencies is:

$$\zeta = \cos\theta \qquad (14)$$

After obtaining the center frequency, the refraction angle θ of the laser which goes through the thin glass plate can be calculated from Equation 14. Then, get the value of θ_o by the laws of refraction. Therefore, according to Equation 9, to obtain the value of K, ultimately, according to Equation 8 to obtain the thickness d.

At the same time, Figure 4 shows the simulation result of the spectrum of FFT of hetero-dyne signal second harmonic, corresponding to the measurement of the glass thickness by the multi-beam laser heterodyne second harmonic measurement by using MATLAB. It shows that with an increase of the thickness of the plate glass, the relative position of the spectrum shifts to high frequency. It is said that the frequency is increasing as a result of the increasing of glass thickness. The reason is that the relationship of frequency f and the thickness d can be shown as $f = Kd$, and when keeping the scale factor K invariant, the frequency f and d present a linear relation. Therefore, as the glass thickness increases, the relative position of the spectrum moves to high frequency. It must be explained that since heterodyne detection is a detection method with near-diffraction limit, the sensitivity of detection is extremely high, and so the Signal-to-noise Ratio (SNR) of the heterodyne signal is extremely high in Figure 3 and Figure 4.

Because engineer major interest in measurement accuracy of glass thickness, as an example of glass thickness measurement to verify the feasibility of this novel method. Eight sets of data are obtained when using the second harmonic measurement of multi-beam laser heterodyne for the glass thickness, and the simulation results of different glass thickness are shown in Table 1.

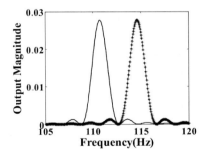

Figure 3. Fourier transform spectrum of normal and oblique incidence heterodyne signals' second harmonic.

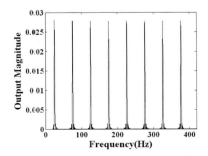

Figure 4. Frequency spectrum of the measurement of different glass thickness.

137

Table 1. The actual values d and simulation measured values d_i of different glass thickness condition.

Time	1	2	3	4
d (mm)	1.000000	3.000000	5.000000	7.000000
d_i (mm)	1.000805	3.000423	5.000044	6.999665
Time	5	6	7	8
d (mm)	9.000000	11.000000	13.000000	15.000000
d_i (mm)	8.999286	11.000891	13.000512	15.000133

It can be found through research that the measuring accuracy of exciting method relative error is higher than 0.3%. Using the simulation data in Table 1, we can calculate that the maximum relative error is less than 0.08%, which can improve by an order of magnitude the measuring accuracy aspect. At the same time, because the reflect lights E_1, E_2, ..., E_m pass through the same external environment, the error caused by the external environment can be removed by the heterodyne method, the error of simulation being mainly owing to the trueness error after FFT and the roundoff error in the processing of the calculation.

4 CONCLUSION

According to precision measurements, plate glass thickness in the engineering needs. This paper proposes a method of second harmonic measurement of multi-beam laser heterodyne for the glass thickness, based on laser heterodyne techniques. The test thickness information of the plate glass is loaded into a frequency difference of the heterodyne signal second harmonic, and demodulated by the Fourier transform. It is easy to measure the thickness information of the plate glass, and obtain high-measurement accuracy. This method is a good non-contact measurement method to measure the thickness of plate glass.

Simulation results show that when measuring different plate glass thickness, measurement error is less than 0.08% using this method, indicating that the method is feasible, reliable and able to meet the requirements of small glass thickness measurements. As many engineering fields provide a good means of measurement, it can be widely used in laser radar, machinery, instruments and electronic products manufacturing, and it has good prospects and value.

ACKNOWLEDGMENTS

The project is supported by the National Natural Science Foundation of China (Grant No. 61505050, 61501176), Heilongjiang Province Natural Science Foundation (Grant No. F2015015), Outstanding Young Scientist Foundation of Heilongjiang University (Grant No. JCL201504), China Postdoctoral Science Foundation (Grant No. 2014M561381), Heilongjiang Province Postdoctoral Foundation (Grant No. LBH-Z14178), Special Research Funds for the Universities of Heilongjiang Province (Grant No. HDRCCX-2016Z10), University Nursing Program for Young Scholars with Creative Talents in Heilongjiang Province (Grant No. UNPYSCT-2017116), Science and Technology Innovation Talent Research Special Fund Project of Harbin (Grant No. RC2017JQ009003), and Postdoctoral research initiation fund project settled in Heilongjiang.

REFERENCES

Li, Y.C., Wang, C.H., Gao, L., Cong, H.F. & Qu, Y. (2012a). Second harmonic multi-beam laser heterodyne measurement for small angle based on oscillating mirror sinusoidal modulation. *Acta Phys. Sin.*, *61*(1), 010601-1–010601-5.

Li, Y.C., Wang, C.H., Gao, L., Cong, H.F. & Qu, Y. (2012b). Multi-beam laser heterodyne measurement with ultra-precision for the glass thickness based on oscillating mirror sinusoidal modulation. *Acta Phys. Sin*, *61*(4), 044207-1–044207-6.

Li, Y.C., Zhang, L., Yang, Y.L., Gao, L., Xu, B. & Wang, C.H. (2009). The method for multi-beam laser heterodyne high-precision measurement of the glass thickness, *Acta Phys. Sin.*, *58*(8), 5473–5478.

Liu, C.H. & Li, Z.H. (2008). Application of the astigmatic method to the thickness measurement of glass substrates. *Appl. Opt.*, *47*(21), 3968–3972.

Nemoto, S. (1992) Measurement of the refractive index of liquid using laser beam displacement. *Appl. Opt.*, *31*(31), 6690–6694.

Novikov, M.A., Tertyshnik, A.D., Ivanov, V.V., Markelov, V.A., Goryunov, A.V., Volkov, P.V., … Mishulin, Y. (2004). Optical interference system for controlling float-glass ribbon thickness at hot stages of production. *Glass and Ceramics*, *61*(1–2), 37–41.

Protopopov, V.V., Cho, S., Kim, K., Lee, S., Kim, H. & Kim, D. (2006). Heterodyne double-channel polarimeter for mapping birefringence and thickness of flat glass panels, *Review of Scientific Instruments*, *77*(5), 1–7.

Tsai, M., Huang, H., Itoh, M. & Yatagai, T. (1999) Fractional fringe order method using Fourier analysis for absolute measurement of block gauge thickness. *Opt. Rev.*, *6*(5), 449–454.

Tsai, M., Tian, R., Huang, H., Itoh, M. & Yatagai, T. (1999). Absolute thickness measurement using automatic fractional fringe order method. SPIE, *3897*, 335–339.

Ulrich, R. & Torge, R. (1973). Measurement of thin film parameters with a prism coupler, *Appl. Opt.*, *12*(12), 2091–2098.

Yan-Chao, L. & Chun-Hui, W. (2012). A method of measuring micro-impulse with torsion pendulum based on multi-beam laser heterodyne. *Chin. Phys. B*, *21*(2), 020701-1–020701-6.

Yan-Chao, L., Chun-Hui, W., Yang, Q., Long, G., Hai-Faing, C., Yan-Ling, Y., … Ao-You, W. (2011). Numerical investigation of multi-beam laser heterodyne measurement with ultra-precision for linear expansion coefficient. *Chin. Phys. B*, *20*(1), 014208-1–014208-7.

Frontier Research and Innovation in Optoelectronics Technology and Industry – Habib & Lewis (Eds)
© 2019 Taylor & Francis Group, London, ISBN 978-1-138-33178-5

Second-harmonic spectral compression using Fresnel-inspired binary phase shaping

Baihong Li
College of Sciences, Xi'an University of Science and Technology, Xi'an, Shanxi Province, China
Key Laboratory of Time and Frequency Primary Standards, National Time Service Center,
Chinese Academy of Sciences, Xi'an, China

ABSTRACT: Spectral compression of Second-Harmonic Generation (SHG) is crucial for selective two-photon microscopy and high-precision nonlinear spectroscopy. In this paper, we report a experimental realization of second-harmonic spectral compression by Fresnel-inspired binary phase shaping for a Gaussian and transformed-limited Fundamental Pulse (FP). Given the FP bandwidth, a narrower SHG bandwidth can be obtained by simply increasing the number of binary phases. The best compression factor achieved in our experiments was 11.3, with a bandwidth (FWHM) of only 0.161 nm and high efficiency of 56%. These results will significantly facilitate the applications of selective two-photon microscopy and spectroscopy.

1 INTRODUCTION

High-peak-power ultrashort pulses are commonly used in selective two-photon microscopy and high-precision nonlinear spectroscopy (Silberberg, 2009), but the corresponding broad bandwidths limit the spectral resolution achieved. A high spectral resolution requires the generation of a strong signal at the desired frequency and the suppression of background at other frequencies, which are enabled by spectral compression. Many studies have been conducted to achieve this goal by using the method of quantum coherent control. Broers et al. (Broers et al., 1992) first demonstrated spectral focusing in second-harmonic generation (SHG) and two-photon absorption (TPA) using binary spectral amplitude modulation. Later, Zheng and Weiner demonstrated the coherent control of SHG and obtained a signal with a high contrast using binary encoded pulses borrowed from communications technology (Zheng et al, 2000; Zheng et al., 2001). Moreover, Dantus and co-workers reported various efforts for spectral compression by taking advantage of multiphoton intrapulse interference (MII) (Walowicz et al., 2002; Lozovoy et al., 2003; Pastirk et al., 2003; Cruz et al., 2004; Lozovoy et al., 2004; Comstock et al., 2004; Lozovoy et al., 2005; Lozovoy et al., 2005; Lozovoy et al., 2006) using various pulse shaping technologies(Weiner, 2011). However, most of these methods involve more complex algorithms and a large number of binary sequences. In 2014, Li et al. (Li et al., 2014; Li et al., 2014) proposed a more exact phase-tailored scheme for spectral compression, named Fresnel-inspired binary phase shaping (FIBPS), based on the work by Broers et al. In this work, we experimentally demonstrate SHG spectral compression for a Gaussian TL fundamental pulse (FP) in the cases of N = 13, 21, and 41 binary phase sequences. The narrowest bandwidth of only 0.161 nm is obtained with a compression factor of 11.3 and high efficiency of 56% using only 41 binary phase sequences. The narrower compressed bandwidth of SHG can be easily achieved by simply increasing the number of binary phase masks.

To obtain the spectral compression of SHG, Li et al. (Li et al., 2014; Li et al., 2014) presented an FIBPS scheme for a chirped pulse with a phase distribution $\phi(\omega) = \alpha(\omega - \omega_0)^2$, where α is the chirp parameter. The SHG intensity can be written as

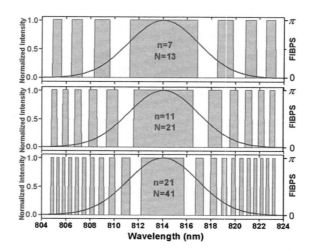

Figure 1.   Shaping scheme for a Gaussian pulse with FIBPS for $n = 7$, 11, and 21, respectively.

$$I_{SHG}(\Omega) \propto \left| \int_{-\Delta\omega/2+\Omega}^{\Delta\omega/2+\Omega} |A(\omega')|^2 e^{i2\alpha\omega'^2} d\omega' \right|^2 \qquad (1)$$

$A(\omega)$ are the spectral amplitude. $\Omega = \omega - \omega_0$ is the frequency deviation from the central frequency ω_0 of the FP. $\Delta\omega$ is the full width of the FP spectrum. The boundary of the nth frequency zone tailored by the FIBPS scheme for an FP spectrum can be expressed as (Li et al., 2014; Li et al., 2014)

$$\pm\Omega_n = \pm\sqrt{(3/2+2(n-1))\pi/\alpha}(n=1,2,3\cdots) \qquad (2)$$

The total number of frequency zones (binary phases) N is $2n–1$. According to Eq. (2), an FIBPS function can be obtained as follows:

$$FIBPS(\Omega) = \frac{\pi}{2}\left(\prod_n \text{sgn}(\Omega_n - |\Omega|) + 1\right) \qquad (3)$$

where sgn is a symbolic function. The above function takes only the value 0 or π for different frequency zones, and the phase difference is π between adjacent frequency zones.

We will demonstrate below that the FIBPS is also valid for a TL FP. Therefore, a chirp is not necessary. For convenience, we will use the above formula with wavelength instead of frequency. The shaping scheme used in our experiment is shown in Fig. 1, where the Gaussian FP spectra are shaped with FIBPS for $n = 7$, 11, and 21, respectively.

2   EXPERIMENT AND RESULTS

2.1   *Experimental setup*

The experimental setup is illustrated in Figure 2. A commercial Ti:sapphire laser (Fusion 100-1200, FEMTOLASERS) centered at 814 nm with a bandwidth of 7 nm, pulse duration of 135 fs, and repetition rate of 75 MHz is used to generate the FP. After being collimated by a pair of lenses, the beam is passed into a pulse shaper (MIIPS-HD, Biophotonic), which is a reflective configured 4-f Fourier pulse shaping system. A half-wave plate HWP1 is used to rotate the polarization of the input pulse field to obtain maximum diffraction efficiency (~92.5%) when the light passes through the polarization-dependent grating G (PC2100 NIR,

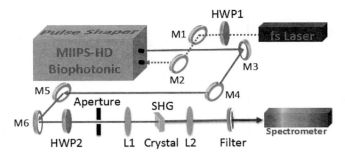

Figure 2.   Experimental setup. L: lens, HWP: half-wave plate, M: mirror.

Spectrogon Inc.). The output efficiency of the shaper system was 73.86% when no phase modulation was applied. A reflection-type liquid crystal on the silicon-spatial light modulator (LCOS-SLM, X10468-2, Hamamatsu) with 792 pixels is placed on the Fourier plane. The different frequencies in the input pulse are scattered in space by the grating and then enter the SLM after being collimated by a cylindrical lens. The spatially dispersed light pulses are phase-modulated along the x-axis needed and then reflected. Thus, the shaped pulse is output above the incident pulse along the vertical direction and then arrives at HWP2 after being reflected four times, as shown in Figure 2. Before the output shaped pulse is focused by L1 into the SHG crystal, an aperture is used to block the unwanted zero-order diffracted light from first-order diffracted light, and HWP2 is used to optimize the SHG efficiency. The generated SHG signals are focused and then sent into the spectrometer (HR4000, Ocean Optics Inc.) after filtering the residual FH light.

The correspondence between the wavelength and pixels is calibrated with the pulse shaper by measuring the FH spectrum before the experiment. To determine the phase of the pulse before it enters the pulse shaper, phase compensation is performed for the input pulse, and its phase information is obtained using the MIIPS method (Lozovoy et al., 2004). In the following experiments, both FIBPS and the compensated phases of the input pulse are introduced by the SLM, and the desired phase functions can be written in the computer to control the phase of pixels in SLM.

2.2   *Experimental results and discussions*

Figure 3(a) shows the experimental results of SHG spectral compression with a 0.5 mm bismuth borate (BIBO) crystal for $n = 7$ (N = 13, red dotted lines), $n = 11$ (N = 21, green lines), and $n = 21$ (N = 41, blue bold lines) with FIBPS. It is found that all SHG spectra are compressed by shaping FH spectrum with FIBPS, and the larger the total number of binary phases N, the narrower the SHG spectrum, but simultaneously the SHG intensity decreases. The SHG spectrum was normalized with respect to the TL SHG intensity for comparison. In those three cases, the compressed bandwidths (FWHM) are 0.393 nm, 0.249 nm and 0.161 nm respectively, and efficiencies are 73%, 66% and 56% respectively. The blue bold line denotes the best result of compression with a compression factor of 11.3. Other experimental parameters are: the center wavelength of the FP is $\lambda_0 = 814$ nm, the bandwidth of shaping FH spectrum is $\Delta\lambda = 18.674$ nm, from 804.648 to 823.322 nm ($\Delta\omega = 5.315 \times 10^{13}$ rad/s). The chirp parameters corresponding to $n = 7$, $n = 11$ and $n = 21$ are $2\alpha_1 = 30023$ fs^2, $2\alpha_2 = 47815$ fs^2, and $2\alpha_3 = 92293$ fs^2, respectively.

To further quantify the results of the SHG spectral compression in Fig. 3, we estimate the contrast ratio C, defined as the ratio between the integrated signal S and integrated background B, according to Ref. (Comstock et al., 2004). We obtained $C_7 = 3.64$, $C_{11} = 1.88$, and $C_{21} = 0.80$. The higher the ratio is, the better is the quality of the signal. It can be seen that the SHG compression bandwidth narrows with increase in N, while the contrast ratio gradually decreases.

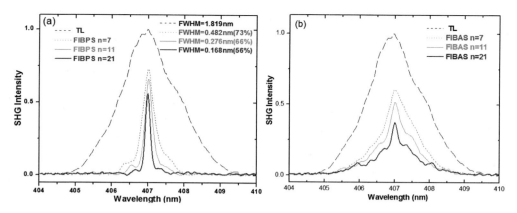

Figure 3. Experimental results of SHG spectral compression with 0.5-mm BIBO crystal for n = 7 (red dotted lines), n = 11 (green lines) and n = 21 (blue bold lines) with FIBPS (a) and FIBAS (b). The SHG spectrum was normalized with respect to the TL SHG intensity (the dashed line).

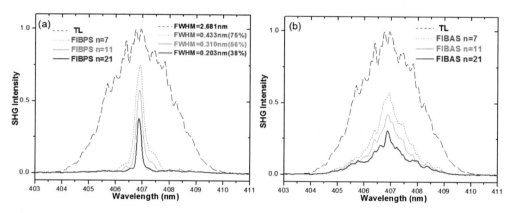

Figure 4. Experimental results of SHG spectral compression with 10-μm BBO crystal for n = 7 (red dotted lines), n = 11 (green lines) and n = 21 (blue bold lines) with FIBPS (a) and FIBAS (b). The SHG spectrum was normalized with respect to the TL SHG intensity (the dashed line).

In addition, we also show compressed results using Fresnel-inspired binary amplitude shaping (FIBAS) as a comparison, as shown in Figure 3(b), where compressed bandwidth and signal intensity are both worse than those in Figure 3(a) with FIBPS. This is a clear fact because that amplitude modulation introduces loss of at least half of the pulse energy.

To illustrate the generality of our method, we also give the experimental results of SHG spectral compression with a 10-μm barium borate (BBO) crystal as shown in Figure 4. In the three cases, the compression bandwidths (FWHM) are 0.433 nm, 0.310 nm, and 0.203 nm respectively, and efficiencies are 75%, 58% and 38% respectively, as shown in Fig.4(a). The blue bold line denotes the best result with a compression factor of 13.2. The chirp parameters corresponding to $n = 7, n = 11$ and $n = 21$ are $2\alpha_1 = 30136$ fs^2, $2\alpha_2 = 47994$ fs^2, and $2\alpha_3 = 92640$, respectively. The estimated contrast ratio C are $C_7 = 4.08$, $C_{11} = 2.47$ and $C_{21} = 1.13$ respectively. Figure 4 (b) shows compressed results with FIBAS as a comparison.

The experimental errors are mainly caused by the following factors: (1) Although the use of the MIIPS method can compensate for the majority of phases of the input pulse, there remains a small residual uncompensated phase, resulting in experimental errors. (2) The BPS we designed does not exactly coincide with the pixels in the SLM because of the limitation of the resolution of this pulse shaper, and this non-correspondence will become more

144

distinct when N increases. This is also one of the main reasons why the compression intensity decreases with n in Fig. 3(a). We also implemented the SHG spectral compression experiment for $n = 31$ (N = 61, not given in the text), where the obtained SHG intensity was lower and the compression bandwidth was even slightly larger than that for $n = 21$, indicating that the non-correspondence between the phase shaping point and SLM pixels has become evident. We believe this error can be gradually eliminated with the improvement in SLM manufacturing technology and pulse shaper resolution. (3) The pixels in the SLM are discrete, and there is a gap (0.4-μm) between adjacent pixels. Since the light field passing through these gaps is not modulated (~2%), this part of the energy will be lost, affecting the compression effect.

When the phase modulation is replaced by amplitude modulation, the compression results of SHG become worse with a broader compressed bandwidth, lower intensity, and larger background. Although the effect of SHG spectral compression can be demonstrated in both crystals with different crystal length, the compressed bandwidth in a short crystal is broader than in a long crystal owing to the broadening of the phase-matching bandwidth. Furthermore, the SHG intensity obtained in a long crystal is higher than that in a short one because the SHG power scales linearly with the crystal length (Zheng et al., 2000). Therefore, in practical applications, longer crystals should be selected as the SHG crystal to obtain a narrower bandwidth with a high signal intensity.

Additionally, FIBPS originates from the quadratic phase factor in Eq. (1), it thus is independent of the shape of the FP spectrum, and is suitable for a variety of commonly used pulses. Moreover, one can also adjust the required amount of dispersion by n based on Eq. (3).

3 CONCLUSION

We have experimentally demonstrated spectral compression in SHG by FIBPS for a Gaussian TL pulse. A narrowest bandwidth (FWHM) of only 0.161 nm was obtained with a compression factor of 11.3 and high efficiency of 56%. For a given FP spectral bandwidth, a narrower bandwidth can be obtained in principle by simply increasing the total number of binary phases N while requiring a higher shaper resolution. Our method provides novel deterministic BPS with a small amount of binary phase sequences for effectively realizing SHG spectral compression without search space maps or any algorithms. These results will be useful in the applications of selective two-photon microscopy and high-precision nonlinear spectroscopy.

ACKNOWLEDGMENT

This work was supported by the National Natural Science Foundation of China (Grant No. 11504292), and Natural Science Basic Research Plan in Shaanxi Province of China (Grant No. 2016 JQ1036).

REFERENCES

Broers, B., Noordam, L.D. & van Linden van den Heuvell, H.B. 1992. Diffraction and focusing of spectral energy in multiphoton processes. *Physical Review A* 46(5): 2749–2756.
Cruz, J.M.D., Pastirk, I., Lozovoy, V.V., Walowicz, K.A. & Dantus, M. 2004. Multiphoton intrapulse interference 3: Probing microscopic chemical environments. *Journal of Physical Chemistry A* 108(1): 53–58.
Comstock, M., Lozovoy, V.V., Pastirk, I. & Dantus, M. 2004. Multiphoton intrapulse interference 6; binary phase shaping. *Optics Express* 12(6): 1061–1066.
Li, B.H., Xu, Y.G., An, L., Lin, Q.L., Zhu, H.F., Lin, F.K., & Li, Y.F. 2014. Quantum focusing and coherent control of nonresonant two-photon absorption in frequency domain. *Optics Letters* 39(8): 2443–2446.
Li, B.H., Xu, Y.G., Zhu, H.F., Lin, Q.L., An, L., Lin, F.K., & Li, Y,F. 2014. Spectral compression and modulation of second harmonic generation by Fresnel-inspired binary phase shaping. *Journal of the Optical Society of America B* 31(10): 2511–2515.

Lozovoy, V.V. & Dantus, M. 2005. Systematic control of nonlinear optical processes using optimally shaped femtosecond pulses. *Physical Chemistry Chemical Physics* 6(10): 1970–2000.

Lozovoy, V.V., Pastirk, I. & Dantus, M. 2004. Multiphoton intrapulse interference. IV. Ultrashort laser pulse spectral phase characterization and compensation. *Optics Letters* 29(7): 775–777.

Lozovoy, V.V., Pastirk, I., Walowicz, K.A. & Dantus, M. 2003. Multiphoton intrapulse interference. II. Control of two- and three-photon laser induced fluorescence with shaped pulses. *Journal of Chemical Physics* 118(118): 3187–3196.

Lozovoy, V.V., Shane J.C., Xu, B. & Dantus, M. 2005. Spectral phase optimization of femtosecond laser pulses for narrow-band, low-background nonlinear spectroscopy. *Optics Express* 13(26): 10882–10887.

Lozovoy, V.V., Xu, B., Shane J.C. & Dantus, M. 2006. Selective nonlinear optical excitation with pulses shaped by pseudorandom Galois fields. *Physical Review A* 74(4): 041805(R).

Pastirk, I., Cruz, J.D., Walowicz, K., Lozovoy, V. & Dantus, M. 2003. Selective two-photon microscopy with shaped femtosecond pulses. *Optics Express* 11(14): 1695–1701.

Silberberg, Y. 2009. Quantum coherent control for nonlinear spectroscopy and microscopy. *Annual review of physical chemistry* 60(60): 277–292.

Weiner, A.M. 2011. Ultrafast optical pulse shaping: A tutorial review. *Optics Communications* 284(15): 3669–3692.

Walowicz, K.A., Pastirk, I., Lozovoy, V.V. & Dantus, M. 2002. Multiphoton intrapulse interference. 1. Control of multiphoton processes in condensed phases. *Journal of Physical Chemistry A* 106(41): 9369–9373.

Zheng, Z. & Weiner, A.M. 2000. Spectral phase correlation of coded femtosecond pulses by second-harmonic generation in thick nonlinear crystals. *Optics Letters* 25(13): 984–986.

Zheng, Z. & Weiner, A.M. 2001. Coherent control of second harmonic generation using spectrally phase coded femtosecond waveforms. *Chemical Physics* 267(1–3): 161–171.

Frontier Research and Innovation in Optoelectronics Technology and Industry – Habib & Lewis (Eds)
© 2019 Taylor & Francis Group, London, ISBN 978-1-138-33178-5

Study of a laser-echo simulation system with high precision and a wide range based on a phase-locked loop frequency multiplier

M. Liu, Z.W. Li & G.Y. Wen
Changchun Observatory, NAO, CAS, Changchun, China

X.N. Yu
NUERC of Space and Optoelectronics Technology, Changcun University of Science and Technology, Changcun, China

Y.Z. Liu
Rocket Force Military Representative Office in Tianjin Area, Tianjin, China

ABSTRACT: A novel laser-echo simulator system with high precision and a wide range is presented in this paper. The system, which is used to detect the ranging performance of a pulse-laser rangefinder, adopts Phase-Locked Loop (PLL) frequency multiplier to analog laser echo. By doubling the frequency of a crystal oscillator clock, the simulation system can achieve both high-resolution precision and a wide dynamic range. The laser-echo simulation system gathers impulse signals extraction, precision time delay and echo signal simulation into an integrated whole, the range limitation is overcome more effectively while maintaining high accuracy by adopting PLL in particular. The technical index of the simulation system discussed in this paper is 0.3 m high precision with a range from 50 m to 10 km. The detailed design of each component of the laser-echo simulator is described. The experimental results indicate that the echo simulation accuracy of the system is better than 0.35 m, meeting the ranging detection requirements of laser rangefinder performance.

Keywords: laser-echo simulator; phase-locked loop frequency multiplier; precision time delay; laser rangefinder performance test; ranging precision detection

1 INTRODUCTION

Pulsed-laser rangefinders have been widely applied in the fields of intelligent transportation, aerospace detection, geodesy, and architectural surveying because of their simple structure, long detection distance, and high-ranging precision (Gao et al., 2012). After many years of research, the current direction of pulsed-laser rangefinder development is embodied in: (1) ranging precision; (2) operating range; (3) ranging frequency (Céspedes et al., 1995). In view of the gradual improvement of these three factors, the testing systems of some parameters at the present stage of development is seriously lagging behind the capabilities of new equipment (Liu et al., 2013), and this has now become an important research topic (Yang et al., 2014).

The quality of automobile collision avoidance detectors in China varies, their range sensitivity and precision repeatability need to be calibrated quickly in the room, and the speed and the distance in a pilotless system need to be detected as well (Zeng et al., 2006). In the space debris monitoring field, it is necessary to detect the performance indexes of laser rangefinders to ensure the accuracy and reliability of astronomical observation data (Sun et al., 2006).

A wide-ranging and high-precision atmospheric laser-echo simulation technology based on Phase-Locked Loop (PLL) frequency doubling is proposed in this paper. The main idea of the method is to create a virtual target in the laboratory to replace the laser echo reflected from the target object. The system can achieve a wider range and higher precision through

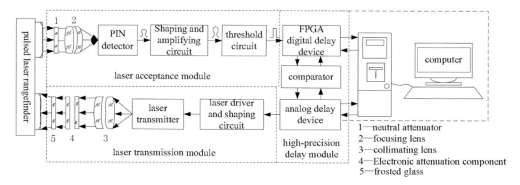

Figure 1.   System components of the detection system.

this method. Once the technology is brought to market, its profitability should be very good. It can also provide reliable technical support for the next generation of driverless systems and space debris detection.

2   COMPONENTS AND PROCESS FLOW OF THE SYSTEM

The basic principle of the system is that a controlled-delay pulsed-laser signal is used instead of the reflected laser echo of the rangefinder. The process flow is as follows:

1. The simulation distance parameter is set to delay counters through the computer's serial port.
2. When the laser rangefinder emits a pulsed laser signal, the optical signal passes through attenuation slices, then through focusing lens components before converging on the high-speed PIN detector; the PIN detector switches the optical signal to an electronic signal.
3. Through an amplification circuit, rectification circuit and threshold circuit, the output signal will become an available analog signal, matching the Transistor–Transistor Logic (TTL) level.
4. A monostable trigger broads the received laser pulse signal immediately as a gate signal of a high-frequency oscillator. The Field-Programmable Gate Array (FPGA) digital delay module is a subtraction counter; when the gate signal is active, the high-frequency oscillator's output pulse signal is used as the delay counter's input pulse. When the delay counter overflows, the analog delay module will be triggered.
5. Once the analog delay module overflows, the radiation module emits a laser beam to the laser rangefinder. The system components are shown in Figure 1.

3   LASER-ECHO SIMULATION SUBSYSTEM DESIGN

3.1   *Laser acceptance module design*

The function of the laser acceptance module is to convert the light pulse emitted by the rangefinder device into an electrical signal, which is used as the precise start-timing signal of the laser-echo simulator. The response time and repeatability of the laser-receiving module directly affect the accuracy of the analog distance measurement. In this paper, an FGA01FC InGaAs photodiode is used as the detector; its performance parameters are shown in Table 1. The spectral response curves and a photograph of the photodiode are shown, respectively, in Figures 2a and 2b.

The photodiode response time and repeatability of the FGA01FC photodiode will directly affect the accuracy and stability of the simulation distance measurement. Its response bandwidth (f_{BW}) and response time (t_R) can be computed with the following formulas:

Table 1. Performance parameters of FGA01FC photodiode.

Response wavelength range	900–1700 nm
Responsivity	1.11 A/W
Effective photoreceptor diameter	120 μm
Rise time (RL = 50 Ω)	300 ps
Dark current (5 V)	0.05 nA
Photodiode capacity (5 V)	2.0 pF

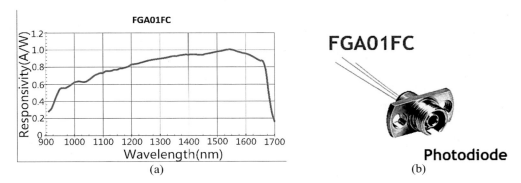

Figure 2. (a) Spectral response curve of the photodiode; (b) photograph of the photodiode.

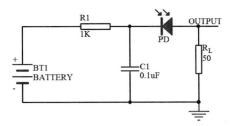

Figure 3. Laser acceptance module filter circuit schematic diagram.

$$f_{BW} = \frac{1}{(2\pi) R_L C_j}, \quad t_R = \frac{0.35}{f_{BW}} \tag{1}$$

The response bandwidth and response time are as follows:

$$t_R = \frac{0.35}{f_{BW}} = \frac{0.35}{1.59(GHZ)} \approx 219.8(ps) \tag{2}$$

In order to meet the requirements for a delay accuracy of a nanosecond, a 220 ps response time by the detector is sufficient. Figure 3 shows a schematic diagram of the filter circuit of the laser acceptance module; R_1 and C_1 are combined as a filter circuit to remove noise interference; D_1 is the detector, R_L is the load.

3.2 *Delay module design based on a phase-locked loop frequency multiplier*

Adopting PLL of FPGA to achieve a high-precision delay, the 50 MHz crystal clock is expanded nine times under frequency doubling to create a high-speed clock of 450 MHz, which is used as the echo simulator delay clock. A single clock cycle is as follows:

$$t = \frac{1}{f} = \frac{1}{450 \times 10^6} \approx 2.22(ns) \tag{3}$$

For the laser-echo simulation, the accuracy of the distance simulation is:

$$\varepsilon = \frac{ct}{2} = \frac{3 \times 10^8 \times 2.22 \times 10^{-9}}{2} \approx 0.33(m) \tag{4}$$

In Equation 4, c is the speed of light. The external clock crystal operates at 450 MHz, which meets the requirement for a 0.3 m laser-echo simulation. The program implements the following function: nine-times frequency doubling by PLL of FPGA is used to achieve a 450 MHz signal (clk_450M), then this high-frequency clock is used as a delay signal clock, when the fire control signal input, precise delay start. Figure 4 shows a Resistor–Transistor Logic (RTL) view of the PLL component.

The PLL program reflects the effect of delay simulation if the high-precision clock meets the requirements. The test clock is 50 MHz; after five cycles, the multiplier effect is shown in Figure 5.

It can be seen in Figure 5 that once the PLL enters the locked state, one main clock contains nine frequency cycles; that is, the frequency after frequency doubling is $50 \times 9 = 450$ MHz and the cycle time is 2222 ps, meeting the 0.3 m high-precision requirement.

The main function of signal conditioning is to convert the signal to be measured by amplifying and filtering it into a standard signal that the acquisition device can recognize. As the front-level processing circuit of the simulator, the conditioning circuit needs high input-voltage sensitivity, a wide voltage-conditioning range and a high input impedance, and must adapt to the periodic and aperiodic requirements of various waveforms; it must also have a

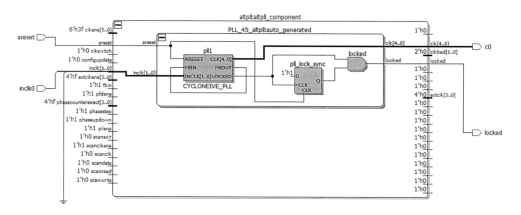

Figure 4.   Schematic RTL diagram of the phase-locked loop component.

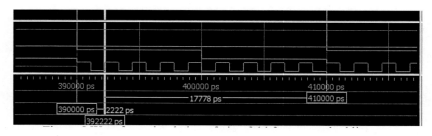

Figure 5.   Waveform simulation of nine-fold frequency doubling.

150

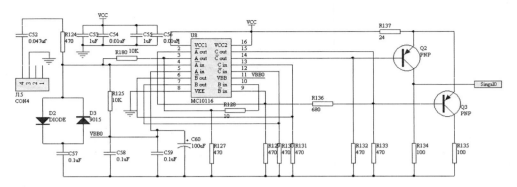

Figure 6. Signal conditioning circuit.

Table 2. Performance parameters of C86119E semiconductor pulse laser.

Performance parameter	Minimum	Typical	Maximum	Unit
Maximum forward current (iFM)			4	A
Pressure drop (pulse width is 100 ns)		2.5	3	V
Luminescence threshold current (ith)		0.25		A
Luminescence response time (t) (10–90%)		<1		ns
Peak radiation wavelength (λm)	1059	1064	1069	nm
Beam divergence angle		4.5° × 38°		Degrees
Luminous surface dimension		2 × 100		μm

good anti-interference capability. The MC10116-based shaping amplifier circuit is shown in Figure 6.

3.3 *Laser radiation module design*

The radiation module uses a C86119E semiconductor pulse laser (PerkinElmer Optoelectronics, Fremont, CA, USA) as the laser-echo radiation device; the response laser light source of the internal is 1.06 μm of laser rangefinder. This type of laser can provide a high-efficiency and low-threshold pulse laser. The laser is packaged with a diode, one silicon photodiode, a semiconductor cooler and a thermistor. The performance parameters of the C86119E semiconductor pulse laser are shown in Table 2.

From Table 2, it can be seen that the driving circuit should accommodate a pressure drop of 2.5–3 V, the luminous threshold current is 0.25 A, the luminous response time is ≤1 ns, the maximum forward current is less than 4 A, and the maximum pulse width of the driving current is less than 200 ns. The rest of the parameters will be used as the design parameters of the laser optical collimation module. In accordance with the driving requirements of the laser, this paper discusses the front drive principle circuit as shown in Figure 7. Figure 8 shows physical circuit boards and test waveforms.

Figure 8b shows that the laser radiation waveform pulse width is about 23 ns, which can be viewed as the system error calibration and this error can be corrected by the oscilloscope.

The function of the laser radiation subsystem is to simulate the power and distribution of the laser-echo energy of the laser rangefinder. Together, this subsystem and the laser-echo distance simulation subsystem complete the echo reproduction, enabling distance simulation and energy to form a simulation loop. The components of the echo energy simulation subsystem are shown in Figure 9.

Figure 10 shows the light spot illumination distribution map of the collimated laser. As we can see from Figure 8b, as the divergent beam passes through the collimator, energy distribution is strong in the middle but weak at the edge, and the uniformity of the spot is more than 94%.

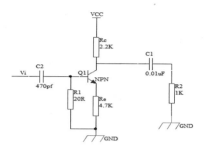

Figure 7.   Front drive principle circuit.

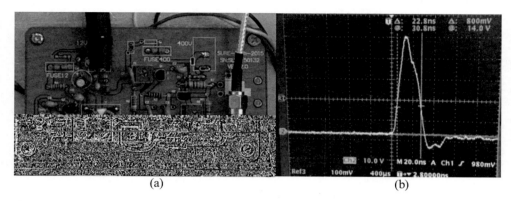

(a)   (b)

Figure 8.   (a) Laser-driven circuit board; (b) laser emission pulse signal test waveform.

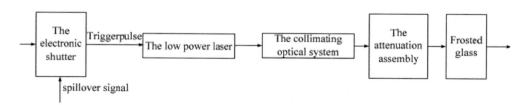

Figure 9.   Components of the laser radiation subsystem.

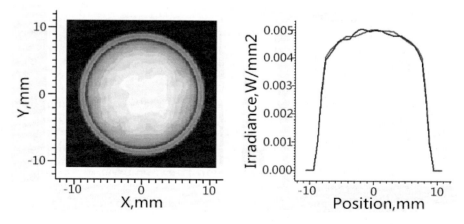

Figure 10.   Photogrammetry distribution.

4   ACCURACY VERIFICATION OF THE LASER-ECHO DISTANCE SIMULATION

This experiment was achieved by calibrating a pulse-laser rangefinder. The performance parameters of the rangefinder were as follows: the wavelength of the laser was 1.064 μm; the output peak power was 2 MW; the divergence angle θ of the laser beam was less than 1 mrad; the ranging accuracy was 1 m. By using a digital oscilloscope with a 1 GHz sampling frequency to detect the delay accuracy of the laser-echo simulation distance, and taking the 1000 m (6667 ns) echo distance as an example, the average distance of 100 repeated experiments of the simulated distance was 1000.16 m, with standard deviation of 0.036 m. A single detection experiment for the distance simulation is shown in Figure 11a, and 100 repeated experimental results are shown in Figure 11b. Figure 12 shows the response distance for the echo signal on the laser-echo simulator.

In a similar fashion, verification experiments on different simulation distances were completed; the results are shown in Table 3.

From Table 3 we can see that the delay error of the laser-echo signal simulator is less than 3 ns; that is, the laser-echo distance simulation accuracy is better than 0.3 m, thus meeting the technical requirements.

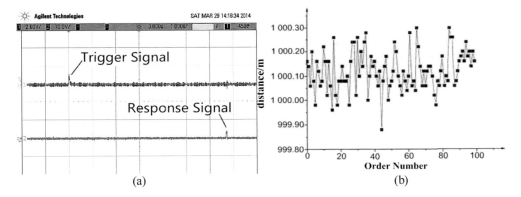

(a)   (b)

Figure 11.   (a) Time delay signal waveform; (b) echo distance simulation statistic curve.

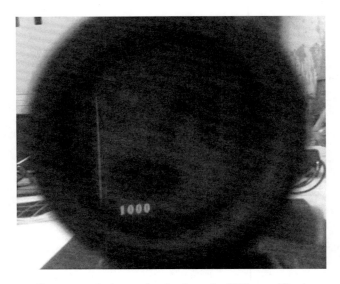

Figure 12.   Response distance on the laser-echo simulator for 1000 m verification test.

Table 3. Verification test data of delay-precision detection.

Target simulation distance (m)	50	100	300	500	1500	2000
Ordinal	Oscilloscope monitoring delay time (ns)					
1	336.13	670.49	2002.42	3332.85	10003.2	13335
2	337.36	665.94	2001.63	3333.14	10003.1	13331.6
3	334.92	670.01	2001.99	3335.13	9999.63	13333.1
4	335.59	666.46	1998.57	3332.25	10000.4	13332.3
5	336.98	669.19	2002.2	3335.94	10003.3	13333.6
6	333.85	667.09	2000.67	3336.02	9999.29	13333.9
7	333.06	670.75	2002.57	3334.74	10000.4	13332.0
8	338.06	671.27	2000.78	3334.67	10000.7	13331.9
9	336.78	668.12	2001.71	3332.72	10002.2	13335.4
10	333.97	671.62	2004.01	3337.01	10002.6	13331.2
Equivalent distance (m)	50.35	+100.364	300.248	500.167	10001.5	1999.95
Equivalent distance error (m)	+0.29	+0.304	+0.248	0.170	+0.22	−0.05

5 CONCLUSIONS

A high-precision and wide-range laser-echo simulation system has been introduced in this paper. The experimental results show that the laser-echo simulation system is of high degree digitization, easy to operate, fast and effective, and its distance simulation error is better than 0.5 m, which means that the time-delay precision has been raised by more than two times compared with other early systems by the method of optical delay (Wang et al., 2017). The whole calibration process for the laser rangefinder could be completed in the laboratory, shortening its development period and reducing the research cost. This device could also be used to detect the accuracy of weapon fire control equipment and to detect the performance of airborne radar ranging. The system has practical significance in the current domestic production and detection of laser rangefinders (Zheng & Voelger, 2001).

REFERENCES

Céspedes, I., Huang, Y., Ophir, J. & Spratt, S. (1995). Methods for estimation of subsample time delays of digitized echosignals. *Ultrasonic Imaging, 17*(2), 142–171.

Gao, Y., Lei, J., Yu, H., Liu, Y. & Liu, J. (2012). Technology of simulation for high-precision echo laser. *Infrared and Laser Engineering, 41*(1), 196–199.

Liu, M., Zhang, G., An, Z., Duan, J. & Lin, Y. (2013). Design of the calibrating system for laser ranging accuracy based on DS1023. In *Proceedings 2013 International Conference on Mechatronic Sciences, Electric Engineering and Computer (MEC)* (pp. 628–631). Piscataway, NJ: Institute of Electrical and Electronics Engineers. doi:10.1109/MEC.2013.6885140

Sun, B., Wan, Q., Zhang, W., Han, L., Wei, S. & Cao, H. (2006). Imitating for laser echo accurately based on CPLD. *Optical Instruments, 28*(3), 8–11.

Wang, X., Wang, J., Li, Q. & Xiao, Y. (2017). Error analysis and calibration of high resolution laser echo simulator. *Journal of Changchun University of Science and Technology, 40*(2), 50–52.

Yang, R., Zhang, G. & Zhang, Z. (2014). Design and experiment of a laser ranging scheme for aerospace applications. *Infrared and Laser Engineering, 11*(3), 700–706.

Zeng, C.-E., Wang, Q., Chang, G., Shan, C.-S. & Wang, Q. (2006). Novel method for maximum range measure of pulse laser range finder. *Infrared and Laser Engineering, 34*(6), 664–668.

Zheng, Y. & Voelger, P. (2001). Monte Carlo simulation of LIDAR return signals in multiple scattering intensities from homogenous haze. *Chinese Journal of Light Scattering, 13*(2), 70–77.

Topological reaction in two superimposed vortex beams

W.R. Miao & X.Y. Pang

School of Electronic and Information, Northwestern Polytechnical University, Xi'an,
Shaanxi Province, China

ABSTRACT: In this paper, the topological characteristics of two non-axial, parallel Gaussian vortex beams, with the same and opposite topological charges, are discussed based on the Huygens–Fresnel principle and Debye approximation. Topological behaviors in both conservation and non-conservation of topological charge and index are analyzed in the focal plane of the composite field. The distribution of newly generated phase singularities is found to be quite different for the vortices of same charge and opposite charge. Comparison and analysis for both cases are also given in this paper.

1 INTRODUCTION

Phase singularity is a substantial characteristic of optical vortex beams' spatial structure (Berry, 1981, 2001; Soskin & Vasnetsov, 2001; Dennis, 2001a, 2001b) and have many potential applications in wireless communication (Wang et al., 2012), microscopy (Spektor et al., 2008; Török & Munro, 2004) and optical trapping (Bishop et al., 2004). Topological reaction in the structured field with phase singularities has been investigated frequently for its fundamental role in studying and applying singular fields (Swartzlander & Maleev, 2003; Soskin et al., 1997; Kotlyar et al., 2007; Pang et al., 2015), including examining the Orbital Angular Momentum (OAM) in such fields (Soskin et al., 1997; Litvin et al., 2011) and dynamics of different singularities (Zhao et al., 2017; Zhao et al., 2018; Pang et al., 2018; Luo & Lü, 2010; Lopez-Mago & Perez-Garcia, 2013).

The optical singularities and their reactions have also attracted a lot of attention in composite fields. In1997, Soskin and colleagues analyzed the properties of light beams carrying phase singularities, they found that in spite of any variation in the number of vortices in a combined beam, the total angular momentum is constant during the propagation (Soskin et al., 1997). The motion of optical phase singularities as the relative phase or amplitude of two interfering collinear non-concentric beams was studied by Swartzlander and Maleev (2003). A superposition of the Bessel and new hypergeometric modes was given as an example of the rotating light beams with zero OAM by Kotlyar et al. (2007). Vasilyeu et al. reported the first experimental generation of the superposition of higher-order Bessel beams (Vasilyeu et al., 2009). In 2010, the composite polarization singularities formed by the transverse and longitudinal electric-field components were investigated by Luo and Lü (2010). Superimposed Bessel beams were created in a laboratory to measure the OAM density by Litvin et al. (2011). The superposition of two off-axis optical vortices with orthogonal polarization states were used to study the dynamics of polarization singularities by Lopez-Mago and Perez-Garcia (2013). However, most works have concentrated on the properties along propagation. For a better understanding of the nature of the superimposed field and the topological reactions of composite phase singularities, we choose to study the topological reaction of two non-coaxial, parallel vortex beams in the focal plane.

In this paper, the topological behaviors of two superimposed Gaussian vortex beams in the focal plane are investigated. The Huygens–Fresnel principle and Debye approximation are used to derive the general form of the vortex beam. The fascinating situation of phase singularities and topological reaction in the focal plane are analyzed in two parts, one being the two beams with identical topological charge, the other with opposite topological charge.

2 FOCUSED FIELD

Consider a scalar focused field with f the focal length and O the geometrical optical focus located at the origin of the Cartesian coordinates (see the lower part in Figure 1). Assume that there is an incident field with the complex amplitude distribution $V_0(\rho, \theta)$, and according to the Huygens–Fresnel principle (Born & Wolf, 1999), the field at observation point P near the focus can be written as:

$$U_1(P) = -\frac{i}{\lambda} \frac{A_1 e^{-ikf}}{f} \iint_W \frac{e^{iks} V_o(\rho,\theta)}{s}\, dS,\tag{1}$$

with λ the wavelength and $k = 2\pi/\lambda$ the wave number. Here, s is the distance between the observation point P and the wave front W, and A_1 is a constant related to the intensity of the incident beam.

Now, if the incident field is a monochromatic Gaussian vortex wave, the complex amplitude distribution can be expressed as:

$$V_o(\rho',\theta) = \rho' e^{-(\rho'/\omega_o)^2} e^{im\theta}\tag{2}$$

with ω_0 the beam waist and m the topological charge. Applying Debye approximation, Equation 1 becomes:

$$U_1(P) = -\frac{i}{\lambda} A_1 \iint_\Omega \rho' \frac{e^{iks-(\rho'/\omega_o)^2} e^{im\theta}}{s}\, dS\tag{3}$$

After some straight calculations, we can derive:

$$U_1(P) = -\frac{i}{\lambda} \frac{a^3 A_1}{f^2} e^{i\left(kz + m\arctan\frac{x}{y}\right)} 2\pi i^m \int_0^1 J_m\left(-k\frac{a}{f}\sqrt{x^2+y^2}\,\rho\right) e^{-\frac{i}{2}k(a/f)^2 \rho^2 z - (a\rho/\omega_o)^2} \rho^2\, d\rho\tag{4}$$

where $J_m(x)$ is the mth order Bessel function of the first kind. Also, $\rho' = a\rho$. Equation 4 is the expression for the field near the focus. If the focusing system is moved a distance b along the positive x axis, we can get another parallel Gaussian vortex beam and its expression can be derived in the same way, as:

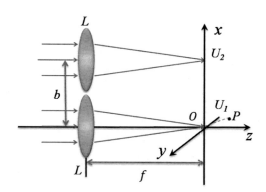

Figure 1. Illustration of two non-coaxial, parallel focusing systems. The origin O of a Cartesian coordinate system is taken at the geometrical focus of the first focused field.

$$U_2(P) = -\frac{i}{\lambda}\frac{a^3 A_2}{f^2} e^{i\left(kz+n\arctan\frac{x-b}{y}\right)} 2\pi i^n \int_0^1 J_n\left(-k\frac{a}{f}\sqrt{(x-b)^2+y^2}\,\rho\right) e^{-\frac{i}{2}k(a/f)^2\,\rho^2 z-(a\rho/\omega_o)^2}\rho^2\,d\rho \qquad (5)$$

where n represents the topological charge of the incident field in the second focusing system. The superimposed field (the system is shown in Figure 1), thus can be expressed as:

$$
\begin{aligned}
U(P) &= U_1(P) + U_2(P)\\
&= -\frac{i}{\lambda}\frac{a^3}{f^2} e^{ikz}\left(A_1 e^{im\arctan\frac{x}{y}}2\pi i^m \int_0^1 J_m\left(-k\frac{a}{f}\sqrt{x^2+y^2}\,\rho\right) e^{-\frac{i}{2}k(a/f)^2\,z\rho^2-(a\rho/\omega_o)^2}\rho^2\,d\rho\right.\\
&\quad \left.+ A_2 e^{in\arctan\frac{x-b}{y}}2\pi i^n \int_0^1 J_n\left(-k\frac{a}{f}\sqrt{(x-b)^2+y^2}\,\rho\right) e^{-\frac{i}{2}k(a/f)^2\,z\rho^2-(a\rho/\omega_o)^2}\rho^2\,d\rho\right)
\end{aligned}
\qquad (6)
$$

3 RESULTS AND DISCUSSION

3.1 *Superposition of two vortex beams with equal topological charge*

For generality and simplicity, we first consider two vortex beams with the same topological charge $m = n = 1$ and it is also assumed that the two beams have identical magnitude $A_1 = A_2$.

The intensity distribution of the composite field in the focal plane is shown in Figure 2(a), where the brightest color indicates the intensity maximum, and vice versa. The vortices, also called the phase singularities can be observed in Figure 2(b) by the intersections of contour lines. By definition, phase singularities only occur where the real part and imaginary part of $U(p)$ equals zero simultaneously. To show the phase singularities more expressly, in Figure 2(c), we use a blue line to indicate the value $\text{Re}[U(p)] = 0$ and a yellow dashed line to represent $\text{Im}[U(p)] = 0$. Thus, phase singularities locate at the intersection points of the two lines. From Figure 2, we can find that, besides the original phase singularities near $(0, 0)$ and $(10\lambda, 0)$, new singularities appear in the midway of two original ones [near $(5\lambda, 0)$] and around them (like near $(-0.5\lambda, -9\lambda)$). These new phase singularities are created by the interference of two beams and such creation does not follow the conservation law of topology. In the following, we will show that these singularities also obey conservation law by gradually changing the distance between the two beams.

The annihilation process of vortices at the transverse focal plane is shown in Figure 3. When the distance between two vortices is 10λ, there are four saddle points, D_1 ($m = 0$, $t = -1$), D_2 ($m = 0$, $t = -1$), D_3 ($m = 0$, $t = -1$), D_4 ($m = 0$, $t = -1$) and three phase singularities, S_1 ($m = 1$, $t = +1$), S_2 ($m = 1$, $t = +1$) and S_3 ($m = -1$, $t = +1$) in our observation region (see (a) of Figure 3). It is well known that the topological charge m and topological index t must be conserved if the field experiences small disturbances. This conservation law was first elucidated by Nye, Hajnal and Hannay (Nye & Wright, 2000; Nye et al., 1988), and can be expressed as a simple equation:

$$V^+ + V^- + 2S \Leftrightarrow 0 \qquad (7)$$

In this equation, V^+ represents a vortex with a positive topological charge, V^- represents a vortex with a negative topological charge and S represents a saddle. In our case, by gradu-

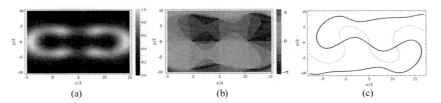

(a)   (b)   (c)

Figure 2. Transverse intensity (a), phase distribution (b) and lines of zero real part and imaginary part of the field (c). Here m = 1, n = 1, z = 0, a/f = 1/10, a/ω_0 = 1, b = 10 λ and $A_1 = A_2$.

157

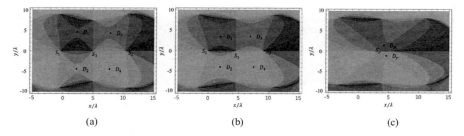

Figure 3. Annihilation process of phase singularities at the transverse focal plane: (a) b = 10λ; (b) b = 9λ; (c) b = 7.7λ. Here m = 1, n = 1, z = 0, a/f = 1/10, a/ω_0 = 1 and $A_1 = A_2$.

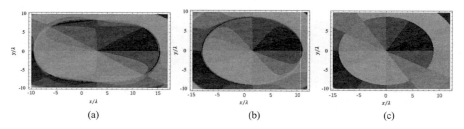

Figure 4. Contours of the phase at the transverse focal plane: (a) $b = 6\lambda$; (b) $b = 3\lambda$; (c) $b = 0$. Here $m = 1$, $n = 1$, $z = 0$, a/f = 1/10, a/ω_0 = 1 and $A_1 = A_2$.

ally decreasing the distance between the two vortices from 10λ to 7.7λ, S_1, S_2 and S_3 start to get close to each other, and annihilated with four saddle points, and at last only one phase singularity with two saddles exists as shown in Figure 3(c). In this progress, the topological charge and index are conserved.

If the distance between the two vortices continues to decrease from 7.7λ, we find that there is always only one phase singularity located at the midway of the two original vortices. That is, when $b = 6\lambda$, it locates at $x = 3\lambda$; when $b = 3\lambda$, it locates at $x = 1.5\lambda$; when $b = 0$, in that the two vortices coincide with each other, the phase singularity is exactly at the original point. Thus, we can say 7.7λ is a critical point, and when b is larger than 7.7λ, there exists an original phase and composite singularities which can be observed to have conserved topological reactions; when b is less than 7.7λ, there is only one phase singularity in the midway of the two original phase singularities which indicates the mergence of two vortex beams. This process is shown in Figure 4.

3.2 Superposition of two vortex beams with opposite topological charge

In this part, we will consider two vortex beams with opposite topological charges, in that $m = +1$ and $n = -1$. Firstly, it is also assumed that the two beams have identical magnitude, in that $A_1 = A_2$. The intensity and phase distribution are shown in Figure 5.

It can be seen that the intensity is zero not only for the point in the midway of the two original phase singularities, but also for the points lying in the mid-line ($x = 5\lambda$) of them, which is quite different from that for the two beams with identical topological charge. This phenomenon is also easily found in plots (b) and (c) in Figure 5. In plot (b), that zero intensity line is actually the line of singularity and the phase crossing this line will have a π phase jump. This line in plot (c) can be observed by the overlying of two lines which represent the zero real part and zero imaginary part of the field, respectively. This behavior is caused by the identical magnitude of the two fields and opposite topological charge of the vortices. If the ratio for magnitudes of the two fields is a little bit changed, this phase singularities line will be destroyed. An illustration is shown in Figure 6, where $A_2/A_1 = 0.9$. It is quite clear that the two lines of the zero real part (blue, solid line) and zero imaginary part (yellow, dashed line) are separated from each other.

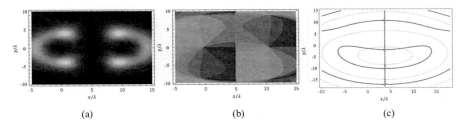

(a)   (b)   (c)

Figure 5.   Transverse intensity (a), phase distribution (b) and lines of zero real part and imaginary part of the field (c). Here $m = 1$, $n = -1$, $z = 0$, $a/f = 1/10$, $a/\omega_0 = 1$, $b = 10\lambda$ and $A_1 = A_2$.

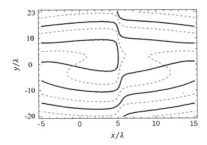

Figure 6.   Lines for the zero real part and zero imaginary part of the field at the transverse focal plane. $Z = 0$ with magnitude ratio 10/9. Here $m = 1$, $n = -1$, $z = 0$, $a/f = 1/10$, $a/\omega_0 = 1$, $b = 10\lambda$ and $A_2/A_1 = 0.9$.

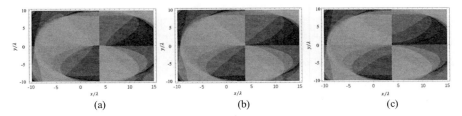

(a)   (b)   (c)

Figure 7.   Contours of the phase at the transverse focal plane: (a) $b = 6\lambda$; (b) $b = 3\lambda$; (c) $b = 0$. Here $m = 1$, $n = -1$, $z = 0$, $a/f = 1/10$, $a/\omega_0 = 1$ and $A_1 = A_2$.

When the separation distance b is decreased, it is also very interesting to find that 7.7λ is still a critical point for the vortex beams with opposite topological charge. That is, when the separation distance is larger than 7.7λ, the original phase singularities can be observed on both sides of the phase singularity line, while, when the distance is less than 7.7λ, the original phase singularities merge with each other and only the phase singularity line exists. This process is shown in Figure 7.

4   CONCLUSION

In this paper, we studied the topological reactions of two non-coaxial, parallel, vortex beams in the focal plane. The topological behaviors in conservation and non-conservation ways both can be observed in the composite field. It was found that there is a critical distance between the two parallel beams which is 7.7λ in our case. When the separation distance is smaller than this value, the two beams will merge with each other and it is hard to observe the topological reaction in the focal plane, while, when the separation distance is greater than the critical distance, there will be rich topological reactions for phase singularities and saddles, like creations and annihilations. It was also found that when the distance is larger than the critical distance, there will be a phase singularity in the midway of the original singularities for the case of identical topological charge, whereas rather than a phase singularity, there will be a singularity line between the two original phase singularities in the case of opposite topological charge.

ACKNOWLEDGEMENTS

This work is supported by National Natural Science Foundation of China (NSFC) (No. 11504296).

REFERENCES

Berry, M.V. (1981). Singularities in waves and rays, in Balian, R. Kléman, M. & Poirier, J.-P. editors, Les Houches Session XXV – *Physics of Defects* (North-Holland).

Berry, M.V. (2001). Geometry of phase and polarization singularities illustrated by edge diffraction and the tides. *Proceedings of SPIE, 4403*, 1–12.

Bishop, A.I., Nieminen, T.A., Heckenberg, N.R. & Rubinszteindunlop, H. (2004). Optical micro rheology using rotating laser-trapped particles. *Physical Review Letters, 92*(19), 198104.

Born, M. & Wolf, E. (1999). Principle of optics: Electromagnetic Theory of Propagation, Interference and Diffraction of Light (Cambridge: Cambridge University Press), seventh(expanded) ed.

Dennis, M.R. (2001a). Topological singularities in wave fields. *University of Bristol.*

Dennis, M.R. (2001b). Phase critical point densities in planar isotropic random waves. *Journal of Physics A: Mathematical and General, 34*(20), L297.

Kotlyar, V.V., Khonina, S.N., Soifer, V.A., & Skidanov, R.-V. (2007). Rotation of laser beams with zero of the orbital angular momentum. *Optics Communications, 274*(1), 8–14.

Litvin, I.A., Dudley, A. & Forbes, A. (2011). Poynting vector and orbital angular momentum density of superpositions of Bessel beams. *Optics Express, 19*(18), 16760–16771.

Lopez-Mago, D., Perez-Garcia, B., Yepiz, A., Hernandezaranda, R.-A. & Gutiérrezvega, J.-C. (2013). Dynamics of polarization singularities in composite optical vortices. *Journal of Optics, 15*(4), 044028.

Luo, Y. & Lü, B. (2010). Composite polarization singularities in superimposed Laguerre-Gaussian beams beyond the paraxial approximation. *Journal of the Optical Society of America A, 27*(3), 578–584.

Maleev, I.D. & Swartzlander, G.A. (2003). Composite optical vortices. *Journal of the Optics Society of America B, 20*(6), 1169.

Nye, J.F., Hajnal, J.V. & Hannay, J.H. (1988). Phase saddles and dislocations in two-dimensional waves such as the tides. *Proceedings of the Royal Society of London A, 417*(1852), 7–20.

Nye, J. and Wright, F.J. (2000). Natural focusing and fine structure of light: Caustics and wave dislocations. *American Journal of Physics, 68*(8), 776–776.

Pang, X. Gbur, G. & Visser, T.D. (2015). Cycle of phase, coherence and polarization singularities in Young's three-pinhole experiment, *Optics Express, 23*(26), 34093–34108.

Pang, X. Cheng, J. & Zhao, X. (2018). Polarization dynamics on optical axis, *Optics Communications, 421*, 50–55.

Soskin, M.S., Gorshkov, V.N., Vasnetsov, M.V., Malos, J.-T. & Heckenberg, N.-R. (1997). Topological charge and angular momentum of light beams carrying optical vortices. *Physical Review A, 56*(5), 4064–4075.

Soskin, M.S. & Vasnetsov, M.V. (2001). Singular optics. *Progress in Optics, 42*(4), 219–276.

Spektor, B., Normatov, A. & Shamir, J. (2008). Singular beam microscopy. *Applied Optics, 47*(4), 78–87.

Török, P. & Munro P. (2004). The use of Gauss-Laguerre vector beams in STED microscopy. *Optics Express, 12*(15), 3605–3617.

Vasilyeu, R., Dudley, A., Khilo, N. & Forbes, A. (2009). Generating superpositions of higher-order Bessel beams. *Optics Express, 17*(26), 23389–23395.

Wang, J., Yang, J.Y., Fazal, I.M., Ahmed, N., Yan, Y., Huang, H., Ren, Y.-X.., Yue, Y., Dolinar, S., Tur, M. & Willner, A.-E. (2012). Terabit free-space data transmission employing orbital angular momentum multiplexing. *Nature Photon, 6*(7), 488–496.

Zhao, X. Cheng, J. Pang, X. & Wan, G. (2017). Properties of a strongly focused Gaussian beam with an off-axis vortex, *Optics Communications, 389*, 275–282.

Zhao, X. Pang, X. Cheng, J. & Wan, G. (2018). Transverse focal shift in vortex beams. *IEEE Photonic Journal, 10*(1), 6500417.

Frontier Research and Innovation in Optoelectronics Technology and Industry – Habib & Lewis (Eds)
© 2019 Taylor & Francis Group, London, ISBN 978-1-138-33178-5

Effects of the electron density on the femtosecond filamentation in gases

X.X. Qi & C.R. Jing
College of Physical & Electronic Information, Luoyang Normal University, Luoyang, China

ABSTRACT: We investigate the effects of electron density on the intense femtosecond laser filamentation in air and argon. We consider the avalanche ionization, attachment and recombination process which could influence electron density. Simulation results show that these effects on the on-axis intensity can be neglected when the pressure is low, and then increase to be almost constant with the increasing pressure. These effects in argon are more obvious than those in air. They should be taken into account in the simulations of the femtosecond filamentation in the high pressure environment larger than 10 atm and 20 atm for argon and air, respectively.

1 INTRODUCTION

Since Braun et al. carried out the experiment with an intense infrared (IR) femtosecond laser pulse demonstrating that the intense ultra-short laser pulse is suited for long range propagation in air (Braun et al. 1995), a large amount of research had been engaged in this propagating type which is called filamentation or self-guided propagation. The physical mechanism of filamentation is interpreted as a dynamic equilibrium between Kerr self-focusing and plasma defocusing in classical model (Couairon & Berge 2000, Kosareva et al. 2011, Kohler et al. 2013, Li et al. 2014) or between Kerr self-focusing and defocusing by higher-order Kerr effect in higher-order Kerrmodel (Qi et al. 2016, Bejot et al. 2011, Carsten et al. 2011, Milchberg 2014, Petrarca et al. 2012, Bejot et al. 2010). In both models, electron density is a key parameter.

Electron density is determined by photo-ionization, avalanche ionization, electron-attachment and the electron-ion recombination (Couairon & Mysyrowicz 2010). The two ionization effects increase the electron density while the attachment and the recombination decrease it. Most of the simulation researches consider the photo-ionization but takes parts of the next three processes into account, especially ignoring the last two processes (Afonasenko et al. 2014, Feng et al. 2015, Deng et al. 2013, Wang et al. 2011, Sprangle et al. 2002). The pulse duration of the majority of femtosecond filamentation research is from 10 fs to 300 fs, the influence of recombination and attachment processes should be discrepant. Furthermore, attachment and recombination can play important roles for high pressure (Fernsler & Rowl 1996, Littlewood et al. 1983).

In this paper, we study the effects of electron density correlative processes on the femtosecond filamentation for varied pressures and pulse durations in air and argon. The content is organized as follows. Section 2 introduces the nonlinear Schrodinger equation for the femtosecond pulse propagation and the computational algorithm employed in the simulations. In section 3, the effects of electron density correlative processes on the filament in air and argon are investigated respectively. The conclusions are summarized in section 4.

2 MODEL AND ALGORITHM

Expressed in the reference frame moving at the group velocity, the propagation equation of a laser pulse with a linearly polarized incident electric filed $E(r,t,z)$ along the propagation direction z reads (Couairon & Mysyrowicz 2010, Xi et al. 2006, Wang et al. 2010, Couairon & Berge 2002):

$$\frac{\partial E}{\partial z} = \frac{i}{2k_0}\left(\frac{\partial^2}{\partial r^2} + \frac{1}{r}\frac{\partial}{\partial r}\right)E - \frac{ik''}{2}\frac{\partial^2 E}{\partial t^2} + \frac{ik_0}{n_0}\Delta n_{\mathrm{kerr}}E - \frac{ik_0}{2}\frac{\omega^2}{\omega_0^2}E - \frac{\beta^{(k)}}{2}|E|^{2K-2}E \qquad (1)$$

where t refers to the retarded time in the reference frame of the pulse ($t \rightarrow t - \frac{z}{v_g}$ with $v_g = \frac{\partial \omega}{\partial k}|\omega_0$ corresponding to the group velocity of the carrier envelope). The first two terms on the right-hand side of Equation (1) are the linear effects, accounting for the spatial diffraction and the second order dispersion. The other three terms represent Kerr effect, plasma defocusing effect and the multiphoton ionization effect, and these terms are the nonlinear effects. In Equation (1), $k_0 = 2\pi/\lambda$ and $\omega_0 = 2\pi c/\lambda$ are the wave number and the angular frequency of the carrier wave, respectively. k'' is the second order dispersion coefficient. The nonlinear index Δn_{kerr} induced by intense femtosecond laser pulses can be written as $\Delta n_{kerr} = n_2|E|^2 + n_4|E|^4 + n_6|E|^6 + n_8|E|^8 + ...$, where n_{2*j} are coefficients that have been reported in references (Loriot et al. 2009, Loriot et al. 2010) are related to $\chi^{(2*j+1)}$ susceptibilities. The plasma frequency is $\omega = \sqrt{\rho e^2/m_e \varepsilon_0}$ with e, m_e, ρ being the electron charge, mass and density. $\beta^{(K)}$ denotes the nonlinear coefficient for K-photon absorption, where K is the minimal number of photons needed to ionize gas.

The evolution of electron density ρ follows the equation (Couairon & Mysyrowicz 2010, Bejot et al. 2007):

$$\frac{\partial \rho}{\partial t} = \frac{\beta^{(K)}|E|^{2K}}{K\hbar\omega_0} + \frac{\sigma}{U_i}\rho|E|^2 - \frac{\beta^{(K)}|E|^{2K}}{K\hbar\omega_0}\frac{\rho}{\rho_{at}} - \alpha\rho^2, \qquad (2)$$

in which the right-hand terms consist of the photo-ionization, the avalanche ionization, the electron attachment and the electron-ion recombination. In Equation (2), U_i is the ionization potential, $\hbar = h/2\pi$ with h the Planck constant, the cross-section σ for inverse bremsstrahlung at 1 atm (standard atmosphere pressure) follows the Drude model (Yablonovitch & Bloembergen 1972) and $\sigma = \frac{k_0 e^2}{\omega_0^2 \varepsilon_0}\frac{\omega_0 \tau_c}{\omega_0^2 \tau_c^2}$ (τ_c = 350 fs is the relaxation time) (Couairon et al. 2006), ρ_{at} denotes the neutral particle density of gas under 1 atm.

The input electric field envelop is modeled by a Gaussian profile with input power P_{in} as

$$E(r,z,t)|_{z=0} = \sqrt{\frac{2P_{in}}{\pi r_0^2}}\exp\left(-\frac{t^2}{t_0^2}\right)\exp\left(-\frac{r^2}{r_0^2}\right), \qquad (3)$$

where r_0 and t_0 denote the radius and duration of the pulse, respectively.

The model is solved via the Split-Step Fourier method (Agrawal 2006), in which all the linear terms are calculated in the Fourier space over a half-step while the nonlinear terms are calculated in the physical space over a second half-step. To integrate the linear part of the Eq. (1) along the propagation axis, we adopt the Crank-Nicholson scheme (Press et al. 2007), which is more stable than the Euler method (Chiron et al. 1999). For Equation (2), the 4th-order Runge-Kutta method is employed.

3 RESULTS AND DISCUSSION

We consider an incident laser pulse with the wavelength $\lambda = 800$ nm and the beam radius $r_0 = 1$ mm. The power of the incident laser pulse is set as $P_{in} = 4P_{cr}$, where $P_{cr} = 3.77\lambda^2/8\pi n_2$ is the critical power. We simulate the propagation of the ultra-short intense laser pulses for five cases:

1. electron density with all the above four processes;
2. electron density without the attachment process;
3. electron density without the recombination process;
4. electron density without the avalanche ionization process;
5. electron density only with the photo-ionization process.

Take the propagation of pulse with duration of 200 fs and pressure of 2 atm as an example, we plot Figure 1 which presents the evolution of laser on-axis intensity and flux for the above five cases. The upper part of each of the sub-figures presents the on-axis intensity and the below one is the flux corresponding with each case. We still can see the discrepancy among them although it's tiny.

Hereinafter, the on-axis intensity and beam radius along the propagating direction are selected to illustrate the discrepancy of the five cases. We define the relative deviations (RDs) of the other cases from case 1 as follows:

$$RD_{(i)} = \frac{\int |I_{case(1)} - I_{case(i)}| dz}{\int I_{case(1)} dz}, \quad (4)$$

where $I_{case(1)}$ is the on-axis intensity in filament zone along the propagation direction z for case 1, $I_{case(i)}$ represents the intensity of other cases sequentially. As stated in Section 1, the duration of the input laser pulses and the pressure can affect the electron density, we choose pressures from 0.3 atm to 40 atm and the durations from 10 fs to 500 fs.

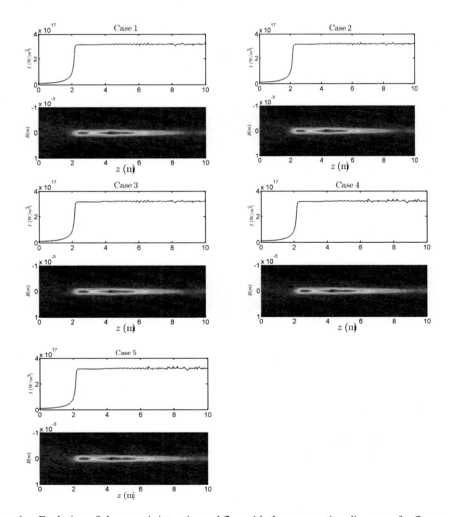

Figure 1.   Evolution of the on-axis intensity and flux with the propagation distance z for five case.

3.1 *Argon medium*

Since the noble gas is mono-atomic and there is no Raman effect nor any complication such as molecular association and fragmentation, many researches choose the noble gas such as argon as the propagating medium (Couairon & Berge 2000, Bejot et al. 2010, Wang et al. 2011, Wolf et al. 2008, Tarazkar et al. 2014, Nakajima et al. 2009). We simulate the filamentation in argon medium whose ionization potential is close to that of the air molecules. The values of parameters we used in simulations are listed in Table 1.

We simulate the five cases with femtosecond laser pulses with duration of 200 fs for varied pressures and show thefour RDs of the on-axis intensity with the pressures in Figure 2.

From Figure 2, we can see that the four *RD*s increase very slowly until about 7 atm and then change greatly which indicating the effects of attachment process, recombination process and avalanche ionization process grow sharply and afterwards become almost invariable with the increasing pressures. Although the *RD*(3) is a bit larger than *RD*(2) by 0.8% after 12 atm, the effects of the electron attachment and the electron-ion recombination are almost the same. The *RD*(4) is larger than *RD*(2) and *RD*(3) by 6% after 12 atm which means the effect of avalanche ionization is greater than these of the next ones. The *RD*(2), *RD*(3) and *RD*(5) are approximately equivalent, this explains indirectly that the electron density is not determined by the simple linear superposition of the four processes and on the contrary, the generation and vanishment of electrons are affected by the nonlinear and complicated interaction among the four processes.

In order to observe the discrepancy of the filaments for the five cases more visually, we plot the evolutions of filament radius with propagating distance at 10 atm in Figure 3 and at

Table 1.   Parameters used in the model for argon medium.

$k''(fs^2\ cm^{-1})$	0.21 p* (Wolf et al. 2008)
$n_2(10^{-19}\ cm^2\ W^{-1})$	1.0 p (Loriot et al. 2010)
$n_4(10^{-33}\ cm^4\ W^{-2})$	$-0.37\ p$
$n_6(10^{-45}\ cm^6\ W^{-3})$	0.4 p
$n_8(10^{-59}\ cm^8\ W^{-4})$	$-1.7\ p$
$n_{10}(10^{-74}\ cm^{10}\ W^{-5})$	8.8 p
K	11 (Wolf et al. 2008)
$\beta^{(11)}\ (m^{-17}\ W^{-9})$	3.32 p 10^{-176}
$U_i(eV)$	15.76
$\sigma(m^2)$	1.01 $\frac{p(1+a_0^2\tau_c^2)}{p^2+a_0^2\tau_c^2}$ 10^{-23} (Couairon et al. 2006)

*p denotes the ratio of the pressure to the standard atmosphere pressure.

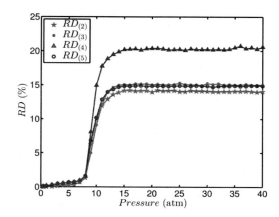

Figure 2.   The evolution of *RD*s of the on-axis intensity with pressures in argon.

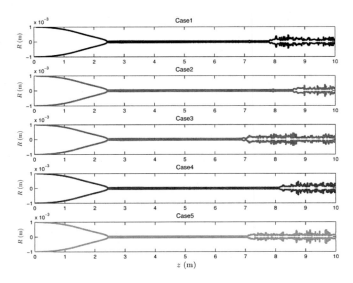

Figure 3.   The evolution of filament radius with distance z for five cases in argon at 10 atm.

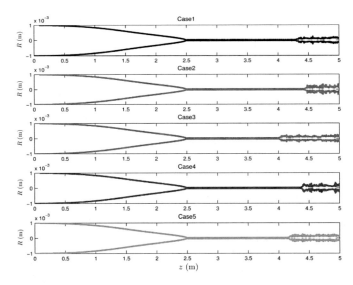

Figure 4.   The evolution of filament radius with distance z for five cases in argon at 30 atm.

30 atm in Figure 4. In these two figures, we can see that the lengths of the filaments for the five cases are not identical yet. For more detailed analysis, we calculate the start point, the end point, the length and the radius of the filament for each case and summarize in Table 2. From the table, we can find that, whether at 10 atm or at 30 atm, the filaments of these cases get the same radius and start point, but the end points are different resulting in the disparate filament lengths. The largest discrepancy among these cases can be up to 27.4%. All the above discrepancies demonstrate that the avalanche ionization, attachment and recombination can not be ignored for filament simulation at high pressures in argon medium. In addition, we can obtain that the length of filament in 10 atm environment are longer than those in 30 atm environment because of the less energy loss induced by the collision, so that the power to generate self-focus decreases more slowly until its value is small than the critical power. We can also get that the radius of filament in 10 atm environment are wider than those

165

Table 2. Properties of filaments for five cases at 10 atm and 30 atm.

10 atm

Case / Filament	Start point (m)	End point (m)	Length (m)	Radius (μm)
Case1	2.434	7.811	5.377	41.016
Case2	2.434	8.590	6.156	41.016
Case3	2.434	6.901	4.467	41.016
Case4	2.434	8.104	5.670	41.016
Case5	2.434	7.058	4.624	41.016

30 atm

Case / Filament	Start point (m)	End point (m)	Length (m)	Radius (μm)
Case1	2.487	4.344	1.798	23.438
Case2	2.487	4.421	1.934	23.438
Case3	2.487	4.002	1.515	23.438
Case4	2.487	4.285	1.857	23.438
Case5	2.487	4.130	1.643	23.438

Table 3. Parameters used in the model for air medium.

$k''(fs^2\ cm^{-1})$	$0.2\ p^*$ (Wolf et al. 2008)
$n_2(10^{-19}\ cm^2\ W^{-1})$	$1.2\ p$ (Loriot et al. 2010)
$n_4(10^{-33}\ cm^4\ W^{-2})$	$-1.5\ p$
$n_6(10^{-45}\ cm^6\ W^{-3})$	$2.1\ p$
$n_8(10^{-59}\ cm^8\ W^{-4})$	$-0.8\ p$
K	10 (Xi et al. 2006)
$\beta^{11)}\ (m^{-17}\ W^{-9})$	$1.27\ p\ 10^{-160}$
$U_i(eV)$	11 (Li et al. 2014)
$\sigma(m^2)$	$5.1\frac{p(1+a_0^2\tau_c^2)}{p^2+a_0^2\tau_c^2}10^{-24}$ (Couairon et al. 2006)

in 30 atm environment because of the following reason. Since the critical power is inversely proportional to Kerr nonlinear index of refraction (Bernhardt et al. 2008) and since Kerr nonlinear index is proportional to the argon density, the critical power increases as the pressure decreases. Hence, to obtain self-focusing at lower pressure, the input peak power of the pulse has to increase. However, since the intensity is clamped at the value under the standard pressure, the higher peak power of the pulse has to be contained inside a region with a larger diameter than that under the standard pressure.

3.2 Air medium

Except the argon medium, we also simulate the femtosecond filamentation in air and calculate the RDs of the on-axis intensity. The properties of the incident laser pulse are identical to the ones in section 3.1. The values of parameters in air medium simulations are listed in Table 3. 3 and the simulation results are plotted in Figure 5. We can get from Figure 5 that the four RDs almost keep constant until 15 atm, then increase precipitously to 25 atm and afterwards maintain invariable. Similarly to Figure 2, the effect of the avalanche ionization for the air medium is also greater than those of the attachment and recombination whose influences are much closer. What's important we'd like to point out is that for the air medium, the four RDs become constant from 25 atm rather than 15 atm for the argon medium which means the effect of the pressure on the on-axis intensity in air medium is less sensitive than that in argon medium.

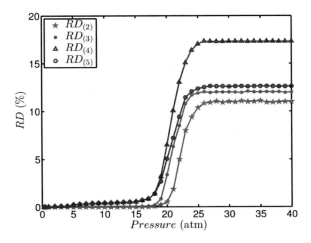

Figure 5. The evolution of RDs of the on-axis intensity with pressures in air.

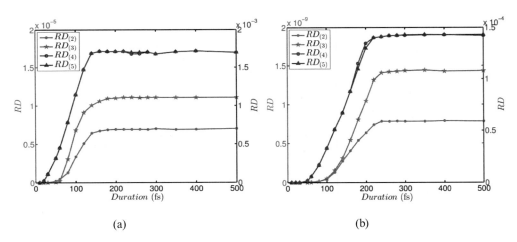

(a)   (b)

Figure 6. The evolution of RDs of the on-axis intensity with pulse durations. (a) in argon; (b) in air.

We also simulate the effects of the three processes on femtosecond filamentation for varied pulse durations from 10 fs to 500 fs at 5 atm in argon and air medium. And we show the results in Figure 6. From the figure, we can see that all the RDs are less than 10^{-3} and even as low as 10^{-9} which indicate the effects can be neglected.

4 CONCLUSIONS

We have investigated the effects of electron density on the intense femtosecond laser filamentation for the air and the argon medium considering the attachment process, the recombination process and the avalanche ionization process. Numerical results show that the effects of these processes increase at first and become constant afterwards with the increasing pressure. Especially for high pressure, the relative deviation can reach about 20% for air and argon, which indicate that the three processes should be taken into account for the filamentation simulation in high pressure environment. On the other hand, these effects on the on-axis intensity in air are less sensitive than those in argon.

ACKNOWLEDGMENTS

The authors thank the reviewer for her/his constructive comments and suggestions on improving the quality of this paper. This work is supported by the National Natural Science Foundation of China Grant No. 11704174.

REFERENCES

Afonasenko, A.V., Apeksimov, D.V., Geints, Y.E., Golik, S.S., Kabanov, A.M., & Zemlyanov, A.A. 2014. Study of filamentation dynamics of ultrashort laser radiation in air: beam diameter effect. Journal of Optics, 16(10), 105204.

Agrawal, G.P. 2006. *Nonlinear Fiber Optics*. Salt Lake City: Academic Press.

Béjot, P., Bonnet, C., Boutou, V., & Wolf, J.P. 2007. Laser noise compression by filamentation at 400 nm in argon. Optics Express, 15(20), 13295–309.

Béjot, P., Kasparian, J., Henin, S., Loriot, V., Vieillard, T., & Hertz, E., et al. 2010. Higher-order kerr terms allow ionization-free filamentation in gases. Physical Review Letters, 104(10), 103903.

Béjot, P., Hertz, E., Kasparian, J., Lavorel, B., Wolf, J.P., & Faucher, O. 2011. Transition from plasma-driven to kerr-driven laser filamentation. Physical Review Letters, 106(24), 243902.

Bernhardt, J., Liu, W., Chin, S.L., & Sauerbrey, R. 2008. Pressure independence of intensity clamping during filamentation: theory and experiment. Applied Physics B, 91(1), 45–48.

Braun, A., Korn, G., Liu, X., Du, D., Squier, J., & Mourou, G., 1995. Self-channeling of high-peak-power femtosecond laser pulses in air. Opt. Lett. 20(1), 73–75.

Carsten B., Ayhan D., & Günter S. 2011. Saturation of the All-Optical Kerr Effect. Physical Review Letters, 106(18), 183902.

Chiron, A., Lamouroux, B., Lange, R., Ripoche, J.F., Franco, M., & Prade, B., et al. 1999. Numerical simulations of the nonlinear propagation of femtosecond optical pulses in gases. The European Physical Journal D – Atomic, Molecular, Optical and Plasma Physics, 6(3), 383–396.

Couairon, A., & Berge, L., 2000. Modeling the filamentation of ultra-short pulses in ionizing media. Phys. Plasmas 7 (1), 193–209.

Couairon, A., & Bergé, L. 2002. Light filaments in air for ultraviolet and infrared wavelengths. Physical Review Letters, 88(13), 135003.

Couairon, A., Franco, M., Méchain, G., Olivier, T., Prade, B., & Mysyrowicz, A. 2006. Femtosecond filamentation in air at low pressures: part i: theory and numerical simulations. Optics Communications, 259(1), 265–273.

Couairon, A., & Mysyrowicz, A. 2010. Femtosecond filamentation in transparent media. Physics Reports, 441(2), 47–189.

Deng, Y., Jin, T., Zhao, X., Gao, Z., & Chi, J. 2013. Simulation of femtosecond laser pulse propagation in air. Optics Laser Technology, 45(1), 379–388.

Feng, Z.F., Li, W., Yu, C.X., Liu, X., Liu, J., & Fu, L.B. 2015. Extended laser filamentation in air generated by femtosecond annular gaussian beams. Physical Review A, 91(3).

Fernsler, R.F., & Rowl, H.L. 1996. Models of lightning-produced sprites and elves. Journal of Geophysical Research Atmospheres, 101(D23), 29653–29662.

Köhler, C., Guichard, R., Lorin, E., Chelkowski, S., Bandrauk, A.D., & Bergé, L., et al. 2013. On the saturation of the nonlinear refractive index in atomic gases. Physical Review A, 87(4), 043811.

Kosareva, O., Daigle, J.F., Panov, N., Wang, T., Hosseini, S., & Yuan, S., et al. 2011. Arrest of self-focusing collapse in femtosecond air filaments: higher order kerr or plasma defocusing? Optics Letters, 36(7), 1035–7.

Li, S.Y., Guo, F.M., Song, Y., Chen, A.M., Yang, Y.J., & Jin, M.X. 2014. Influence of group-velocity-dispersion effects on the propagation of femtosecond laser pulses in air at different pressures. Phys. rev.a, 89(89), 023809.

Littlewood, I.M., Cornell, M.C., Clark, B.K., & Nygaard, K.J. 1983. Two- and three-body electron-ion recombination in carbon dioxide. Journal of Physics D Applied Physics, 16(11), 2113–2118.

Loriot, V., Hertz, E., Faucher, O., & Lavorel, B. 2009. Measurement of high order kerr refractive index of major air components. Optics Express, 17(16), 13429–34.

Loriot, V., Hertz, E., Faucher, O., & Lavorel, B. 2010. Measurement of high order kerr refractive index of major air components. Optics Express, 18(3), 3011–12.

Milchberg, H. 2014. The extreme nonlinear optics of gases and femtosecond optical/plasma filamentation. Physics of Plasmas, 21(10), 47–189.

Nakajima, T., Song, Z., & Zhang, Z. 2009. Transverse-mode dependence of femtosecond filamentation. Optics Express, 17(15), 12217–29.

Petrarca, M., Petit, Y., Henin, S., Delagrange, R., Béjot, P., & Kasparian, J. 2012. Higher-order kerr improve quantitative modeling of laser filamentation. Optics Letters, 37(20), 4347–9.

Press, W.H., Flannery, B.P., Teukolsky, S.A., & Vetterling, W.T. 2007. *Numerical recipies*. New York: Cambridge University Press.

Qi, X., Ma, C., & Lin, W. 2016. Pressure effects on the femtosecond laser filamentation. Optics Communications, 358, 126–131.

Sprangle, P., Peñano, J.R., & Hafizi, B. 2002. Propagation of intense short laser pulses in the atmosphere. Physical Review E Statistical Nonlinear & Soft Matter Physics, 66(2), 046418.

Tarazkar, M., Romanov, D.A., & Levis, R.J. 2014. Higher-order nonlinearity of refractive index: the case of argon. Journal of Chemical Physics, 140(21), 605.

Wang, H., Fan, C., Zhang, P., Qiao, C., Zhang, J., & Ma, H. 2010. Light filaments with higher-order kerr effect. Optics Express, 18(23), 24301–6.

Wang, Z., Zhang, C., Liu, J., Li, R., & Xu, Z. 2011. Femtosecond filamentation in argon and higher-order nonlinearities. Optics Letters, 36(12), 2336–2338.

Wolf, J.P., Kasparian, J., & Béjot, P. 2008. Dual-color co-filamentation in argon. Optics Express, 16(18), 14115–14127.

Xi, T.T., Lu, X., & Zhang, J. 2006. Interaction of light filaments generated by femtosecond laser pulses in air. Physical Review Letters, 96(2), 025003.

Yablonovitch, E., & Bloembergen, N. 1972. Avalanche ionization and the limiting diameter of filaments induced by light pulses in transparent media. Physical Review Letters, 29(14), 907–910.

Frontier Research and Innovation in Optoelectronics Technology and Industry – Habib & Lewis (Eds)
© 2019 Taylor & Francis Group, London, ISBN 978-1-138-33178-5

A novel method for evaluating the consistency of PZTs with folded-cavity He-Ne laser

Y.C. Quan, Z.Q. Tan, J.P. Liu & Y. Huang
Department of Optoelectronic Engineering, College of Advanced Interdisciplinary Studies,
National University of Defense Technology, Changsha, Hunan, China

ABSTRACT: To evaluate the consistency of PZTs, a novel method based on folded-cavity He-Ne laser has been proposed and demonstrated. In this method, the laser's intensity and frequency are modulated by the voltages put on the PZTs beside the reflecting mirrors. Because of the nonlinear responses and different piezoelectric coefficients of PZTs, the cavity length could not stay unchanged under the same driving voltages, neither could the laser intensity and frequency. Thus, the inconsistency of PZTs to voltages can be evaluated through these parameters of the laser. A nearly double difference on piezoelectric coefficients between the two used PZTs are detected in both approaches, intensity modulation and frequency-beat system. The first approach provides a quick response method to evaluate the consistency of PZTs and the second approach provides a high precision method, whose resolution can reach 0.026 nm/V. And the second approach shows two orders of magnitude higher resolution than the first one.

Keywords: He-Ne laser, PZT, folded cavity

1 INTRODUCTION

Piezoelectric ceramic transducer (Pr-Zr-Ti, PZT) is a functional material which can realize the interconversion of electric charge and deformation, and the piezoelectric effect can be divided into direct piezoelectric effect and inverse piezoelectric effect (Hagood, N.W. & Flotow, A.V. 1991). This kind of material is widely used in communication, satellite broadcasting, bioscience and some high technology application such as aviation, becoming an indispensable component of industry and research. America and Japan, as the two of the largest producers of top grade electronic components, have already put much energy and money on the research and manufacture (Liu, X.J. et al. 2006), intending to produce higher Curie temperature, bigger piezoelectric coefficients and higher mechanical intensity PZTs. The parameters can reach extremely high for a single PZT, thus users can exploit new applications which were restricted before.

PZT thin films are used widely for actuators and sensors of Micro Electro Mechanical System (MEMS) because of the excellent performance in deformation and response speed. But it shows a complicated hysteresis behavior, which means it is a nonlinear response. PZT is often used as driving machine to the interference instruments in the measurement of optical, then the nonlinearity should affect the precision of phase shift directly. The relation between the macroscopic hysteresis response and microscopic domain switching and structural phase transition were investigated (Wakabayashi, H. & Tsuchiya, K. 2017). Besides, kinds of methods for the hysteresis behavior were proposed to analyze this complicated character, including phase-shifting phase retrieval based on time-domain Fourier Transform (Zhang, W.P. et al. 2015), modeling and feed-back controlling for PZT micro-displacement (Liu, B. et al. 2013) and adaptive control (Sabanovic, A. et al. 2006). Researchers have paid much attention to the hysteresis and nonlinearity of PZT, so that a few cares about the consistency of a pair of PZTs. So called consistency means the different PZTs should have the same displacement performance under the same motivation voltage. In other words, the displacement to voltage coefficients of different PZTs are not exactly the same in practical applications. Undoubtedly, the consistency of PZTs is much more important than the nonlinearity in some special applications, such as synchronous control, synchronous displacement actuate and so on.

The corresponding evaluating methods for the nonlinear corresponding parameters now can reach a high precision. While for synchronous applications, it may not reach our expected result if the two PZTs are not match, even if each PZT performs excellently when works alone. For example, the cavity length control in square ring resonator (Chen, M.X. & Yuan, J. et al. 2011), which needs to change two mirrors at the same time. So, the evaluating method for the consistency of PZTs also needs to be developed.

In this paper, we will present a folded He-Ne laser device to evaluate the consistency of PZTs by detecting the intensity and frequency of the outer laser. The principle used is the inverse piezoelectric effect, which means that the deformation of PZT is modulated by the outer electric field (Zhou, K.F. et al. 2006). The displacement of PZT is very tiny for its size (Shieh H. et al. 2004), so it possesses a great advantage in displacement resolution. And we prove the feasibility experimentally and found that it provides a high precision approach to monitor the synchronism of PZT's displacement.

2 EXPERIMENTAL SETUP

To evaluate the consistency of two piezoelectric transducers of the same size, we propose a folded-cavity He-Ne laser device, which contains a He-Ne gain medium channel, three high reflective films and two PZTs of the same size, as presented in Fig. 1. The whole cavity is made of ceramic-glass, which has a coefficient of thermal expansion of $2 \times 10^{-8}/°C$, possessing an extraordinary thermostability (Zhang, B. & Long, X.W. 2011), so the physical cavity length remains unchanged in a very high accuracy at room temperature. And the gas tightness of the cavity is better than $1 \times 10^{-12} \ Pa \cdot m^3/s$. In addition, the high voltage laser driver (DW-GSC12 A, current stability: 1 μA) can offer constant current, so the index of the inside He-Ne gas still stays unchanged. The effective cavity length is quite stable, so is the output laser. After preheating for one hour, the frequency fluctuation of the output laser is less than 1 MHz.

Although the cavity length is about 10.95 centimeters, the high reflectivity (≈99.99%) of the films guarantees the gain is larger than the cavity loss, making up the short gain length. The length of the cavity can be slightly changed by varying the motivation voltage of PZTs outside the films, so do the intensity and the resonance frequency of the laser. We measured the mechanical displacement of the piezoelectric transducers by detecting the variation of the laser intensity or frequency, thus the consistency of the two piezoelectric transducers can be evaluated.

The evaluation system is shown in Fig. 2. When pumped, the laser device carries two same out lasers at the folded mirror, and the laser parameters, including frequency and intensity

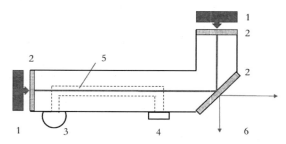

Figure 1.   The laser device: 1. PZT. 2. The high reflective film. 3. Cathode tube. 4. Anodic tube. 5. The gain medium (helium-neon gas). 6. The outer laser.

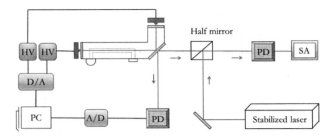

Figure 2. The experimental setup of laser beat-frequency system. SA: accept the beat note signal from our laser device and stabilized laser. PD: detect the direct laser intensity. HV: high voltage.

are modulated by the movement of the two outer piezoelectric transducers. One of the lasers is directly detected by photo detector and the intensity fluctuation is recorded and sent to PC by the data collector. The other outer laser is used to beat with a frequency-stabilized laser, whose short-term frequency stability can reach 9.4 kHz, provided by National Institute of Metrology (NIM), generated by an iodine stabilized He-Ne laser (AI-05-01, NIM). The frequency difference is measured by spectrum analyzer. The different displacement responses to voltage of the two PZTs can be analyzed through the intensity and frequency fluctuation.

3 RESULTS AND DISCUSSIONS

3.1 *Test one PZT*

Scanning a single PZT outside the reflecting mirror changes the effective cavity length, and the best working point is achieved when the cavity length matches with the gain medium. And PZT had a special frequency response characteristic, thus the testing result varied with the chosen motivating frequency, which was chosen at 50 Hz properly.

A 100 V DC bias is loaded onto the left side PZT to provide an initial cavity length, also to match with the gain medium. The working cavity length changes with the driving triangular wave accordingly, whose peak-to-peak value is not beyond 200V because the negative voltage has no impact on the effective cavity length. Fig. 3 shows the outer laser intensity fluctuation with different driving triangular waves (the data is received from photo detector, so the unit is volt). It can be concluded that the resonant frequency modulated by the changing cavity length is all in the gain bandwidth, as shown in Fig. 3(a), and the intensity curve also reflects the gain spectrum of He-Ne gas.

Two longitudinal modes appear when the amplitude of the triangular gets larger, along with dark field between two longitudinal modes, as shown in Fig. 3(b). With the increase of peak-to-peak value, more longitudinal modes appear but the number of modes is not linear to the driving voltage, proving the nonlinear response of PZT.

3.2 *Testing two PZTs*

The initial cavity length is designed to fit the gain medium by adjusting the dc bias voltages applied to the PZTs, making the laser intensity be the largest and the device has the highest sensitivity and the lowest possibility to work in dark fields. The frequency (10 Hz) of driving triangular wave loaded onto the left is the same with the right, but the phase is opposite, and the amplitude is tunable independently.

The left triangular wave, whose peak-to-peak value is 50V, stays unchanged, and the right voltage increases gradually. The corresponding laser intensity is listed in Fig. 4. When the right voltage is relatively small, the effective cavity length is modulated with the left driving voltage, shown in Fig. 4(a), (b). But Fig. 4(e), (f) demonstrate that the modulation becomes anti-phase when the right voltage gets relatively larger. The phase of laser intensity

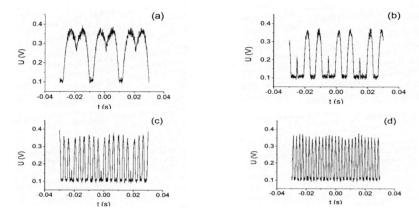

Figure 3. Intensity fluctuation under different driving triangular wave. The voltage of triangular: (a) –20V, (b) –70V, (c) –150V, (d) –200V.

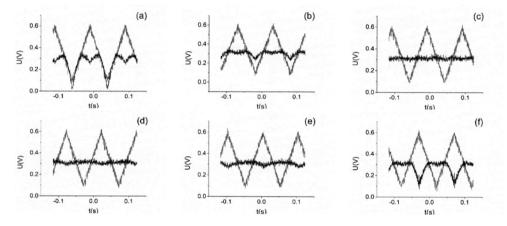

Figure 4. Intensity fluctuation with the increase of right triangular wave (10 Hz). The blue curve represents left triangular wave, whose peak-to-peak value is 50V, and the right: (a) –60V, (b) –80V, (c) –92V, (d) –94V, (e) –100V, (f) –120V. The red line represents the left driving voltage and the blue line represent the laser intensity.

fluctuation keeps the same with the triangular wave which causes larger displacement. While the actual displacements in both sides caused by the two triangular waves are the same or have little differences, the effective cavity length hardly varies, so the outer laser intensity stays relatively stable, as shown in Fig. 4(c), (d). We can conclude from intensity fluctuation that this trend can prove that the PZTs have different displacement responses to the motivation voltages even though the voltages are the same, and we can evaluate this difference, resulting from circumstance, installation technology and so on, in this intensity modulation approach.

It proves that the consistency of PZTs can be evaluated experimentally through the intensity fluctuation when the two PZTs are phase-reversely driven at the same time, but we could not distinguish the intensity fluctuation obviously even the driving voltage has varied from 92 V to 94 V, as shown in Fig. 4(c), (d). It works well for some usual applications with such resolution,but it is not enough for high-precision applications. We design a laser beat system, as shown in Figure 2, to improve the resolution by measuring the laser frequency variation while the triangular waves are loaded onto the PZTs. To get more details and observe easily,

Table 1. Frequency fluctuation caused by different driving triangular wave.

Group	Left (V)	Right (V)	Frequency fluctuation (MHz)
1	0	0	<1
2	0	0.5	20
3	1	1.95	2–3
4	5	9.7	2–3
5	23	43.8	5

we adjust the driving frequency from 10 Hz to 1 Hz, and the experimental procedure is as the same as the intensity modulation approach. The driving frequency difference will not impact the evaluation results, and the frequency response is not much important in this work, which just provides and demonstrates a feasible method.

Table 1 displays the result we get at NIM. We adjust the voltages put on the right and left side PZTs to find out which pair of combination can stabilize the cavity length. Unlike the laser intensity, directly detected through a simple photo detector, which stays unchanged when the cavity length varies a bit, the laser frequency can sense a tiny cavity length change. We care the frequency fluctuation instead of the absolute frequency. The laser beat-frequency system can measure the frequency variations caused by the cavity changes, so the different displacement responses to voltage between the two PZTs can be evaluated.

The revolution of the beat-frequency system is constructed by the stability of the ceramic-glass laser. From group 1, we can conclude that the frequency stability of the laser device is quite excellent. When the voltage is not applied to PZTs, the frequency fluctuation of the laser is less than 1 MHz. That is to mean, the frequency fluctuation below 1 MHz caused by the different responses of PZTs could not been measured. The frequency fluctuation can reach 20 MHz even the voltage variate is only 0.5 V, as shown in group 2. Compared with the first solution we take, which hardly distinguishes 2 V voltage difference, the resolution of beat-frequency system is two orders of magnitude higher. Group (3), (4) display the least frequency variation we can take. However, the driving voltage is too small to reflect the characters of the whole PZT but a single point. The frequency spacing of two adjacent longitudinal modes in the cavity is matched with a half wavelength's displacement, so 1 MHz frequency fluctuation is corresponded to displacement of 0.23 nm. The limit displacement resolution can reach 0.026 nm/V when the voltage loaded to the right PZT is 43.8 V, shown as group (5). There are two longitudinal modes when the voltage is 70 V, shown as Fig. 3(b), so scanning the PZT with 43.8 V is enough for laser applications. Furthermore, if the initial dc bias voltage can be adjusted more precisely and the spectrum analyzer has a higher revolution, a better revolution of this beat-frequency system can be presented. When the voltage gets larger, the least frequency fluctuation reaches 5 MHz, exposing the non-linearity of PZT, as shown in Group (5).

Adjusting the voltages applied on the PZTs, and keeping the frequency fluctuation at the same level, more details about the differences between the two PZTs can be concluded. We can also draw a common conclusion from the two experiments above: the piezoelectric coefficient of the left-side PZT is about twice as large as that on the right (as shown in Table 1).

4 CONCLUSION

In conclusion, a PZT evaluation system is presented using a folded He-Ne laser, and the consistency of a pair of PZTs can be easily got by this system. We provide two approaches: the intensity modulation approach for simple system and the beat-frequency system for high-precision applications. The first approach provides a quick solution to evaluate the consistency of PZTs and the second approach provides a high precision method, whose resolution can reach 0.026 nm/V. Both have concluded that the piezoelectric coefficient of left side PZT is about twice as large as that of the right side. They show a quite excellent ability to identify the different displacement responses to voltages between PZTs. This laser method for the consistency of PZTs is quite new and has a referential significance in synchronous applications.

ACKNOWLEDGEMENT

This work is sponsored by the Research Project of National University of Defense Technology (No. ZK16-03-21).

REFERENCES

Chen, M.X. & Yuan, J. et al. 2011. Backscattering coupling effect in square ring resonator [J] Acta Optica Sinica, 31(s1): s100522.

Hagood, N.W. & Flotow A.V. 1991. Damping of structural vibrations with piezoelectric materials and passive electrical networks [J]. Journal of Sound and Vibration, 146(2): 243–268.

Liu, B. et al. 2013. Modeling and control for PZT micro-displacement actuator [J]. optics and precision engineering, 21(6):1503–1509.

Liu, X.J. et al. 2006. Development and new application of piezoelectric ceramics at home and abroad [J]. Bulletin of the Chinese Ceramic Society, 25(4), 101–107.

Sabanovic, A. et al. 2006. Sliding Mode Adaptive Controller for PZT Actuators [C]. IEEE Industrial Electronics, IECON 2006 -, Conference on. IEEE, 5209–5214.

Shieh H. et al. 2004. Adaptive tracking control solely using displacement feed-back for a piezo-positioning mechanism [J]. Journal of Intelligent & Robotic Systems, 151(5): 653–660.

Wakabayashi, H. & Tsuchiya, K. 2017. Multiscale numerical study on ferroelectric nonlinear response of PZT thin films [C]. Society of Photo-Optical Instrumentation Engineers. Society of Photo-Optical Instrumentation Engineers (SPIE) Conference Series, 1024603.

Zhang, B. & Long, X.W. 2011. Lamb-Dip frequency-stabilized He-Ne laser with an integrated cavity made of zerodur (I): structure and techniques [J]. Acta Optica Sinica, 8, 31.

Zhang, W.P. et al. 2015. Generalized Phase-Shifting Phase Retrieval Approach Based on Time-Domain Fourier Transform [J]. Chinese laser, 42(9), 262–268.

Zhou, K.F. et al. 2006. High frequency piezoelectric micromachined ultrasonic transducers for imaging applications [J]. Micro and Nanotechnologies for Space Applications, 25(62): 2230–2235.

Frontier Research and Innovation in Optoelectronics Technology and Industry – Habib & Lewis (Eds)
© 2019 Taylor & Francis Group, London, ISBN 978-1-138-33178-5

Numerical analysis of the thermal effect of a detector irradiated by a repetition frequency pulsed laser

X.D. Ren & W.H. Lei
School of Electronic Countermeasures, National University of Defense Technology, Hefei, China

L.Q. Zeng
The Troops of 32026, Wulumuqi, China

ABSTRACT: Based on heat conduction theory, a theoretical model of the mercury cadmium telluride detector irradiated by repetition frequency pulsed laser was built. The method of applying the heat source on the focal plane of the detector was used to analyze the thermal effect of the mercury cadmium telluride (HgCdTe) detector. The difference of the thermal effect was compared under different repetition frequencies of the laser. The results show that the energy of the laser was mainly absorbed by the HgCdTe photosensitive layer and the high temperature field was mainly distributed near the interface between the HgCdTe layer and the cadmium zinc telluride (CdZnTe) layer when the HgCdTe detector was irradiated by a repetitive frequency pulse laser. There was almost no temperature accumulation effect when the frequency was 1 KHz and 10 KHz. When the frequency is 100 KHz, the peak temperature increased with the increase in the number of pulses, and there was an obvious temperature accumulation effect.

Keywords: pulsed laser, HgCdTe detector, temperature field

1 INTRODUCTION

Infrared detectors are key components of space-based infrared systems, having a very strong target detection ability and constituting an important threat to military action. It is possible to achieve the purpose of satellite countermeasures by interfering or damaging its detectors under the operation conditions of satellites. This has important reference value for laser blinding weapons and the anti-laser performance of the space-borne detector (Shao et al., 2006). The optical system of the detector has a high gain performance and the detector's destruction by high-power laser is mainly caused by the heating of optical elements and target materials, resulting in thermal ablation and thermal stress destruction (Li et al., 2012). It also makes it extremely vulnerable to interference. Meanwhile, with the development of high-power laser technology, it is possible to disturb space-borne detectors using laser weapons. Tang et al. (2013) studied the thermal effects of the mercury cadmium telluride (HgCdTe) crystal irradiated by single pulsed and the high repetition frequency laser. Wei et al. (2014) analyzed the thermodynamic effects of the silicon based positive-intrinsic negative (PIN) detector irradiated by millisecond pulsed laser with the finite element method. Lei et al. (2013) carried out an experiment on the interference and damage effects of the HgCdTe image sensors irradiated by pulsed laser. Zhao et al. (2017) studied the temperature distribution of a photodiode irradiated by a pulse series laser. However, very few people have modeled and simulated the HgCdTe detector irradiated by different repetition frequency lasers.

In this paper, the theoretical model of the mercury cadmium telluride detector irradiated by repetition frequency pulsed laser is built based on heat conduction theory. The method of applying the heat source on the focal plane of the detector is used to analyze the thermal effect of the HgCdTe detector under different repetition frequencies of the laser.

2 THEORETICAL ANALYSIS

2.1 *Physical model of the HgCdTe detector*

HgCdTe detectors are mostly used in the field of military space. The performance is superior and the cost is expensive. This paper mainly focuses on the investigation of HgCdTe detectors. The COMSOL Multiphysics software is used to construct the geometric model (Zheng et al., 2017; Su et al., 2016; Wang et al., 2010) for the multilayer structure of the HgCdTe detector. The detector is divided into four layers, as shown in Figure 1. The first layer is cadmium zinc telluride (CdZnTe) with a thickness of 50 μm and it is the best substrate material of the HgCdTe detector. The second layer is the photosensitive layer of HgCdTe with a thickness of 10 μm. In the third layer, the indium columns (In) and the epoxy resin are arranged in line with each other. The indium column connects the photosensitive chip and the read-out circuit with a thickness of 10 μm and width of 10 μm. The width of the epoxy resin is 40 μm. The fourth layer is the silicon (Si) base, with a thickness of 50 μm, which is the read-out circuit part of the detector. Its two-dimensional simplified model is shown in Figure 1.

2.2 *Theoretical analysis of the temperature field*

The study of the temperature field of the materials produced by laser irradiation is a basic work for the failure mechanism and damage effect of the interaction of laser and material (Zhang et al., 2013; Huang et al., 2016; Duan et al., 2004). In the thermal coupling process of the material in each structure layer between the laser and the detector, the material of each layer absorbs laser energy into heat to increase the temperature, and then the temperature field is formed through heat conduction between layers. As a result, the temperature field changes the properties and structure of the materials, and even causes the failure of the material function. In the three-dimensional rectangular coordinate system, the general form of the partial differential equation (Zhu et al., 2012; Wang et al., 2016) for heat conduction is:

$$\frac{\partial}{\partial x}\left(K_i \frac{\partial T}{\partial x}\right) + \frac{\partial}{\partial y}\left(K_i \frac{\partial T}{\partial y}\right) + \frac{\partial}{\partial z}\left(K_i \frac{\partial T}{\partial z}\right) + q(x,y,z,t) = \rho_i c_i \frac{\partial T}{\partial t} \qquad (1)$$

In Equation 1, c_i is the specific heat capacity of the layer i, K_i is the thermal conductivity of the layer i, $q(x, y, z, t)$ represents the laser deposition energy per unit time and unit volume, which is expressed as (Zhu et al., 2012):

$$q(x,y,z) = I(x,y,t)e^{-\alpha z}\alpha(1-R) = I_0 f(x,y)g(t)e^{-\alpha z}\alpha(1-R) \qquad (2)$$

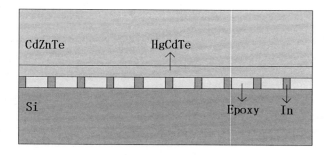

Figure 1.   The two-dimensional model of the HgCdTe detector.

where $I(x, y, t)$ is the function of the pulsed laser and α represents the absorption coefficient. For multilayer structure detectors, the absorption coefficient of different structural materials is different: R is the surface reflection coefficient, I_0 is the peak power density of the incident laser and is expressed as:

$$I_0 = \frac{E}{\pi w_0^2 \tau} \tag{3}$$

where τ is the pulse width, w_0 is the radius of the laser spot, E is the incident laser energy and $f(x, y)$ represents the spatial distribution of the pulsed laser beam. For the TEM_{00} mode pulse laser, the spatial distribution of the laser beam can be expressed as:

$$f(x, y) = \exp\left(-\frac{2(x^2 + y^2)}{w_0^2}\right) \tag{4}$$

In Equation 2, $g(t)$ represents the time distribution of the pulsed laser beam. For a repetition frequency laser with the frequency f and the pulse width τ, $g(t)$ can be described as:

$$g(t) = \begin{cases} 1, & n/f \leq t \leq n/f + \tau \\ 0, & n/f + \tau \leq t \leq (n+1)/f \end{cases} \tag{5}$$

where $n = 0, 1, 2, \ldots$ in Equation 5. In the thermal analysis, the radiation and convection heat loss on the upper surface of the sample are ignored. The initial temperature is set to 77 K and the bottom surface is cooled by the Dewar cold head and the temperature is kept at 77 K. The boundary conditions can be expressed as:

$$T(x, y, z, t = 0) = 77\text{K} \tag{6}$$

$$T(x, y, z = 0, t) = 77\text{K} \tag{7}$$

3 RESULTS

In the process of solving the heat conduction equation, an important problem is to deal with the thermophysical parameters of the material. In the process of laser irradiation, the thermal physical parameters of the materials in each layer of the HgCdTe detector are regarded as a constant. The thermal parameters of each layer of the detector (Lei et al., 2013) are shown in Table 1. The laser spot radius, after focusing, is set to 15 μm which corresponds to the single pixel size of the detector (about 30 μm). The peak power density is 15.5 MW/cm². The wavelength of the laser source is 10.6 μm and the pulse width τ is set to 200 ns. Experiments show that mercury will precipitate at the temperature of 350 K, destroying the crystal structure of HgCdTe and influencing the device performance. The indium column may fall off when the temperature of the indium column reaches melting point which is about 426 K. The detector will be completely destroyed when the temperature of the HgCdTe layer reaches melting point which is about 990 K.

Table 1. The parameters of each layer of the HgCdTe detector.

	CdZnTe	HgCdTe	In	Epoxy	Si
$\rho(\text{kg} \cdot \text{m}^{-3})$	5680	7600	7310	1250	2330
$C(\text{J} \cdot \text{Kg}^{-1} \cdot \text{K}^{-1})$	159	150	237.6	1560	550
$K(\text{W} \cdot \text{m}^{-1} \cdot \text{K}^{-1})$	0.97	20	82.06	0.2	250

The laser beam focused on the focal plane of the detector was assumed to be Gaussian distributed with a peak power density of 20 MW/cm². When the repetition frequencies are 1 KHz, 10 KHz and 100 KHz, the temperature distribution of the center point between every two adjacent layers of the detector were analyzed.

Figure 2 shows the temperature distribution of the center point of the interface between the HgCdTe layer and CdZnTe layer during five pulse laser irradiation. It can be seen from the diagram that the temperature of the central point rises rapidly to a very high peak temperature during the laser pulse, exceeding the melting point of the HgCdTe, and the temperature decreases rapidly when the pulse laser is finished. When the frequency is 1 KHz and 10 KHz, the temperature dropped to the initial temperature before the next pulse, and there is almost no cumulative effect. When the frequency is 100 KHz, the temperature of the center point cannot reduce to the initial temperature before the next pulse arrives, so there is a certain temperature accumulation effect. During the five pulse irradiation, the peak temperature gradually increases from 1400 K to 1500 K. At the same time, the analysis shows that the center point of the interface has recovered from the peak temperature to 80 K when $t \approx 0.07$ ms after the first pulse. Therefore, the temperature accumulation effect will be achieved when the repetition frequency of the pulse laser is greater than 15 KHz. From the point of view of laser damage, it is necessary to ensure that a single pulse irradiation can achieve laser damage when the detector is irradiated by a low frequency pulse laser. When the frequency is high enough, the temperature accumulation effect can achieve laser damage when the irradiation time is sufficient.

Figure 3 shows the temperature distribution of the center point of the interface between the HgCdTe photosensitive layer and the In column layer irradiated by different repetition frequencies. When the frequency is 1 KHz and 10 KHz, the temperature of the center point rises from 77 K to about 133 K. The temperature of the center point of the irradiation center

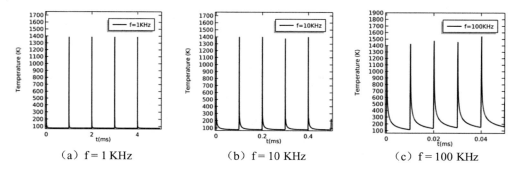

Figure 2. Temperature distribution at the center point of HgCdTe and CdZnTe interfaces at different reputation frequency: (a) f = 1 KHz; (b) f = 10 KHz; (c) f = 100 KHz.

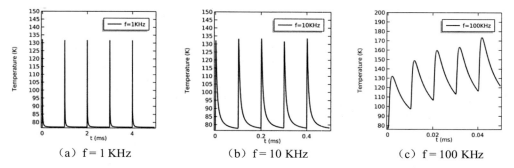

Figure 3. Temperature distribution at the center point of HgCdTe and In interfaces at different reputation Frequency: (a) f = 1 KHz; (b) f = 10 KHz; (c) f = 100 KHz.

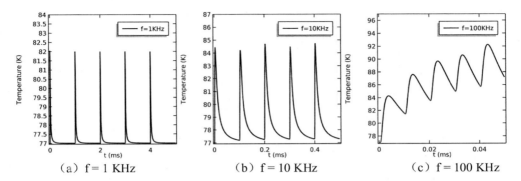

Figure 4. Temperature distribution at the center point of HgCdTe and In interfaces at different reputation frequency: (a) f = 1 KHz; (b) f = 10 KHz; (c) f = 100 KHz.

drops to the initial temperature before the next pulse arrives under the effect of heat conduction and heat diffusion. When the frequency is 100 KHz, the cumulative effect of the temperature rise is obvious, the peak temperature gradually increases, with the peak temperature increasing to 174 K at the end of the fifth pulse. It is known that the indium column will fall off when the temperature reaches 426 K. In contrast to Figure 2, it is known that the melting damage of the HgCdTe layer occurs prior to that of the In column layer.

Figure 4 shows that the temperature distribution of the center point of the interface between the interface of the In column layer and the Si layer, and the temperature rise trend is in accordance with the temperature distribution at the center point of the interface between the HgCdTe and the In column layer, but the temperature rise is less.

The analysis shows that the laser energy is mainly absorbed by the HgCdTe photosensitive layer, and the laser energy transmitted to the Si layer is very limited. Therefore, it can be concluded that the melting damage first appears at the interface of the HgCdTe layer and the CdZnTe substrate layer.

4 CONCLUSION

This paper built a theoretical model of the HgCdTe detector irradiated by repetition frequency pulsed laser based on heat conduction theory by applying the heat source to the focal plane of the detector. The thermal effect was compared under different repetition frequencies of laser. The results show that the energy of the laser is mainly absorbed by the HgCdTe photosensitive layer and the high temperature field is mainly distributed near the interface between the HgCdTe layer and the CdZnTe layer when the detector is irradiated by the laser source. The melting damage will first appear at the interface between the HgCdTe layer and the CdZnTe substrate layer. There is almost no temperature accumulation effect when the frequency is 1 KHz and 10 KHz. However, when the frequency is 100 KHz, the peak temperature increases with the increase of the pulse number, and there is an obvious temperature accumulation effect.

REFERENCES

Duan XF, Wang YF, Niu YX. (2004). Analytic calculation and evaluation of thermal and mechanical damage in optical materials induced by laser. *Chinese Journal of Lasers*, 12(31), 1455–1459.
Huang F, Niu YX., Wang YF, Duan XF. (2016). Calculation of thermal and mechanical effect induced by laser in optical window materials. *Acta Optica Sinica*, 4(26), 576–580.
Li DW, Meng W, Ma LH. (2012). Finite element study on heat damage of moving target under the high energy laser irradiating. *Optical Technique*, 5(38), 624–629.

Lei P, Li H, Bian JT, Nie JS. (2013). Experimental study of HgCdTe imaging sensor irradiated by TEA-CO2 laser. *Acta Optica Sinica*, 2(33), 021400201–021400205.

Shao L, Li SG, Sun XQ. (2006). Analysis of the missile early warning satellites detecting principle and discussion of attack-defense measure. *Infrared Technology*, 4(28), 43–46.

Su XL, Niu CH, Ma MY. (2016). Research on the thermal damage of HgCdTe infrared detector under laser irradiation of 10.6 um wavelength. *Infrared Technology*, 1(38), 6–9.

Tang W, Ji TB, Guo J, Shao JF, Wang TF. (2013). Numerical analysis of HgCdTe crystal damaged by high repetition frequency CO2 laser. *Chinese Optics*, 5(6), 736–742.

Wang SW, Li Y, Guo LH. (2010). Analysis on the disturbance of CO_2 laser to long-wave infrared HgCdTe detector. *Journal of Infrared & Millimeter Waves*, 2(29), 102–104.

Wang YB, Jin GY, Zhang W. (2016). Temperature and thermal stress analysis of aluminum alloy plate irradiation by long pulsed laser. *Chinese Journal of Lasers*, 4(8), 124–131.

Wei Z, Jin GY, Peng B, Zhang XH, Tan Y. (2014). Numerical simulation of thermal and stress field in silicon-based positive-intrinsic-negative photodiode irradiated by millisecond-pulsed laser. *Acta Physica Sinica*, 5(63), 1940205–1940210.

Zhao HY, Wei Z, Jin GY. (2017). The damage numerical research of pulse sequence laser irradiation photodiode. *Laser & Optoelectronics Progress*, 4(54), 040401–040412.

Zhang YC, Shen H, Zhu RH. (2013). Three-dimensional temperature field of material irradiated by continuous wave laser. *Chinese Journal of Lasers*, 8(40), 1–5.

Zheng YL, Hu YH, Zhao NX, Ren XD. (2017). Analysis of the influence of pulse width and repetition frequency on damage threshold of HgCdTe detector. *Infrared Technology*, 2(6), 11–18.

Zhu ZW, Cheng XA, Huang LJ, Liu ZJ. (2012). Lightfield intensification induced by nanoinclusions in optical thin-films. *Applied Surface Science*, 258(12), 5126–5130.

Frontier Research and Innovation in Optoelectronics Technology and Industry – Habib & Lewis (Eds)
© 2019 Taylor & Francis Group, London, ISBN 978-1-138-33178-5

Analysis of thermal-gradient-induced aberration of coolant in direct liquid-cooled thin-disk laser

X. Ruan
School of Information Science and Technology, Fudan University, Shanghai China
Institute of Applied Electronics, China Academy of Engineering Physics, Mianyang China
Key Laboratory of Science and Technology on High Energy Laser, China Academy of Engineering Physics, Mianyang China

B. Tu, J.L. Shang, J. Wu & X.C. An
Institute of Applied Electronics, China Academy of Engineering Physics, Mianyang China
Key Laboratory of Science and Technology on High Energy Laser, China Academy of Engineering Physics, Mianyang China

H. Su
Key Laboratory of Science and Technology on High Energy Laser, China Academy of Engineering Physics, Mianyang China

J.Y. Yi, H.X. Cao, Q.S. Gao, K. Zhang & C. Tang
Institute of Applied Electronics, China Academy of Engineering Physics, Mianyang China
Key Laboratory of Science and Technology on High Energy Laser, China Academy of Engineering Physics, Mianyang China

ABSTRACT: In direct-liquid-cooled thin-disk lasers, the laser passes through the coolant several times during the oscillation process. The temperature distribution of the coolant has a great influence on the laser wavefront. Comparison of the thermal-gradient-induced aberration of liquid has been conducted using D_2O and CCl_4 as coolants. The influence of the thermal conductivity and specific heat capacity of the coolant on the laser wavefront is analyzed too. The results show that liquids with poor thermal properties and high thermo-optic coefficients such as CCl_4 cause serious wavefront aberration.

1 INTRODUCTION

Laser systems with high average power, high efficiency and high beam quality have been attractive for a broad range of medical, commercial, scientific and military applications. For a Solid-State Laser (SSL) with 100 kW output power, the thermal effect is the main limiting factor for further raising of the output power, and may bring serious aberration and the risk of fracturing the gain medium (Giesen & Speiser, 2007; Bruesselbach & Sumida, 2005; Sazegari et al., 2010). With its efficient thermal management, the Direct-Liquid-Cooled thin-disk Laser (DLCL) is one of the most important technologies in solving the thermal problem in high-power lasers (Wang et al., 2016).

In DLCLs, tens of disk pieces integrate into one gain module and the circulating liquid flows over the largest surface of every disk to extract the heat. Due to their efficient thermal management, DLCLs have become extremely attractive in the high-power laser field (Li et al., 2013; Nie et al., 2014; Okada et al., 2006; Gong et al., 2013). However, the beam quality of DLCLs is not as good as other solid-state lasers (Fu et al., 2014; Tu et al., 2016). This is because the laser passes through the coolant several times during the oscillation, with the temperature distribution of the coolant having a great influence on the laser wavefront.

Based on the characteristics of the laser, a convective cooling model of DLCLs is established. A large eddy simulation model is used to analyze the aberrations, such as the tilt aberration, defocus aberration and high-order aberrations with high frequency that were introduced in the cooling process. The cooling effect and wavefront aberration are compared when using different liquids as coolants. The influence of the thermal conductivity and specific heat capacity are also analyzed. The results show that the specific heat capacity, thermal conductivity and thermo-optic coefficient of the coolant have a significant effect on wavefront aberration. The Peak-to-Valley (PV) of the aberration is almost ten times greater when using CCL_4 as the coolant in comparison with H_2O.

2 THEORETICAL MODEL

A schematic diagram of a DLCL is shown in Figure 1. Tens of thin Neodymium-doped Yttrium Aluminum Garnet (Nd:YAG) disks are integrated into the Gain Module (GM). Four cubes with a thickness of 0.5 mm are spliced on the corners of the thin disk to provide the cooling channel. Coolant flows over the largest surface of every disk to remove the heat. There is a section of YAG without doping above and below the Nd:YAG, which is used to homogenize the flow field. Two dichroic mirrors (M1 and M2) are used to detect the wavefront aberration of the GM.

The convective cooling model of DLCLs is shown in Figure 2. Coolant flows between the two plates in the y direction. The laser is in the z-axis. In this model, the length d of the two plates is 10×10 mm. The thickness of coolant is 0.5 mm.

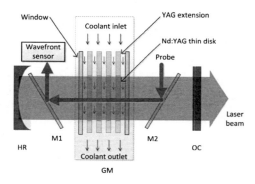

Figure 1. Configuration of a direct-liquid-cooled Nd:YAG multi-disk laser resonator: HR – high reflector; OC – output coupler; M1, M2 – dichroic mirrors; GM – gain module.

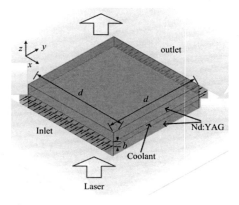

Figure 2. Convective cooling model of Nd:YAG thin-disk laser.

184

3 RESULTS AND DISCUSSION

Through use of ANSYS engineering simulation software (Ansys, Inc., Canonsburg, PA), a large eddy simulation model is used to analyze the temperature distribution of the coolant under uniform thermal loading. Then the wavefront aberration is calculated based on the thermo-optic effect. The heat flux on the wall of coolant is 1.8×10^5 W/m^2. For the inlet, the average flow rate is 2 m/s and the temperature is 290 K.

3.1 Flow field at the inlet

In this section, the flow model at the YAG extension is first established, and the flow velocity at the entrance of the Nd:YAG is calculated. Figure 3 shows the velocity of the coolant at the entrance of the Nd:YAG. In the thickness direction (z direction), because of the small scale, the boundary layer effect is obvious. That is, the flow velocity in the center is high and the velocity near the wall is low.

3.2 Aberration with D_2O as coolant

Due to its excellent thermal properties, D_2O is commonly used as a coolant. First, we analyze the wavefront aberration when D_2O is used as the coolant. Figure 4 shows the temperature of the coolant along the direction of flow. The maximum temperature rise of the liquid is 7.99 K. In the middle of the flow channel, the temperature changes little, indicating that the thickness of the flow channel can be reduced. At the boundary layer, the temperature rises obviously.

Figure 5 shows the wavefront aberration caused by the fluid when D_2O is used as the coolant. The flow direction of the liquid corresponds to the negative direction of the y-axis in Figure 5. As can be seen from Figure 5, the main wavefront aberration is tilt aberration along the flow direction, and its PV accounts for 88.8% of the total aberration. The coolant absorbs

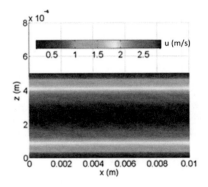

Figure 3.   Velocity of the coolant at entrance of the Nd:YAG.

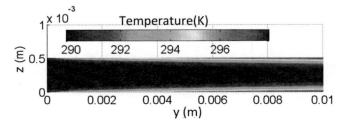

Figure 4.   Temperature of the coolant along the flow direction.

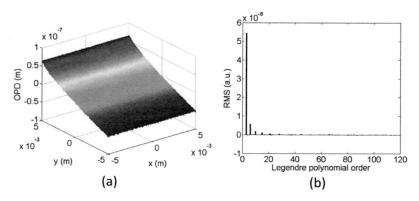

Figure 5. Wavefront aberration caused by the coolant when D_2O is used: (a) total aberration; (b) Legendre polynomial expansion of the aberration.

the heat flux and temperature rise during the flow, which causes the tilt aberration. Compensation of the tilt can be realized by use of two identical gain modules with opposite flow directions. In addition the tilt aberration, followed by the defocus aberration, its PV accounts for 6.9% of the total aberration. The wavefront aberration at the inlet is larger than in other places because the temperature of the boundary layer at the inlet is low, the convection heat transfer coefficient is large, and the temperature rises quickly when the coolant reaches the Nd:YAG region. During the cooling process, the temperature gradient gradually decreases and the wavefront gradient decreases too.

3.3 *Aberration with CCl_4 as coolant*

The absorption coefficient of CCl_4 at 1064 nm is smaller compared to that of D_2O. The use of carbon tetrachloride as coolant can significantly reduce cavity loss. However, the simulation results show that the maximum temperature rise of the coolant was 25.9 K, which is much higher than the 7.99 K seen with D_2O due to the poor thermal characteristics of CCl_4.

Figure 6 shows the wavefront aberration caused by the fluid when CCl_4 is used as the coolant. The velocity and the heat flux are both the same as in Figure 5 (for D_2O). As shown in Figure 6, when CCl_4 is used as the coolant, the wavefront aberration caused by the fluid under the same heat loading is about ten times that of D_2O. This is mainly because the specific heat capacity and thermal conductivity of CCl_4 are less than that of heavy water, but the thermo-optic coefficient is higher than that of D_2O.

3.4 *Temperature versus specific heat capacity and thermal conductivity*

The maximum temperature versus specific heat capacity and thermal conductivity are analyzed. The heat flux on the wall of coolant is 1.8×10^5 W/m². For the inlet, the average flow rate is 2 m/s and the temperature is 290 K. Figure 7a shows the maximum temperature versus specific heat capacity. The thermal conductivity is 0.11 W/m·K. The other parameters of the liquid are the same as the water. It can be seen that the maximum temperature has a negative exponential relationship with the specific heat capacity. Figure 7b shows the maximum temperature versus thermal conductivity. The specific heat capacity is 886 J/kg·K. The maximum temperature also has a negative exponential relationship with the thermal conductivity. In order to obtain a better cooling effect, a liquid with specific heat capacity of more than 4000 J/kg·K and thermal conductivity of more than 0.6 W/m·K should be used.

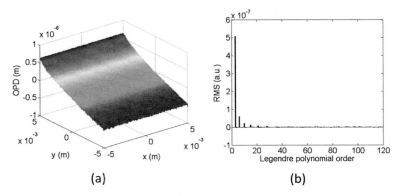

(a)   (b)

Figure 6.   Wavefront aberration caused by the coolant when CCl$_4$ is used: (a) total aberration; (b) Legendre polynomial expansion of the aberration.

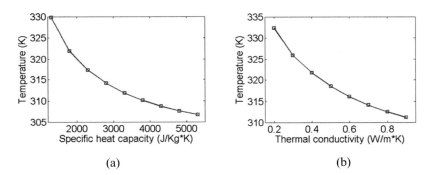

(a)   (b)

Figure 7.   Maximum temperature of coolant: (a) maximum temperature vs specific heat capacity; (b) maximum temperature vs thermal conductivity.

4   CONCLUSION

Based on the characteristics of the laser, a convective cooling model of DLCLs is established. A large eddy simulation model was used to analyze the temperature of the coolant. The cooling effect and wavefront aberration are compared when using different liquids as the coolant. The result shows that, under uniform thermal loading, the main aberration of the liquid is the tilt and defocus along the direction of flow. The specific heat capacity, thermal conductivity and thermo-optic coefficient of the coolant have a significant effect on wavefront aberration. The maximum temperature rise of the coolant reaches 25.9 K when CCl$_4$ is used as the coolant, which is much higher than the 7.99 K rise seen with D$_2$O. The PV of aberration is almost ten times greater when using CCl$_4$ as the coolant, compared with H$_2$O.

REFERENCES

Bruesselbach, H. & Sumida, D.S. (2005). A 2.65-kW Yb:YAG single-rod laser. *IEEE Journal of Selected Topics in Quantum Electronics*, *11*(3), 600–603.

Fu, X., Li, P., Liu, Q. & Gong, M. (2014). 3 kW liquid-cooled elastically-supported Nd:YAG multi-slab CW laser resonator. *Optics Express*, *22*(15), 18421–18432.

Giesen, A. & Speiser, J. (2007). Fifteen years of work on thin-disk lasers: Results and scaling laws. *IEEE Journal of Selected Topics in Quantum Electronics*, *13*(3), 598–609.

Gong, M., Li, P., Liu, Q. & Fu, X. (2013). Analysis of wavefront aberration induced by turbulent flow field in liquid-convection-cooled disk laser. *Journal of the Optical Society of America B, 30*(8), 2161–2167.

Li, P., Liu, O., Fu, X. & Gong, M. (2013). Large-aperture end-pumped Nd:YAG thin-disk laser directly cooled by liquid. *Chinese Optics Letters, 11*(4), 47–50.

Nie, R.Z., She, J.B., Zhao, P.F., Li, F.L. & Peng, B. (2014). Fully immersed liquid cooling thin-disk oscillator. *Laser Physics Letters, 11*(11), 115808.

Okada, H., Yoshida, H., Fujita, H. & Nakatsuka, M. (2006). Liquid-cooled ceramic Nd:YAG split-disk amplifier for high-average-power laser. *Optics Communications, 266*(1), 274–279.

Sazegari, V., Milani, M.R. & Jafari, A.K. (2010). Structural and optical behavior due to thermal effects in end-pumped Yb:YAG disk lasers. *Applied Optics, 49*(36), 6910–6916.

Tu, B., Liu, C., Tang, C., Wang, K., Gao, Q. & Cai, Z. (2016). Kilowatt-level direct-'refractive index matching liquid'-cooled Nd:YLF thin disk laser resonator. *Optics Express, 24*(2), 1758–1772.

Wang, K., Tu, B., Jia, C., Shang, J., An, X. & Liao, Y. (2016). 7 kW direct-liquid-cooled side-pumped Nd:YAG multi-disk laser resonator. *Optics Express, 24*(13), 15012–15020.

Frontier Research and Innovation in Optoelectronics Technology and Industry – Habib & Lewis (Eds)
© 2019 Taylor & Francis Group, London, ISBN 978-1-138-33178-5

Detecting collagen fibrils' structure in ovarian cancer using second harmonic generation microscopy

Jun-Fang Wu, Xi-Da Li & Chao Li

School of Physics and Optoelectronic Technology, South China University of Technology, Guangzhou, Guangdong, China

ABSTRACT: Ovarian cancers have the highest mortality rate of all gynecologic cancers. Many researchers have found significant modifications in the morphology of collagen fibers in ovarian tumors. Second Harmonic Generation (SHG) is considered to have the potential to probe the structure of collagen fibrils within tissues. However, how the structure of collagen fibrils impacts the SHG remains to be solved. In this paper, Fourier transformation was adopted to analyze the quasi-phase matching, which is critical to the efficient generation of SHG. Then, the intensities of forward and backward SHG can be calculated. Moreover, two types of distribution models of collagen fibrils were built for normal ovaries and high-grade serous tumors. The intensity ratios of forward and backward SHG calculated from our models were a good fit to the measured ones. Therefore, our models are helpful for understanding the structure of ovaries at the fibril level.

1 INTRODUCTION

Ovarian cancer is the leading cause of death of all gynecologic cancers. Early ovarian cancer usually has no obvious symptoms, which is why the disease is typically diagnosed at an advanced stage. Women diagnosed with localized-stage ovarian cancer have more than a 90% five-year survival rate. For late-stage ovarian cancer, the five-year survival drops to 29%. Therefore, early detection of this disease is very important in order to improve the survival rate.

Although screening tests are usually used to detect disease, there is no effective screening test for ovarian cancer. The two tests used most often to screen for ovarian cancer are Transvaginal Ultrasound (TVUS) and the cancer antigen (CA)-125. TVUS can help to find a mass in the ovary, but most of the masses found are not cancer. CA-125 is used as a tumor marker. Levels of CA-125 in many ovarian cancer patients are high, while a high CA-125 level is more often caused by other conditions.

Extracellular Matrix (ECM) is an important component of the microenvironment for normal tissues and tumors. ECM varies in the progress of invasion and metastasis of ovarian cancer. Many researchers are trying to find the relationship between ECM remodeling and the growth of ovarian tumors. For example, a semi-quantitative polymerase chain reaction was used to detect levels of TGF-β1 in the cytoplasm and ECM (Keming et al., 2014). The TGF-β1 levels were found to be significantly different between epithelial ovarian cancer and the corresponding normal ovarian tissue. Ricciardelli observed that in the metastasis of ovarian cancer, a number of ECM proteins including fibronectin, TGF-β1 and periostin play some roles in the ovarian cancer–peritoneal cell interaction (Ricciardelli et al., 2016). Li found that highly expressed ITGBL1, a kind of protein in ECM, could promote ovarian cancer cell migration (Suna et al., 2016). Collagens are the most abundant fibrous proteins of the ECM. Much focus has been placed on collagens.

Increased collagens are found in ovarian cancers compared to healthy ovaries. Imaging of collagens in the ECM of malignant ovarian tissues reveals highly regular helical structures of collagen fibers. In contrast, collagen fibers are cross-hatched in normal tissues (Oleg et al.,

2010). Many researchers have found significant differences in the distribution and organization of collagen fibers in ovarian tumors as compared with normal tissues (Adur et al., 2014).

By common consent, structural modifications have taken place in collagens of ovarian cancer. However, the quantitative changes in diseased collagens' structure, which are crucial for understanding the mechanism of cancer progression, are difficult to measure. In recent years, Second Harmonic Generation (SHG) has emerged as a powerful technique to probe the structure of collagen fibrils within tissues. SHG is a coherent non-linear process, in which two lower energy photons are up-converted to twice the incident frequency of an excitation laser. Because of its underlying physical origin, SHG is highly sensitive to not only the collagen fibril structure, but also the changes occurring in it. Some researchers paid attention to SHG directionality, which is measured through the ratio of forward to backward SHG (F/B). They thought SHG directionality was affected by collagen fibril diameter, spacing and disorders of fibril packing within a fiber. Therefore, the microstructure of collagen can be measured by the F/B (Kathleen et al., 2015; Houle et al., 2015). Polarization-resolved SHG microscopy has been noted to provide valuable information on collagen arrangement not available with intensity measurements alone (Danielle et al., 2015; Hristu et al., 2017). The laser beam polarization components are utilized in polarization-resolved SHG microscopy, which enables measurements of the main orientation and angular dispersion of collagen fibrils. Then, the anisotropy in collagen fibrils is used to distinguish healthy from diseased. The SHG images are also analyzed in the frequency domain. In Paolo et al. (2009), the SHG images were analyzed by means of discrete Fourier transform, auto-correlation and entropy analyses. Therefore, the disorder induced within the irradiated area of corneal stroma was quantified.

Although SHG has the potential to provide insight into collagen fibrils, how the organization and size of collagen fibrils impact the SHG remains to be solved. In this paper, we propose a method to calculate SHG intensity. Besides, the intensity ratios of forward and backward SHG are utilized to build proper models of collagen fibrils.

2 THE THEORY OF SHG

2.1 The generation of SHG

The SHG emitted by collagen fibrils results from the non-linear polarization of collagen fibrils illuminated by exciting laser.

The second-order polarization $\mathbf{P}_2$ generated by the exciting electric field $\mathbf{E}_1$ is in the form:

$$\mathbf{P}_2 = 2\varepsilon_0 d\mathbf{E}_1^2 \tag{1}$$

where ε_0 is the permittivity of free space, d is the non-linear susceptibility. Then, the second-order polarization acts as the source of the second harmonic generation. The wave equation about the second harmonic wave is:

$$\nabla^2 \mathbf{E}_2 - \frac{\varepsilon_r}{c^2} \frac{\partial^2 \mathbf{E}_2}{\partial t^2} = \frac{1}{\varepsilon_0 c^2} \frac{\partial^2 \mathbf{P}_2}{\partial t^2} \tag{2}$$

We consider some simplified conditions. Assume both the exciting beam and the second harmonic wave propagate in the +z direction. The electric field strength of the exciting wave ($i = 1$) and the second harmonic wave ($i = 2$) are represented as:

$$\mathbf{E}_i = A_i e^{i(k_i z - \omega_i t)} \tag{3}$$

The amplitude of $\mathbf{P}_2$ can then be written as:

$$P_2 = 2\varepsilon_0 d A_1^2 e^{i \cdot 2k_1 z} \tag{4}$$

By substituting Equations 1, 3 and 4 into Equation 2, while utilizing the slowly varying amplitude approximation, we obtain:

$$\frac{dA_2}{dz} = i\frac{2\omega_2^2}{k_2c^2}A_1^2de^{i(2k_1-k_2)z} \tag{5}$$

In the case of $2k_1 - k_2 = 0$, which is known as the perfect phase matching, the amplitude and the intensity of the second harmonic wave will reach the maximum.

2.2 Quasi-phase matching

For biological tissues, the phase matching condition is difficult to satisfy. The efficient generation of SHG depends on quasi-phase matching. The non-linear susceptibility d has different values for collagen fibrils and their surroundings. Assume d varies in the period of Λ. Then the spatial variation of d can be described in terms of a Fourier series as:

$$d(z) = \sum_{n=-\infty}^{\infty} c_n e^{i n \omega_0 z} \tag{6}$$

where $\omega_0 = \frac{2\pi}{\Lambda}$ and $c_n = \frac{1}{\Lambda}\int_0^\Lambda d(z)e^{-i n \omega_0 z}dz$.

Equation 6 means $d(z)$ in biological tissues can provide many phase components, such as $\pm\omega_0, \pm2\omega_0, \dots$. Of all these components, one particular component is assumed to dominantly prompt SHG. We denote the particular component as $N\omega_0$. Equation 5 can be rewritten by replacing d with its component $N\omega_0$, yielding:

$$\frac{dA_2}{dz} = i\frac{2\omega_2^2}{k_2c^2}A_1^2 \cdot c_N \cdot e^{i(2k_1-k_2+N\omega_0)\cdot z} \tag{7}$$

Quasi-phase matching means $2k_1 - k_2 + N\omega_0 = 0$. For biological tissues, the value of $2k_1 - k_2 + N\omega_0$ should be as close to zero as possible. The amplitude of the second harmonic wave is given by integrating Equation 7 from $z = 0$ to $z = L$, yielding:

$$A_2(L) = \int_0^L \frac{dA_2(L)}{dz}dz = i\frac{2\omega_2^2}{k_2c^2}A_1^2 \cdot c_N \cdot \frac{e^{i(2k_1-k_2+N\omega_0)L}-1}{i(2k_1-k_2+N\omega_0)} \tag{8}$$

The intensity of the second harmonic wave is then given by:

$$F_{SHG} = 2n_2\varepsilon_0 c|A_2(L)|^2 = \frac{2\omega_2^2 I_1^2}{n_1^2 n_2 \varepsilon_0 c^3}\cdot|c_N|^2 \cdot L^2 \sin c^2\left[\frac{(2k_1-k_2+N\omega_0)L}{2}\right] \tag{9}$$

2.3 The ratio of forward and backward SHG

The second harmonic wave propagating in the $+z$ direction is known as the forward SHG. The second harmonic wave can also emit in the $-z$ direction, which is known as the backward SHG. The backward SHG is similar to the forward SHG in the requirement of quasi-phase matching. The quasi-phase matching for the backward SHG takes the following form:

$$2k_1 + k_2 + M\omega_0 = 0 \tag{10}$$

where $M\omega_0$ is the particular Fourier component of $d(z)$, which dominantly prompts the backward SHG. The intensity of the second harmonic wave propagating along the backward direction can be deduced in the same way as the deduction of Equation 9. The intensity of the backward SHG is given by:

$$B_{SHG} = \frac{2\omega_2^2 I_1^2}{n_1^2 n_2 \varepsilon_0 c^3}\cdot|c_M|^2 \cdot L^2 \sin c^2\left[\frac{(2k_1+k_2+M\omega_0)L}{2}\right] \tag{11}$$

The intensity ratio of forward and backward SHG, R, is calculated by using Equations 9 and 11, yielding:

$$R = \frac{F_{SHG}}{B_{SHG}} = \frac{|c_N|^2 \cdot \sin c^2 \left[\dfrac{(2k_1 - k_2 + N\omega_0)L}{2} \right]}{|c_M|^2 \cdot \sin c^2 \left[\dfrac{(2k_1 + k_2 + M\omega_0)L}{2} \right]} \tag{12}$$

3 EXPERIMENTAL RESULTS AND ANALYSIS

3.1 *The measured intensity ratio of forward and backward SHG*

Campagnola's lab measured the intensity ratios of forward and backward SHG for normal stroma and High-Grade Serous (HGS) ovarian tumor, which is the predominant subtype of cancer in patients (Karissa et al., 2017). The intensity ratios measured at different excitation wavelengths are shown in Figure 1.

What we want to do is not just measure the intensity ratio. The more important thing is to understand how collagen fibril assembly decides the intensity ratio. The intensity ratio does not give collagen fibril assembly directly. We can build some distribution models of collagen fibrils for normal stroma and HGS tumors, and calculate the intensity ratios of these models. The models from which the calculated intensity ratios best fit the measured ratios are considered to be the closest assemblies to the real ones.

3.2 *The models of collagen fibrils built for normal stroma and HGS tumors*

The differing intensity ratios for normal stroma and HGS tumors are determined by changes in collagen fibril assemblies, such as fibril size, fibril morphology and fibril density. We built two types of distribution of collagen fibrils for normal stroma and HGS tumors. Most parameters, including fibril diameter D, distance among fibrils t and excitation wavelength w, in our two model types are almost the same. The only difference lies in the distribution of d. In other words, we focused on the intensity ratio dependence on the distribution of d.

In our models, a fibril is represented by a cylinder. All fibrils are parallel. The sign of d can change from one fibril to another. The positive ds and negative ds in the second harmonic wave generating domain distribute as evenly as possible. The normal stroma model adopts the radiated distribution of d, where one fibril acts as the radiant center and other fibrils surround the center fibril with different radii. The HGS tumor model adopts the parallel distribution of d, where fibrils are arranged in an array of a parallelogram. Nine and five models were built for normal and HGS tumors, respectively. Figure 2 displays some examples of these models projected on the cross sections perpendicular to fibril axes. A white circle in Figure 2 represents a fibril with positive d, while a black circle represents a fibril with negative d.

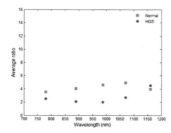

Figure 1.   The measured intensity ratios of forward and backward SHG for normal stroma and HGS tumor.

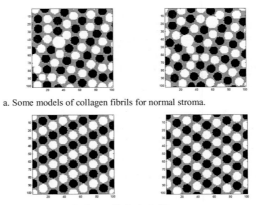

a. Some models of collagen fibrils for normal stroma.

b. Some models of collagen fibrils for HGS tumor.

Figure 2.   The distribution types of collagen fibrils.

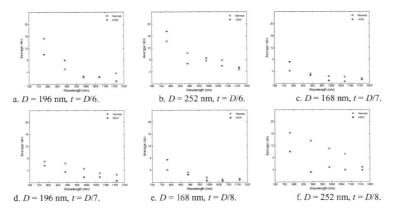

a. D = 196 nm, t = D/6.   b. D = 252 nm, t = D/6.   c. D = 168 nm, t = D/7.

d. D = 196 nm, t = D/7.   e. D = 168 nm, t = D/8.   f. D = 252 nm, t = D/8.

Figure 3.   The average intensity ratios of forward and backward SHG simulated for normal stroma and HGS tumor with different sets of parameters.

3.3   *The calculated intensity ratios and analysis*

The structures of biological tissues have much flexibility, for example, the assembly patterns of collagen fibrils in real ovaries are not as regular as our models, which allow more randomness, and the shapes of each fibril are not ideal cylinders. Therefore, unreasonable models often lead to large deviations from measurements, which can usually be two or three orders of magnitude.

If our distribution models for normal stroma and HGS tumors are reasonable, there will be a set of parameters, D and t, which makes the calculated intensity ratios close to the measurements. We experiment with four values of D, which are 168 nm, 196 nm, 224 nm and 252 nm. For each value of D, t has been experimented with three values, which are D/6, D/7 and D/8. So there are twelve sets of parameters (D, t) in total.

As mentioned previously, there were nine distribution models for normal stroma. We calculated their intensity ratios of forward and backward SHG according to a set of parameters (D, t). The average of these ratios was regarded as the final intensity ratio for normal stroma under the special set of parameters. The similar computations were done to get the final ratio for HGS tumors. Therefore, we obtained twelve comparisons between the two kinds of tissues. Figure 3 shows some of these comparisons.

As for our models, the simulated intensity ratios of forward and backward SHG achieved the same order of magnitude as that of the measured ones. This means the distribution types of collagen fibrils in our models are close to real fibrils. In the theoretical analysis, a fibril can be modeled by a cylinder. The signs of ds in real fibrils should distribute as evenly as possible, so as to make the whole bulk of collagen fibrils into a condition of electrical neutrality.

Furthermore, we can determine the most reasonable set of parameters for normal stroma and HGS tumors by comparing Figure 1 and Figure 3. The result shown in Figure 3(d) has the best fit to the measurements. Therefore, the corresponding set of parameters, $D = 196$ nm and $t = D/7$, can give us an insight into the structures of normal ovaries and HGS tumors at the fibril level. The average diameter of fibrils in ovaries is about 196 nm. The distance between the centers of adjacent fibrils is about 8/7 of the fibril diameter.

4 CONCLUSIONS

SHG excited by laser with proper wavelength can provide information about tissue structures at the fibril level. SHG has been found to have superior power to distinguish normal ovaries and ovarian cancer. How to interpret the relationship between SHG emission directions and inner structure of collagen fibrils of ovaries is critical for SHG to become a clinical screening tool.

The efficient generation of SHG in biological tissues depends on quasi-phase matching. In this paper, Fourier transformation was adopted to analyze the possible phase components which collagen fibrils can provide. One particular component was assumed to dominantly prompt SHG according to quasi-phase matching. Then, the intensities of the forward and backward SHG, as well as their ratio, can be calculated.

We also built distribution models of collagen fibrils for normal ovaries and high-grade serous tumors. The comparison between the simulated and measured ratios of forward and backward SHG revealed that the assembly patterns of fibrils in our models are reasonable. In addition, the proper parameters for the models were found, which made the simulated ratios meet the measurements with varying wavelengths of the excitation beam. All of these help us have an insight into the collagen fibrils in the human ovary.

REFERENCES

Adur, J., Pelegati, V.B., Thomaz, A.A., Baratti, M.O., Andrade, L.A., Carvalho, H.F., Bottcher-Luiz, F. & Cesar, C.L. (2014). Second harmonic generation microscopy as a powerful diagnostic imaging modality for human ovarian cancer. *Journal of Biophotonics*, 7(1–2), 37–48.

Danielle, T., Richard, C. & Ahmad, G. et al. (2015). Ultrastructural features of collagen in thyroid carcinoma tissue observed by polarization second harmonic generation microscopy. *Biomedical Optics Express*, 6(9), 3475–3480.

Houle, M.A., Couture, C.A. & Bancelin, S. et al. (2015). Analysis of forward and backward second harmonic generation images to probe the nanoscale structure of collagen within bone and cartilage. *Journal of Biophotonics*, 8, 993–1001.

Hristu, R., Stanciu, S. & Tranca, D. et al. (2017). Improved quantification of collagen anisotropy with polarization-resolved second harmonic generation microscopy. *Journal of Biophotonics*, 10(9), 1171–1179.

Karissa, B.T., Kirby, R.C., Kevin, W.E., Sana, M.S., Manish, P. & Paul, J.C. (2017). Stromal alterations in ovarian cancers via wavelength dependent second harmonic generation microscopy and optical scattering. *BMC Cancer*, 17, 102.

Kathleen, A.B., Ryan, P.D. & Mehar, K.C. et al. (2015). Second-harmonic generation scattering directionality predicts tumor cell motility in collagen gels. *Journal of Biomedical Optics*, 20(5), 051024.

Keming, C., Hua, W., Shengrong, L. & Cunjian, Y. (2014). Expression and significance of transforming growth factor-β1 in epithelial ovarian cancer and its extracellular matrix. *Oncology Letters*, 8, 2171–2174.

Oleg, N., Ronald, B.L., Molly, A.B. & Paul, J.C. (2010). Alterations of the extracellular matrix in ovarian cancer studied by second harmonic generation imaging microscopy. *BMC Cancer*, 10, 94.

Paolo, M., Fulvio, R. & Francesca, R, et al. (2009). Photothermally-induced disordered patterns of corneal collagen revealed by SHG imaging. *Optics Express*, 17(6), 4868–4878.

Ricciardelli, C., Lokman, N.A., Ween, M.P. & Oehler, M.K. (2016). Ovarian cancer–peritoneal cell interactions promote extracellular matrix processing. *Endocrine-Related Cancer*, 23(11), T155–T168.

Suna, L., Defeng, W., Xiaotian, L., Lingling, Z., Hui, Z. & Yingjie, Z. (2016). Extracellular matrix protein ITGBL1 promotes ovarian cancer cell migration and adhesion through Wnt/PCP signaling and FAK/SRC pathway. *Biomedicine & Pharmacotherapy*, 81, 145–151.

Frontier Research and Innovation in Optoelectronics Technology and Industry – Habib & Lewis (Eds)
© 2019 Taylor & Francis Group, London, ISBN 978-1-138-33178-5

Effect of laser intensity on the energy of accelerating electron in vacuum

Liqiang Xin
Bell Honor School, Nanjing University of Posts and Telecommunications, Nanjing, People's Republic of China

Youwei Tian
College of Science, Nanjing University of Posts and Telecommunications, Nanjing, People's Republic of China

ABSTRACT: The influence of laser intensity on relativistic motion and the energy gains from electron oscillations driven by circularly polarized tightly focused laser pulses have been investigated theoretically and numerically using a single electron model. Ponderomotive force accelerates an electron at the focus of diverse intensity tightly focused short-pulse laser is considered. Due to the asymmetry of acceleration and deceleration, the accelerated electron is extracted from the tightly focused laser pulse by the longitudinal ponderomotive force. Since the distance of the election movement much less than the beam waist radius, the laser pulse acting on the election is close to the plane wave during the interaction, so the electron can not be accelerated. By using MATLAB to simulate the electron trajectory and dimensionless particle normalized energy, found that when pulse width and beam waist are constant, the energy of the electron is increasing with the increasing of laser pulse intensity.

Keywords: laser intensity; electron control; circularly polarized laser pulses; tight focus

1 INTRODUCTION

In 1995, Hartemann proposed the extraction mechanism of nonlinear ponderomotive force scattering. Due to the obvious asymmetry of the focused Gauss pulses, when electrons meet the laser pulses, electrons can be scattered by the ponderomotive force of the laser field, and some energy (Luo W et al., 2016) could be obtained from it. In the course of the study of relativistic effects in the laser pulse and electron interaction, highly nonlinear physical phenomena are revealed. So far, many authors have studied the interaction between free electrons and intense laser fields. Various nonlinear phenomena have been found (He F et al., 2004). The main motivation to study these phenomena comes from the desire to produce the desktop particle acceleration and the X ray source. We hope to miniaturize the particle accelerator.

In order to explore the precise control of the electronic movement by laser further, this paper puts forward a scheme that puts a still electronic at the cross position between the focal plane of the laser pulse and the propagation direction. We go to observe the electron trajectory and the electronic energy when different amplitude laser pulse interact with the electron. We will explore inflection point when will the electron gain obvious energy.

2 THE ACCELERATION MODEL

For a circularly polarized tightly focused laser pulse, the vector potential can be expressed as:

$$a = a_0 \exp\left(-\frac{\eta^2}{L^2} - \frac{x^2 + y^2}{b^2}\right) \times \left(1 + z^2/z_f^2\right)^{-1/2} \hat{a}, \tag{1}$$

where L and b are the width of the laser pulse and the radius of the waist. a_0 is the peak amplitude normalized by mc^2/e. $\eta = z-t$; z and t are the coordinate position and time after the normalization by k_0^{-1} and ω_0^{-1}, k_0 and ω_0 represent the wavenumber and the circular frequency of the tight focusing laser pulse. $b = b_0\left(1 + z^2/z_f^2\right)$, b is the waist radius of the ultrashort and super intense focused laser pulse propagating to Z, while b_0 is the minimum radius of the laser pulse. It has been normalized by k_0^{-1}. $z_f = b_0^2/2$ means the Rayleigh length of the laser pulse. m is the quality of the electron. When the waist radius changes slightly, the intensity of the laser pulse changes accordingly. The waist radius of the pulse is the smallest at the focal plane, so the intensity of the laser is the largest here. The laser beam radius is much larger than at the focal plane when it is far away from the focus, and the laser intensity is obviously smaller than at the focal plane, too.

The electron is still placed in focus center of the tight focusing laser pulse, when the laser pulse propagation from left to right, the rising edge of the laser pulse accelerate the electron. Due to very close to the focus, the waist radius of the laser pulse is really small, so the laser pulse energy all convergence here. It is good for electron acceleration effect. The ponderomotive force pushes the electron away from the focus, and the radius of the laser beam is increasing, and the intensity of the laser is gradually reduced. When the laser decline interacts with the electron, because the laser is away from the focal plane, according to the characteristics of tightly focused laser beam, the beam radius is larger than before. It is the asymmetric phenomenon of electron acceleration and deceleration that makes the electron gain a certain energy after the laser pulse separated from the electron.

The schematic geometry of laser-electron interaction is depicted in Fig. 1. Here, we assume that the laser pulse propagates along the $+\hat{z}$ axis and an electron is initially stationary at the origin of coordinates. The electron is accelerated at the focus, then the energetic electron emits radiation in the direction $\mathbf{n} = \sin\theta\cos\phi\hat{x} + \sin\theta\sin\phi\hat{y} + \cos\theta\hat{z}$ [Fig. 1]. In this expressions θ and ϕ have the usual meanings as per the spherical coordinate system with the $\hat{z}$ axis aligned along the electron longitudinal motion.

3 THE INTERACTION FUNCTION

The motion of electrons in the electromagnetic field can be described by the Lagrange equation and the energy equation of the electron:

$$d_t(p - a) = -\Delta_a(u \cdot a) \tag{2}$$

$$d_t\Upsilon = u \cdot \partial_t a \tag{3}$$

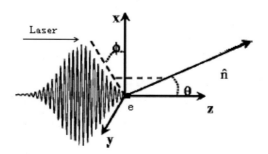

Figure 1. Schematic diagram showing the interaction of an incident circularly polarized tightly focused laser pulse with a stationary electron, we assume the laser field propagation along +z axis.

u is the velocity of the electron. $p = \Upsilon u$ is the momentum of the electron. a is the vector potential. They are separately normalized by c, mc, mc^2. $\Upsilon = \left(1 - u^2\right)^{-1/2}$ is a relativity factor, and it is also an electron energy normalized by mc^2.

(1) is the laser formula expressed in Descartes coordinates, then each component of the light field is written as:

$$a_x = a_L \cos(\eta), a_y = a_L \sin(\eta) \tag{4}$$

which

$$a_L = a_0 \exp\left(-\frac{\eta^2}{L^2} - \frac{x^2 + y^2}{b^2}\right) \times \left(1 + z^2 / z_r^2\right)^{-1/2}$$

The equation (4) is brought into the equation (2) and (3), and the following equations are obtained after the simplification.

$$\Upsilon d_t u_x = \left(1 - u_x^2\right)\partial_t a_x + u_y\left(\partial_y a_x - \partial_x a_y\right) + u_z \partial_z a_x - u_x u_y \partial_t a_y \tag{5a}$$

$$\Upsilon d_t u_y = \left(1 - u_y^2\right)\partial_t a_y + u_x\left(\partial_y a_x - \partial_x a_y\right) + u_z \partial_z a_y - u_x u_y \partial_t a_x \tag{5b}$$

$$\Upsilon d_t u_z = -u_x \partial_z a_x - u_y \partial_z a_y - u_z\left(u_x \partial_t a_x + u_y \partial_t a_y\right) \tag{5c}$$

$$d_t \Upsilon = u_x \partial_t a_x + u_y \partial_t a_y \tag{5d}$$

In which, u_x, u_y, u_z are the velocity components of electrons in the direction respectively. By solving these differential equations, we can get the curves of electron trajectories and electron energy gains over all the time.

4 RESULTS AND DISCUSSION

Figure 2 shows the electron trajectory [Fig. 2(a)] in 3D and the dimensionless particle normalized energy $\Delta\gamma$ [Fig. 2(b)]. The peak amplitude of laser pulse is $a_0 = 4.0$ (blue line), $a_0 = 5.0$ (red line), and $a_0 = 6.0$ (yellow line). The energy gain $\Delta\gamma$ is about 0.0585 ($a_0 = 4$), 0.332 ($a_0 = 5$) and 1.5618 ($a_0 = 6$). From Fig. 2(a), it can be seen that when the electron is separated from the laser, it will get a certain amount of energy. It is running along the direction of the laser propagation, and the collimation is better. With the increase of the laser pulse energy, the

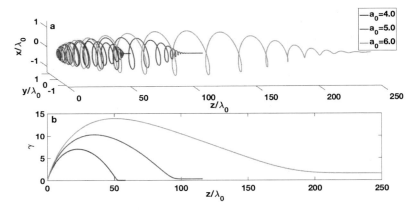

Figure 2. Electron trajectory in a circularly polarized tightly focused Gaussian laser pulse (a) and the dimensionless particle normalized energy gain $\Delta\gamma$(b) when a_0 is 4,5,6.

maximum distance of the electron radial motion increases and the acceleration distance is longer. From Fig. 1(b), we can also see that the acceleration distance of electrons is different under three different laser intensity pulses. The larger the laser intensity is, the faster the acceleration will be, and the larger energy gain of electron.

Figure 3 shows the 3D electron trajectories and the electronic dimensionless normalized energy when the peak amplitude of laser pulse is 7, 8 and 9 respectively. The blue line, the red line and the yellow line are the peak ranges of the laser pulse respectively. The energy gains $\Delta\gamma$ during the interaction shown in Fig. 3(b) is about 6.3738 ($a_0 = 7$), 16.1574 ($a_0 = 8$) and 26.8036 ($a_0 = 9$). When a focused laser pulse intensity increases up to 8, the deceleration of electron is not obvious. From Fig. 2(a), we can see that when the electron is separated from the laser, it gets some energy, but it doesn't move along the direction of laser propagation. It moves spiral around the axis of laser propagation. The scattering angle of the spiral motion is not easy to calculate exactly, which increases the difficulty of accurate control of high energy electrons after acceleration.

According to the theoretical results, it is known that the energy obtained after the electron accelerated is strongly dependent on the laser intensity. Fig. 4 shows the relation between the peak amplitude a_0 of the circular polarized laser and the energy gain $\Delta\gamma$ of the electron. The parameters of the laser pulse used in the simulation are: laser pulse width $a_0 = 5.0$, waist radius $b_0 = 5.0$. It can be seen from the diagram that the amplitude of the laser is different and the energy gain of the electron at the stationary point is definitely different. As a whole, the larger the intensity of the laser, the more the energy gain of the electrons. When the peak amplitude of the laser pulse is less than 4, the electron energy gain is almost zero. When it is more than 6, the electronic energy increases rapidly.

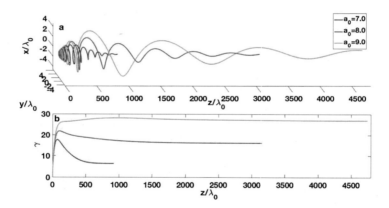

Figure 3. Electron trajectory in a circularly polarized tightly focused Gaussian laser pulse (a) and the dimensionless particle normalized energy gain $\Delta\gamma$ (b) when a_0 is 7,8,9.

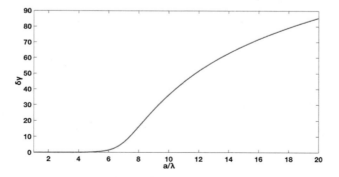

Figure 4. The relationship between the energy of the electron ejection and the laser pulse energy when a_0 is between zero and twenty.

5 CONCLUSIONS

In conclusion, the dynamics and full spatial distribution of radiated emission has been investigated numerically with a single electron model in the cases of different laser intensity of driving circularly polarized femtosecond tightly focused laser pulses. It shows that the angular distribution and the electron motion depends sensitively on the laser intensity of the driving tightly focused laser pulse. The full spatial distribution of radiation is different for different tightly focused laser pulse intensity.

The stationary electron trajectories and energy changes which is accelerated by tightly focused laser pulses are discussed in detail. When the peak amplitude of laser pulse is less than 5, we found that the stationary electron first accelerate and then decelerate, and ultimately do not gain energy. With the increasing of laser pulse intensity, the electron's energy is also increasing. When the peak amplitude of laser pulse is over 12, the increasing speed of the electronic energy gain slows down. It should be remembered in mind that if we want to control the electron accurately, not only take into account its energy, but also should consider the scattering angle of the electron. These issues will be discussed in a separate paper.

ACKNOWLEDGMENTS

This work has been supported by the National Natural Sciences Foundation of China under Grant No. 10947170/A05 and No. 11104291, Natural science fund for colleges and universities in Jiangsu Province under Grant No. 10KJB140006, Natural Sciences Foundation of Shanghai under Grant No. 11ZR1441300 and Foundation of NJUPT under Grant No. NY212080 and sponsored by Jiangsu Qing Lan Project and STITP Project under Grant No. XYB2013012.

REFERENCES

Guo H, Liu T, 1997. Relativistic plasma and laser wakefield accelerate nonlinear optics in electrons. Laser & Optoelectronics Progress. 11: 32–34.

He F, Yu W, Lu P X, Yuan X, Liu J R, 2004. Electron acceleration by a tightly focused femtosecond laser beamin vacuum. Acta Physica Sinica. 01: 165–170.

He F, Yu W, Xu H, Lu P X, 2005. Electron preheated acceleration by a relativistic femtosecond laser beam in vacuum. Acta Physica Sinica. 09: 4203–4207.

Huang S H, Wu F M, 2008. Focused laser pulse vacuum accelerated electron scheme for external electrostatic. Acta Physica Sinica. 57(12):7680–7684.

Kong Q, Zhu L J, Wang J X, Huo Y K, 1999. Dynamic characteristics of electrons in an ultea intense laser field. Acta Physica Sinica. 04: 93.

Luo W, Huang Y, Sun Y, An N, 2016. Developments of electron and proton acceleration using Laser-plasma Interaction. Opt Instr. 38(03): 278–282.

Tian Y W, Yu W, Lu P X, He F, Ma F J, Xu H, Jing G L, Qian L J, 2005. Electron capture and violent acceleration by a tightly focused ultra short ultra intense laser pulse in vacuum. Acta Physica Sinica. 09: 4208–4212.

Wang Z J, Chen S H, 1982. Talk about electron accelerated by laser. Chinese Science Bulletin. 13:780–783.

Xu T, B Shen, J Xu, S Li, Y Yu, 2016. Ultrashort megaelectron positron beam generation based on laser accelerated electrons. Physics of Plasmas. 23(3): 457.

Yin F, Tao X Y, 2011. Acceleration of relativistic electrons in Gauss laser field. Laser Technology. 35(03): 384–387.

Frontier Research and Innovation in Optoelectronics Technology and Industry – Habib & Lewis (Eds)
© 2019 Taylor & Francis Group, London, ISBN 978-1-138-33178-5

Infrared laser stealth based on anomalous reflection by phase gradient meta-surfaces

Cuilian Xu, Shaobo Qu, Yongqiang Pang, Jiafu Wang, Mingbao Yan, Yongfeng Li, Jun Wang & Hua Ma
Department of Basic Sciences, Air Force Engineering University, Xi'an, China

ABSTRACT: In considering 10.6 µm laser guidance stealth, this paper presents the design and characterization of a phase gradient meta-surface. Simulation results demonstrate that this design can redirect an impinging light into a single anomalous reflection beam with the same polarization. Compared to previous gradient meta-surfaces in infrared regimes, our samples exhibit much higher conversion efficiency (~84%) in the anomalous reflection mode at normal incidence and retain unchanged light polarization after the anomalous reflection. This design has potential applications in infrared laser stealth.

1 INTRODUCTION

With the rapid development of advanced detectors and precision guidance technology, stealth technology has become the most important and effective technique in penetration tactics (Cai et al., 2007; Grant et al., 2011; Z.X. Wang et al., 2014; W. Wang et al., 2015; Costantini et al., 2015). Stealth technology can effectively avoid aspects of exposure of weapon equipment targets to detectors. Stealth technology can be subdivided into several different types, including radar stealth, laser stealth, visible light stealth and infrared stealth. Laser detection actively detects and identifies targets using its own laser echo, and so to achieve laser stealth, materials with low or anomalous reflection can reduce signal detection.

Phase Gradient Meta-surfaces (PGMs) (Sun et al., 2012; Farmahini-Farahani & Mosallaei, 2013; Azad et al., 2013; Ding et al., 2013; Meinzer et al., 2014; Li et al., 2014, 2016) are artificial sub-wavelength-thickness anisotropic metallic arrays engineered with a spatial reflection/refraction phase response achieved by spatially arranging different subunit resonators. This response is realized according to the generalized version of reflection and refraction law (Snell's law). PGMs have attracted considerable research in recent years. Many novel and interesting phenomena have been demonstrated within the ultra-thin plane structure, such as anomalous reflection/refraction, surface wave excitation, flat lenses and waveplates. Using a PGM, the wavefront and polarization of the reflected/refracted waves can be manipulated with greater freedom. The key issues for designation of a substance as a PGM are phase response control and amplitude manipulation. Polarization conversion meta-surfaces, electric/magnetic resonators, and other resonators are employed to serve as the subunit resonators that manipulate the amplitudes and phases of the reflected/refracted waves in both terahertz and microwave regimes. In the polarization conversion meta-surface, a 2π phase manipulation can be easily achieved, and a dispersionless phase gradient can be derived over a wide operational bandwidth. However, achievement of anomalous reflection and transmission are usually accompanied by polarization conversion for the anomalously reflected/refracted waves, and this is undesirable for some specific applications.

Currently, the operating wavelengths of common military neodymium-doped Yttrium Aluminum Garnet (Nd^{+3}:YAG) lasers and CO_2 lasers are 1.06 µm and 10.6 µm, respectively, corresponding to near infrared and far infrared regions. At present, laser guidance missiles usually rely on 10.6 µm laser beams to search for objects. Aimed at 10.6 µm laser guidance stealth, our work proposes a polarization-independent resonator for the design of a reflective PGM. This can achieve wideband high reflectivity, and a 2π phase shift can be easily realized by changing its geometrical parameters. Based on this resonator, a co-polarization PGM is designed. The distributions of the electric field for anomalous reflection at wavelength $\lambda = 10.6$ μm are simulated. The proposed design has potential applications in 10.6 μm infrared laser stealth.

2 A HIGH-EFFICIENCY AND POLARIZATION-INDEPENDENT RESONATOR

The proposed resonator is illustrated in Figure 1, in which Figure 1a is a perspective view and Figure 1b the front view. The resonator depicted consists of a metallic pattern, a metal groundsheet and a dielectric spacer. The metallic layer employed was gold, following the Drude model for dielectric function (Xu et al., 2018) of $\varepsilon_m(\omega) = 1 - \omega_p^2/\omega(\omega + i\omega_c)$, with plasma frequency of $\omega_p = 2\pi \times 2.175 \times 10^3$ THz and collision frequency of $\omega_c = 2\pi \times 6.5$ THz. A MgF$_2$ layer is chosen as the dielectric spacer with dielectric constant of $\varepsilon = 1.892$. The repetition period of the unit is a, the thickness of the dielectric substrate is d, and the width of the metallic square patch is b.

For the specific metallic pattern, all the resonance frequencies are approximately linearly distributed about the frequency. The reflection phase of the resonator can be manipulated by changing the value of the geometrical parameter b. To investigate the resonance characteristics of the proposed resonator, numerical simulations were performed using CST Microwave Studio commercial software (Dassault Systèmes, Paris, France) to simulate the distributions of the surface currents (Ma et al., 2013; Liang et al., 2013; Kim et al., 2015). The values of the geometrical parameters for the unit cell used in the simulations were as follows: $a = 2.5$ μm, $b = 3.8$ μm and $d = 0.6$ μm. The distributions of the surface currents under y- and x-polarized wave normal incidence at wavelength $\lambda = 9.4$ μm are shown in Figure 2, where Figure 2a illustrates y-polarized wave incidence and Figure 2b depicts x-polarized wave incidence. From the figures, it is found that the surface currents are highly enhanced in the directions of the inci-

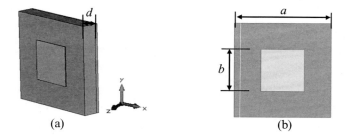

(a)   (b)

Figure 1. Schematic views of the proposed resonator: (a) perspective view; (b) front view.

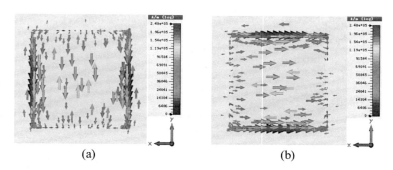

(a)   (b)

Figure 2. Surface current distributions of the resonator under: (a) y-polarized wave normal incidence at $\lambda = 9.4$ μm; (b) x-polarized wave normal incidence at $\lambda = 9.4$ μm.

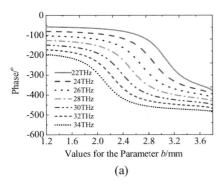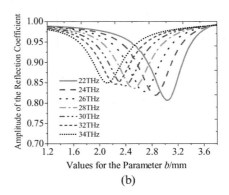

(a) (b)

Figure 3. Values of the parameter b for x-polarized waves normally illuminating the proposed resonators versus: (a) simulated phases of the reflection coefficients; (b) amplitudes of the reflection coefficients.

dent electric fields. The surface currents on the metallic pattern and the metal groundsheet form opposing current loops. Accordingly, strong magnetic resonances dominate. Moreover, due to the symmetrical characteristic of the structure, it has almost the same response to Transverse Electric (TE) and Transverse Magnetic (TM) polarized light for normal incidence.

Keeping the geometrical parameters a and d of the resonator unchanged, the resonance frequency will be altered by changing geometrical parameter b. To study the phase and amplitude modulation of the reflection coefficient for electromagnetic wave normal incidence, we simulated the reflection coefficients versus parameter b from 1.2 to 3.8 μm at different frequencies of x-polarized wave normal incidence. The values of the geometrical parameters a and d are 2.5 and 0.6 μm, respectively. The simulated results are shown in Figure 3, where Figure 3a shows the phases of the reflection coefficients versus the values of the parameter b at frequencies $f = 22, 24, 26, 28, 30, 32$ and 34 THz, and the corresponding amplitudes of the reflection coefficients are shown in Figure 3b. It is demonstrated that the phase coverage is different at different frequencies, with the least phase coverage being below 289° at $f = 34$ THz, and the widest phase coverage being up to 320° at $f = 22$ THz. Most importantly, for the resonators with different parameter b, the amplitudes for the reflection coefficients are all greater than 80% over a wide frequency range, from 22 to 34 THz. Therefore, a 2π phase manipulation and high-efficiency reflection can be simultaneously achieved using this resonator. The working frequency can move to a lower-frequency band as the dielectric spacer thickness d is increased.

3 DESIGN PRINCIPLES OF THE PHASE-MODULATED META-SURFACE

The quasi-periodic micro-structure units of the phase gradient meta-surface produce a linear distribution "phase change". The phase gradient $d\Phi/dx$ is added to the incident wave in the x-direction by the phase gradient meta-surface. This additional "artificial" phase gradient phase-modulates the reflected waves so that they no longer adhere to the traditional law of reflection and must establish generalized reflection laws (Snell's law) (Yu et al., 2011). As shown in Figure 4, a beam from point Q in medium 1 incident on the interface of mediums 1 and 2 with an incidence angle θ_i is reflected to reach point Q' in medium 1, where n_i and n_t are the refractive indices of mediums 1 and 2, respectively. There is a phase gradient meta-surface on the interface between the mediums. According to Fermat's principle, the light propagation path is the path which requires least time. By selecting two beam paths that are infinitely close to the actual beam path, as shown by the blue and red arrowed lines in the figure, the phase accumulation of the two beam paths is equal:

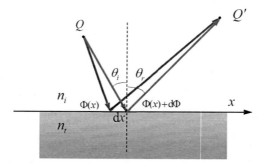

Figure 4.   Schematic diagram to describe the principle of PGM.

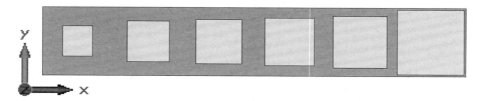

Figure 5.   Super unit of the designed reflective PGM.

$$\left(k_0 n_i \sin\left(\theta_i \right) dx + \Phi\left(x \right) + d\Phi \right) - \left(k_0 n_i \sin\left(\theta_r \right) dx + \Phi\left(x \right) \right) = 0 \qquad (1)$$

The general law of reflection can be obtained:

$$\sin\left(\theta_r \right) - \sin\left(\theta_i \right) = \frac{1}{k_0 n_i} \frac{d\Phi}{dx} \qquad (2)$$

Thus, the reflected wave can be anomalously reflected in any direction via accurate phase gradient design.

4   PGM DESIGNED FOR HIGH-EFFICIENCY CO-POLARIZATION ANOMALOUS REFLECTIVITY

According to the above analysis, a 2π phase manipulation and high-efficiency co-polarization reflectivity can be simultaneously achieved across a wide wavelength range using the designed resonator. Using the resonators as subunit cells, we design a PGM for wideband co-polarization anomalous reflection based on the generalized version of reflection law. The super unit of the PGM (Figure 5) is composed of six resonators with different geometrical parameters b. The phase differences $\Delta\Phi$ between adjacent subunit resonators are 60° at the wavelength $\lambda = 10.6$ μm. The simulated phases and amplitudes of the reflection coefficients for the six different resonators are presented in Figure 6, in which the reflection phase is an approximate linear function about the geometrical parameter b. High-efficiency reflections above 84% for all the resonators are observed, as shown in Figure 6b. The phase gradient of the PGM can be calculated by $\nabla\Phi = \Delta\Phi / a = \pi / 3_a$ on the basis of the generalized version of the reflection law:

$$\sin\theta_r = \frac{k_i \sin\theta_i + \nabla\Phi}{k_i} \qquad (3)$$

204

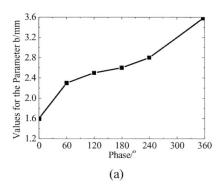
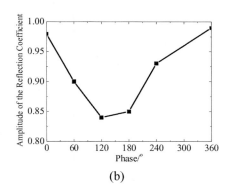

(a)   (b)

Figure 6.   Phase gradient design of the reflective PGM: (a) the six resonators used; (b) amplitudes of their reflection coefficients.

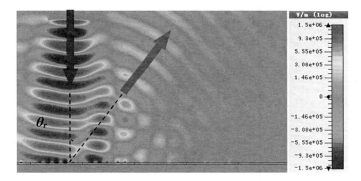

Figure 7.   Simulated distributions of the electric field for x-polarized wave normal incidence at wavelength $\lambda = 10.6$ μm.

where θ_i and θ_r are the incident and reflected angles, respectively, and k_i is the wave number of the incident wave $k_i = \lambda/2\pi$, in which λ is the wavelength of the incident wave. The value of the geometrical parameter a is 2.5 μm. At wavelength $\lambda = 10.6$ μm, the calculated anomalous reflected angle is $\theta_r = 33°$ for y- and x-polarized wave normal incidence.

To visualize the anomalous reflection of the designed PGM, numerical simulation is performed for the PGM with finite dimension (10×12 super unit arrays). The boundary conditions in x- and y-directions are all set as open. The x-polarized wave, which is achieved by using a finite port, normally illuminates the PGM from direction $+z$. The simulated distributions of the electric field component for x-polarized wave normal incidence at $\lambda = 10.6$ μm are illustrated in Figure 7, from which the anomalous reflection phenomenon can be clearly observed. In the figures, the propagation direction of the reflected wave is depicted by the red directional arrow. The anomalous reflection angle is about 33°, which is consistent with the theoretically calculated result.

5   CONCLUSIONS

In summary, a polarization-independent resonator is proposed to serve as the subunit resonators for the design of a reflective PGM, aimed at use in 10.6 μm laser guidance stealth. For this resonator, a total 2π phase shift and high-efficiency co-polarization reflectivity can be easily realized by changing its geometrical parameters. Based on this resonator, the phase profile design is only required for the reflective PGM designation, and as an example a reflective PGM is designed. At wavelength $\lambda = 10.6\,\mu m$, the electric field distribution is simulated. The simulation results show promising performance in high-efficiency co-polarization anomalous reflection. The proposed structure shows promising potential for application in infrared stealth.

ACKNOWLEDGMENT

The authors would like to thank the National Nature Science Foundation of China (nos. 61331005, 60501502 and 61501497).

REFERENCES

Azad, A.K., Taylor, A.J., Dalvit, D.A.R. & Chen, H.T. (2013). Terahertz metamaterials for linear polarization conversion and anomalous refraction. *Science*, *340*(6138), 1304–1307.

Cai, W., Chettiar, K.U., Kildishev, A.V. & Shalaev, V.M. (2007). Optical cloaking with metamaterials. *Nat Photonics*, *1*(4), 224–227.

Costantini, D., Lefebvre, A., Coutrot, A.-L., Moldovan-Doyen, I., Hugonin, J.-P., Boutami, S., … Greffet, J.-J. (2015). Plasmonic metasurface for directional and frequency-selective thermal emission. *Phys Rev Applied*, *4*(1), 014023.

Ding, K., Xiao, S.Y. & Zhou, L. (2013). New frontiers in metamaterials research: Novel electronic materials and inhomogeneous metasurfaces. *Front Phys*, *8*(4), 386–393.

Farmahini-Farahani, M. & Mosallaei, H. (2013). Birefringent reflectarray metasurface for beam engineering in infrared. *Opt Lett*, *38*(4), 462–464.

Grant, J., Shi, X., Alton, J.D. & Cumming, R.S. (2011). Terahertz localized surface plasmon resonance of periodic silicon microring arrays. *J Appl Phys*, *109*(5), 257401.

Kim, Y.J., Yoo, Y.J., Kim, K.W., Rhee, J.Y., Kim, Y.H. & Lee Y.P. (2015). Dual broadband metamaterial absorber. *Opt Express*, *23*(4), 3864–3868.

Li, Y.F., Zhang, J.Q., Qu, S.B., Wang, J.F., Chen, H.Y., Xu, Z. & Zhang, A.X. (2014). Wideband radar cross section reduction using two-dimensional phase gradient metasurfaces. *Appl Phys Lett*, *104*, 221110.

Li, Y.F., Zhang, J.Q., Zhang, Y.D., Chen, H.Y. & Fan, Y. (2016). Wideband, co-polarization anomalous reflection metasurface based on low-Q resonators. *Appl Phys A*, *122*(9), 1–6.

Liang, Q.Q., Wang, T.S.H., Lu, Z.W., Sun, Q., Fu, Y.Q. & Yu, W.X. (2013). Metamaterial based two dimensional plasmonic subwavelength structures offer the broadest waveband light harvesting. *Adv Opt Mater*, *1*(1), 43–49.

Ma, W., Wen, Y.Z.H. & Yu, X.M. (2013). Broadband metamaterial absorber at mid-infrared using multiplexed cross resonators. *Opt Express*, *21*(25), 30724–30730.

Meinzer, N., Barnes, W.L. & Hooper, I.R. (2014). Plasmonic meta-atoms and metasurfaces. *Nat Photonics*, *8*(12), 889–898.

Sun, S.L., Yang, K.Y., Wang, C.M., Juan, T.K., Chen, W.T., Liao, C.Y., … Tsa, D.P. (2012). High-efficiency broadband anomalous reflection by gradient meta-surfaces. *Nano Lett*, *12*(12), 6223–6229.

Wang, W., Fang, S.J., Zhang, L.P. & Mao, Z.P. (2015). Infrared stealth property study of mesoporous carbon-aluminum doped zinc oxide coated cotton fabrics. *Text Res J*, *85*(10), 1065–1075.

Wang, Z.X., Cheng, Y.Z., Nie, Y., Wang, X. & Gong, R.Z. (2014). Design and realization of one-dimensional double hetero-structure photonic crystals for infrared-radar stealth-compatible materials applications. *J Appl Phys*, *16*(5), 054905.

Xu, C.L., Qu, S.B., Pang, Y.Q., Wang, J.F., Yan, M.B., Zhang, J.Q., …Wang, W.J. (2018). Metamaterial absorber for frequency selective thermal radiation. *Infrared Phys. Techn*, *88*, 133–138.

Yu, N., Genevet, P., Kats, M.A., Aieta, F., Tetirnne, J.-P., Capasso, F. & Gaburro, Z. (2011). Light propagation with phase discontinuities: Generalized laws of reflection and refraction. *Science*, *334*(6054), 333–337.

Frontier Research and Innovation in Optoelectronics Technology and Industry – Habib & Lewis (Eds)
© 2019 Taylor & Francis Group, London, ISBN 978-1-138-33178-5

A novel frequency-locking scheme for optical-feedback cavity-enhanced absorption spectroscopy

Zanran Xu, Zhongqi Tan, Zhifu Luo, Jiayi Li & Siqi Liu
College of Advanced Interdisciplinary Studies, National University of Defense Technology, Changsha, Hunan, China

ABSTRACT: To deal with the frequency locking between the lasers with optical passive cavity in Cavity-Enhanced Absorption Spectroscopy (CEAS), a novel scheme is proposed and analyzed based on optical amplification technique and self-injection frequency-locking technique. A simplified model is established for this scheme, and the relation between the intensity of resonance light and the absorption coefficient of the gas sealed in the cavity is obtained in steady-state status. By comparing the simulation results of 100 ppm concentration CH_4 with the theoretical results, the feasibility of the proposed scheme is verified.

Keywords: cavity-enhanced absorption spectroscopy; optical amplification; self-injection frequency-locking

1 INTRODUCTION

Traditional chemical detection methods have been widely applied to gas concentration detection in practice, such as monitoring of atmospheric environment, detection of industrial gas emission and diagnosis of disease although (Lexing Wang & Yangqing Liu, 2002). However, on the other hand, a newly-emerging methd, i.e. laser spectroscopy is developing rapidly in recent years (Lili Feng, 2015). The laser spectroscopy techniques are based on the detection of the interaction between the material and the light field. With these techniques, the composition and content of atoms (molecules) can be obtained by analyzing the absorption and emission spectra of electromagnetic waves (E. Wahl, 2006 & D. Yamano & C. Wang, 2009 & B. Gao, 2010). The direct absorption techniques, i.e. cavity ring-down spectroscopy (CRDS) and its extended techniques have significantly higher sensitivity than conventional absorption spectroscopy, and they have been extensively researched and widely applied in many fields (Shixin Pei, 2005).

Compared with CRDS, cavity-enhanced absorption spectroscopy (CEAS) has more advantages in terms of high-spectral resolution and simple structure. The essential difference between CRDS and CEAS is that the resonance between the laser and cavity mode is never broken for CEAS. In CEAS technique, the laser is injected into a high-finesse optical resonance cavity. The intensity of resonance light can be recorded during the scanning, and the absorption spectrum of gaseous species sealed in the cavity is obtained. As mentioned above, it can be found that the frequency locking between the laser and the optical cavity is critical to CEAS, and the traditional scheme uses the way of changing the laser wavelength or scanning the cavity length. In 1999, D. Romanini et al. proposed a novel frequency-locking scheme using the inject-locking effect of the V-shaped cavity, and this scheme had significantly improved the CEAS performance of apparatus. In fact, when the inject-locking effect occurs, laser line width is narrowed and its frequency is locked on the resonance frequency of V-shaped cavity. The resonance status of the locked laser frequency can maintain for a short time, so the recorded signal has a higher signal to noise ratio.

Therefore it is quite significant to find the way to lock the frequency of laser with cavity mode and increase the resonance time in CEAS technique. In this paper, a novel frequency-locking scheme is proposed and analyzed based on optical amplification technique and self-injection frequency locking technique. And the stimulation results provide the theoretical basis and technical preparation for the feasibility of high-resolution CEAS.

2 EXPERIMENTAL SCHEME

Based on the principle of self-injection frequency-locking technique and optical amplification technique, a novel frequency-locking scheme for CEAS is proposed. The schematic of the proposed scheme is shown in Fig. 1.

It can be found from Fig. 1 that a distributed feedback (DFB) laser is used as the light resource and the V-shaped cavity acts as the absorption cell, so that the advantages of the optical-feedback CEAS are retained. In particular, the V-shaped cavity has the advantages of high coupling efficiency and can avoid the false lock between the frequency of laser and the mode of resonant cavity. Obviously, V resonator is still used as the core component in this scheme (Zhongqi Tan, 2009). Furthermore, to further improve the performance of frequency locking between the laser and the cavity, an optical amplifier is introduced into the scheme. To realize the self-injection and the amplification of the laser, an optical switch is applied to switch the optical-feedback resonance light and optical amplifier resonance light.

The specific workflow of this scheme is as follows. Firstly, with the control of a computer, the optical switch brings the light from the narrow lin-width DFB laser (2 MHz) into the V-shaped cavity. In this case, the DFB laser can be considered as the "ignition" light source. When the laser is coupled into the resonator successfully via scanning the length of the V-shaped cavity by the piezoelectric ceramic, the optical-feedback effect will occur. After that, the amplitude of the resonant light is amplified by the optical amplifier, but the laser wavelength remains unchanged. The optical switch leads the output of optical amplifier into the V-shaped resonant cavity within tens of nanoseconds, so an active ring resonant optical path emergs. In order to make the resonant light of the V-shaped cavity satisfy the resonance condition of the annular optical path, a Piezo actuator is adopted which also controls the phase of feedback light in the optical feedback mode. Finally, the absorption spectrum of the gas sealed in the V-cavity is determined by measuring the intensity of the resonant light of V-shaped cavity, and the scanning of laser wavelength is achieved by changing DFB laser or scanning the length of V-cavity. In this scheme, to ensure that the light in laser will not fade away before the optical-feedback frequency locking is formed (the light switch conversion is formed), the decay time of the V cavity is required to be greater than that of the optical switch, which is also an indispensable condition for the formation of a stable resonance.

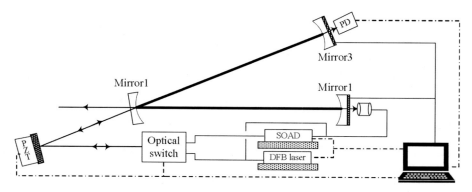

Figure 1. Schematic diagram of self-injection locked CEAS.

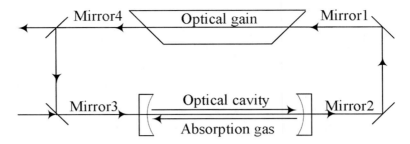

Figure 2. Simplified model of the proposed scheme based on optical amplification.

3   SIMULATION MODEL

The simplified model of the proposed scheme based on optical amplification technique is established, as shown in Fig. 2. The model is composed of a resonator optical amplifier, a reflector, and an auxiliary optical path.

Four mirrors are adopted here to investigate the influence of other optical elements in the system on the intensity of light. Their reflectivity reflects the ratio of the beam splitters, the coupling ratio of the laser to the resonant cavity when the laser enters the resonator, the coupling ratio of the laser from the cavity ejection to the optical amplifier, and the losses caused by roads and devices. The relation between the amplitude of the incident light and the amplitude of the emitted light is calculated using the principle of multi-beam interference

$$E = \frac{E_0 \cdot (1 - R) \cdot e^{-\alpha \frac{l}{2}}}{1 - R^2 \cdot e^{-2\alpha l}} \tag{1}$$

where E_0 = the light that enters the cavity for the first time when the light intensity is stable; R_i = the reflectivity of mirror i; l = the length of the cavity; and α = the absorption coefficient of the gas sealed in the cavity.

A is the coefficient of optical amplifier. I and I_s represent the input light intensity and its saturated light intensity, respectively. The small signal gain is g_0. The laser amplification under steady conditions can be obtained by:

$$A = \frac{g_0}{1 + I/I_s} \tag{2}$$

Considering of the various losses in the optical path, we replace the various losses as $R_1 \sim R_4$. In this case, the self-consistent equation can be obtained by passing a whole round in the optical path, and the function of resonant output light intensity of the ring optical path can be expressed as:

$$I = \frac{I_s \cdot g_0 \cdot (1 - R)^2 \cdot e^{-\alpha l} \cdot R_1 \cdot R_2 \cdot R_3 \cdot R_4}{(1 - R^2 \cdot e^{-2\alpha l})^2} - I_s \tag{3}$$

Using Eq. (3), we can analyze the variation in the steady light intensity I with the absorption coefficient of the medium in the cavity.

4   SIMULATION RESULTS AND DISCUSSIONS

Simulation experiment was conducted on CH_4. The absorption coefficient of CH_4 with a concentration of 100 ppm in the range of 6600~6620 cm^{-1} was acquired from the HITRAN database. Fig. 3 shows the changing of absorption coefficient with the laser's wave-number. It can be found that CH_4 has two strong absorption peaks at the wave-numbers of 6605.4 cm^{-1} and 6612.6 cm^{-1}.

Based on Eq. (3), the CEAS of CH_4 was simulated in this case. Firstly, some relevant parameters were initialized. To be more specific, the saturation output power of optical amplifier was about 10 dbm at the wavelength of 1512 nm, and the gain of the small signal was about 25 dB (nearly 320 times). Therefore, I_s and g_0 in the theoretical model were 0.01 W and 320, respectively. In the simplified model, R_1, R_2, R_3 and R_4 were set to 0.9, 0.8, 0.8, and 0.8, respectively, and the ring resonance optical path length was set to 0.08 m.

The relation between the steady-state light intensity and the laser wave-number was obtained, and is the result is shown in Fig. 4. It can be found that there are two absorption peaks in the simulated steady-state CEAS result under the same condition, and it is completely consistent with the results by HITRAN database.

Generally, the absorption coefficient of gas depends on specific state. In practical applications, the absorption spectrum of the gaseous medium in V-shaped cavity can be obtained by measuring the resonant intensity of the cavity with the change of the laser wavelength using CEAS technique, and then the concentration can be determined. Therefore, according

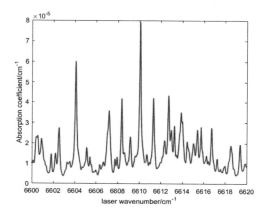

Figure 3. Absorption coefficient of CH_4 with a concentration of 100 ppm in the range of 6600~6620 cm^{-1}.

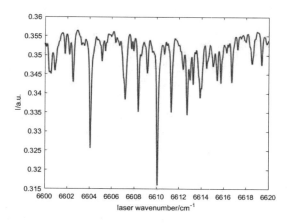

Figure 4. Simulated CEAS result of 100 ppm CH_4.

to the relation between the resonant signal intensity and the laser wave-number obtained in simulation, it can be concluded that the proposed scheme can effectively measure the CEAS of gaseous species sealed in V-shaped cavity, and obtain the gas concentration based on the measured results.

In order to verify the effectiveness of the model and the reliability of the obtained results, an extreme case is considered. That is, the concentration of CH_4 is set as zero, which means there is no absorption loss in the V-cavity. Then, the obtained simulation results in this case will not show an absorption curve. In fact, the simulation results are true, and CEAS curve is a straight line. These results verify the feasibility of the proposed scheme from the other perspective.

5 CONCLUSION

To improve the performance of optical-feedback CEAS, a novel frequency-locking scheme based on optical amplification technique is proposed and analyzed. A simplified model is established to verify the feasibility of the proposed scheme. The results indicate the potential application of the proposed scheme in CEAS measurement.

ACKNOWLEDGEMENT

This work is sponsored by the Research Project of National University of Defense Technology (No. ZK16-03-21).

REFERENCES

Gao, B., W. Jiang, and A. Liu et al 2010, Ultrasensitive near-infrared cavity ring-down spectrometer for precise line profile measurement, Rev. Sci. Instrum, 81, 043105.

Le-xin Wang, Zhi-ming Zhao, Hong-bing Yao, Yu-ming Chen, Lei Shi, and Yong Hao, 2002, Analysis and application of infrared absorption spectra of blood, Spectroscopy and Spectral Analysis, 22(6):980.

Li-li Feng, 2015, Cavity fading spectroscopy enhancement technology and its application research, xi 'an optical precision machinery research institute,Master's Thesis.

Shi-xin Pei, Xiao-ming Gao, Fen-ping Cui, Wei Huang, Jie Shao, Hong Fan, and Wei-jun Zhang, 2005, Study on cavity enhanced absorption and high sensitive absorption spectra of CO2, Spectroscopy and Spectral Analysis, 25(12): 1908–1911.

Wahl, E., S. Tan, S. Koulikov, B. Kharlamov, C. Rella, E. Crosson, D. Biswell, and B. Paldus, 2006, Ultra-sensitive ethylene post-harvest monitor based on cavity ring-down spectroscopy, Opt. Express, 14, 1673–1684.

Wang, C. and P. Sahay, 2009, Breath analysis using laser spectroscopic techniques: breath biomarkers, spectral fingerprints, and detection limits, Sensors,9, 8230–8262.

Yamano, D., Y. Sakamoto, A. Yabushita, M. Kawasaki, I. Morino, and G. Inoue, 2009,Buffer-gas pressure broadening for the 2v3 band of methane measured with continuous-wave cavity ring-down spectroscopy," Appl. Phys. B. 97, 523–528.

Yang-qing Liu, Rong Yuan, Hui-hui Zhao, Gong Bai, Hui-fang Yang, 2002, The Comparison of infrared absorption characteristics between chicken bile and common animal bile, Spectroscopy and Spectral Analysis,33(3):264–266.

Zhong-qi Tan, 2009, Research and application of new laser absorption spectroscopy technology based on high quality optical passive cavity, National University of Defense Technology.

Frontier Research and Innovation in Optoelectronics Technology and Industry – Habib & Lewis (Eds)
© 2019 Taylor & Francis Group, London, ISBN 978-1-138-33178-5

Optimal design of the emission device used for an anti-tank missile laser simulator

P.P. Yan, Y.Q. Yang & W.J. She
The Photoelectric Measurement and Control Technology Research Department, Xi'an Institute of Optics and Precision Mechanics, Chinese Academy of Sciences, Xi'an, China

ABSTRACT: A laser has the characteristics of good monochromaticity, good directivity, good coherence, and high-brightness. Because of these, laser guidance technology plays a more and more important role in modern war. Besides, with the improvement of science and technology, laser simulators can simulate the targeting effects of the real firing weapons as exactly as possible. They have not only achieved their purpose in training, but have also cut down the expenses. In order to achieve the actual effect and imitate the fighting process in practical training, we need to develop a realistic simulator. This paper designs an emission device of an anti-tank missile laser simulator. It consists of a fiber-coupled laser diode with 915 nm wavelength, a collimating optical system which can compress the divergence angle from 218 mrad to 6.5 mrad, a zoom collimating and beam expanding system, and a rotary dual-wedge system. The advantages of this system are that it is simple, and can be easily installed and adjusted. It can simulate the return signals of the pulse laser simulator when it finds the range for the targets with different characteristics at different distances under various atmospheric conditions.

Keywords: laser simulator, emission device, optical design, collimating and beam expanding

1   INTRODUCTION

Science and technology have been developing rapidly in modern times. Meanwhile, the modernization of national defense has been developing continuously. Along with the extensive use of advanced and new technology, such as automatic control and computer techniques in the military training field, laser simulators have been born which are applied to various kinds of weapons (Simpson, 2008; Arcus, 2003; Chen & Speyer, 2008). Because of the improvement of science and technology, laser simulators can simulate the targeting effects of the real firing weapons as exactly as possible. They can not only achieve their purpose in training, but also cut down the expenses. In order to achieve the actual effect and imitate the fighting process in practical training, we need to develop a realistic simulator.

By directing at the actual needs, this paper designs and realizes an anti-tank missile laser simulator. It can solve the problem that anti-tank missile training counterwork cannot be popularized at present. It also provides the basis for evaluating the attack ability of the weapons scientifically, so as to provide exact inspection results for the training. This simulator is stable and safe. It is close enough to the operating mode and the attack effect of a real weapon. It roundly improves the actual combat training level.

2   THE EMISSION DEVICE

The emission device is a very important part of the laser simulator. Figure 1 is the schematic diagram of a laser simulator (Yan et al., 2010). It consists of a fiber-coupled laser diode with

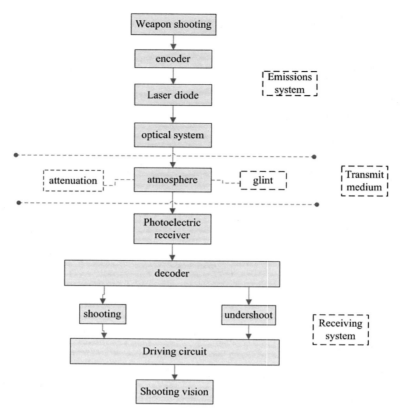

Figure 1.   The schematic diagram of a laser simulator.

a 915 nm wavelength, a collimating optical system which can compress the divergence angle from 218 mrad to 8 mrad, a zoom collimating and beam expanding system which can keep the beam size the same at different distances, and a rotary dual-wedge system which can actualize ballistic trajectory simulation.

2.1   *The laser diode*

The laser is the core part of the whole simulator. According to its working principle, it can be divided into several categories: solid state, gaseous state, liquid state, and semiconductor. To contrast between a fiber-coupled laser diode and a pulsed laser, here we choose the fiber-coupled laser diode. It has better performance in peak power, average power of duty cycle, and heat generation. The wavelength is 915 nm. The output power is 10 W. The N.A. is 0.22. The diameter of the fiber is 105 um. The layout of the laser diode is as follows.

2.2   *The collimating optical system*

The beam which is emitted from the laser diode is not completely parallel light. It has a certain divergence angle. The divergence angle is usually as the important parameter of laser diode. The divergence angle of the laser is 25°. Based on the required precision and actual detection background, it should be reduced to less than 8 mrad. So, we need a collimating optical system. The collimating optical systems mainly include single-lens, lens groups, reflection methods, and diffraction methods. In comparison, the single-lens method has a simpler structure and a lower cost. It can be used in laser simulators.

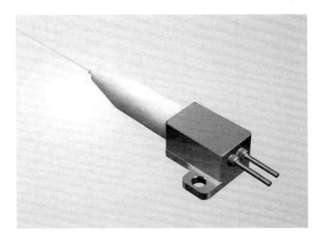

Figure 2.   Layout of the laser diode.

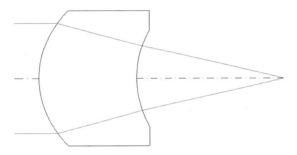

Figure 3.   Layout of the collimating optical system.

The Gaussian beam can be used to assume the beam emitted by the laser diode (Yan et al., 2010; Zhong & Du, 2000; Zhou et al., 2007; Yushu, 1999). According to the characteristics of the Gaussian beam, the beam is divergent during propagation. The half divergence angle is shown in Equation 1.

$$\theta = \frac{dW(Z)}{dZ} = \frac{\lambda^2}{\pi W_0} \frac{1}{\sqrt{\pi^2 W_0^4 / Z^2 + \lambda^2}} \tag{1}$$

where Z is the transmitting distance, W(Z) is the spot radius, λ is the wavelength, and W_0 is the Gaussian beam waist radius. When the transmitting distance $Z \to \infty$,

$$\theta' = \frac{\lambda}{\pi} \sqrt{\frac{1}{W_0^2}\left(1 - \frac{d_1}{f}\right)^2 + \frac{1}{f^2}\left(\frac{\pi W_0}{\lambda}\right)^2} \tag{2}$$

where d_1 indicates the position of the beam waist on the optical axis. Equation 3 shows that the collimation effect is good when the back focus of the lens and beam waist position are equal. Therefore, Equation 3 is an important basis for designing a laser transmitter optical collimation device.

Figure 3 is the single-lens collimating system. We use an aspherical lens. The focal length is 8 mm. After passing through the designed aspherical lens, the 25° divergence angle of the laser diode is compressed to 6.5 mrad respectively.

The Modulation Transform Function (MTF) is perhaps the most comprehensive of all optical system performance criteria. The MTF tells us how well the modulation in the object is transferred to the image by the lens. Figure 4 shows the MTF curve of this optical system. We can see that the quality of the image approaches the diffraction limited.

2.3 *The zoom collimating and beam expanding system*

Although the divergence angle is reduced to 8 mrad, the beam size will increase as the distance increases. We should keep the beam size the same at different distances. Compared with other beam expanding systems, the zoom beam expanding system has the advantage of an adjustable beam expanding ratio, so it has a wide range of applications in fields such as holography, laser ranging, and optical communication.

The zoom beam expanding system can be divided into a two-element zoom lens, a three-element zoom lens, and a four-element zoom lens, according to the number of elements (Kong & Hao, 2001; Liu et al., 2010). After analyzing the two-element, three-element, and four-element zoom systems, we choose the three-element zoom system, consisting of the zoom group, fixed group, and compensation group. The advantages of this system are that the continuous zooming can be achieved, system aberrations are small and the total system length is short. The specific system structure is shown in Figure 5.

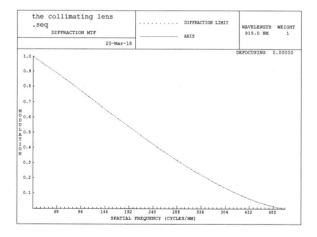

Figure 4.   Modulation transform function curve of the collimating optical system.

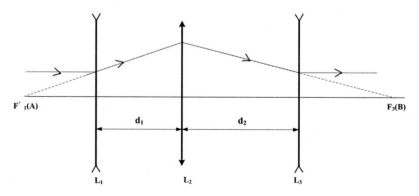

Figure 5.   The zoom lens with three elements.

Where L_1 is the variator, L_2 is the fixed group, L_3 is the compensation group, d_1 is the distance between the main surface of L_1 and L_2, d_2 is the distance between the main surface of the L_2 and L_3. When collimated parallel beams are incident on L_1, if the image points of L_2 and the front focal point F_3 of L_3 coincide, the outgoing light beam is parallel to the optical axis. In order to change the aperture of the outgoing beam to achieve the purpose of changing the zoom ratio, it is necessary that if the moving distance of L_1 is q, then the compensation group should move distance e accordingly to make the front focus F_3 of L_3 and the image point B' of L_2 superposed, as Figure 6 shows.

By analysis of the three-element zoom lens, the movement of the variator, fixed group and compensator are derived. We design a zoom beam expander which has the magnification from $1 \times$ to $10 \times$, as shown in Figures 7 and 8. The theoretical results are tested. Using the optimization function of the CODEV, the aberration of the system can be satisfied.

The shift magnification curves of L_1 and L_3 are shown in Figure 9. The ordinate is the expansion ratio and it also represents the position of L_2. The abscissa is the distance between L_3 and L_2, and the distance between L_1 and L_2 at different expansion ratios. As can be seen from Figure 3, the movement of L_1 is linear motion and the movement of L_3 is non-linear motion.

By analyzing the wave-front aberration (OPD) of the system at different expansion ratios M, the RMS wave aberration of the system is 0.0017λ when M = 1 and 0.0021λ when M = 10. All of the RMS are less than $\frac{\lambda}{4}$. The image quality of the system is good.

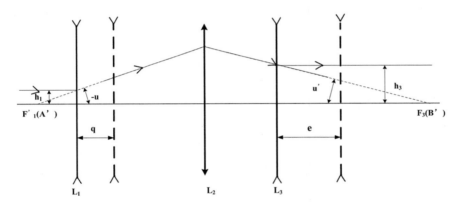

Figure 6.   The theory of a zoom lens with three elements.

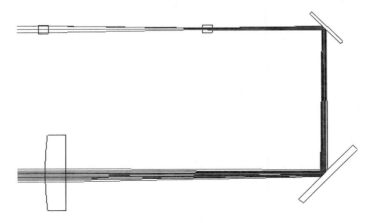

Figure 7.   The 2D layout of the zoom system (M = 1).

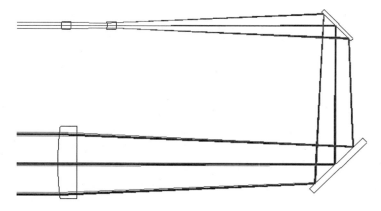

Figure 8. The 2D layout of the zoom system (M = 10).

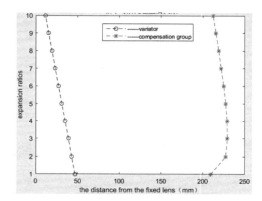

Figure 9 The practical curve of L_1 and L_3's movement.

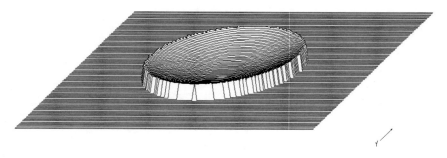

Figure 10. The pupil map of the zoom system (M = 1).

Figure 12 shows the 3D layout of the whole optical system. Figure 13 is the prototype of the emission device. Figure 14 is the testing process of the emission device. We can learn from the final result that the structure of this system is simple. It has a high performance which could be used in practical applications.

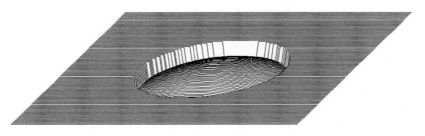

Figure 11.　The pupil map of the zoom system (M = 10).

Figure 12.　The 3D layout of the optical system of the emission system.

Figure 13.　The prototype of the emission device.

Figure 14.   The testing process of the emission device.

3   CONCLUSION

This paper designs a prototype of an emission device of an anti-tank missile laser simulator. We discuss the compartments of the emission device and analyze every part of it. It consists of a fiber-coupled laser diode, a collimating optical system, a zoom collimating and beam expanding system, and a rotary dual-wedge system. Here we analyze the collimating principle. According to it, we design a single-lens collimating system. In order to achieve the expansion of the laser beam, we give the analysis of the three-element zoom lens. By analyzing the movement of the variator, fixed group and compensator, we design a zoom collimating and beam expanding system which has a magnification from $1 \times$ to $10 \times$. The advantages of this system are that it is simple, and can be easily installed and adjusted.

REFERENCES

Arcus, P. (2003). American national missile defense system, *Space Policy*, *19*(1), 7–13.
Chen, R.H. & Speyer. J. (2008). Terminal and boost phase intercept of ballistic missile defense. *Navigation and Control Conference and Exhibit, Beijing* (pp. 28–39).
Kong, X. & Hao, P. (2001). Optical design of zoom beam system. *Applied Optics*, *22*(5), 7–11.
Liu, H, An, Z. & Gao, Y. (2010). Design of a varifocal beam expander system. *Journal of Changchun University of Science and Technology (Natural Science Edition)*, *33*(4), 19–20.
Simpson, J. (2008). Shoot-down key to survival for airborne laser program. *Inside the Air Force*, *19*(13), 10–12.
Yan, M.L., Shan, X.Q. & Qu, Z. (2010). Laser active imaging-guided anti-tank missile system small-scale integration design. The 5th International Symposium on Advanced Optical Manufacturing and Testing Technologies-Optoelectronic Materials and Devices for Detector, Imager, Display, and Energy Conversion Technology, Dalian.
Yushu, Z.X.Y.C.Z. (1999). Collimation of semiconductor laser beams. *ACTA OPTICA SINICA, 3*.
Zhong, C.X. & Du, C.L. (2000). Optical design of collimating and focusing optical system for semiconductor diode array. *Optical Instruments*, *22*(6), 25–29.
Zhou, B.K., Gao, Y.Z. & Chen, C.R. (2007). Laser principle. *Beijing National Defence Industry Press*.

Frontier Research and Innovation in Optoelectronics Technology and Industry – Habib & Lewis (Eds)
© *2019 Taylor & Francis Group, London, ISBN 978-1-138-33178-5*

Mode effects on supercontinuum generation from 15/130 large mode area fiber

Xin Yang, Sheng-Ping Chen & Zong-Fu Jiang
College of Advanced Interdisciplinary Studies, National University of Defense Technology, Changsha, Hunan, China

ABSTRACT: A 15/130 Large Mode Area (LMA) fiber amplifier is constructed for supercontinuum generation with a mechanical Long Period Grating (LPG) to control the transverse modes. LP_{01} and LP_{11} modes can be generated respectively from this amplifier. Obvious difference of supercontinuum generation between the two modes is experimentally observed, due to their different dispersion characters. Strong mode interaction is observed when exciting both of the two modes, which generates a series of new spectrum components due to intermodal nonlinear interaction, providing a way to flatten the gaps between different orders of Raman peaks so as to get a flatter supercontinuum spectrum. Supercontinuum covering the near infrared range from 800 nm to 2000 nm is obtained in this case.

1   INTRODUCTION

The nonlinear effects in fibers have gathered an increasing attention from fundamental research to practical applications. Dispersion interaction, stimulated Raman scattering (SRS), self-phase modulation (SPM), four waves mixing (FWM), modulation instability (MI), are widely used in fiber lasers and supercontinuum generation (SCG). However, these nonlinear effects are concentrated on the fundamental mode. Recently, the nonlinear phenomena within high order modes in few-mode fiber as well as multi-mode fiber has attracted considerable attention owing to their promising applications in space-division multiplexing (SDM) (DJ. Richardson, 2013) in optical communication, creation of optical vortices (J. Hanmazal, 2010), and power scaling (S. Ramachandran, 2006) of fiber lasers. The multimode solitons (LG. Wright), SRS of high order modes (I. Flammer, 2004), intermodal FWM (C. lesvigne, 2007 & A. Bendahmane, 2018), cross-phase modulation (T. Chaipiboonwong, 2007), Kerf-crselleaning (R. Guenard, 2017 & Liu Z, 2016), to name just a few, are considered when exciting high order modes. However, most of them are performed in multi-mode fiber with mixing modes pumping (MA. Eftekhar, 2017 & ZS. Ezmaveh, 2017), focusing on exploring the intermodal effects. As for the nonlinear effects of a single high order mode, most schemes are performed in PCFs which have special dispersion and nonlinear characters (R. Cherif, 2008 & Liu T, 2014). Besides, the output power is milliwatt level because of the low coupling efficiently by off-set pumping to excited different high order modes. Relatively less effort was made to exploit the clear process of nonlinear effects in common step-index fibers. The difference between the fundamental mode and high order modes in dispersion characters, the effective mode field area and cut-off wavelength, all play significant roles in laser propagation in fibers.

The purpose of this study concentrates on the mode effects on supercontinuum generation in 15/130 large mode area (LMA) step index fibers. In order to investigate the mode effects, a mechanical long period grating (LPG) is used as a mode converter. LP_{01} and LP_{11} modes can be generated respectively from this mode converter. Obvious difference of supercontinuum generation between the two modes is experimentally observed and analyzed.

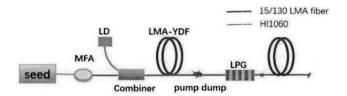

Figure 1. Schematic of the experiment. MFA, mode field adapter. LD, laser diode. LMA-YDF, largemodearea ytterbium doped fiber. LPG, long period grating.

2 EXPERIMENTAL SETUP

The schematic diagram of the experimental setup is shown in Fig. 1. The seed laser is a picosecond fiber laser with pulse width of ~330 ps, central wavelength of 1060 nm, and repetition frequency of 1 MHz. The average output power of the seed is 23 mW. The pigtail fiber of this seed laser is HI1060 fiber, which is a single-mode fiber at the operating wavelength. A mode field adapter with input fiber of HI1060 and output fiber of 15/130 is arranged between the seed laser and the fiber amplifier to ensure LP_{01} mode seeding of the amplifier. By using a combiner, the seed and pump laser are coupled in to a 4-mlong 15/130 LMA cladding-pumped ytterbium doped fiber (YDF) which is used as an amplifier to provide sufficient signal power for supercontinuum generation. After the YDF, there is a 1-mlong 15/130 LMA passive fiber that can be stressed by the mechanical LPG to realize mode conversion (I. Giles, 2012). By applying an appropriate pressure to the mechanical LPG, the fundamental LP_{01} mode can be coupled into the LP_{11} mode effectively when the modal beat length between the LP_{01} and the LP_{11} modes $L_B = \frac{2\pi}{\beta_{LP_{01}} - \beta_{LP_{11}}}$ is equal to the Bragg wavelength of the mechanical grating, that is $\Lambda = L_B$. Compared with other mode selection elements such as polarization component, spatial light modulators and the photonic lanterns, the mechanical LPG is easy to be fabricated and has high transfer efficiency between LP_{01} and LP_{11} modes. A pump dumping section is arranged between the YDF and the 15/130 passive fiber to let the residual pump light out. Another section of 15/130 LMA passive fiber with length of 11 m and 170 m is fusion spliced respectively to the LPG fiber to enhance the nonlinear effects. The fiber output end is angle-cleaved to avoid harmful back-reflection.

3 RESULTS AND DISCUSSION

The switching between different modes is implemented by applying/releasing the pressure on the mechanical LPG. It can be switchable between the LP_{01} and the LP_{11} mode. The output power is almost the same between the two modes, which means the high transfer efficiency of the mode converter (Liu T 2016). The average output power measured after the LPG-reaches8.18 W at the maximum pump power, corresponding to a peak power of about 20 kW. Because of the high peak power, nonlinear effects occur, leading to the broadening of the signal spectrum. Fig. 2 shows the laser spectrum at the output port of the LPG. Clear Raman peak can be seen when releasing the pressure to get the fundamental mode output. While the spectrum becomes smoother on both sides of the central wavelength when adjusting the mechanical LPG to get the LP_{11} mode.

In this study, large-mode-area double-clad fibers (LMA-DCFs) are utilized as the nonlinear medium. The core diameter, numerical aperture (NA), and normalized V parameter at the operating wavelength (1060 nm) of the LMA fibers are 15 μm, 0.08 and 3.56 respectively. Both the LP_{01} and the LP_{11} mode are supported in such few-mode fibers. The cut-off wavelength of the LP_{11} mode is 1625 nm. The group velocity dispersion (GVD) profiles and the group velocity curves of the LP_{01} and the LP_{11} modes in the near infrared ranges are numerically calculated by using the fiber intrinsic equation, which is shown in Fig. 3. It can be found that the zero dispersive wavelengths (ZDW) of the LP_{01} mode and the LP_{11} mode

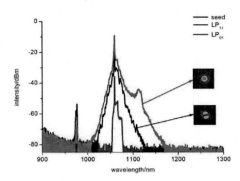

Figure 2. Laser spectrum after the LPG of the amplifier at 8.18 W average output power of LP01 and LP11 mode. The inset shows the output facula of both modes.

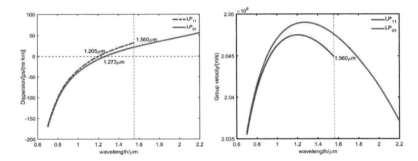

Figure 3. The group velocity dispersion (GVD) profiles (a) and the group velocity curves (b) of the LP_{01} and LP_{11} modes in the near infrared ranges.

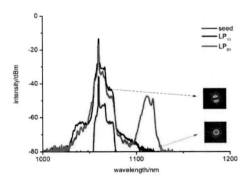

Figure 4. The output spectra of the LP_{01} and the LP_{11} mode at the same exciting power (average power:0.995 W) after 11 m 15/130 LMA-GDF transmission.

is 1205 nm and 1273 nm respectively. The effective mode field area and nonlinear coefficient of both modes are calculated. The second-order mode has a relatively larger effective mode field area of 205 μm^2, comparing to the value 136 μm^2 of the fundamental mode, therefore the LP_{11} mode has a lower nonlinear coefficient of 0.869 W/km compared to1.306 W/km of the fundamental mode. The differences in dispersion and nonlinear coefficient between the two modes suggest their performance in the fiber would be different, which will be illustrated in the followed paragraphs.

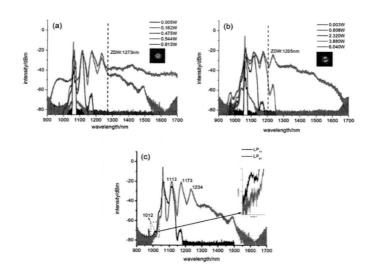

Figure 5. The output spectral of the LP01 (a) and LP11 (b) mode at the different exciting power after 170 m 15/130 LMA-GDF transmission. (c) is the comparison diagram of the LP01 and LP11 mode when the anti-stokes begin to appear, the average exciting power is 0.544 W and 2.320 W respectively.

The experiments were performed by changing the length of the GDF. Firstly, an 11 m 15/130 LMA-GDF is fusion spliced to the pigtail fiber of the mode converter. In order to reduce the bending loss, the fiber is coiled at a large diameter of 18 cm. The bending loss for such length of fiber in this condition is estimated to be neglectable by using the method in ref (Liu T, 2016). Fig. 4 shows the output spectral of both the LP_{01} and LP_{11} mode at the same exciting power. Comparing the nonlinear effects between the LP_{01} mode and the LP_{11} mode, obvious difference can be observed. For the LP_{01} mode, the Raman peak around 1113 nm is rather high at about 1 W exciting power. While for the LP_{11} mode, there is nearly no Raman peak at 1 W exciting power. This means that the power needed to generate the first order Raman peak is much higher for the LP_{11} mode as compared with the LP_{01} mode, in other words, LP_{11} mode has a higher Raman threshold. We have measured the output power when the first Raman peak appears to get a rough calculation of the Raman threshold. The Raman threshold of the LP_{11} mode is about 1.684 W, while it is 0.47 W for the LP_{01} mode. Although there may exist some deviations between the real Raman threshold and the calculation, the result is still useful, which is roughly consistent with the results of the effective mode field area and nonlinear coefficient calculated above.

In this condition, the output power of the LP_{11} mode after the 11 m-long passive fiber is lower than that of the LP_{01} mode. From the production information table of the 15/130 LMA-GDF, the fiber loss is 6 dB/km for the LP_{01} mode at 1060 nm. So, the loss of the 11 m long fiber for the LP_{01} mode can be neglected. While for the LP_{11} mode, the loss is measured by using the cut back method to be about 30 dB/km, which is much larger than that of the fundamental mode.

Then the 11 m passive fiber is replaced by a same kind of fiber with 170 m length for further investigation. The fiber is coiled in a loop with 16 cm diameter. The bending loss of both the LP_{01} and the LP_{11} mode can be neglected. In this experiment, we can see the difference of spectrum broadening at the short wavelength range between the two modes. When exciting the LP_{01} mode, we can see three orders of Raman peaks at very low output power, after that the short wavelength around 1012 nm begin to appear, while the short wavelength generation of LP_{11} mode is just after the second Raman peak at a certain power, as shown in Fig. 5. This phenomenon can be explained by the different dispersion characters of the two modes. The LP_{11} mode exhibits a shorter ZDW, which is helpful to the short wavelength generation when the spectrum broads to a certain long wavelength near the ZDW. It is consistent with the result of the calculated ZDW of each mode as shown in Fig. 3, which is 1205 nm for the LP_{11} mode and 1273 nm for the LP_{01} mode. The spectrum peak at 1012 nm for the LP_{11} mode is the anti-stokes light corresponding to the stokes light (the first Raman peak) at 1113 nm.

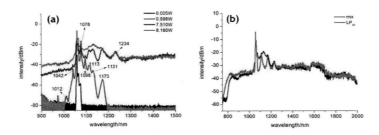

Figure 6.   The output spectrum when both LP_{01} and LP_{11} modes are excited in the 170 m long 15/130 LMA-GDF (a) and Comparision of the spectrum of only LP_{01} mode is excited and both of the LP_{11} and the LP_{01} mode are excited in 170 m 15/130 LMA-GDF at maximum average power of 8.18 W (b).

It is interesting that a series of spectrum peaks appear when both modes are excited in such long fiber by adjusting the pressure of the mechanical grating. The output spectrum is shown in Fig. 6. Except the Raman peaks in the longer wavelength range, the 1113 nm peak (the first order Raman peak), the 1173 nm peak (the second Raman peak), and the 1234 nm peak (the third order Raman peak), the other peaks appear only when both of the two modes are excited. This phenomenon can be also verified by bending the 30 cm long passive fiber after the mode converter into small coils with diameter of about 3 cm to guarantee that the input mode is only the fundamental mode when both two modes are excited in the mode converter. There is a pair of narrow lines with equal frequency shift from the central wavelength, at the 1042 nm and the 1078 nm respectively. According to Fig. 3, there exists phase mismatching between 1042 nm and 1078 nm, which excludes the dispersive wave and four waves mixing (FWM) effect. The group-velocity mismatch is helpful for modulation instability. Also, it can affect the gain spectrum of modulation instability of cross phase modulation between the two modes in terms of its shape, peak value, and position (Zhong X, 2007), which is also called the intermodal modulation instability (IM-MI). Both of the two peaks contain the LP_{11} mode and the LP_{01} mode (R. Dupiol, 2017). Also, the 1098 nm peak can be regarded as a sideband of 1060 nm related to the 1030 nm peak which is covered by the ASE spectrum of the YDF. As the 1078 nm peak has a relative high power, the first order Raman peak at 1131 nm of it appears.

Two optical spectrum analyzers (OSAs with measuring range of 600–1700 nm; and 1200–2400 nm) were adopted to analyze the output spectrum of the system. The SC spectrum was collected by a multimode fiber and recorded by the two different OSAs. In addition, a long-pass filters with 1200 nm cutoff wavelength is used when measuring the 1600–2000 nm spectra to eliminate the measuring errors caused by high order diffraction.

These intermodal nonlinear effects make the spectrum more complex than the condition of only one mode exists in such length of 15/130 fiber. The gaps between each order of Raman peaks are somewhat filled by these new spectrum components produced by inter-modal modulation nonlinearities, as shown in Fig. 6(b). As the pump power increases to the maximum value of over 8 W, the supercontinuum spectrum covers the near infrared range from 800 nm to 2000 nm with 4.77 W output power. Compared with the output spectrum of the pure LP_{01} mode, the gap between the laser wavelength and the first order Raman peak almost disappears, as well as the gap between the first and the second Raman peaks.

4   CONCLUSION

In conclusion, we report on the conclusion experimental observation of the LP_{11} mode pulse laser transmission in 15/130 LMA fiber to explore the nonlinear effects of this mode. Compared with the fundamental mode, the LP_{11} mode has much larger effective mode field area and lower nonlinear coefficient, resulting in an increase of Raman threshold. This character allows us to increase the peak power of the ability of fiber-based devices and systems to transport information or deliver energy. The 11 m 15/130 LMA fiber is changed into 170 m to explore the effect of dispersion and the cut-off wavelength of the LP_{11} mode on the spectrum broadening. As the length of the passive fiber increases, the LP_{11} mode perform differently in short wavelength generation when compared with the LP_{01} mode, due to their different dispersion characters. The anti-stokes light appears just after the second order Raman peak appears due to the relatively short ZDW of it, while for the LP_{01} mode the anti-stokes light appears after the third Raman peak. As the pump power increases, it can be coupled into the fundamental mode when broadens to the wavelengths longer than the cut-off wavelength.

By adjusting the pressure of the mechanical grating, both of the two modes can be excited into the 170 m passive fiber. Strong mode interaction is observed when exciting both of the two modes, producing a way to flatten the gaps between different orders of Raman peaks so as to get a flatter supercontinuum spectrum. A spectrum covering the near infrared range from 800 nm to 2000 nm is obtained in this case. Because of the large mode area and low nonlinear coefficient of the high order mode, the power can be increased in future studies to generate high power supercontinuum in high order mode in passive fiber. Compared with previously published supercontinuum generation from ytterbium-doped fiber amplifiers (Song R, 2013), the phenomenon of supercontinuum generation in this work is more complex due to the intermodal nonlinearities between the two modes, and the spectrum broadens more into the short wavelength range from 1000 nm to 800 nm.

REFERENCES

Bendahmane A. and K. Krupa, 2018, Seeded intermodal four-wave mixing in a highly multimode fiber, Journal of Optical Society of America B, 36(2):295–301.

Chaipiboonwong, T., P. Horak, JD. Mills, and WS. Brocklesby, 2017, Numerical study of nonlinear interactions in a multimode waveguide, Optics Express, 15(14):9040–47.

Cherif, R., M. Zghal, L. Tartara, and V. Degiorgio, 2008, Supercontinuum generation by higher-order mode excitation in a photonic crystal fiber, Optics Express, 16(3):2147–52.

Dupiol, R., A. Bendahmane, K. Krupa, J. Fatome, A. Tonello, M. Fabert et al, 2017, Intermodal modulational instability in graded-index multimode optical fibers, Opt Lett., 42(17):3419–22.

Eftekhar, MA., LG. Wright, MS. Mills, M. Kolesik, RA. Correa, FW. Wise et al, 2017, Versatile supercontinuum generation in parabolic multimode optical fibers, Optics Express, 25(8):9078–87.

Ezmaveh ZS. and M.A. Eftekhar, 2017, Tailoring frequency generation in uniform and concatenated multimode fibers, Opt Lett., 42(5):1015–18.

Flammer I. and P. Sillard, 2004, Calculation of Raman gain when pumping with higherorder order modes, Optical Society of America.

Giles, I., A. Obeysekara, C. Rongsheng, D. Giles, F. Poletti F, and D. Richardson, 2012, Fiber LPG Mode Converters and Mode Selection Technique for Multimode SDM, IEEE Photonics Technology Letters, 24(21):1922–5.

Guenard, R., K. Krupa, R. Dupiol, M. Fabert, A. Bendahmane, V. Kermene et al, 2017, Kerr self-cleaning of pulsed beam in an ytterbium doped multimode fiber, Optics Express, 25(5):4783–92.

Hamazakiand J. and R. Morita, 2010, Optical-vortex laser ablation, Optics Express, 18(3):2144–51.

Lesvigne, C., V. Couderc, and A. Tonello, 2007, Visible supercontinuum generation controlled by intermodal four-wave mixing in microstructured fiber, Opt Lett., 32.

Liu T, Chen S, Qi X, and Hou J, 2016, High-power transverse-mode-switchable all-fiber picosecond MOPA, Optics Express, 24(24):27821–7.

Liu T, Chen S, Zhang B, Qi X, Liu W, and Hou J, 2014, Visible supercontinuum generation through hollow beams in a two-mode photonic crystal fiber, Applied Physics Express, 7(6).

Liu Z, LG. Wright, DN. Christodoulides, and FW. Wise, 2016, Kerr self-cleaning of femtosecond-pulsed beams in graded-index multimode fiber, Opt Lett., 41(16):3675–8.

Ramachandran S. and JW. Nicholso, 2006, Light propagation with ultralarge modal areas in optical fibers, Opt Lett., 31(12):1797–99.

Richardson DJ. and JM. Fini, 2013, Space-division multiplexing in optical fibres, Nature Photonics, 7:354.

Song R, Hou J, Chen SP, Yang WQ, Liu T, and Lu QS, 2013, Near-infrared supercontinuum generation in an all-normal dispersion MOPA configuration above one hundred watts, Laser Physics Letters, 10(1):015401.

Wright, LG. and WH. Renninger, DN. Christodoulides, and FW. Wise, 2015, Spatiotemporal dynamics of multimode optical solitons, Optics Express, 23(3):3492–506.

Zhong X, 2007, Effects of group-velocity mismatch and cubic-quintic nonlinearity on cross-phase modulation instability in optical fibers, Chinese Optics Letters, 5(9):534–7.

Frontier Research and Innovation in Optoelectronics Technology and Industry – Habib & Lewis (Eds)
© *2019 Taylor & Francis Group, London, ISBN 978-1-138-33178-5*

Design of performance parameters for a missile-borne laser altimeter

H. Yang
Department of Information System, Dalian Naval Academy, Liaoning Dalian, China

Y.F. Zhang
Department of Missile and Naval Gun, Dalian Naval Academy, Liaoning Dalian, China

ABSTRACT: The laser altimeter with infrared/blue–green dual-wavelength output cannot only detect terrestrial terrain, but also near-shore seabed terrain. Therefore, it can be applied to the terrain matching guidance system for cruise missiles. In order to design a missile-borne laser altimeter, it is necessary to accurately determine the performance parameters of the laser altimeter. Based on laser altimetry/bathymetry equations and the output Signal-Noise Ratio (SNR) expression, the digital simulation calculation was used to assess the performance of the laser altimeter. Further, such parameter values could be determined, including peak power, pulse energy, pulse width, pulse repetition rate and so on.

Keywords: laser altimeter, altimetry/bathymetry, peak power, pulse repetition frequency

1 INTRODUCTION

In recent years, laser altimeters have been widely used in various airborne scanning altimeters, lunar exploration systems and oceanographic radar systems. The laser altimeter features high detection accuracy and strong anti-electromagnetic interference. Laser of a certain wavelength (such as blue–green laser) has a strong capability for seawater penetration (Zhao et al., 2017). Therefore, the laser altimeter with near-infrared/blue–green dual-wavelength output cannot only detect terrestrial terrain by infrared laser, but also near-shore seabed terrain by blue–green laser, which has been widely used in the terrain matching guidance for cruise missiles (Zhang & Yang, 2007) and other fields (Tang et al., 2017; Yu et al., 2013). To design a missile-borne laser altimeter, the key performance parameters of the altimeter components must be accurately determined in advance. The laser altimeter performance indices include size, weight, power consumption, cost and so on, which must meet the tactical and technical requirements for cruise missiles. To determine the performance parameters, first the measuring model of a laser altimeter should be built. Then, the digital simulation method can be applied to assess the altimeter performance (Zhu et al., 2017). Finally, the appropriate performance parameter values can be determined. The simulation models include the laser ranging equations and the output SNR expression of photo-detectors.

2 OVERALL PLAN FOR A MISSILE-BORNE LASER ALTIMETER

A diagram of terrain matching guidance for a cruise missile is shown in Figure 1. When the cruise missile is carrying out terrain matching navigation, it is necessary to determine the plane coordinates of the route matching area, the height from the land surface or the depth of the seabed, that is, determine the X, Y and Z values of navigation points in three-dimensional coordinates. The plane coordinate values could be determined by the missile's inertial navigation system (INS). The height/depth values are mainly measured by the laser ranging system. The infrared or blue–green laser from the altimeter is collimated and then emitted to the ground or sea surface. The infrared laser reflected by the ground is received by the infrared radiation (IR) receiving system. The blue–green laser can penetrate seawater and reach the seabed, which is scattered and received by the blue-green radiation (GR) receiving system. According to the scanning angle, the missile's flight height and other parameters, such as the height/depth values along the flight route could be calculated.

Due to the bad installing environmental conditions on the cruise missile, such as small space and high flight speed of the missile, the laser altimeter equipped to the missile must feature small size, light weight, long range, large detection area, high detection sensitivity and strong anti-background interference. In theory, the higher the output pulse energy is, the narrower the pulse width is and the greater the pulse peak power is, the stronger the detecting capability of the altimeter is. However, in practice, restricted to state-of-the-art technology, the output pulse energy and pulse peak power is limited. In general, the detection height to land of the laser altimeter is less than 1.5 km, the detection depth to the seabed is less than 50 m. The detection accuracy to the land is ± 0.2 m, and the detection accuracy to the seabed is ± 0.5 m. Therefore, we should consider these factors in designing the perfect altimeter.

The schematic layout of a laser altimeter is shown in Figure 2, which can satisfy the requirements of tactical and technical indices for cruise missiles. The designed laser altimeter comprises of a laser transmitter, photoelectric detector, transmitting and receiving optical system, A/D sampling and data storage system with high speed, integrated signal processing system and power supply. The laser transmitter is the core of laser altimeter. The laser transmitter should output dual-wavelength laser beams, such as an infrared beam at a wavelength of 1064 nm and

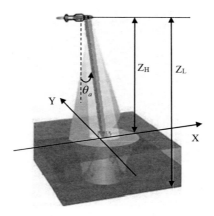

Figure 1.   Diagram of terrain matching quidence for cruise missle.

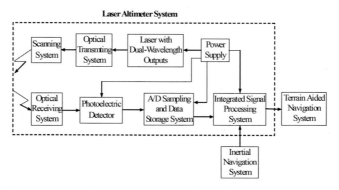

Figure 2.   Schematic layout of the laser altimeter for cruise missiles.

228

blue–green beam at a wavelength of 532 nm. The scanning system can carry out scanning measures through a rotating reflector and changing of the zenith angle. The photoelectric detector is used to receive infrared pulse returns from land or sea surface, and to receive blue–green pulse returns from seabed. The function of A/D sampling and the data storage system is sampling echo signals with high speed and storing data with large capacity. The INS can provide attitude and position information for cruise missiles. The signal processing system integrates all inputting information, namely, the sampling echo signals, missile's attitude and position information and scanning angle information. Based on this information, the values of height and depth can be calculated accurately. Finally, the information of land and seabed terrain contour can be provided for cruise missiles to carry out terrain matching navigation.

3   LASER RANGING EQUATION

The laser altimeter adopts the technology of pulsed ranging to measure distance. The measuring range is an important technical index for designing and developing the laser altimeter. The performance of ranging can be described by the laser ranging equation. The ranging equation describes the relationship between received power and transmitting power, optical transmission rate, viewing angle of receiver, atmospheric or seawater attenuation coefficient, and target reflectance. Because the designed laser altimeter cannot only detect the land terrain but also detect the seabed terrain, it is necessary to evaluate the ranging performance of the altimeter from altimetry and bathymetry.

3.1   Laser altimetry equation

The altimetry equation of the laser altimeter for land terrain can be expressed as follows (Wei et al., 2005):

$$P_r = \frac{P_t K_t K_r A_r \rho}{\pi R^2} e^{-2\mu R} \tag{1}$$

In Equation 1, P_t is the transmitting peak power, P_r is the received power, K_t is the transmitting optical transmissivity, K_r is the received optical transmissivity, A_r is the received area of the target, ρ is the diffuse reflection coefficient, R is the measuring distance, μ is the atmospheric attenuation coefficient. In general, the atmospheric attenuation coefficient can be calculated simply as follows:

$$\mu = (3.91/V) \tag{2}$$

In Equation 2, V (km) is atmospheric visibility.

3.2   Laser bathymetry equation

Similar to Equation 1, the bathymetry equation of the laser altimeter can be expressed as follows (Jiang et al., 2005):

$$P_r = \frac{P_t K_t K_r T_w^2 A_r \rho}{\pi (h + Z/n)^2} e^{-2(\mu h + \Gamma Z)} \tag{3}$$

In Equation 3, T_w is the transmissivity of sea water, h is the flight height of the missile from the sea surface, Z is the depth from sea surface to seabed, n is the refractivity of sea water and Γ is the effective attenuation coefficient of sea water.

It is necessary to explain the effective attenuation coefficient: some of the photons transmitted in the sea are scattered by the water particles and deviate from the optical axis. However, through multiple scattering, the photons can re-enter the optical axis and be received by the detection system again. Therefore, the underwater energy attenuation for a collimated laser beam cannot simply be expressed by the attenuation coefficient μ_a of seawater, which should be expressed by the effective attenuation coefficient Γ. Obviously, the scattering loss is present, but the scattering loss is lower than that of the collimated beam. That is, $\kappa \leq \Gamma \leq \mu_a$. κ is the diffuse attenuation coefficient. Γ could be described as follows (Chen, 1996):

$$\Gamma = \mu_a(1 - 0.832\omega_0) \tag{4}$$

In Equation 4, ω_0 is the single scattering rate.

4   OUTPUT SNR EXPRESSION OF DETECTOR

The echo signals received by the altimeter are usually mixed with noises. To calculate the minimum detectable signal power $P_{r\min}$, the SNR of the altimeter must be determined. The altimeter noises include two parts, that is, thermal noise and quantum noise caused by the dark current. Under the condition of strong background interference, the output SNR expression of Si-APD or photomultiplier is as follows:

$$S/N = \cfrac{MP_r\eta e/h\nu}{\left[\cfrac{4KT \cdot \Delta f}{R_L} + 2eM^2F_n\left(\cfrac{\eta e}{h\nu}P_r + \cfrac{\eta e}{h\nu}P_b + I_{Dd}\right) \cdot \Delta f\right]^{1/2}} \tag{5}$$

In Equation 5, M is the detector current gain, P_r is the signal power, P_b is the background power, η is the quantum efficiency, e is the electronic charge $(1.602 \times 10^{-19}C)$, h is the Planck constant $(6.626 \times 10^{-34}J \cdot s)$, ν is the incident laser frequency (Hz), K is the Boltzmann constant $(1.38 \times 10^{-23}J/K)$, T is the absolute temperature, Δf is the system noise bandwidth, R_L is the conductor impedance, F_n is the noise multiplication factor, I_{Dd} is the direct current (DC) component of noise current generated by the dark current.

The factors that affect the background power include direct sunlight and the reflected sunlight from the ground, cloud and moon. The background power received by the laser altimeter is approximated as follows:

$$P_b = E_b\Delta\lambda K_r A_r \theta_r \tag{6}$$

In Equation 6, E_b is the background spectral radiance $(W/m^2 \cdot sr \cdot \mu m)$, $\Delta\lambda$ is the bandwidth of narrow-band filter and θ_r is the received field of view.

5   SIMULATION ANALYSIS

5.1   *Altimetry simulation of laser altimeter*

Supposed the maximum height measured by the laser altimeter to the land surface is 1.5 km with 1064 nm infrared laser output, the detector adopts Si-APD. The performance parameters of the laser altimeter are shown in Table 1. According to Equation 1 and Equation 5, the relationship between the received SNR and emitting peak power under different working conditions can be determined by digital simulations. Supposed the background radiance is 50 $W/m^2 \cdot sr \cdot \mu m$, the background radiation power is calculated by Equation 6, and the atmospheric transmittance is calculated by Equation 2.

The simulation result shows that SNR varies with the transmitting peak power under different visibility, as shown in Figure 3. We can conclude that if the peak power of infrared laser reaches 50 kW, the desired height value can be measured by the altimeter under the condition of better visibility. However, if the visibility is low, the SNR will be very low. Low visibility is usually caused by fog and other climatic phenomena. If the altimeter is applied in harsh climates, the peak power of the laser should be increased.

Table 1. Performance parameters of the laser and detector in altimetry.

Parameter	Value	Parameter	Value
Atmospheric visibility V	20 km	Detector receiving calibre D	60 mm
Transmitting optical transmissivity K_t	0.9	Received optical transmissivity K_r	0.48
Laser emission angle ϕ	45°	target reflectivity ρ/Ω	$0.15/\pi\,sr^{-1}$
Filter bandwidth $\Delta\lambda$	5 nm	dark current I_{Dd}	100 nA
Excess noise factor F_n	4.5	Internal gain M of APD	100
Noise bandwidth Δf	10 MHz	Load resistance R_L	50 Ω
quantum efficiency η	40%	Conductor temperature T	25°C

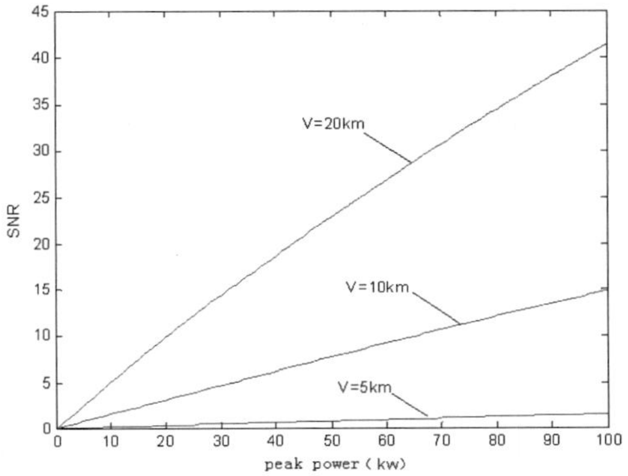

Figure 3. The curve of SNR with peak power under different visibility.

Table 2. Performance parameters of laser and detector in bathymetry.

Parameter	Value	Parameter	Value
Missile flight height h	30 m	Detector receiving calibre D	60 mm
Transmitting optical transmissivity K_t	0.9	Received optical transmissivity K_r	0.48
Seabed reflectivity ρ	0.2	Sea surface reflectivity ρ_W	0.02
Transmissivity of sea water T_W	0.98	Refractivity of sea water n	1.333
Atmospheric attenuation coefficient μ	1.96×10^{-4} m^{-1}	Attenuation coefficient of sea water μ_a	1 m^{-1}
Single scattering rate ω_0	0.8	Laser emission angle ϕ	45°
Filter bandwidth $\Delta\lambda$	5 nm	dark current I_{Dd}	300 nA
Excess noise factor F_n	7.5	Internal gain M	200
Noise bandwidth Δf	20 MHz	Load resistance R_L	100 Ω
quantum efficiency η	50%	Conductor temperature T	25°C

5.2 Bathymetry simulation of laser altimeter

As seawater attenuation is much larger than atmospheric attenuation, the depth value of seabed terrain measured by the altimeter should not be too large when the cruise missile is fighting across the sea. The altimeter outputs blue–green laser with 532 nm wavelength to detect seabed terrain. The detector adopts a photomultiplier tube, and the related performance parameters are shown in Table 2. Using the laser sounding Equation 3 and the

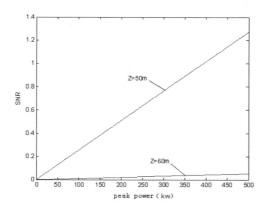

Figure 4.   The curve of SNR with peak power under different depth.

signal-to-noise-ratio calculation Equation 5, we can simulate the relationship between the signal-to-noise-ratio and the peak power in different water depths. Supposed the seabed reflectivity is 0.2 and the background radiance is still $50W/m^2 \cdot sr \cdot \mu m$.

The variation of SNR with laser peak power under different water depths is simulated, as shown in Figure 4. The simulation results show that if the depth is about 50 m, SNR of the altimeter is generally greater than one, and the peak power of blue–green laser is greater than 450 KW. From the simulation results of Figure 4, we can see that the detection depth is only increased by 10 m, and the SNR decreases rapidly, indicating that the attenuation of seawater to laser energy is very serious. Therefore, under the premise that the peak power of the laser emission peak is no more than 500 kW, when the terrain matching guidance is carried out for cruise missiles, the selected area of the matching area should not be too deep and generally should not exceed the 50 m.

6   OTHER PERFORMANCE PARAMETERS OF A MISSILE-BORNE LASER ALTIMETER

In addition to the peak power, other performance parameters of a laser altimeter also need to be determined, including the emission pulse energy, the pulse width, the pulse repetition frequency and so on.

6.1   The emission pulse energy W_T

From the formulas of laser altimetry and bathymetry, the maximum value of height and depth measured by a laser altimeter is related to the emission pulse energy. The greater the emission pulse energy, the greater the detection distance is. However, detection distance is not significantly increased with the emission pulse energy, so it is not ideal to increase the detection distance by improving the emission pulse energy. The emission pulse energy of the altimeter cannot be too large. Because the installation space on the cruise missile is limited, the power of the battery cannot be too large. Considering the performance parameters of some laser detection systems abroad, the output energy of the single pulse of the 1064 nm infrared laser is 0.25 mJ, and the output energy of the 532 nm blue–green laser is 2.25 mJ.

6.2   The pulse width τ

The emission pulse width is related to the gain of the laser rod, the rise time of the Q switch and the geometric parameters of the resonator. The pulse width can be estimated by using the emission pulse energy and peak power. Supposed the output energy of 1064 nm infrared

232

laser and 532 nm blue–green laser is 0.25 mJ and 2.25 mJ, respectively, the peak power is 50 kW and 450 kW, then the pulse width can be calculated as 5 ns. Therefore, the designed pulse width of the infrared laser and the blue–green laser is 5 ns.

6.3 The pulse repetition frequency f_n

Supposed the flight height h of a cruise missile and the scanning angle θ of a laser altimeter are definite values, the width of scanning mowing can be determined. To keep the desired width of scanning mowing, the detection density will be determined by the missile's flight speed and laser pulse repetition frequency. The calculation formula is as follows. Supposing the flight height of a cruise missile is 1000 m, the scanning angle of a laser altimeter is 30°, and the width of the scanning mowing is $2 \times 1000 \times tg30° = 1155\,m$. If the cruise missile's flight speed is 270 m/s, the size of the terrain matching mesh is $25 \times 25\,m$, that is, the detection density is one point per 25 m^2, and the distance between two points is 5 m, then the laser pulse repetition frequency is $f_n = (1155 \times 270)/(25 \times 25) = 500Hz$. The high pulse repetition frequency is also dependent on the performance of the laser. For the diode-pumped solid-state laser, a pulse repetition frequency of $500Hz$ can be achieved.

7 CONCLUSIONS

According to the formulas for laser altimetry/bathymetry and the output SNR expression of the detector, the detection performance of a missile-borne laser altimeter with infrared/blue–green dual-wavelength output is evaluated by digital simulations. The performance parameters of a laser altimeter are finally determined, including peak power, single pulse energy, pulse width and pulse repetition frequency. The simulation results provide a design basis for the development of a missile-borne laser altimeter.

REFERENCES

Chen. W.G. (1996). The effective attenuation coefficient of airborne oceanic lidar system. *Acta Electronica Sinica, 24*(6), 47–50.
Jiang, L., H. Zhu, & S. Li. (2005). The relationship between max survey depth of airborne ocean lidar and secchi depth. *Laser & Infrared, 35*(6), 397–399.
Tang, X.M., J.F. Xie, X.K. Fu, F. Mo, S.N. Li & X.H. Dou. (2017). ZY3–02 laser altimeter on-orbit geometrical calibration and test. *Acta Geodaetica et Cartographica Sinica, 46*(6), 714–723.
Wei, D.Z., J. Xiao, D.Y. Zhang & K. Wang. (2005). The research for raindrop attenuation characteristic of laser signal. *Guidance & Fuze, 36*(4), 9–12.
Yu, Z.Z., X. Hou & C.Y. Zhou. (2013). Progress and current state of space-borne laser altimetry. *Laser & Optoelectronics Progress, 50*(2), 52–61.
Zhang Y.F. & H. Yang. (2007). Laser altimeter technology and its applications. *Aerospace China, 12,* 19–23.
Zhao, J.H., Y.Z. Ouyang & A.X. Wang, (2017). Status and development measurement technology tendency for seafloor terrain. *Acta Geodaetica et Cartographica Sinica, 46*(10), 1786–1794.
Zhu, D.Z., Y.J. Zhang, X. Li, W. Wei, G.G. Qiu, Q.Q. Guo & W.X. Zhao. (2017). Design of ground calibration simulation system of laser altimeter. *Journal of Atmospheric and Environmental Optics, 12*(4), 313–320.

Frontier Research and Innovation in Optoelectronics Technology and Industry – Habib & Lewis (Eds)
© 2019 Taylor & Francis Group, London, ISBN 978-1-138-33178-5

Assessing the dynamic characteristics of a femtosecond laser micro plasma expansion process with an optical fiber sensing probe

D. Zhong & Z. Li
School of Electronics and Information, Hubei University of Science and Technology, Xianning, Hubei, China

ABSTRACT: The acoustic signal of the laser micro plasma expansion for the femtosecond laser when ablating a silicon wafer has been detected using the whole optical fiber sensing probe. Meanwhile, a high-speed CCD camera has been applied to observe the expansion process of the laser micro plasma. In comparison, the frequency and amplitude of the acoustic emission spectrum has been analyzed, and the dynamic characteristics of the femtosecond laser micro plasma expansion have been researched. The results show that the detected acoustic emission signal frequency spectrum pattern is fixed with changing the laser energy. The amplitude of the acoustic emission spectrum decreases along with the increasing of the detection distance. Further analysis of the continuous 10 ms acoustic emission signal frequency spectrum found that the acoustic emission signal amplitude of the plasma dynamic expansion process increased first, and then decreased under the same conditions of the femtosecond laser energy and the detection distance. The rule conforms to the dynamic process from the high-speed CCD.

1 INTRODUCTION

The diversity of phenomena that arise during the development of the interaction between lasers and materials inducing plasma has promoted research within this discipline in recent years (Eliezer, 2002). A significant amount of laser plasma research has involved the characterization of the transient response of laser-generated plasma, primarily associated with the ablation of target materials using short-pulse, high peak-power lasers (Bulgakova et al., 2000; Kabashin et al., 1998). Laser plasma research has found a wide range of applications. The study of laser plasma dynamic characteristics provides a new development direction for laser shock strengthening processing, laser shock forming, laser cleaning, laser-induced fusion, laser propulsion, laser medicine and other fields. At present, a series of researches have been carried out, both at home and abroad, into the dynamic characteristics of the laser plasma expansion process. These researches have especially made great achievements in the theoretical study of the dynamic characteristics of nanosecond-pulsed laser plasma expansion. The simulations by the existing nanosecond or longer-pulse laser ablation target material are mostly based on the classical theory of heat transfer. The establishing of a dynamic model of laser ablation plasma plume steady expansion was mainly through the related research into the transmission of pulse laser (Bogaerts et al., 2011). On the femtosecond scale, nanosecond and long-pulse laser ablation model is no longer applicable (Lv et al., 2009). In order to obtain the dynamic characteristics of the plasma expansion process induced by femtosecond laser ablation target material, laser-induced plasma diagnostic techniques are used in experimental research. Garnov et al. used over-speed spectroscopy to observe the process of plasma formation, progress and ionization at the early stage (Garnov et al., 2009). Liu et al. researched the influence of liquid environments on the femtosecond laser ablation of silicon (Liu et al., 2010). Gao et al. studied the plasma space and time-resolved emission spectra by the femtosecond-pulse laser ablation silicon (111). Gao et al. summarized the evolution of plasma plume expansion space emission wavelength shift and spectral intensity process (Gao et al., 2011). Odachi et al. studied the work on the ablation of crystalline silicon by femtosecond laser pulses in air and vacuum (Odachi et al., 2013). However, these techniques are mainly concerned on molecular spectroscopy. They cannot comprehensively monitor the dynamic characteristics of the plasma expansion process.

Some scholars had provided great value in exploring the theoretical mechanism of the dynamic characteristics and related detection methods of the nanosecond and long-pulse laser plasma expansion process. Zhang et al. obtained the relationship between the impulse coupling coefficient and laser power density by using the pendulum and photoelectric velocity measurement method (Zhang et al., 2009). The characteristics of the emission spectra and hologram of the plasma plume have been investigated during femtosecond laser ablation (Amer et al., 2015; Hu et al., 2009). Based on this, in the present paper we present the application of an all-fiber non-gel microstructure probe F-P acoustic emission sensing measurement system to detect the acoustic emission signal of the femtosecond laser plasma expansion process, using a femtosecond laser with the center wavelength of 780 nm and pulse width of 180 fs ablation of silicon in air. The dynamic characteristics of the femtosecond laser micro plasma expansion process were obtained by analysis. At the same time through high-speed CCD, the motion characteristics of the femtosecond laser micro plasma expansion process were observed and analyzed.

2 EXPERIMENTAL

The schematic of the experimental setup is shown in Figure 1. A commercial Ti: sapphire femtosecond laser with 180 fs laser pulses and 780 nm central wavelength, 500 Hz repetition rate was used in the micromachining system (Japanese Cyber Laser Company; LS-IF-FW-C-401). The pulse energy could be continuously varied by a variable neutral density filter and a mechanical shutter could turn the laser on and off. Before femtosecond laser irradiation, the silicon wafers were cleaned by ultrasonic bath for 15 min in acetone and rinsed in deionized water. The sample was positioned horizontally on a three-dimensional computer-controlled stage. For the F-P optical fiber sensing system, a tunable semiconductor laser with 1,545 nm wavelength and 0.2 mw power was used as the light source (Santec Company). The multi-channel data acquisition card of NI Company was used for the data collection. The highly efficient signal processing circuit is developed independently by our laboratory. The high-speed CCD used in the experiment was a type of PCO. dimax HD high-speed digital camera, produced by PCO Company of Germany; the exposure time was 1.5 μs–40 ms.

The experiments were carried out as follows. A tunable semiconductor laser was used as the light source in the measurement process. A femtosecond laser beam with a high power density and a short pulse was focused using a quartz lens. As the laser beam irradiated the silicon wafers in air, the laser micro plasmas could be formed, which led to an acoustic signal

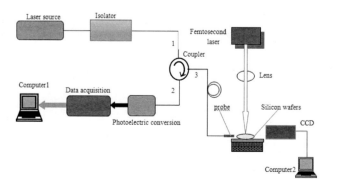

Figure 1.   Schematic layout of the experimental setup.

due to the expansion of the laser micro plasma. The signals were detected by the whole optical fiber sensing probe and collected by the data acquisition module after passing through a coupler. Meanwhile, the CCD was connected to the computer by a 1394 interface and finished shooting photo storage through the MATLAB software programming to call CCD own function package. The high-speed CCD camera had been applied to observe the expansion process of the laser micro plasma. In comparison, the frequency and amplitude of the acoustic emission spectrum has been analyzed, and the dynamic characteristic of the femtosecond laser micro plasma expansion has been researched.

3 RESULTS AND DISCUSSION

In the air background, the 180 fs femtosecond laser pulse with pulse frequency of 500 Hz was applied to irradiate the silicon wafers under the different femtosecond laser energies (110 mw, 230 mw, 340 mw) and the different detection distances (0.5 mm, 1 mm, 1.5 mm, 2 mm, 3 mm). The acoustic emission signal of the femtosecond laser micro plasma was observed. We tried to find the relationship between the acoustic signal frequency, or signal strength, and the different femtosecond laser detection distances or energies. The results are shown in Figure 2 and Figure 3. Figure 2 shows the typical acoustic emission signal frequency spectrum of the

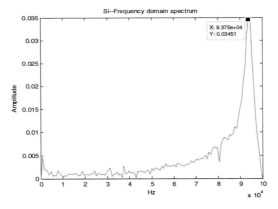

Figure 2. The typical acoustic emission signal frequency spectrum of femtosecond laser micro plasma produced by a femtosecond laser ablation of silicon wafer in air background.

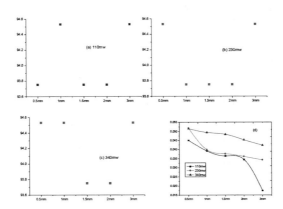

Figure 3. The change rules of femtosecond laser micro plasma acoustic emission signal frequency and signal strength with different detection distances and different femtosecond laser pulse energies.

femtosecond laser micro plasma, produced by the femtosecond laser ablation of a silicon wafer in air. Figure 3 shows the change rules of the femtosecond laser micro plasma acoustic emission signal frequency and signal amplitude with different detection distances and different femtosecond laser pulse energies. Figure 3(a) shows the change rules of the femtosecond laser micro plasma acoustic emission signal frequency with different detection distances under the femtosecond laser pulse energy of 110 mw. Figure 3(b) shows the change rules of the femtosecond laser micro plasma acoustic emission signal frequency with different detection distances under the femtosecond laser pulse energy of 230 mw. Figure 3(c) shows the change rules of the femtosecond laser micro plasma acoustic emission signal frequency with different detection distances under the femtosecond laser pulse energy of 340 mw. Figure 3(d) shows the change rules of the femtosecond laser micro plasma acoustic emission signal amplitude with different detection distances under different femtosecond laser pulse energies. Because the acoustic emission signal first peak has the logo for the femtosecond laser plasma acoustic emission signal, we therefore take the first signal peak as being the plasma acoustic emission signal.

4 CONCLUSIONS

In summary, space-selective precipitation of Si nanocrystals was achieved in borosilicate glass using 500 Hz femtosecond laser pulse irradiation. An increased refractive index of 8.7% was observed from the crystallization region owing to the high refractive index of the Si nanocrystals. A Z-scan measurement reveals that the precipitated nanocrystals can greatly enhance the third-order optical nonlinearities. These results may find some application for the fabrication of ultra-fast optical switches and diffractive optical devices.

ACKNOWLEDGMENTS

The authors gratefully acknowledge the financial support for this work provided by the Program of the Natural Science Foundation of Hubei Province, Grant No. 2016CFB515, the Humanities and Social Science project of Hubei Education Department, No. 16Q259 and No. 18Q164, the National Natural Science Foundation of China (NSFC) under the Grant No. 61575148, the Dr. Start-up Fund of Hubei University of Science and Technology No. BK1524, the colleges and universities of Hubei Province innovation and entrepreneurship training plan, No. 201710927006Z, the Team Plans Program of the Outstanding Young Science and Technology Innovation of Colleges and Universities in Hubei Province, Grant No. T201817, and the teaching reform program for Chinese and foreign cooperation of Hubei University of science and technology, No. 2017-ZYA-002.

REFERENCES

Amer, E., Gren, P. & Sjödahl, M. (2015). Laser-ablation-induced refractive index fields studied using pulsed digital holographic interferometry. *Optics and Lasers in Engineering*, 47(6), 793–799.

Bogaerts, A., Aghaei, M., Autrique, D., Lindner, H., Chen, Z.Y. & Wendelen, W. (2011). Computer simulations of laser ablation plume expansion and plasma formation. *Advanced Materials* Research, 76(12), 1–10.

Bulgakova, N.M., Bulgakov, A.V. & Bobrenok, O.F. (2000). Double layer effects in laser-ablation plasma plumes. *Phys. Rev. E*, 62(3), 5624–5635.

Eliezer, S. (2002). The interaction of high-power lasers with plasma. *Institute of Physics Publishing, Bristol and* Philadelphia, 45(3), 105–127.

Gao, X., Song, X., Guo, K., Tao, H. & Lin, J. (2011). Research on the emission spectrum of femtosecond laser ablation of silicon surface plasma generation. *Phys. Lett.*, 60(2), 025203.

Garnov, S.V., Bukin, S.V., Malyutin, A.A. & Strelkov, V.V. (2009). Ultrafast space-time and spectrum-time resolved diagnostics of multicharged femtosecond laser microplasma. In *AIP Conf. Proc. 1153*(1), 37–48. AIP.

Hu, H., Wang, X., Li, Z., Zhang, N. & Zhai, H. (2009). The femtosecond laser ablation of aluminum ultrafast pulse digital holographic diagnosis ejecta. *Journal of Physics, 58*(11), 662–667.

Kabashin, A.V., Nikitin, P.I., Marine, W. & Sentis, M. (1998). Electric fields of a laser plasma formed by optical breakdown of air near various targets. *Quantum Elect, 28*(5), 24–28.

Liu, H., Chen, F., Wang, X., Yang, Q., Bian, H., Si, J. & Hou, X. (2010). Influence of liquid environments on femtosecond laser ablation of silicon. *Thin Solid Films, 518*(43), 5188–5194.

Lv, D., Tong, X., Li, L., Lin, K. & Liu, Z. (2009). Research progress of bio-engineering materials laser processing organization. *Laser Journal, 30*(4), 7–9.

Odachi, G., Sakamoto, R., Hara, K. & Yagi, T. (2013). Effect of air on debris formation in femtosecond laser ablation of crystalline Si. *Applied Surface Science, 282*, 525–530.

Zhang, H., Lu, J. & Ni, X. (2009). Optical interferometric analysis of colliding laser produced air plasmas. *Journal of Applied Physics, 106*(6), 063308.

Zhu, J., Yang, Y., Yang, B., Shen, Z., Lu, J. & Ni, X. (2016). The experimental study of laser plasma transfer on impulse. *Journal of Nanjing University of Aeronautics & Astronautics, 39*(5), 628–632.

Frontier Research and Innovation in Optoelectronics Technology and Industry – Habib & Lewis (Eds)
© 2019 Taylor & Francis Group, London, ISBN 978-1-138-33178-5

Combining principle component analysis and a back-propagation network for classifying plant water-stress based on images of chlorophyll fluorescence

Chunyan Zhou, Jian Dong Mao, Hu Zhao & Bai Zhang
School of Electrical and Information Engineering, North Minzu University, Yinchuan (Ningxia), China

ABSTRACT: The objective of this research was to classify plant water-stress levels using PCA (Principle Component Analysis) and a Back-Propagation (BP) neural network via chlorophyll fluorescence induction profiles. This was achieved using chlorophyll fluorescence images combined with the chlorophyll fluorescence kinetics parameters. The images were recorded at 690 nm with a high-resolution imaging device consisting of Light Emitting Diodes (LEDs) for an excitation at 460 nm, and an EMCCD camera. Then, the chlorophyll fluorescence kinetics curve was obtained and chlorophyll fluorescence kinetics parameters could be calculated. Furthermore, the characteristics of chlorophyll fluorescence kinetics curve were extracted and their dimensions were reduced by PCA. According to the accumulative variance contribution rate, three principle components were selected to replace the character parameters. We clustered and classified different water-stress levels with these principle components, using a BP neural network. The result showed that the accuracy rate of clustering water-stress on different levels could reach up to 92%. This technique has the potential to monitor the water-stress condition of plants non-destructively and simply.

Keywords: PCA, BP, Chlorophyll Fluorescence, Water-Stress

1 INTRODUCTION

A suitable water supply is a most important limiting factor for crop production. It is necessary to monitor the variation of plant water conditions. So, development of a fast, non-destructive and real-time system for monitoring plant growing conditions is desirable, to reduce resource waste and to promote crop production. Chlorophyll Fluorescence (ChlF), as an indicator of plant photosynthesis, could make stress symptoms apparent to the naked eye. Therefore, chlorophyll fluorescence has been widely used to monitor all kinds of stress state in the plant-growing process (Gorbe & Calatayud, 2012; Lichtenthaler & Babani, 2000; Lichtenthaler et al., 1996). The study of ChlF response in plants dates back to 1931. H. Kautsky discovered that when a plant was dark-adapted and then exposed to light, its chlorophyll fluorescence displayed characteristic changes in intensity corresponding to the induction of photosynthetic activity. This is known as the Kautsky effect (Kautsky & Hirsch, 1931). Chlorophyll fluorescence Induction (CFI) is a sensitive indicator of photosynthesis, and it has been found to be extremely suitable for the screening of the physiological parameters of plants (Govindje, 1995). The fluorescence measurements performed in the last decades concentrated on the measurements of single leaf points. This means that per each measurement, the ChlF of only one rather small leaf point (leaf spot) is excited and sensed. However, such single leaf point measurements provide only limited information on the photosynthetic apparatus, since single leaf spots often are not representative of the photosynthetic activity of the whole leaf. For this reason, high-resolution chlorophyll fluorescence induction imaging techniques have been developed (Lichtenthaler et al., 2005).

Many scientists have done a lot of researches on the kinetic images of chlorophyll fluorescence. They have set up various parameters to monitor the status of plants, such as frequently used R_{fd}-value images, which can provide quick information on the active photosynthetic performance of all leaf parts, and also allow the study of the successive loss of photosynthetic activity of leaves under stress conditions (Takayama et al., 2011). One study created an index named PFI, by illuminating tomato plants, and acquired images by CCD at 15 fps (Hsiao et al., 2010). Another study developed the DFI index, using the slow transience of chlorophyll fluorescence induction (Blumenthal et al., 2014). However, most studies point out the effect of stresses, but less study is carried out on the classification of plant stress levels. One researcher had clustered plant stress level using a hidden Markov model based on all points of the curve, but he just used the slow parts of the chlorophyll fluorescence transient curve, and this method involves a large amount of computing (Seuret et al., 2017). So, it is necessary to search for a new method to classify the water condition level more accurately, and to compress the data volume and reduce the dimension of the characteristics value.

As an important statistical analysis method, Principal Component Analysis (PCA) can reveal the dependency and relationship among multiple random variables. It also provides a way to reduce high-dimensional data to low-dimensional data and simplify data structures, with minimal loss of the original data information (Vankayala & Rao, 1993). Artificial Neural Networks (ANNs) are used to handle experimental data in various fields of technology and science, and their ability to tackle complex calculation issues, have been more and more recognized (Kirova et al., 2009; Huang & Dong, 2002). The Back-Propagation (BP) neural network is an influential neuron model, which could fit any given function, and could implement forecasts by training the sample data. It has been widely used in such areas as classification, approximation, regression and compression (Gottumukkal & Asari, 2004). This paper tries to use PCA combined with a BP neural network, reducing the dimension of feature points of a chlorophyll fluorescence kinetics curve, and then classifying the level of plant water-stress.

2 MATERIAL AND METHOD

2.1 *Plant material*

Plants of *Scheffera octophylla* (Lour.) Harms, which react quickly to environmental changes, were used for this study. For the drought stress treatment, 30 pots plants were divided into three groups, not watered for two weeks, and subjected to normal daylight conditions and room temperature. Then, three sets of plants were watered with 15 ml, 50 ml, and 80 ml respectively every three days. About 25 leaves of each set were picked, from which chlorophyll fluorescence profiles were collected and processed.

2.2 *Measuring protocol and data handling*

Figure 1 shows a schematic of the chlorophyll fluorescence induction imaging system, consisting of four major components: an array of LEDs (460 nm), an EMCCD camera (iXon Ultra 897, Andor), an interference filter (690FS10–50, Andover Co. Ltd., USA) and a computer. A blue LED panel provided excitation light to irradiate the leaves of the plants, which were dark-adapted at least 20 minutes before the experiment. The interference filter is mounted between the lens and the CCD-chip of the camera. The EMCCD with 512×512 pixels was used in the 2×2 binning mode for higher sensitivity and higher frame rate.

The acquisition mode of the EMCCD was set into kinetics mode, the cycle time was 0.5 s, and 500 pictures were collected during 250 s. The fluorescence intensity of each image averaged and were arranged according to the time order, so the chlorophyll fluorescence induction kinetics curves were obtained for each different water-stress (15 ml, 50 ml, 80 ml), as shown in Figure 2. It can be seen from the diagram that the decline rate of the fluorescence kinetics curve, and other fluorescence kinetic parameters such as the maximum fluorescence and steady fluorescence, obviously changed with the level of water-stress.

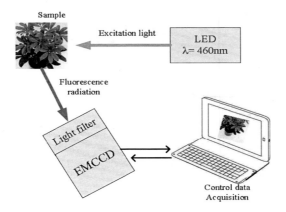

Figure 1.   The chlorophyll fluorescence induction imaging system.

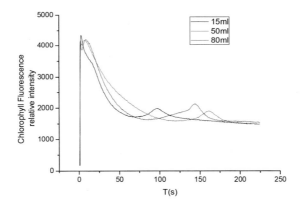

Figure 2.   The chlorophyll fluorescence transient variation curve for different water-stress levels.

2.3   *Characteristic values extraction*

The background data was removed before the maximum value from the kinetics curve. Then eight characteristic parameters were extracted, such as the fluorescence maximum value (F_{max}), the minimum value (F_{min}), the secondary maximum value (F_{max}'), the arise time of the secondary maximum ($T_{Fmax'}$), and the average intensity of the steady-state ($F_{average}$). At the same time, due to the rate of the downward gradient being diverse under different levels of water-stress, so a fitted curve is shown in Figure 3 as a green line, and its quadratic coefficient (A), monomial coefficient (B) and constant term (C) were extracted. Part of the original data is shown in Table 1. There were a total of 83 samples; 80% are used for modeling, and 20% for the model evaluation.

2.4   *Introduction to the PCA and BP neural network*

The major factor considered when selecting the initial characteristics is the separability of samples. So, the original features were interrelated to a large extent. A large data set would increase the burden of computing, and therefore some characteristics which had little contribution for

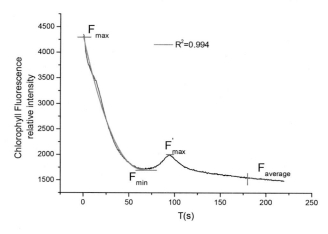

Figure 3. Characteristic values of chlorophyll fluorescence transient variation curve.

Table 1. Initial data of characteristic values.

Num	F_{max}	F_{min}	$F_{max'}$	$T_{Fmax'}$	Average	A	B	C
1	4,886.9	1,721.7	2,033.8	333.0	1,709.1	0.025	−26.5	4,909.8
2	4,719.7	1,723.5	2,136.3	235.0	1,612.6	0.155	−44.1	4,763.2
3	4,913.9	1,716.7	2,147.8	251.0	1,599.5	0.156	−45.7	4,917.0
4	4,753.3	1,781.0	2,139.7	238.0	1,661.2	0.187	−47.1	4,762.1
5	5,267.2	1,960.7	2,288.5	277.0	1,834.2	0.180	−49.2	5,327.8
6	5,123.4	1,828.8	2,241.5	301.0	1,788.4	0.109	−39.9	5,049.4
7	4,843.8	1,774.6	2,130.4	286.0	1,728.4	0.181	−46.2	4,819.3
8	4,264.7	1,619.3	2,181.1	281.0	1,607.4	0.114	−38.5	4,422.8

classification could be discarded in order to get a smaller set. Principal component analysis finds a few comprehensive factors to take the place of many of the original features, which can reflect the information of the original variables as much as possible, and these factors are unrelated to each other. Thus, the purpose of reducing the original data dimension can be achieved. Let vector $X = [x_1, x_2, ..., x_p]$ with the sample number n and the variable number p. Firstly, the correlation coefficient r_{ij} (i,j = 1,2,...,p) between x_i and x_j is calculated. Secondly, the eigenvalue λ_i and the corresponding eigenvector e_i are solved by the equation $|\lambda I - R| = 0$. Thirdly, the first m (m ≤ p) principal component F_i corresponding to λ_i(I = 1, 2,..., m), whose cumulative contribution rate reaches 85–95%, is selected to meet the analysis demand (Yin et al., 2010; Pan et al., 2009). A BP neural network, which is a more representative and prevalent model among ANNs, is a multi-layer and feed-forward network with the network training based on error back-propagation, and weight or threshold adjustment.

In this study, we adopt a three-layer network with input layer, hidden layer and output layer to cluster water-stress.

3 RESULTS AND DISCUSSION

3.1 *Preprocessing by the PCA method*

If the initial eight characteristic parameters were all used to model, the data dimension would be high and the computing required increased. Therefore, we used the PCA algorithm to reduce the dimension. The eigenvalues and corresponding eigenvectors of the correlation coefficient matrix R and the contribution rate were calculated, and the principal components were extracted. Table 2 shows that the characteristic value of the first principal component is 4.218; the variance accounted for 52.7% of the total variance, which is the main performance characteristic of the water-stress. The first three principal components' variance accounted for 94.5% of the total variance; according to the selection criteria of the principal component, the original eight characteristics can instead be replaced by these three principal components.

The principal component factor loading matrix is shown in Table 3. According to the PCA matrix model, the relationship between Y1, Y2, Y3 and original variables could be obtained, and the factor expressions could be derived.

$$Y1 = -0.454X1 - 0.398X2 - 0.415X3 - 0.277X4 - 0.449X5 - 0.019X6 + 0.076X7 - 0.423X8$$
$$Y2 = 0.161X1 - 0.203X2 - 0.186X3 - 0.097X4 - 0.145X5 + 0.628X6 - 0.63X7 + 0.276X8$$
$$Y3 = 0.034X1 - 0.45X2 - 0.312X3 + 0.824X4 + 0.109X5 - 0.051X6 + 0.064X7 + 0.05X8$$

3.2 Water-stress degree classified based on PCA combining with BP network

The principal components 1, 2, and 3 are set as input parameters, according to the water quantity. The degree of water-stress could be divided into three groups as an output parameter: the first class (15 ml), the second class (50 ml), and the third class (80 ml). About 80% of the samples were set as training data. Using MATLAB to create the BP neural network, tansig was set as the transfer function between the neurons of the hidden layer and the input layer, and trainlm was set as the training function. The maximum training number was set at 2,000 times, the training error was 0.0001, the learning rate was 0.1, and the rest of the parameters were default values. The confusion matrix for the network classifier is as shown in Figure 4, and the classified accuracy of the training samples is 96.7%.

Table 2. Principle components and their contribution rates.

Num	Eigenvalue	Contribution Rates/%	Accumulated Contribution Rates/%
1	4.218	52.721	52.721
2	2.419	30.239	82.961
3	0.926	11.575	94.535
4	0.237	2.967	97.502
5	0.134	1.678	99.18
6	0.056	0.702	99.882
7	0.008	0.095	99.977
8	0.002	0.023	100

Table 3. Factor load matrix of principle components.

varname	Y1	Y2	Y3
Fmax	−0.454	0.161	0.034
Fmin	−0.398	−0.203	−0.45
Fmax'	−0.415	−0.186	−0.312
TFmax'	−0.277	−0.097	0.824
Average	−0.449	−0.145	0.109
A	−0.019	0.628	−0.051
B	0.076	−0.63	0.064
C	−0.423	0.276	0.05

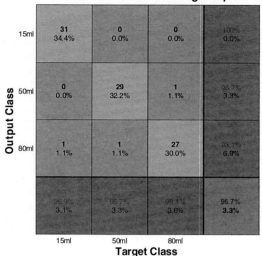

Figure 4.   Confusion matrix for the training sample set.

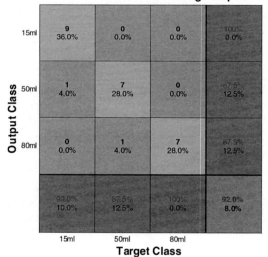

Figure 5.   Confusion matrix of testing sample set.

3.3   *Evaluation of classified result*

In order to test the classification effect of the classifier, 20% of the samples were used to validate and evaluate the BP network model. The confusion matrix of the testing sample set is as shown in Figure 5. The average classification accuracy is 92%. We discovered that the higher the plant water condition (80 ml), the higher of sample recognition accuracy (100%), but the first class and the second class had wrongly classified each other. This is partly because there were fewer samples, and also because 15 ml and 50 ml water had similar effects on the plant growth, so they had similar fluorescence characteristics. As expected, because water is the substance directly involved in the photosynthesis of plants, the chlorophyll fluorescence showed a strong relation to different water conditions. So, the water condition of the plant is related to the chlorophyll fluorescence, and, on the other hand, the water-stress of the plant could be classified by ChF accurately.

4 CONCLUSION

In this paper, the chlorophyll fluorescence kinetics curve was obtained on the basis of fluorescence imaging, eight characteristic parameters of which were extracted for analysis. Principal component analysis was used for data dimension reduction, and three principal component factors were extracted. The classifiers were designed using a BP neural network to classify plants for different water quantity. The results showed that the average classification accuracy of the training sample set is 96.7%, and of the testing sample set is 92%. The accuracy of 80 ml water quantity is 100%, but less water (15 ml and 50 ml) tends to have a small wrong recognize rate. This can be improved by increasing the sample numbers and widening the distance of water content between samples. Due to the complexity of plant chlorophyll fluorescence, which affected by such factors as the growth environment, nutrition, and water status? The next step of this work is to optimize the model and to control the part influence factors, in order to achieve finer classification results. Further, more plant species should be experimented with to determine the practicality and universality of the model.

ACKNOWLEDGMENTS

This work was supported by the National Natural Science Foundation of China (NSFC) under Grant 61565001, 61765001, 61450007, 51505005 and The Research Project of North Minzu University under Grant 2017DX008.

REFERENCES

Blumenthal, J., Megherbi, D.B. & Lussier, R. (2014). Unsupervised machine learning via hidden Markov models for accurate clustering of plant stress levels based on imaged chlorophyll fluorescence profiles & their rate of change in time. 76–81.
Gorbe, E. & Calatayud, A. (2012). Applications of chlorophyll fluorescence imaging technique in horticultural research: A review. *Scientia Horticulturae, 138*, 24–35.
Gottumukkal, R. & Asari, V.K. (2004). An improved face recognition technique based on modular PCA approach. *Pattern Recognition Letters, 25*(4), 429–436.
Govindje, E. (1995). Sixty-three years since Kautsky: Chlorophylla fluorescence. *Australian Journal of Plant Physiology, 22*(2), 131–160.
Hsiao, S.C., Chen, S., Yang, I.C., Chen, C.T., Tsai, C.Y., Chuang, Y.K., … Lo, Y.M. (2010). Evaluation of plant seedling water stress using dynamic fluorescence index with blue LED-based fluorescence imaging. *Computers and Electronics in Agriculture, 72*(2), 127–133.
Huang, S.W. & Dong, M.L. (2002). Application of adaptive variable step size BP network to evaluate water quality. *Journal of Hydraulic Engineering, 10*(10), 119–123.
Kautsky, H. & Hirsch, A. (1931). Neue Versuche zur Kohlensäureassimilation. *Naturwissenschaften, 19*(48), 964–964.
Kirova, M., Ceppi, G., Chernev, P., Goltsev, V. & Strasser, R. (2009). Using artificial neural networks for plant taxonomic determination based on chlorophyll fluorescence induction curves. *Biotechnology & Biotechnological Equipment, 23*(sup1), 941–945.
Lichtenthaler, H.K. & Babani, F. (2000). Detection of photosynthetic activity and water stress by imaging the red chlorophyll fluorescence. *Plant Physiology and Biochemistry, 38*(11), 889–895.
Lichtenthaler, H.K., Lang, M., Sowinska, M., Heisel, F. & Miehé, J.A. (1996). Detection of vegetation stress via a new high resolution fluorescence imaging system. *Journal of Plant Physiology, 148*(5), 599–612.

Lichtenthaler, H.K., Langsdorf, G., Lenk, S. & Buschmann, C. (2005). Chlorophyll fluorescence imaging of photosynthetic activity with the flash-lamp fluorescence imaging system. *Photosynthetica*, *43*(3), 355–369.

Pan, G., Yan, G., Song, X. & Qiu, X. (2009). BP neural network classification for bleeding detection in wireless capsule endoscopy. *Journal of Medical Engineering & Technology*, *33*(7), 575.

Seuret, M., Alberti, M., Liwicki, M. & Ingold, R. (2017). PCA-initialized deep neural networks applied to document image analysis. In *Document Analysis and Recognition (ICDAR), 2017 14th IAPR International Conference* (Vol. 1, pp. 877–882).

Takayama, K., Nishina, H., Iyoki, S., Arima, S., Hatou, K., Ueka, Y. & Miyoshi, Y. (2011). Early detection of drought stress in tomato plants with chlorophyll fluorescence imaging–practical application of the speaking plant approach in a greenhouse–. *IFAC Proceedings Volumes, 44*(1), 1785–1790.

Vankayala, V.S. & Rao, N.D. (1993). Artificial neural networks and their application to power systems. *Proceedings—IEEE International Symposium on Circuits and Systems, 12*, 67–69.

Yin, S., Steven, X.D., Naik, A., Deng, P. & Haghani, A. (2010). On PCA-based fault diagnosis techniques. In *Control and Fault-Tolerant Systems, 2010* (pp. 179–184). IEEE.

Optical communications

Frontier Research and Innovation in Optoelectronics Technology and Industry – Habib & Lewis (Eds)
© 2019 Taylor & Francis Group, London, ISBN 978-1-138-33178-5

An optical fiber ultrasonic sensor based on cascaded Fiber Bragg Grating Fabry-Perot probe

X.H. Bai, M.L. Hu, T.T. Gang, J. Wang & L.G. Dong
Key Laboratory of Photoelectric logging, School of Physics, Northwest University, Xi'an, China

ABSTRACT: A fiber-optic ultrasonic sensor based on cascaded Fiber Bragg Grating Fabry-Perot (FBG-FP) is proposed and experimentally demonstrated. An air bubble with a small size is formed on the end of FBG when the corrosion of FBG and Single Mode Fiber (SMF) are spliced with fiber fusion splicer. Because of the synergic effect of the cascaded FBG-FP, the sensor performs 67 dB Signal-to-Noise Ratio (SNR) to high frequency ultrasound. In addition, two models are designed to verify the chromatography ability of the sensor. The results of the present study indicate that the sensor could distinguish the interface of the water and the models within a range.

Keywords: Fiber optic sensor; FBG-FP; ultrasonic sensor

1 INTRODUCTION

The fiber-optic sensor has been widely used for detecting ultrasound (Park S. et al. 2012). Compared with the traditional piezoelectric transducer (PZT), it has some advantages such as high sensitivity, good electromagnetic interference immunity, wide dynamic range, small volume, and multiplex capability (Yamashita K. et al. 2015, Zhou J.C. et al. 2015, Rong Q.Z. et al. 2014). As one kind of the fiber-optic sensors, optical means based on FBGs with its unique advantages, such as absolute parameter, strong anti-interference ability, high reusability and so on, are noticed widely by many researchers. Furthermore, it can realize quasi-distributed sensing networks by cascaded fiber gratings (Yu Z. et al. 2016, Li Z.X. et al. 2012, Rong Q.Z. et al. 2017).

By contrast, the sensor based on interferometric fiber-optic sensor (IFOS) is more compact, especially the sensor based on Fabry-Perot interferometer (FPI). It has wide bandwidth and multi ways of the demodulation method. However, inevitably IFOS also has high sensitivity of other physical variables at the same time, such as temperature, low frequency strain and other parameters (Feng X. et al. 2014, Shinoda M. et al. 2013, Wu C.et al. 2011).

In this paper, we propose an ultrasonic sensor which combined with FBG and FPI. Unlike most papers though, the role of the FP is for modulating the FBG, rather than for measuring other parameters. In addition, the FP is designed to a small size (90-μm-overall length) to avoid the interference of other physical variables (for example temperature, low frequency strain). Experiments show that the sensor based on cascaded FBG-FP performs a high signal-to-noise ratio (SNR) to the high frequency ultrasound after FP modulation.

2 EXPERIMENTAL DESIGN

The schematic diagram of the proposed sensor is shown in Fig. 1(a). The FP with a 66-μm-long and 24-μm-thick is cascaded on the end of the FBG. When the UW is applied on the sensor, not only will the center wavelength of the FBG shift, the reflection spectrum of the FP will also shift. Therefore, the synergic effect of the cascaded FBG-FP will make the sensor perform a high sensitivity. Although the wavelength shift of the FP is not so much owing to the small size of the air bubble, the interference of other physical variables can be avoided.

The FBG-FP-making process is shown in Fig. 1(b). Both FBG and SMF are placed into hydrofluoric acid solution to be etched for 45 min. Because the cladding and the fiber core have different corrosion rate, there is a pinhole on the end of FBG and SMF respectively. Then an air bubble forms at the end of the FBG after the splicing with fiber fusion splicer. In order to improve the sensitivity of the sensor to the ultrasound, some grooves are etched on the air bubble by using 800 nm FemtoSecond laser. Finally, we can get the cascaded FBG-FP structure with the precise cleaving setup as shown in Fig. 1(c). Meanwhile, the sensitivity of the sensor to ultrasound is improved further due to the small size.

With a broadband source and optical spectrum analysis, the reflection spectrum of the sensor can be obtained, as shown in Fig. 2(a). The rad, green and blue curve demonstrate the reflection spectrum of the sensor at the first measurement, after a month later and after cleaning respectively. As can be seen from the diagram, the shape of the reflection spectrum has not changed, the energy of the reflected spectrum just decreases slightly after a month later, and it increases after cleaning. This indicates that the sensor has the advantage of good stability. But we should pay attention to keeping the sensor and the placement environment clean. Figure 2(b) shows the larger version of the reflection spectrum. The spectral bandwidth

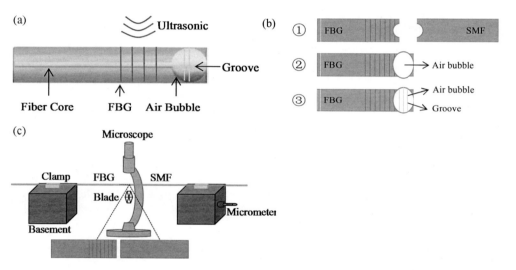

Figure 1.   (a) The schematic diagram of the sensor; (b) The production process of the sensor; (c) Schematic diagram of the precise cleaving setup.

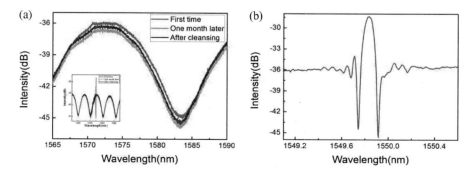

Figure 2.   (a) The reflection spectrum of the cascaded FBG-FP; (b) The larger version of the reflection spectrum.

just is 0.17 nm, which is very important for the side-band filtering technology-based intensity interrogation.

3 EXPERIMENT RESULTS AND DISCUSSION

The schematic diagram of the measurement device using the proposed sensor is shown in Fig. 3. The light, emitted by a tunable laser (Santec, 710) with a 100 KHz linewidth and a 0.1 pm tunable resolution, is launched into the sensor by a circulator. The light reflected by the sensor is guided to a photoelectric detector (New Focus, 2053) with a bandwidth of 10 MHz at 0 dB gain. A piezoelectric transducer (PZT) driven by a function generator is employed to generate UW. All of the connecting lines are connected to the shielded connector block. So we can record the signals in real-time by the PCI data acquisition card. In order to move the PZT and the sensor conveniently, we controlled the scanning platform by using a personal computer (PC) with RS232-USB. In addition, the PZT, the sensor and the models are placed in water because of the great loss of UW in air.

With the measurement device, we detected the continuous sinusoidal UW firstly. A continuous sinusoidal wave of 1 MHz with a voltage of 10 Vpp is used for driving the PZT and the responses of the sensor are shown in Fig. 4(a). The voltage of the signal (as the green

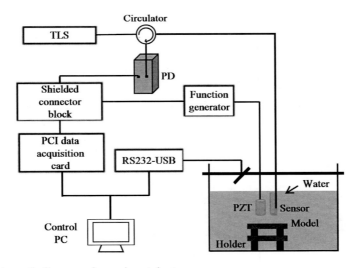

Figure 3. Schematic diagram of experimental set-up.

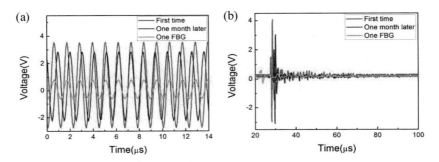

Figure 4. (a) The responses of the sensor to continuous UW in first time and a month later; (b) The responses of the sensor to pulse UW in first time and a month later.

253

curve shows), obtained with single FBG ultrasonic sensor, is about 1.42 V while the voltages of the signals (as the red curve and blue curve show) are 6.34 V and 5.07 V, respectively. Obviously, the synergic effect of the FBG-FP can be observed when it is used for the ultrasonic detection, as the FBG-FP shows superior effect to the single operation. The only difference between two signals observed for the first time and a month later is the 1.27 V change of the voltage, which is caused by the change of the reflected spectrum energy. There for, the higher the reflected spectrum energy, the better the sensor performs. Figure 4(b) shows the responses of the sensor to pulse-wave. The difference between these signals is similar with continuous sinusoidal UW. And the SNR of the signal is 67 dB, which is higher than previous sensors (53 dB).

Two models are designed to verify the chromatography ability of the sensor. As shown in Fig. 5(a), three glass slides are equally distributed at regular intervals (10 mm). The difference between Fig. 5(a) and Fig. 5(b) is that the distance between the second slide and the third slide changes from 10 mm to 20 mm. The corresponding responses of the sensor are shown in Fig. 6(a) and 6(b).

From Fig. 6(a), three signals corresponding to the first slide, second slide and third slide respectively could be found and the time differences are 6.01 μs, and 6.12 μs respectively. The velocity of sound in water is taken to be 1550 m/s. According to the range formula,

$$s = v \cdot t \tag{1}$$

where s, v, t represent distance, speed and time respectively, we can obtain that the corresponding distances are 9.32 mm, and 9.49 mm. Compared with actual distance, the relative errors are 6.8% and 5.1%. In addition, the water surface signal and bottom surface of water tank signal also could be found in the far left and right in Fig. 6(a).

In Fig. 6(b), the time differences are 6.16 μs and 12.96 μs respectively. With the same method, the corresponding distances are 9.55 mm and 20.09 mm respectively the relative

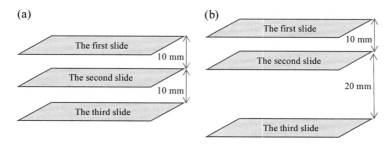

Figure 5.   The abridged general views of (a) Model no. 1 and (b) Model no. 2.

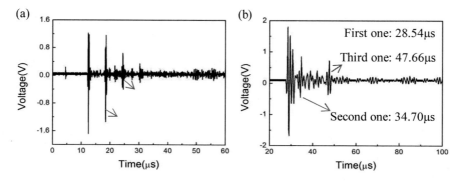

Figure 6.   Real-time measurement results of (a) Model no. 1 and (b) Model no. 2.

errors are 4.5% and 0.5%. In fact, there will be error during the making process of the models. And the impurities in the water will cause a change of the velocity. Overall, the sensor has the chromatography ability and the average measuring error is 4.2%, which is important for us to use it for chromatography detection.

4 CONCLUSION

In this paper, we proposed and experimentally demonstrated a fiber-optic ultrasonic sensor based on cascaded fiber Bragg grating Fabry-Perot (FBG-FP). Because the difference of corrosion rate between the cladding and the fiber core, an air bubble with a small size was formed on the end of FBG when the corrosion of FBG and single mode fiber (SMF) were spliced with fiber fusion splicer. The role of the FP is for modulating the FBG. Due to the small size, the interference of other physical is avoidable. In addition, to verify the chromatography ability of the sensor, two models were designed and tested using the sensor and the average measuring error is 4.2%.

FUNDING

This work is supported by the National Natural Science Foundation of China, Natural Science Foundation of Shaanxi Province, the Natural Science Foundation of Northwest University.

REFERENCES

Feng, X. 2014. Fiber-optic acoustic pressure sensor based on large-area nanolayer silver diaghragm, *Opt. Lett*: 39, 2838–2840.

Li, Z.X. 2012. Analysis of ultrasonic frequency response of surface attached fiber Bragg grating, *Appl. Opt*: 51(20), 4709–4714.

Park, S. 2012. Standing wave brass-PZT square tubular ultrasonic motor, *Ultrasonics*: 52, 880–889.

Rong, Q.Z. 2014. Orientation-dependent fiber-optic accelerometer based on grating inscription over fiber cladding, *Opt. Lett*: 39, 6616–6619.

Rong, Q.Z. 2017. Ultrasonic imaging of seismic physical models using fiber Bragg grating Fabry-Perot probe, *IEEE J. Sel. Top. Quantum Electron*: 23(2), 5600506.

Shinoda, M. 2013. Vibration measurement of structures under railway bridge by use of small oscillator and optical fiber sensor, *J. Civil Eng. Jpn*: 69,40–56.

Wu, C. 2011. High-pressure and high-temperature characteristics a Fabty-Perot interferometer based on photonic crystal fiber, *Opt. Lett*: 36, 412–414.

Yamashita, K. 2015. Intrinsic stress control of sol-gel derived PZT films for buckled diaphragm structures of highly sensitive ultrasonic microsensors, *Eurosensors*: 120, 1205–1208.

Yu, Z. 2016. Theoretical and experimental investigation of fiber Bragg gratings with different lengths for ultrasonic detection, *Photonic Sensors*: 6(2), 187–192.

Zhou, J.C. 2015. Water temperature measurement using a novel fiber optic ultrasound transducer system, *2015 IEEE International Conference on*.

Frontier Research and Innovation in Optoelectronics Technology and Industry – Habib & Lewis (Eds)
© *2019 Taylor & Francis Group, London, ISBN 978-1-138-33178-5*

Nonlinear optical properties of water-soluble PbS and Ag$_2$S quantum dots under femtosecond pulses

Q. Chang, X.D. Meng & X.W. Liu
Heilongjiang University, Harbin, China

ABSTRACT: The nonlinear optical properties of Ag$_2$S and PbS quantum dots were studied through a top-hat Z-scan technique under femtosecond laser pulses. The open Z-scan results show that Ag$_2$S and PbS quantum dots both have a valley value, which indicates that they have strong reverse-saturation absorption characteristics. The nonlinear absorption coefficients of Ag$_2$S and PbS quantum dots are 4.17×10^{-12} and 8.38×10^{-13} esu, respectively. The results of closed Z-scans show that Ag$_2$S and PbS quantum dots all have self-focusing properties, and their nonlinear refractive index coefficients are 3.06×10^{-12} and 4.48×10^{-12} esu, respectively. Using transmittance curve fitting, the third-order nonlinear polarizabilities of Ag$_2$S and PbS quantum dots are calculated as 1.15×10^{-13} and 1.07×10^{-13} esu, respectively. The nonlinear absorption characteristics of an Ag$_2$S quantum dot are significantly better than those of a PbS quantum dot. The physical mechanism of the nonlinear optical response of Ag$_2$S and PbS quantum dots is discussed. The results show that the nonlinear refraction of both samples is enhanced with an increase of excitation energy, but the difference is that the refractive coefficient of an Ag$_2$S quantum dot increases with an increase of energy, while the nonlinear refractive coefficient of a PbS quantum dot decreases with an increase of energy. These two materials are of great significance for research into optoelectronics, biosensors, optical limiting and adaptive techniques.

1 INTRODUCTION

Semiconductor quantum dot materials have quantum confinement effects, surface effects and strong nonlinear optical effects. As a result, they have very broad application prospects in the fields of optical switching, optical limiting and optical communication (O'Flaherty et al., 2003). PbS is a direct wide-band-gap semiconductor of the IV-VI category, and its forbidden bandwidth is 0.41 eV at room temperature. Ag$_2$S is an I-VI semiconductor material with a forbidden bandwidth of 1.19 eV at room temperature. In recent years, researchers have studied the transient kinetics and the excited-state absorption process of SiO$_2$ sol-gel film composed of PbS nanoparticles and PbS nanocrystals doped in silicate glass (Saraidarov et al., 2005; Li et al., 2006; Dementjev & Gulbinas, 2009). The ultra-fast carrier relaxation and multi-exciton generation processes of PbS quantum dots have been studied, which prove that photo induced absorption and photo bleaching occur simultaneously in the excited-state absorption process, and there is a very intense competition process (Wheeler et al., 2011); when the photon energy is greater than 2.7 times that of the band gap, a PbS quantum dot will have multiple exciton effects (Shen et al., 2012). The optical nonlinearity of a PbS quantum dot is greatly enhanced in the strong confinement region. Based on this feature, it has broad application prospects in optical information storage and fast communication switching devices (Capoen et al., 2003; Liu et al., 2015; Aleali & Mansour, 2016). The strong absorption characteristics of an Ag$_2$S quantum dot give it a superior nonlinear optical effect. Therefore, there are additional in-depth research opportunities and important applications in the fields of sensors, biomedical and liquid lubrication (Kitova et al., 1994;

Hull et al., 2002). In this paper, the nonlinear characteristics of PbS and Ag$_2$S quantum dots under excitation by different laser energies are compared through femtosecond laser pulses.

2 THEORETICAL ANALYSIS

The Gaussian beam emitted by the light source passes through the diaphragm A_1 to form a top-hat beam, which is concentrated by L_1. The sample moves along the Z-axis on the stepping motor. After the top-hat beam has passed through the sample, the background light can be eliminated by A_2; then the light passing through the sample is split into two beams by a beam splitter. Finally, it is transmitted to the detector D through L_2, as shown in Figure 1. This device allows more high-frequency light signals to pass through the baffle, increasing the sensitivity of the measurement.

The transverse light field distribution of the top-hat beam after convergence can be written as (Zhao & Palffy-Muhoray, 1994):

$$E(r,z,t) \propto \exp\left[-\frac{t^2}{2\tau^2} \right] \frac{2J_1(\pi r / \omega_0)}{\pi r / \omega_0} \otimes \frac{\exp(i\pi r^2 / \lambda z)}{\lambda z} \tag{1}$$

where r is the lateral polar coordinate and J_1 is the first-order Bessel function. Therefore, the beam's waist radius adopts the same expression formula as the literature (Shen et al., 2008; Zhao & Palffy-Muhoray, 1993), that is, $\omega_0 = \lambda f / d$, the diffraction length of the beam $z_0 = \pi \omega_0^2 \lambda$.

Here, for the sake of discussion, we only consider the third-order nonlinear refraction. The relationship between the light intensity and the phase of the sample is as follows:

$$\frac{d\Delta\varphi}{dz'} = kI\gamma \tag{2}$$

where $\Delta\varphi$ is the phase change, z' is the depth of propagation of the beam in the sample, I is the intensity of the light in the sample, and γ is the nonlinear refractive index of the sample.

The light field distribution in the sample is:

$$E(r,z,t) = E_0(r,z,t)\exp(-\alpha L)\exp[i\Delta\varphi(r_1,t)] \tag{3}$$

The distribution of the light field at the baffle is defined as $E_d(r_2,z,t)$; under linear conditions, the light field distribution at the baffle can be defined as $E_d'(r_2,z,t)$. The light field distribution at the stop A_2 and the baffle can be obtained by two Fresnel diffraction integrals. Therefore, the beam intensity in D can be obtained by integrating the light intensity distribution $|E_d|^2$ at the baffle with the space and pulse time. Through normalization, we can calculate nonlinear transmittance:

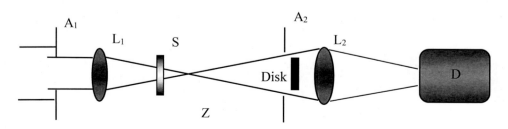

Figure 1.   Experimental device for a top-hat Z-scan.

$$T(z) = \frac{\int_{-\infty}^{+\infty} \int_{r_d}^{+\infty} 2\pi r_2 \,|\, E_d \,|^2 \, dr_2 dt}{\int_{-\infty}^{+\infty} \int_{r_d}^{+\infty} 2\pi r_2 \,|\, E_d' \,|^2 \, dr^2 dt} \qquad (4)$$

Thus, by fitting the curve of the nonlinear transmittance $T(z)$, we can obtain the nonlinear refractive index coefficient γ of the sample.

3 EXPERIMENT

Material preparation methods are detailed in the literature (Chang et al., 2017). The incident source was provided by a mode-locked **PHAROS Yb:KGW** laser (Light Conversion, Vilnius, Lithuania) with a pulse width of 190 fs (full width at half-maximum) and a repetition rate of 10 Hz. The experimental setup is shown in Figure 1.

In order to compare the nonlinear absorption and the nonlinear refraction intensity of water-soluble Ag_2S and PbS quantum dots, we measured them under the same experimental conditions. The laser wavelength used in the experiment was 400 nm, the waist radius of the beam was 86 μm. The experimental results of the two samples under 110 nJ laser energy are shown in Figures 2 and 3. At the same time, experiments were carried out on the effects on the samples of different energies. The pulse width was 190 fs, the excitation wavelength was 515 nm, the beam waist radius was 11.1 μm, and the laser energies used were 0.1, 0.15 and 0.2 μJ. The experimental results are depicted in Figures 4 and 5.

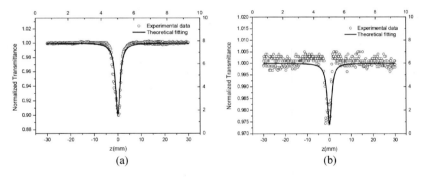

Figure 2.   Open-aperture Z-scan normalized transmittance curves of water-soluble quantum dots of: (a) Ag_2S; (b) PbS (190 fs, 400 nm).

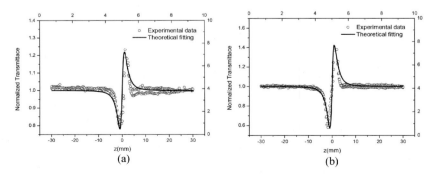

Figure 3.   Closed-aperture Z-scan normalized transmittance curves of water-soluble quantum dots of: (a) Ag_2S; (b) PbS (190 fs, 400 nm).

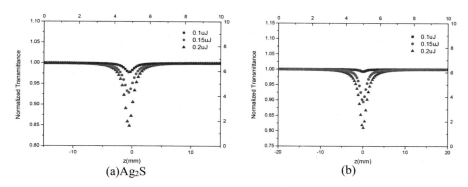

(a)Ag₂S   (b)

Figure 4. Open-aperture Z-scan normalized transmittance curves of water-soluble quantum dots under different energies: (a) Ag$_2$S; (b) PbS (190 fs, 515 nm).

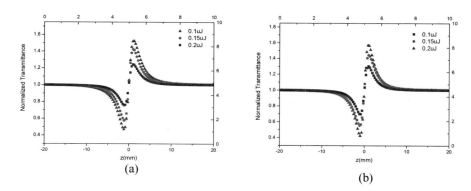

(a)   (b)

Figure 5. Closed-aperture Z-scan normalized transmittance curves of water-soluble quantum dots under different energies: (a) Ag$_2$S; (b) PbS (190 fs, 515 nm).

Table 1. Third-order nonlinear characteristic parameters of water-soluble Ag$_2$S and PbS quantum dots under a femtosecond pulse.

	Ag$_2$S	PbS
Nonlinear refractive index coefficient n_2 (esu)	3.06×10^{-12}	4.48×10^{-12}
Nonlinear absorption coefficient β (m/W)	4.17×10^{-12}	8.38×10^{-13}
Third-order nonlinear polarizability $\chi^{(3)}$	1.15×10^{-13}	1.07×10^{-13}

4 RESULTS AND DISCUSSION

Figure 2 shows us that the open Z-scan normalized transmittance curves of the two samples are all in the shape of a valley, so their nonlinear absorption exhibits a reverse-saturation absorption effect. Similarly, from Figure 3, we can see that the closed Z-scan normalized transmittance curves of the two samples are also the same, both having valley peaks first, so their nonlinear refraction exhibits a self-focusing effect and the nonlinear refractive coefficients are positive. By fitting the experimental results of the Z-scan and bringing the data into the formula, we can obtain the nonlinear absorption coefficients, nonlinear refractive coefficients and third-order nonlinear polarization of the two samples. The third-order nonlinear characteristic parameters of water-soluble Ag$_2$S and PbS quantum dots are shown in Table 1.

Table 1 shows that the nonlinear absorption of a water-soluble Ag$_2$S quantum dot is strong, and that of a water-soluble PbS quantum dot is weak. The nonlinear refractive effect of a water-soluble Ag$_2$S quantum dot is similar to that of a PbS quantum dot.

We compare and analyze the nonlinear properties of water-soluble Ag_2S and PbS quantum dots at different laser energies. The normalized transmittance curves of the open and closed Z-scans of the two samples are shown in Figures 4 and 5, respectively.

Figure 4 reflects the effect of different energies on the nonlinear absorption of the samples. By comparing the normalized open-hole Z-scan transmission curves of PbS and Ag_2S quantum dots, it can be seen that the corresponding valleys of the two samples are sequentially deepened as the energy increases. The intensity of the light at the focus is:

$$I_0 = 2E_0 / (\pi^{3/2} \omega_0^2 \tau) \tag{5}$$

where E_0 is the pulse energy, that is, when the beam waist radius ω_0 and the pulse width τ are constant, the light intensity I_0 is proportional to the pulse energy E_0. Therefore, the light intensity at the focus increases as the energy increases. It can be seen from Figure 6 that as the light intensity increases, the transmittance of the two samples decreases in turn, which is a reverse-saturation absorption (Liaros et al., 2013). The nonlinear absorption coefficients of water-soluble PbS and Ag_2S quantum dots at different energies are shown in Figure 6, which shows that the greater the energy of the excitation light the stronger the anti-saturation absorption effect of the two samples.

Figure 7 reflects the effect of different energies on the nonlinear refraction of the two samples. It shows that the peak-to-valley values of the closed Z-scan transmittance curves

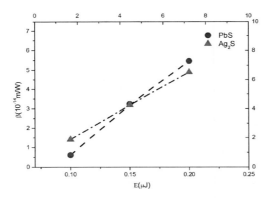

Figure 6.   Nonlinear absorption coefficient of water-soluble Ag_2S and PbS quantum dots under different energies (190 fs, 515 nm).

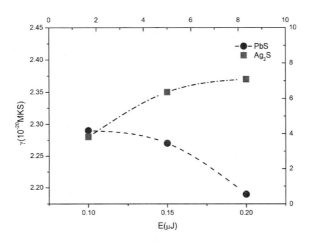

Figure 7.   Nonlinear refraction coefficient of water-soluble Ag_2S and PbS quantum dots under different energies (190 fs, 515 nm).

of the two samples are proportional to the energy, that is, the greater the light intensity, the more obvious the phenomenon of refraction. The difference is that the nonlinear refractive coefficient of an Ag_2S quantum dot increases with an increase of energy, and the nonlinear refractive coefficient of a PbS quantum dot decreases with an increase of energy.

5 CONCLUSIONS

The nonlinear absorption and refraction properties of water-soluble Ag_2S and PbS quantum dots were studied by a femtosecond top-hat Z-scan technique. The experimental data shows that as the energy increases, the transmittance of the two samples decreases, which is consistent with the anti-saturation absorption effect. The water-soluble Ag_2S and PbS quantum dots all exhibit a self-focusing effect. The nonlinear refractive coefficient of an Ag_2S quantum dot increases with an increase of energy, whereas the nonlinear refractive coefficient of a PbS quantum dot decreases with an increase of energy, but these two materials have a wide range of potential applications in optoelectronics, biosensors and optical limiting.

REFERENCES

Aleali, H. & Mansour, N. (2016). Nanosecond high-order nonlinear optical effects in wide band gap silver sulfide nanoparticles colloids. *Optik, 127*(5), 2485–2489.

Capoen, B., Martucci, A., Turrell, S. & Bouazaoui, M. (2003). Effects of the sol-gel solution host on the chemical and optical properties of PbS quantum dots. *Journal of Molecular Structure, 651–653,* 467–473.

Chang Qing, Gao Ying, Liu Xiwen, Chang Cun. (2017). Nonlinear properties of water-soluble Ag_2S and PbS quantum dots under picosecond laser pulses. *IOP Conference Series: Earth and Environmental Science.*

Dementjev, A. & Gulbinas, V. (2009). Excited state absorption of PbS nanocrystals in silicate glass. *Optical Materials, 31*(4), 647–652.

Hull, S., Keen, D.A., Sivia, D.S., Madden, P.A. & Wilson, M. (2002). High temperature structural study of Ag_2S. *J Phys Condens Matter, 14*(1), 19–17.

Kitova, S., Eneva, J., Panov, A. & Haefke, H. (1994). Infrared photography based on vapor-deposited silver sulfide thin films. *J Imaging Sci Technol, 38*(5), 484–488.

Liaros, N., Aloukos, P., Kolokithas-Ntoukas, A., Bakandritsos, A., Szabo, T., Zboril, R. & Couris, S. (2013). Broadband optical power limiting of graphene oxide colloids in the picosecond regime. *J Phys Chem C, 117,* 6842–6850.

Liu, L.W., Hu, S.Y., Dou, Y.P., Liu, T.H., Lin, J.Q. & Wang, Y. (2015). Nonlinear optical properties of near-infrared region Ag2S quantum dots pumped by nanosecond laser pulses. *Beilstein J Nanotechnol, 6,* 1781–1787.

Li, D., Liang, C.-J. & Qian, S.-X. (2006). The femtosecond nonlinear optical properties of PbS nanoparticles composite sol-gel films. *Chinese Journal of Luminescence, 27*(4), 614–616.

O'Flaherty, S.M., Hold, S.V., Brennan, M.E., Cadek, M., Drury, A., Coleman, J.N. & Blau, W.J. (2003). Nonlinear optical response of multiwalled carbon-nanotube dispersions. *J Opt Soc Am B, 20*(1), 49–58.

Saraidarov, T., Reisfeld, R., Sashchiuk, A. & Lifshitz, E. (2005). Nanocrystallites of lead sulfide in hybrid films prepared by sol-gel process. *Materials Science, 34*(2), 137–145.

Shen, Q., Katayama, K., Sawada, T., Hachiya, S. & Toyoda, T. (2012). Ultrafast carrier dynamics in PbS quantum dots. *Chemical Physics Letters, 542,* 89–93.

Shen, X., Chen, W. & Lu, M. (2008). Wireless sensor networks for resources tracking at building construction sites. *Tsinghua Science and Technology, 13*(S1), 78–83.

Wheeler, D.A., Fitzmorris, B.C., Zhao, H.G., Ma, D.L. & Zhang, J.Z. (2011). Ultrafast exciton relaxation dynamics of PbS and core/shell PbS/CdS quantum dots. *Sci China Chem, 54*(12), 2009–2015.

Zhao, W. & Palffy-Muhoray, P. (1993). Z-scan technique using top-hat beams. *Appl Phys Lett, 63,* 1613–1615.

Zhao, W. & Palffy-Muhoray, P. (1994). Z-scan measurement of $\chi^{(3)}$ using top-hat beams. *Appl Phys Lett, 65,* 673–675.

Frontier Research and Innovation in Optoelectronics Technology and Industry – Habib & Lewis (Eds)
© 2019 Taylor & Francis Group, London, ISBN 978-1-138-33178-5

A three-dimensional indoor positioning system based on visible light communication

Beibei Chen, Junyi Zhang, Wei Tang, Yujun Liu & Shaohua Liu
School of Electronic Engineering, Beijing University of Posts and Telecommunications, Beijing, China

Yong Zuo & Yitang Dai
State Key Laboratory of Information Photonics and Optical Communications, Beijing University of Posts and Telecommunications, Beijing, China

ABSTRACT: Visible Light Positioning (VLP) has drawn more and more attention in recent years. In this paper, a novel three-dimensional indoor positioning algorithm is proposed, which is suitable for both the small and large scene. Then, the position of the terminal can be estimated based on the Received Signal Strength Indication (RSSI), a three-dimensional positioning algorithm which combines the Minimum Triangle (MT) method and Weighted Centroid (WC) method (MT-WC-RSSI). The simulation results show that the proposed positioning algorithm has an average positioning error of about 4.11 cm and an average height error of about 1.78 cm.

Keywords: VLC, VLP, three-dimensional positioning, RSSI, minimum triangle method

1 INTRODUCTION

Visible Light Communication (VLC) has becoming a popular research area in recent years (Tang et al., 2017). As a part of visible light communication, Visible Light Positioning (VLP) is considered a promising technology in the field of indoor positioning (Zhao & Chi, 2015). Traditional techniques based on Radio Frequency (RF) like Wi-Fi, Bluetooth, Radio Frequency Identification (RFID) and Ultra-Wide Band (UWB) have some limitations (Thomas et al., 2013; Stefania & Gianluigi, 2014; Abderrahmen et al., 2015). Apart from high cost, these methods are relatively poor in positioning accuracy and can produce electromagnetic interference that is harmful to the human body (Yang et al., 2012). Compared with these methods, VLP has several advantages. It can improve the utilization efficiency of LED, for LED can be used for both illumination and communication. Besides, VLP has no electromagnetic interference, better positioning accuracy and a lower cost (Sevil & Taner, 2015; Armstrong et al., 2013; Vellambi et al., 2014). Three-dimensional positioning is a new direction in the field of indoor positioning and there are few algorithms for indoor three-dimensional positioning. This paper proposed a novel three-dimensional indoor positioning algorithm (MT-WC-RSSI).

This paper is organized as follows. In Section 2, the channel model and indoor positioning algorithm are described. Section 3 describes the simulation results and analysis. Section 4 gives the conclusion of this paper.

2 SYSTEM DESIGN AND ALGORITHM

2.1 VLC channel model

An indoor VLC system model with a size of 5 m × 5 m × 3 m is established in this paper. Four LEDs are evenly distributed on the ceiling, playing the roles of transmitters as well as positioning reference nodes. A photo detector is used as a receiver and its position is that to be estimated in the following part. Because the attenuation of the Non-Line Of Sight (NLOS) links is far greater than the Line Of Sight (LOS) links, only LOS links are considered in this paper. In the case of the LOS optical link, the channel gain between the transmitter and the receiver can be expressed as:

$$H_{TR} = \begin{cases} \dfrac{(m_S+1)A}{2\pi D_d^2} \cos^{m_S}(\varphi)\cos(\psi)T_s(\psi)g(\psi) & 0 \le \psi \le \Psi_c \\ 0, & \psi > \Psi_c \end{cases} \qquad (1)$$

where A is the detector's physical area, D_d is the distance between the transmitter and the receiver, φ is the irradiation angle of source, ψ is incidence angle of detector, $Ts(\psi)$ is the gain of an optical filter, $g(\psi)$ is the gain of an optical concentrator, Ψc denotes the width of a Field Of Vision (FOV) at the receiver, m_s is the Lambertian parameter and is calculated by:

$$m_s = -\frac{\ln 2}{\ln \cos \varphi_{1/2}} \qquad (2)$$

where φ is the half power angle of the transmitter.

The power of the P_r is defined as:

$$P_r = H_{TR} \bullet P_t \qquad (3)$$

where P_t is the power of the transmitter. In a VLC system, noise is unavoidable, and it is also the cause of positioning error. Generally speaking, noise is composed of shot noise, thermal noise and Inter-Symbol Interference (ISI). However, as ISI can be eliminated by modulation, only shot noise and thermal noise are considered in this paper:

$$\sigma_{noise}^2 = \sigma_{shot}^2 + \sigma_{thermal}^2 \qquad (4)$$

$$\sigma_{shot}^2 = 2q\gamma P_r B + 2qI_{bg}I_2 B \qquad (5)$$

$$\sigma_{thermal}^2 = \frac{8\pi k T_k}{G} \eta A I_2 B^2 + \frac{16\pi^2 k T_k \Gamma}{g^m} \eta^2 A^2 I_3 B^3 \qquad (6)$$

where q is the elementary charge, B is the equivalent noise bandwidth of receiver circuit, I_{bg} is the background current, I_2 and I_3 are the noise bandwidth factors, k is the Boltzmann's constant, T_k is the absolute temperature, G is the open loop voltage gain, η is the fixed capacitance of unit area of detector, Γ is the FET channel noise factor and g^m is the FET transconductance.

2.2 The Minimum Triangle (MT) method

As Equation 1 described, the distance Dd between the transmitter and the receiver can be calculated as:

$$D_d = \sqrt{\frac{(m+1)A\cos^m(\phi)\cos(\psi)T_s(\psi)g(\psi)P_t}{2\pi P_r}} \qquad (7)$$

The transmitter plane and receiver plane are parallel, so the horizontal distance of the reception is estimated as:

$$d_{xy} = \sqrt{\sqrt[\frac{m+3}{2}]{\frac{(m+1)AT_s(\psi)g(\psi)P_t h^{(m+1)}}{2\pi P_r}} - h^2} \qquad (8)$$

264

However, as both d_{xy} and h are unknown in a three-dimensional indoor positioning system, in order to get h, a minimum triangle method is proposed. It takes only three LEDs to determine the h value of a point, as shown in Figure 1(b). The receiver plane is divided into four symmetric areas, points in the blue area are related to LED1, LED2 and LED3 (LED.123), similarly, points in the pink, yellow and green areas correspond to LED.124, LED.134 and LED.234, respectively. Taking the blue area as an example, the minimum triangle method is illustrated in Table 1.

The area of the triangle can be obtained by Heron's formula, which is expressed as:

$$S = \sqrt{p(p-a)(p-b)(p-c)} \tag{9}$$

where a, b, c represent the length of triangle sides, p is the semi-perimeter of the triangle, which is calculated by:

$$p = \frac{a+b+c}{2} \tag{10}$$

The estimated height h corresponds to the minimum triangle. Taking a test point whose real terminal position is (1.8, 2.4, 0.85) as an example, the minimum triangle is 0.0115 cm^2 when the estimated height h is 0.87 m.

A method based on the total area of the three circles overlap parts to obtain h was proposed in a previous paper (Guan et al., 2017). However, this method is only suitable for a small part of the test points. As shown in Figure 2(a) and (b), when the real terminal position

Table 1. The principle of the minimum triangle method.

1. Calculate the value of d_{xy} of LED.123 according to Equation 8 and record as d_{xy1}, d_{xy2}, d_{xy3} respectively while $h = 0$;
2. Plot circles whose center is the position of each LED;
3. If circles are intersecting, find these six intersection points.
4. Eight triangles can be obtained from six intersection points, calculate their areas and get the minimum;
5. Insert the minimum and its corresponding height h into an array; increasing the height of receiver h by 1 cm;
6. Repeat step 2 to step 5 until $h = 1.5$;
7. Compare values of areas in the array and find the height h corresponding to the minimum.

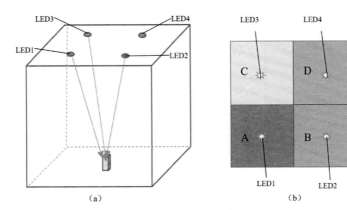

(a) (b)

Figure 1. (a) The scene of indoor visible light communication; (b) division of the receiving area.

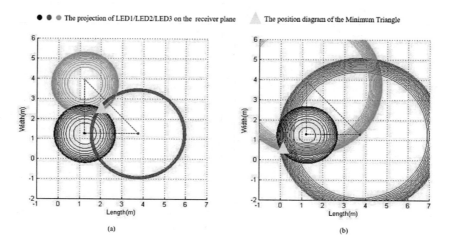

Figure 2. The change in radius when the real terminal position is (a) (1.8, 2.4, 0.85) and (b) (0.2, 0.6, 0.85).

is (1.8, 2.4, 0.85), the minimum overlap area method is applicable, but it is not applicable while the position is (0.2, 0.6, 0.85). In a small room with a small number of light sources, Guan's method cannot locate the receiver close to the edge of the room. In general, Guan's method is not suitable for positioning in a small scene like a meeting room or classroom.

In light of this situation, this paper presents a minimum triangle method, which is applicable to all test points. Compared with Guan's method, the proposed algorithm has wider applicability. It enables the positioning of the whole space in both a small and large scene.

2.3 The Weighted Centroid (WC) method

Multiple LEDs can avoid interference and improve accuracy effectively, so a weighted centroid method is introduced in this paper. In the above part, any three LEDs are combined in four ways (LED.123, LED.124, LED.134 and LED.234). Four estimated coordinates expressed as (x_{R1}, y_{R1}), (x_{R2}, y_{R2}), (x_{R3}, y_{R3}) and (x_{R4}, y_{R4}) are obtained by using the least square estimation. Then, a weighted centroid algorithm is used and expressed in the following form:

$$\begin{cases} x_R = w_1 x_{R1} + w_2 x_{R2} + w_3 x_{R3} + w_4 x_{R4} \\ y_R = w_1 y_{R1} + w_2 y_{R2} + w_3 y_{R3} + w_4 y_{R4} \end{cases} \quad (11)$$

where w_i represents the weighted value and can be obtained by Tang's method (Tang et al., 2017).

To measure the three-dimensional positioning performance, the positioning error can be defined as:

$$error = \sqrt{\left(x_R - x_{real}\right)^2 + \left(y_R - y_{real}\right)^2 + \left(h - z_{real}\right)^2} \quad (12)$$

where $(x_{real}, y_{real}, z_{real})$ are the real coordinates of the receiver.

3 SIMULATION AND RESULT

The parameters of the system are listed in Table 2; 676 test points were taken in each horizontal plane with different heights.

Suppose that the height of the area of human activity is at 0–1.5 m in the actual application environment as a room with a size of 5 m × 5 m × 3 m; the results of three-dimensional positioning error distribution are shown in Figure 3(a–c). The average error of the test plane with a height of 0.5 m, 0.85 m and 1.25 m is 3.81 cm, 3.96 cm and 4.56 cm, respectively. We can see that the maximum error always exists at the room's corner, and test points near the connection of two LEDs also have larger errors, which is a result of edge effect of the minimum triangle method. The simulation results demonstrated that the maximum position error is less than 12.14 cm, the minimum position error is 1.0 cm and the average error distance is 4.11 cm. Apart from a broader application scene of the proposed algorithm, the position accuracy of the minimum triangle method is as high as that of Guan's method.

Table 2. Parameters of the positioning scheme.

Transmitter	Position	LED1(1.25, 1.25, 3) m LED2(1.25, 3.75, 3) m LED3(3.75, 1.25, 3) m LED4(3.75, 3.75, 3) m
	Half power angle	60deg
	Power of each LED	15W
Receiver	Height	0–1.5 m
	Field of view	90deg
	Area	1 cm²
	Conversion efficiency	0.40 [A/W]
	Concentrator gain	1
	Filter gain	1
	Numbers	676/side

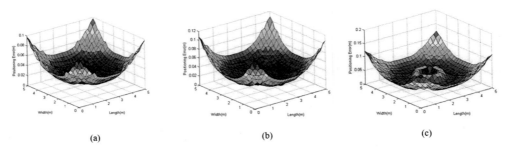

(a) (b) (c)

Figure 3. The distribution of the three-dimensional positioning error: (a–c) represent the results in 0.5 m, 0.85 m and 1.25 m high position points, respectively.

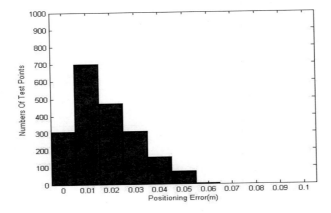

Figure 4. The statistical histogram of the height error.

267

Figure 4 is the statistical histogram of the height error. All the test points ($676 \times 3 = 2028$) are considered. As can be seen from the histogram, the average height error is 1.78 cm, the maximum height error is 6 cm, and the minimum error is 0 cm. Ninety percent of test points have a height error less than 5 cm, and the height error of most points is distributed in the interval of 0–2 cm. As a general rule, the accuracy is related to the height step, so the height error obtained by the minimum triangle method is adjustable.

4 CONCLUSION

In this paper, we proposed a novel three-dimensional indoor positioning algorithm, which has broad application prospects. The scheme of realizing three-dimensional indoor positioning by the minimum triangle method is universal as well as flexible, which can achieve the positioning of the whole space in both a small and large scene. The proposed positioning algorithm was analyzed in an indoor optical wireless environment with dimensions of $5 \text{ m} \times 5 \text{ m} \times 3 \text{ m}$ by computer simulations. The simulation results show that the proposed positioning algorithm has an average positioning error of about 4.11 cm and an average height error of about 1.78 cm.

ACKNOWLEDGMENTS

This work was supported in part by NSFC Program (61471065) and NSFC Program (61335002).

REFERENCES

Abderrahmen, M., Khaled, A.H. & Mohamed, A. (2015). Exploiting social information for dynamic tuning in cluster based WiFi localization. *11th International Conference on Wireless and Mobile Computing, Networking and Communications (WiMob)*, October 19–21, 2015, Abu Dhabi. Piscataway: IEEE.

Jean Armstrong, Ahmet Y. Sekercioglu, Adrian Neild, (2013). Visible light positioning: A roadmap for international standardization. *IEEE Communications Magazine*, 51(12), 68–73.

Jiaqi Zhao, Nan Chi, (2015). Comparative research on several key indoor positioning technologies based on LED visible light communication. *Light and Lighting*, 1, 34–41.

Se-Hoon Yang, Deok-Rae Kim, Hyun-Seung Kim, (2102). Indoor positioning system based on visible light using location code. *Fourth International Conference on Communications and Electronics (ICCE)*, August 1–3, 2012. Piscataway: IEEE.

Sevil, T. & Taner, T. (2015). Indoor localization with Bluetooth technology using artificial neural networks. *19th International Conference on Intelligent Engineering Systems*, September 3–5, 2015, Bratislava, Slovakia. Piscataway: IEEE.

Stefania, M. & Gianluigi, F. (2014). An experimental model for UWB distance measurements and its application to localization problems. *International Conference on Ultra-Wideband*, September 1–3, 2014, Paris, France. Piscataway: IEEE.

Thomas, Q., Wang, Y. & Ahment, S. et al. (2103). Analysis of an optical wireless receiver using a hemispherical lens with application in MIMO visible light communications. *Journal of Lightwave Technology*, 31(11), 1744–1754.

Vellambi, B. N., Muhammad Yasir, Siuwai Ho, (2014). Indoor positioning system using visible light and accelerometer. *IEEE Journal of Lightwave Technology*, 32(19), 3306–3316.

Wei Tang, Junyi Zhang, Beibei Chen, et al, (2017). Analysis of indoor VLC positioning system with multiple reflections. *16th International Conference on Optical Communications and Networks (ICOCN)*, August 7–10, 2017, Wuzhen, China. Piscataway: IEEE.

Weipeng Guan, Yuxiang Wu, Shangsheng Wen, (2017). A novel three-dimensional indoor positioning algorithm design based on visible light communication. *Optics Communications*, 392, 282–293.

Frontier Research and Innovation in Optoelectronics Technology and Industry – Habib & Lewis (Eds)
© *2019 Taylor & Francis Group, London, ISBN 978-1-138-33178-5*

Theoretical investigation of quality factor and light-speed reduction in a plasmon-induced transparency device

P. Dara & H. Kaatuzian
Photonics Research Laboratory, Department of Electrical Engineering, Amirkabir University of Technology, Tehran, Iran

ABSTRACT: In this study, the group index and the quality factor of a plasmon-induced transparency structure, introduced recently, have been investigated in detail with the aim of narrowing the frequency transparency window and obtaining a higher quality factor. The structure consists of an Al_2O_3 grating and metallic film (silver or gold) coated onto SiO_2/TiO_2/SiO_2 layers. The induced transparency occurred due to the wave vector matching between the surface plasmon polaritons and guided mode in the high refractive index layer (TiO_2) in a terahertz regime. This phenomenon can reduce the group velocity of the light. Group index at its highest has been estimated theoretically to be about 298 and 393, respectively, using gold and silver as metals in such an induced transparency system.

1 INTRODUCTION

An Electromagnetically Induced Transparency (EIT) effect is an optical nonlinear phenomenon that is produced by the quantum interference that minimizes the effect of an optical medium on incident light, eliminating absorption and making the medium transparent over a narrow spectral window, which is called the transparency window (Harris, 1997). The high dispersion within this transparency window can reduce group velocity (Hau et al., 1999). This phenomenon has many applications in nonlinear optics, optical storage, optical switching and, generally, all-optical integrated circuits (Lukin & Imamoglu, 2001; Fleischhauer et al., 2005). In order to achieve an all-optical integrated circuit, both the quality factor and slow-down factor have significant roles. In recent years, Surface Plasmon Polaritons (SPPs) have attracted lots of attention because of their interesting physical characteristics, such as confining light at sub-wavelength scale and overcoming the diffraction limit of light, so that using SPPs to fabricate optical storage, optical buffers, terahertz switches and, in general, optical integrated circuits has been the subject of many studies. An analog of the EIT effect was observed in various plasmonic systems such as planar plasmonic metamaterials and a large-area of hybrid plasmon-waveguide systems. The main characteristic of an EIT-like response in plasmonic systems is the appearance of a sharp and narrow transparency window in the absorption spectra, along with a high reduction of light speed, making it perfect for fabricating all-optical integrated circuits (Song et al., 2015; Chen & Fan, 2015; Chai et al., 2016; Hassani Keleshtery et al., 2018; Lu et al., 2018). In this paper, both of these properties have been deeply investigated in a proposed sample plasmonic system, introduced by Lu et al. (2018). The plasmonic system model and the properties of different material layers of the device will be introduced in Section 2. The quality factor will be estimated theoretically in Section 3 and the slow-down factor will be investigated in Section 4, with the aim of improving the technical characteristics of such a plasmonic device.

2 PLASMONIC SYSTEM MODELING

2.1 Structural characteristics

Figure 1 shows the schematic of the plasmonic device composed of an Al_2O_3 grating and SiO_2/TiO_2/SiO_2 layers coated with metallic film (silver or gold). In this system, p, w and h are the pitch, width and height of the Al_2O_3 grating, respectively; t and g represent the thicknesses of the metal and TiO_2 layers and d is the thickness of the SiO_2 spacer between the metal and TiO2 layers. The refractive indices of SiO_2, Al_2O_3 and TiO_2 are $n_s = 1.45$, $n_a = 1.76$ and $n_t = 2.13$, respectively (Lu et al., 2018). Other dimensional features are described in more detail in each section.

2.2 Modeling

The relative permittivity of the metal can be calculated by the Drude dispersion model described in Equation 1 (He et al., 2018; Lu et al., 2018; Kaatuzian & Naseri, 2015):

$$\varepsilon_D = \varepsilon_\infty - \frac{\omega_p^2}{\omega^2 + \Gamma^2} + i\frac{\omega_p^2\Gamma}{\omega(\omega^2 + \Gamma^2)} \tag{1}$$

where ε_∞ is the permittivity limit at high frequencies, ω_p denotes the plasma frequency and Γ is the damping coefficient. The Drude model parameters for gold and silver are listed in Table 1; ω is the angular frequency of light and light is assumed to be normally incident in the y direction.

To simulate the optical response of the plasmonic system, the Finite-Difference Time-Domain (FDTD) method was utilized. In FDTD simulations, the perfectly matched layer absorbing boundary conditions are set at the top and bottom and the periodic boundary conditions are set on the right and left sides of the unit cell. Figure 2 shows the field distribution at the absorption wavelength for gold in the simulated unit cells where it is assumed

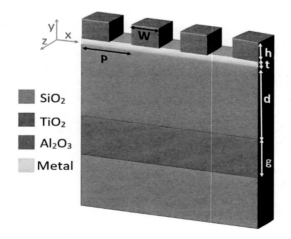

Figure 1. Schematic of the plasmonic system (not to scale).

Table 1. Drude model parameters for gold and silver (He et al., 2018; Lu et al., 2018).

	Gold	Silver
ε_∞	1	3.7
ω_p (rad/s)	1.37e16	1.38253e16
Γ (rad/s)	4.08e13	27.347e12

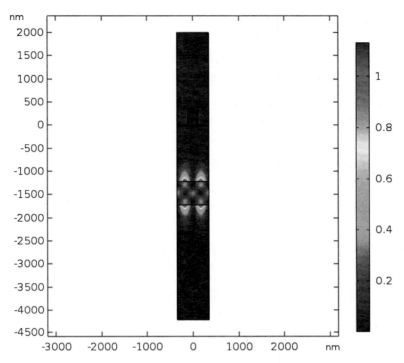

Figure 2. Field distribution at the absorption wavelength for gold in the simulated unit cells when $g = 505$ nm, $t = 19$ nm, $h = 270$ nm, $p = 700$ nm, $w = 220$ nm and $d = 1200$ nm.

that $g = 505$ nm, $t = 19$ nm, $h = 270$ nm, $p = 700$ nm, $w = 220$ nm and $d = 1200$ nm. Figure 2 illustrates that the induced transparency appears when the guided mode is shaped in the High Refractive Index (HRI) layer.

3 QUALITY FACTOR

In this section the quality factor will be studied by structural layer size and material-type alterations. In order to achieve the wave vector matching condition between SPPs and the guided mode in the high refractive index layer (TiO_2) the dimensional properties are set as $g = 495$ nm, $t = 20$ nm, $h = 250$ nm, $p = 700$ nm and $w = 200$ nm, as obtained by Lu et al. (2018) as optimum for silver, and $g = 505$ nm, $t = 19$ nm, $h = 270$ nm, $p = 700$ nm and $w = 220$ nm for gold, with d varying from 1000 to 1400 nm in both cases. Figures 3a and 3b show the absorption spectra of the structure using gold and silver, respectively, as the metal. It can be concluded that by applying these dimensions in both cases (gold or silver) the quality factor will be increased when d increases. It is also observed that by using gold and adjusting the dimensions, the transparency window will be narrower and the quality factor will be increased but, as can be concluded from Figure 3c, the transparency frequency will be decreased by replacing silver with gold and adjusting the dimensions. Table 2 shows the calculated quality factors for gold and silver as a result of changing the dimension of d.

4 SLOW-DOWN FACTOR

In this section, the slow-down factor or group index will be investigated. The group index is calculated via Equation 2 (Zentgraf et al., 2009; Zhang et al., 2018):

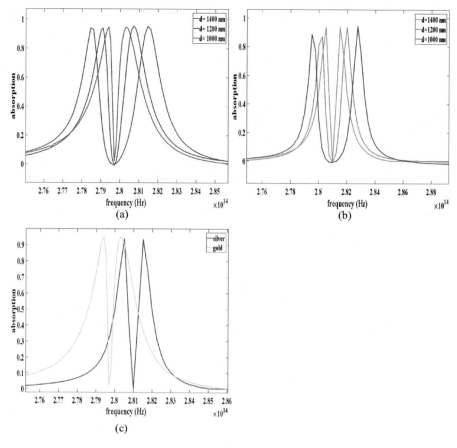

Figure 3. Absorption spectra using: (a) gold when $g = 505$ nm, $t = 19$ nm, $h = 270$ nm, $p = 700$ nm, $w = 220$ nm; (b) silver when $g = 495$ nm, $t = 20$ nm, $h = 250$ nm, $p = 700$ nm, $w = 200$ nm and d varies from 1000 to 1400 nm; (c) gold and silver when $g = 505$ nm, $t = 19$ nm, $h = 270$ nm, $p = 700$ nm, $w = 220$ nm and $d = 1400$ nm using gold as the metal, and $g = 495$ nm, $t = 20$ nm, $h = 250$ nm, $p = 700$ nm, $w = 200$ nm and $d = 1400$ nm using silver as the metal.

Table 2. Quality factor of structure for gold and silver.

d (nm)	Gold	Silver
1400	1398.5906.45	
1200	266.38255.78	
1000		174.81133.33

$$n_g = \frac{C_0}{v_g} = \frac{C_0}{L} \frac{d\varphi(\omega)}{d\omega} \qquad (2)$$

where C_0 is the speed of light in vacuum, v_g is the group velocity, L is the length of the plasmonic system, ω is the angular frequency of light and $\varphi(\omega)$ stands for the transmission phase shift from the light source to the monitor. Table 3 shows the highest slow-down factor for gold and silver by changing the dimension of d.

Figure 4 shows the group index (n_g) using (a) gold and (b) silver as the metal with the dimensional parameters set as in Section 3. It can be concluded that the group index will be increased as d increases.

Table 3.   Highest slow-down factor for gold and silver.

d (nm)	Gold	Silver
1400	298	393
1200	280	128
1000	176	60

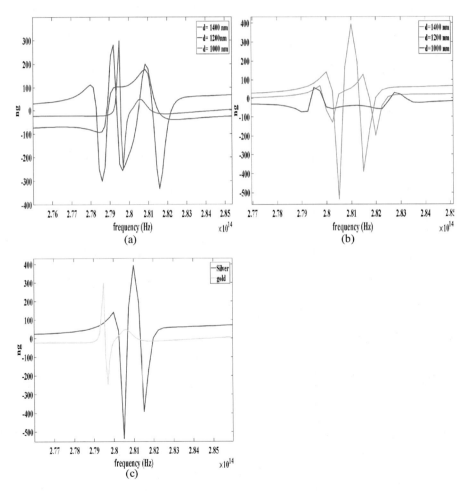

Figure 4.   Group index using: (a) gold when $g = 505$ nm, $t = 19$ nm, $h = 270$ nm, $p = 700$ nm, $w = 220$ nm; (b) silver when $g = 495$ nm, $t = 20$ nm, $h = 250$ nm, $p = 700$ nm, $w = 200$ nm and d varies from 1000 to 1400 nm; (c) gold and silver when $g = 505$ nm, $t = 19$ nm, $h = 270$ nm, $p = 700$ nm, w = 220 nm and $d = 1400$ nm using gold as the metal, and $g = 495$ nm, $t = 20$ nm, $h = 250$ nm, $p = 700$ nm, $w = 200$ nm and $d = 1400$ nm using silver as the metal.

5 CONCLUSION

It is concluded that by using gold instead of silver in the proposed sample plasmonic system, introduced by Lu et al. (2018), and adjusting the dimensional parameters, the transparency window will get narrower and therefore the quality factor of the plasmonic system will be increased, but the frequency of the transparency window will be decreased. Further, it is concluded that the system can reduce the speed of light. The highest group index was estimated to be 298 and 393, respectively, when gold and silver were used as the metal. It is also observed that in both cases (gold and silver) the quality factor and slow-down factor are increased when d increases.

REFERENCES

Chai, Z., Hu, X., Yang, H. & Gong, Q. (2016). All-optical tunable on-chip plasmon induced transparency based on two surface plasmon-polaritons absorption *Applied Physics Letters*.

Chen, X. & Fan, W.-H. (2015). Plasmon-induced transparency in terahertz planar metamaterials. *Optics Communications*, *356*, 84–89.

Fleischhauer, M., Imamoglu, A. & Marangos, J.P. (2005). Electromagnetically induced transparency: Optics in coherent media. *Reviews of Modern Physics, 77*, 633–673.

Harris, S.E. (1997). Electromagnetically induced transparency. *Physics Today*, *50*(7), 36–42.

Hassani Keleshtery, M., Mir, A. & Kaatuzian, H. (2018). Investigating the characteristics of a double circular ring resonators slow light device based on the plasmonics-induced transparency coupled with metal-dielectric-metal waveguide system. *Plasmonics*, *13*(5), 1523–1534.

Hau, L.V., Harris, S.E., Dutton, Z. & Behroozi, C.H. (1999). Light speed reduction to 17 ms^{-1} in an ultra-cold atomic gas. *Nature*, *397*, 594–598.

He, Z., Ren, X., Bai, S., Li, H., Cao, D. & Li. G. (2018). Λ-type and V-type plasmon-induced transparency in plasmonic waveguide systems. *Plasmonics*. doi:10.1007/s11468-018-0746-y.

Kaatuzian, H. & Naseri Taheri, A. (2015). Applications of nano scale plasmonic structures in design of stub filters: A step towards realization of plasmonic switches. In A. Bananej (Ed.), *Photonic crystals*. London, UK: IntechOpen.

Lu, H., Gan, X., Mao, D., Jia, B. & Zhao, J. (2018). Flexibly tunable high-quality-factor induced transparency in plasmonic systems. *Scientific Reports*, *8*, 1558. doi:10.1038/s41598-018-19869-y.

Lukin, M.D. & Imamoglu, A. (2001). Controlling photons using electromagnetically induced transparency. *Nature*, *413*(6853), 273–276.

Song, J., Liu, J., Song, Y., Li, K., Zhang, Z., Xu, Y., ... Song, G. (2015). Plasmon-induced transparency and dispersionless slow light in a novel symmetry-broken metamaterial. *IEEE, Photonics Technology Letters*.

Zentgraf, T., Zhang, S., Oulton, R.F. & Zhang, X. (2009). Ultranarrow coupling-induced transparency bands in hybrid plasmonic systems. *Physical Review B*.

Zhang, Z., Yang, J., He, X., Han, Y., Zhang, J., Huang, J., ... Xu, S. (2018). Active enhancement of slow light based on plasmon-induced transparency with gain materials. *Materials (Basel)*, *11*(6), 941. doi:10.3390/ma11060941.

Frontier Research and Innovation in Optoelectronics Technology and Industry – Habib & Lewis (Eds)
© 2019 Taylor & Francis Group, London, ISBN 978-1-138-33178-5

Improved method for seeking null frequencies in a Sagnac distributed optical fiber sensing system

N. Fang & L.T. Wang
Key Laboratory of Specialty Fiber Optics and Optical Access Networks, Joint International Research Laboratory of Specialty Fiber Optics and Advanced Communication, Shanghai Institute for Advanced Communication and Data Science, Shanghai University, Shanghai, China

ABSTRACT: In order to accurately locate external disturbances in a distributed fiber sensing system based on a Sagnac interferometer, a null-frequencies-seeking method of calculating autocorrelation of frequency spectrum of the interference signal is proposed and demonstrated. The average value of delay frequency intervals between the adjacent autocorrelation peaks of the interference signal spectrum is used as the null-frequency interval. The disturbance can be located by the null-frequency interval. A simulation of a Sagnac fiber-ring interferometer sensing system is built using OptiSystem software. Various disturbances acting on different positions of the sensing fiber are simulated and located. The maximum and minimum location errors within a 1 km sensing fiber are 4.5 m and 0 m, respectively. The simulation results verify that the proposed method is accurate and simple.

Keywords: distributed fiber sensing, Sganac interferometer, null frequencies, autocorrelation of frequency spectrum

1 INTRODUCTION

The distributed optical fiber sensing system based on a Sagnac interferometer configuration has been widely researched in recent years (Huang et al., 2007; Kumagai et al., 2012; Wang et al., 2014). It can be used to detect and locate issues such as pipeline leakage, illegal intrusion, and deformation of large structures. Null frequencies of the frequency spectrum of the phase-change signal induced by an external disturbance are used to locate the disturbance. However, it is often difficult to locate null frequencies due to the influence of system noise. An improved algorithm named "twice-FFT" (double-FFT) has been proposed for multi-point intrusion location in distributed Sagnac sensing systems (Wang et al., 2014). However, this is algorithmically complex and inexplicit in its physical meanings.

In this paper we propose an improved null-frequencies-seeking method, by calculating autocorrelation of the frequency spectrum of the interference signal. The method is demonstrated by a Sagnac fiber-ring interferometer sensing system built using OptiSystem software. The simulation results verify that the proposed method is simple and effective.

2 SYSTEM STRUCTURE AND PRINCIPLES

The system structure of the Sagnac interfering-ring distributed optical fiber sensing system is shown in Figure 1. It consists of an optical source, a 3dB fiber coupler, a standard single-mode fiber, a photodetector (PD) and a signal-processing unit. Half of the fiber acts as the sensing fiber, while the other half acts as the delay fiber. The light from the source is divided into two beams by the coupler and then propagates along counter-directions in the ring. When the beams meet in the coupler, interference occurs.

If an external sinusoidal disturbance acts on the sensing fiber at a point with a distance R from the coupler the AC component of the interference signal power measured by the PD will be given by:

$$P(t) = -\frac{P_0}{2}\cos\left[\Delta\phi_1 - 2\varphi_0\cos\omega_s\left(t - \frac{\tau}{2}\right)\sin\frac{\omega_s\Delta\tau}{2}\right] \tag{1}$$

where P_0 is the power transmitted into the coupler, $\Delta\phi_1$ is the phase bias, φ_0 and ω_s are, respectively, the amplitude and angle frequency of the sinusoidal phase-change signal induced by the external disturbance, τ is the time taken by light propagating once in the ring, $\Delta\tau = n$ $(L-2R)/c$ is the time difference between the propagating time of the two light beams from the disturbance point to the coupler along the counter-directions, n is the refractive index of the fiber core, c is light speed in a vacuum, and L is the ring length.

Assuming φ_0 is small, $\Delta\phi_1$ can be treated as $\pi/2$ (Hoffman & Kuzyk, 2004); thus Equation 1 can be rewritten as:

$$P(t) \approx -P_0\varphi_0\cos\phi_s\left(t - \frac{\tau}{2}\right)\sin\frac{\omega_s\Delta\tau}{2} \tag{2}$$

If the disturbance is broadband, we can say that the powers of the interference signal will be equal to zero when $\omega_s\Delta\tau = 0, 2\pi, \ldots 2N\pi$, where N is an integer, as shown in Figure 2. These frequencies are called null frequencies and are represented as $f_{null} = \omega_{s,null}/(2\pi) = N/\Delta\tau = Nc/[n(L-2R)]$. The frequency intervals of adjacent null frequencies are expressed by $\Delta f_{null} = c/[n(L-2R)]$. Therefore, the disturbance position can be determined by:

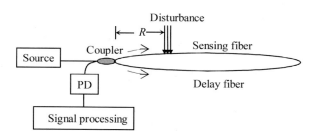

Figure 1. System structure.

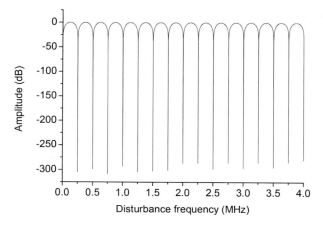

Figure 2. The Fourier response of the interferometer with a ring length of 2000 m to a broadband disturbance at 600 m from the coupler.

276

$$R = \frac{L}{2} - \frac{c}{2n\Delta f_{null}} \tag{3}$$

When the system noise is large, the null frequencies will be overwhelmed. However, as indicated by Figure 2 we can regard the frequency spectrum as a periodic signal. The period is the interval of adjacent null frequencies (Δf_{null}). Therefore, by calculating the autocorrelation of the frequency spectrum of the interference signal, Δf_{null} can be identified. This is a similar process to distinguishing the period from a noisy periodic signal.

3 SIMULATION SYSTEM AND RESULTS

We employed OptiSystem software (Optiwave Systems Inc., Ottawa, Canada) to build the simulation of a Sagnac interferometer fiber-ring distributed sensing system to demonstrate the proposed method of seeking null frequencies. The simulated system is shown in Figure 3. The optical source is a Continuous Wave (CW) laser at 1550 nm. The delay fiber is a 1 km bidirectional optical fiber. The sensing fiber consists of two segments of bidirectional optical fiber. The external disturbance signal acts on the sensing fiber by inserting a phase modulator between the two segments of sensing fiber. By changing the two sensing fibers' lengths we can adjust the location of the disturbance applied to the sensing fiber. After the DC component of interference signal power measured by the PD is blocked, the power data are collected by an oscilloscope.

Because the null-frequency location method is applicable for only broadband disturbance signals, a Gaussian pulse with a bandwidth of 20 MHz serves as the external disturbance signal. Figure 4 presents its temporal waveform and frequency spectrum. When the signal acts on the sensing fiber at a distance of 0.7 km from the coupler and a phase deviation of 5° is induced, the interference signal waveform is as shown in Figure 5.

The frequency spectrum of the interference signal data is obtained from a Fourier transform using MATLAB software (MathWorks Inc., Natick, MA, USA), as shown in Figure 6, from which it can be seen that it is difficult to determine the null frequencies. The autocorrelation of the frequency spectrum data in Figure 6 is then calculated and the autocorrelation curve is as shown in Figure 7. The frequency interval of the adjacent autocorrelation peaks is easily established using a MATLAB program and has a Δf_{null} of 334.28 kHz. According to Equation 3, the disturbance position is 700.9 m. The error compared with the actual disturbance point of 700 m is 0.9 m and the relative error is 0.13%.

The simulation experiments for different disturbance positions were carried out by shifting the disturbance point in the sensing fiber by intervals of 100 m from 0 m to 900 m. The location errors are shown in Figure 8. The maximum error is 4.5 m at 200 m. The minimum error can be as small as zero, such as at 500 m, 600 m and 800 m. The errors are not uniformly

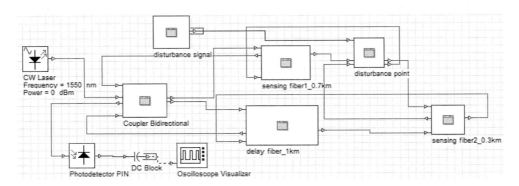

Figure 3.   Simulated system.

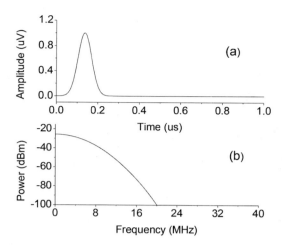

Figure 4.   External disturbance signal: (a) temporal waveform; (b) frequency spectrum.

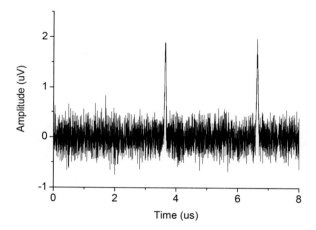

Figure 5.   Temporal waveform of interference signal.

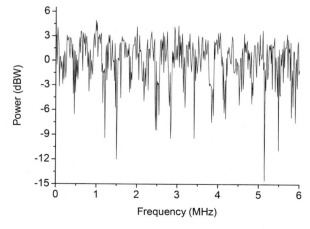

Figure 6.   Frequency spectrum of interference signal.

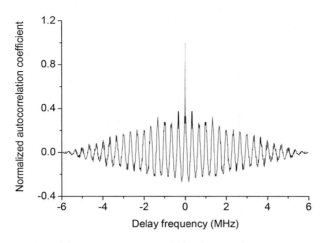

Figure 7.   Autocorrelation of frequency spectrum of interference signal.

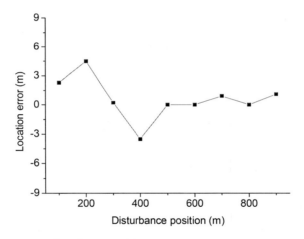

Figure 8.   Location error vs. disturbance position.

distributed along the whole sensing fiber. Moreover, in these positions, the longer the lengths of the frequency spectrum data involved to calculate the autocorrelation, the smaller the location errors.

4 DISCUSSION

The horizontal axis of the autocorrelation of the frequency spectrum is the delay frequency. However, the meaning of the horizontal axis of the second Fast Fourier Transform (FFT) is not clear. Moreover, the relationship of frequency axis of the twice-applied FFT is determined constrainedly. In general, the autocorrelation curve of the spectrum has multiple quasi-periodic peaks. Therefore, we can use the average value of delay frequency intervals between the adjacent autocorrelation peaks as the null-frequency interval (Δf_{null}). The average interval is more accurate than that obtained by the delay frequency difference of two adjacent autocorrelation peaks. However, only one frequency value can be achieved by the twice-FFT method, as shown in Figure 9. The random error is often large.

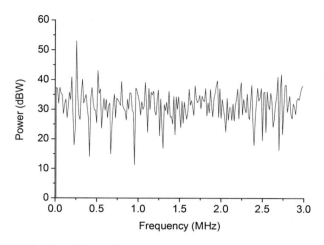

Figure 9.   Double-FFT of interference signal.

5   CONCLUSION

A null-frequencies-seeking method is proposed and demonstrated by the autocorrelation of the frequency spectrum of the Sagnac interference signal. The experimental results of a software simulation show that the proposed method is feasible. Compared to the twice-FFT method (Wang et al., 2014), the method is simple, accurate, and explicit in its physical meanings.

ACKNOWLEDGMENTS

This work is sponsored by the National Natural Science Foundation of China (grant no. 61377082, as well as grant no. 61108004) and Shanghai Pujiang Program (grant no. 14PJD017).

REFERENCES

Hoffman, P.R. & Kuzyk, M.G. (2004). Position determination of an acoustic burst along a Sagnac interferometer. *Journal of Lightwave Technology*, *22*(2), 494–498.

Huang, S.C., Lin, W.W., Tsai, M.T. & Chen, M.H. (2007). Fiber optic in-line distributed sensor for detection and localization of the pipeline leaks. *Sensors and Actuators A: Physical*, *135*(2), 570–579.

Kumagai, T., Sato, S. & Nakamura, T. (2012). Fiber-optic vibration sensor for physical security system. In *Proceedings of 2012 IEEE International Conference on Condition Monitoring and Diagnosis, Bali* (pp. 1171–1174). doi:10.1109/CMD.2012.6416369.

Wang, H., Sun, Q.Z., Li, X.L., Wo, J.H., Shum, P.P. & Liu, D.M. (2014). Improved location algorithm for multiple intrusions in distributed Sagnac fiber sensing system. *Optics Express*, *22*(7), 7587–7597.

Frontier Research and Innovation in Optoelectronics Technology and Industry – Habib & Lewis (Eds)
© 2019 Taylor & Francis Group, London, ISBN 978-1-138-33178-5

A bandwidth-reconfigurable optical filter based on a stimulated Brillouin scattering effect using an optical frequency comb

Jingwen Gong, Xiaojun Li, Qinggui Tan, Wei Jiang & Dong Liang
China Academy of Space Technology (Xi'an), Xi'an, Shaanxi Province, China

ABSTRACT: Due to the fact that the single passband of a microwave photonic filter based on Stimulated Brillouin Scattering (SBS) in fiber is formed by mapping the Brillouin gain spectrum, bandwidth-reconfiguration can be implemented by changing the Brillouin gain linewidth. In this paper, a bandwidth-tunable and reconfigurable microwave photonic filter is proposed and experimentally demonstrated Brillouin gain spectrum narrowing and broadening by using an optical frequency comb. Experiments have shown that the 3 dB and 20 dB bandwidth of the filter are 202 MHz and 314 MHz, respectively and its 20-dB shape factor is 1.56. The passband ripple is ~2.5 dBm with a stop-band rejection of 30 dBm under 15 dBm optical pump power.

Keywords: Microwave photonic filter, stimulated Brillouin scattering, optical frequency comb, bandwidth-reconfigurable

1 INTRODUCTION

With growth of the signal frequency range and bandwidth, the signal composition and the interference information caused by the background become more complicated and changeable. In applications such as telecommunications, radar and electronic warfare systems, both receivers and transmitters need to use bandpass filters with different bandwidths in different center frequency simultaneously to extract desired microwave signals for further signal processing. The reconfigurable and tunable advantages of the microwave photonic filter exactly meet this need. There are currently many categories of microwave photonic filter. The Phase-Shifted Fiber Bragg-Grating (PS-FBG) can obtain a bandwidth of hundred-MHz-scales, but its manufacturing process is complicated and the cost is high. Additionally, it has a narrow tuning range and the bandwidth is not reconfigurable (Guo et al., 2014). The Fabry–Perot (FP) filter has a bandwidth of a few GHz while its form factor is poor. Before the analog-to-digital conversion, an anti-aliasing filter is required to improve the performance by an electric bandpass filter (Wang et al., 2015). The silicon-based micro-ring resonators have a wide range of adjustable bandwidths while having a complex fabrication process. A random drift of resonant peaks random can be caused by width fluctuations of micro-ring resonator, and silicon-based integrated devices are highly susceptible by temperature (Dahlem et al., 2011). Thanks to the narrow SBS linewidth, the SBS-based MPFs have a very high discrimination resolution. The filter tunability is easily realized by modulating a single tone signal at a different frequency or by using a second laser source. The filter selectivity can be further increased by using a phase modulated probe or double sideband cancellation. The filter bandwidth is reconfigurable by broadening the Brillouin pump with different modulation schemes. It becomes one of the most promising solutions for microwave photonic filters (Yi et al., 2017).

Several SBS-based MPFs have been reported in the past decade. The main research trends are summarized as follows. First, is the bandwidth reconfigurable? Various schemes have been proposed, such as controlling the amplitude phase of the combs by using digital programs (Yi et al., 2016), expanding the spectrum by the use of multi-source spectra (Sakamoto et al., 2008), extending the spectrum of the pump modulated by pseudo-random codes (Yi et al., 2011) or a pulse coded by a PSK signal (Yi et al., 2006). The second is frequency tunability. We can control the center frequency of the SBS gain spectrum by changing the wavelength of a continuous wave as the pump (Wei et al., 2017; Yi et al., 2016), changing the frequency of the Carrier-Suppressed Single-Sideband (CS-SSB) signal as the pump (Yi et al., 2016) and controlling the polarization angle of a Polarization Modulation (PolM) (Zhang et al., 2013). Additionally, a rational selection in a multi-pump's frequency to cancel out the gain and loss spectrum can increase the tuning range (Xiao et al., 2015).The last is high selectivity. By splitting the pump into two stages and amplifying the signal twice successively to improve the filter selectivity (Wei et al., 2015) and cascaded four-wave mixing simultaneously broadens and smooths the comb spectra, resulting in an extremely high (> 60 dB) main lobe to side lobe suppression ratio (Supradeepa et al., 2011).

Here, we demonstrate experimentally a wide bandwidth SBS single pass band microwave photonics filter that operates with a really low pump power (3.93 dBm), while maintaining high, reconfigurable resolution (202 MHz) and a high stop-band rejection of 30 dBm.

We achieved this performance through an optical frequency comb as a multi-pump which can overlay the gain spectrum, therefore broadening the spectrum width of the filter. The results presented here point to new possibilities for creating high performance SBS-based reconfigurable MWP filters that will play a key role in modern RF systems for next generation radar (Ghelfi, 2014) and electronic warfare systems.

2 PRINCIPLE

Stimulated Brillouin scattering is a non-linear effect caused by the interaction between a light wave and an acoustic wave in optical fibers. When a pump light with a frequency f_p is injected into an optical fiber, it generates a gain in the vicinity of the frequency $f_p - f_B$ (f_B: Brillouin frequency shift) in the counter-propagating direction and its gain spectrum is given by:

$$H(\omega) = \exp\left(\frac{(\Gamma_B/2)^2 g_0 I_P L/2}{(\omega-\omega_r)^2 + (\Gamma_B/2)^2}\right) \bullet \exp\left(i\frac{(\omega-\omega_r)^2 g_0 I_P L \Gamma_B/4}{(\omega-\omega_r)^2 + (\Gamma_B/2)^2}\right) \quad (1)$$

Figures 1(a) and 1(b) show the diagram and MATLAB simulation of SBS gain spectra, respectively.

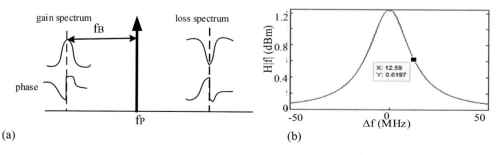

Figure 1. Diagram (a) and simulation (b) of the SBS gain spectrum.

In communication systems, the channel bandwidth often needs to be adjusted according to the actual communication capacity. The passband bandwidth of the filter requires flexible control to achieve its purpose. Therefore, we need to expand the narrow SBS natural gain bandwidth (11.8 MHz). To obtain a broad and flat SBS gain, we must inject a pump with a broad and flat spectrum. However, as shown in Figure 2, we can also obtain a flat SBS gain spectrum by using a pump light with flat multi-line spectra that are equally spaced. This technique does not require large pump power because the pump consists of discrete line spectra, and not a continuous spectrum. We used an Optical Frequency Comb (OFC) which is a comb-like optical spectrum with equally spaced spectral lines to generate a pump that met the above requirement.

A variety of schemes have been developed for OFC generation. The method based on laser mode-locking requires sophisticated feedback controls to attain stable operation, and the comb line spacing is hard to tune. Fiber non-linearity, such as a Four-Wave-Mixing (FWM) effect, is also used for OFC generation. This method needs high power amplifiers and a complicated system designed to achieve high FWM efficiency. In contrast, generation of OFC by external modulation of a Continuous Wave (CW) light using external modulators offers advantages of low complexity, high stability and good frequency tenability (Chen et al., 2013). Here, we propose a novel scheme to generate OFC as shown in Figure 3(a), a spectrally flattened optical frequency comb generation technique employing a CW laser and a conventional Dual-Drive Mach–Zehnder Modulator (DDMZM). The output light field of the DDMZM is $E_{out}(t) = A \sum_{n=-\infty}^{+\infty} j^n J_n(m) \exp[j(\omega_c + n\omega_1)t]$. By controlling the modulation depth m, an optical frequency comb with a number of 2n + 1 and frequency spacing of f_2 can be obtained.

Twenty five flat comb lines with a power deviation within 0.6 dBm and basic line spacing of 20 MHz are successfully generated, which is demonstrated by the Optisystem 15.0 software as shown in Figure 3(b).Compared with other schemes for generating OFC based on external modulators, this program is novel, unique, simple in structure and easy to control.

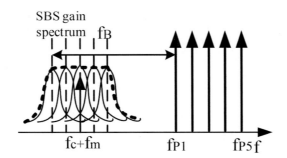

Figure 2.   Schematic diagram of an expanded spectrum by multi-pump.

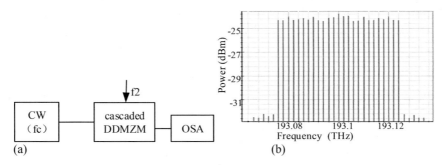

Figure 3.   Generator (a) and spectrum (b) of the optical frequency comb.

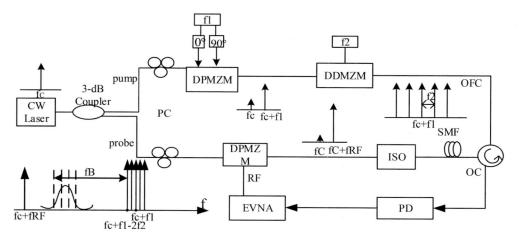

Figure 4. Experimental setup.

3 EXPERIMENT SETUP

Figure 4 shows the experimental setup of the SBS filter. A single optical carrier is divided into two branches by a 3 dB coupler. In the upper branch, the optical carrier goes to a Dual-Parallel Mach–Zehnder Modulator (DPMZM) fed by a microwave oscillator with a frequency of f_1 to generate a Single-Sideband-Suppressed Carrier (SSB-SC) modulation. The SSB-SC signal goes to a DDMZM fed by another microwave oscillator with a frequency of f_2 to generate an OFC. After boosting by a high power Erbium-Doped Fiber Amplifier (EDFA), the OFC signal is sent into a 20 km long Single Mode Fiber (SMF) acting as the Brillouin pump. In the lower one, a swept signal covering the whole SBS gain region was generated by an EVNA. It modulated the CW light in a DDMZM to generate a Single-Sideband (SSB) signal to avoid the combination of the passband and notch Brillouin responses and as the probe signal. Then, the probe signal detected by a photodiode (PD) was sent into the EVNA, where the amplitude and phase response were measured. The SBS gain spectra were obtained by comparing the results when the SBS pump was on and off. A Polarization Controller (PC) was used to maintain the SBS gain at the maximum value.

Here, the frequency of OFC is controlled to let the modulated signal fall within the Brillouin gain region, which can expressed by $f_c + f_{RF} = f_c + f_1 - f_B$ and thus, the filter center frequency is $f_{filter} = f_1 - f_B$ (here $f_1 \geq f_B$). By adjusting N, the number of OFC and comb spacing f_2, the filter bandwidth can be reconfigured. The center frequency of the filter can be tuned by f_1.

4 SIMULATION AND EXPERIMENT RESULTS

The tunability can be easily realized by changing the center frequency of the optical comb. Therefore, we concentrated on how to realize the reconfigurability of the bandwidth. Firstly, the Optisystem 15.0 software was used to verify the SBS superposition effects with 25 pumps at spacings of 20 MHz, 15 MHz and 10 MHz, respectively, as shown below. We compared and analyzed the filter's frequency responses in Figure 5 and concluded its performance in Table 1. As we know, we need a filter with low ripple in-band an ideal rectangular 20 dB shape factor close to one and high gain power.

The above table shows that the optical comb with a spacing of 20 MHz obtains the best performance. Therefore, an experiment was set to verify it: a DFB laser emits a wavelength of 1551.12 nm and was modulated by a tunable microwave source using a DPMZM. The two sub-modulator voltages of DPMZM were set at the minimum transmission points (−21.336V and +7.881V, respectively), while its main modulator was set at the quadrature transmission point

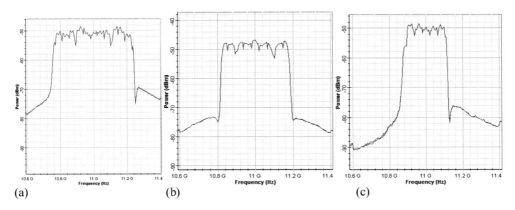

(a)   (b)   (c)

Figure 5.   The SBS gain spectrum at spacings of 20 MHz, 15 MHz and 10 MHz, respectively.

Table 1.   Performance comparison of different comb spacings.

Spacing (MHz)	ω_{3dB} (MHz)	ω_{3dB} (MHz)	Shape factor Q	Gain (dBm)	Tone-to-noise (dBm)	Ripple (dBm)
20	457.9	508.6	1.11	−48.0	25.0	6.23
15	342.2	383.4	1.12	−47.1	27.3	6.16
10	213.8	257.5	1.20	−48.5	33.4	4.30

(+4.714V) to get the upper sideband. The SBS-based filter's tunability can be easily achieved by changing the frequency of the tunable microwave source, which is not verified due to the limitation of the 3dB hybrid. The upper sideband with a frequency of 193609.2 GHz modulated by a DPMZM directly comes into the 2 0 km SMF to obtain the SBS gain spectrum with a 3 dB bandwidth of 11.8 MHz as shown in Figure 6(a) below. The influence of the input optical power on the SBS gain is given as shown in Figure 6(b): the threshold power of SBS effect is less than 3.93 dBm and SBS gain power is nearly proportional to the pump power.

Here, we use an OFC to obtain a wider bandwidth of the SBS gain spectrum as shown in Figure 7(a). A comb with five lines, a spacing of 100 MHz and a flatness of 1.01 dB was experimentally obtained by setting the bias voltage of the DDMZM at +1 6.583 V as shown in Figure 7(b). Using a five-line comb with a spacing of 20 MHz experimentally obtains a 33 MHz 3 dB bandwidth of the SBS gain spectrum. To expand the spectrum bandwidth, a comb with 25 lines and a spacing of 20 MHz is proposed by cascading two DDMZM. Setting the bias voltages of the two cascaded DDMZM at +16.583 V and 14.000 V, respectively, the spectrum of a 25-line comb can be experimentally obtained as in Figure 7(c). Here, we only need to control the two DC bias voltages of the cascaded DDMZM to produce a 25-line excellent performance optical comb, which is simpler in operation and structure than traditional multiple IM and PM cascading schemes. A 25-line comb cannot be clearly observed due to the resolution limitations of the OSA. After boosting by a high power EDFA, the 25-line comb signal is sent into a 20 km long SMF acting as the Brillouin pump and we will obtain its SBS gain spectrum in opposite direction of fiber. Figure 7(d) shows the resulting SBS gain spectrum with a 3 dB bandwidth of 202 MHz and the ripple in-band is less than 2.5 dBm. As it is very difficult to precisely control the pump spectrum, the exact flat top and steep edges for the ideal rectangular filter can hardly be achieved. We obtained a 20 dB shape factor of 1.56 by precisely controlling the bias voltages of the two cascaded DDMZMs, which can precisely control the pump spectrum. This excellent microwave photonics filter technique find versatile applications in telecommunications, radar and electronic warfare systems.

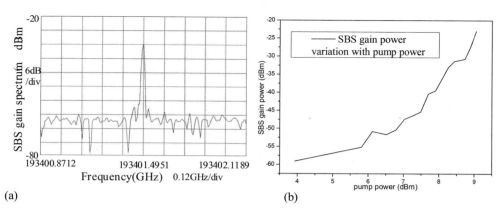

(a)   (b)

Figure 6.   (a) The SBS gain spectrum and (b) SBS gain power variation with pump power.

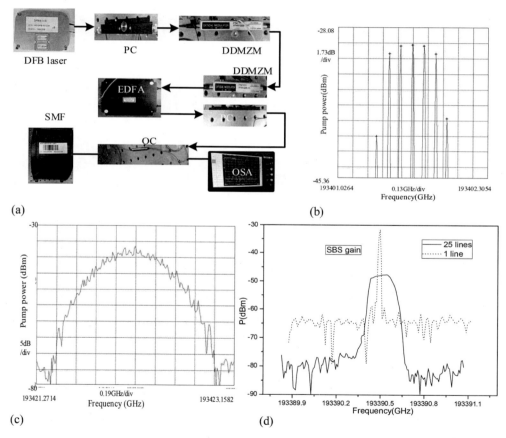

(a)   (b)

(c)   (d)

Figure 7.   (a) Experimental setup, (b) comb with five lines, (c) comb with 25 lines and (d) SBS gain spectrum with 25 lines.

5   CONCLUSIONS

We demonstrated a bandwidth-tunable microwave photonic filter based on stimulated Brillouin scattering in fiber with a bandwidth from 11.8 MHz to 202 MHz by using a 25 line comb with a spacing of 20 MHz obtained by two cascaded DDMZMs. Experiments have shown that the 3 dB and 20 dB bandwidth of the filter is 202 MHz and 314 MHz, respectively and its 20 dB shape factor is 1.56. The passband ripple is ~2.5 dBm with a stop-band rejection of 30 dBm. The experimental result was very close to the simulation result, and the difference was due to the imperfection of the devices we used. Although the ability of the optical comb to spread the spectrum was proven experimentally, the disadvantages of combs such as high flatness and extreme instability caused by bias point drifting severely restrict its application. Therefore, a pseudo-random bit sequence pulse modulation is proposed to extend the pump bandwidth in the future, then the SBS gain spectrum width would be extended to spread the SBS filter bandwidth.

ACKNOWLEDGMENTS

This work was supported in part by the National Advanced Research Foundation of China under Grant 614241105010717.

REFERENCES

Chen, C., Zhang, F. & Pan, S. (2013). Generation of seven-line optical frequency comb based on a single polarization modulator. *IEEE Photonics Technology Letters*, *25*(22), 2164–2166.

Dahlem, M.S., Holzwarth, C.W., Khilo, A., et al. (2011). Reconfigurable multi-channel second-order silicon microring-resonator filterbanks for on-chip WDM systems. *IEEE Photonics Global Conference*, 1–3.

Ghelfi, P. (2014). A fully photonics-based coherent radar system. *Nature*, *507*, 341–345.

Guo, J., Xue, S., Zhao, Q., et al. (2014). Ultrasonic imaging of seismic physical models using a phase-shifted fiber Bragg grating. *Optics Express*, *22*(16),19573–19580.

Sakamoto, T., Yamamoto, T., Shiraki, K., et al. (2008). Low distortion slow light in flat Brillouin gain spectrum by using optical frequency comb. *Optics Express*, *16*(11), 8026–8032.

Supradeepa, V.R., Long, C.M., Wu, R., et al. (2011). Comb-based radio frequency photonic filters with rapid tunability and high selectivity. *Nature Photonics*, *6*(3), 186–194.

Wang, P., Zhao, H., Liu, J., et al. (2015). Dynamic real-time calibration method for fiber Bragg grating wavelength demodulation system based on tunable Fabry–Perot filter. *Acta Optica Sinica*, *35*(8), 0806006.

Wei, W., Yi, L., Jaouën, Y., et al. (2015). Brillouin rectangular optical filter with improved selectivity and noise performance. *IEEE Photonics Technology Letters*, *27*(15), 1593–1596.

Wei, W., Yi, L., Jaouën, Y., et al. (2017). GHz-wide arbitrary-shaped microwave photonic filter based on stimulated Brillouin scattering using directly-modulated laser. *International Topical Meeting on Microwave Photonics*, 1–3.

Xiao, Y., Wang, X., Zhang, Y., et al. (2015). Bandwidth reconfigurable microwave photonic filter based on stimulated Brillouin scattering. *Optical Fiber Technology*, *21*(21), 187–192.

Yi, L., Jaouën, Y. & Gabet, R. (2011). 10-Gb/s slow-light performance based on SBS effect in optical fiber using NRZ and PSBT modulation formats. *Optical Communication*, 1–2.

Yi, L., Wei, W. & Jaouën, Y. (2017). Software-defined microwave photonic filter with high reconfigurable resolution. *The International Multidisciplinary Conference on Optofluidics*, 4449.

Yi, L., Wei, W., Jaouën, Y., et al. (2016). Polarization-independent rectangular microwave photonic filter based on stimulated Brillouin scattering. *Journal of Lightwave Technology*, *34*(2), 669–675.

Yi, L., Wei, W., Shi, M., et al. (2016). Design and performance evaluation of narrowband rectangular optical filter based on stimulated Brillouin scattering in fiber. *Photonic Network Communications*, *31*(2), 336–344.

Yi, L., Zhan, L. & Su, Y. (2006). Delay of RZ PRBS data based on wide-band SBS by phase-modulating the Brillouin pump. *IEEE European Conference on Optical Communications*, 1–2.

Zhang, C., Yan, L.S., Pan, W., et al. (2013). A tunable microwave photonic filter with a complex coefficient based on polarization modulation. *IEEE Photonics Journal*, *5*(5), 5501606–5501606.

Frontier Research and Innovation in Optoelectronics Technology and Industry – Habib & Lewis (Eds)
© 2019 Taylor & Francis Group, London, ISBN 978-1-138-33178-5

Design of an integrated primary mirror and support structure for a space-borne laser communication terminal

Xiang Li, Xiaoming Li & Jiaqi Zhang
National and Local Joint Engineering Research Center of Space and Optoelectronics Technology, Changchun University of Science and Technology, Changchun, Jilin, China

ABSTRACT: In order to reduce the effect of the in-orbit space environment on the optical system of a laser communication terminal, thus improving communication quality and tracking accuracy, an integrated primary mirror structure made of high-volume-fraction SiC/Al is proposed. It resolves a problem of stress concentrations arising as a result of using materials of mismatched coefficients of thermal expansion for the mirror and its support. The temperature stability of the primary mirror surface shape has been improved. On the basis of this work, a lightweight optical system has been achieved. Simulation analysis shows that when the temperature of the working environment changes by $20 \pm 5°C$, the Peak-to-Valley (PV) value of the surface-shape error is $\lambda/16$ and the Root Mean Square (RMS) value of the surface-shape error is $\lambda/67$. In a scenario in which the loading forces of gravity and temperature are coupled, the PV value of the surface-shape error is $\lambda/16$, and the RMS value of the surface-shape error is $\lambda/63$. The base frequency of the integrated primary mirror is 436 Hz, and it has improved dynamic stiffness. The structure can satisfy the requirements of the RMS value of the surface-shape error, $\lambda/50$, of a laser communication system, and provides a technical basis for follow-up research work.

Keywords: optics, laser communication, primary mirror and support, integrated structure

1 INTRODUCTION

Laser communication will become the main approach for information transmission in space in the future, because it has the advantages of a high transmission rate, strong anti-interference ability and high security. In order to establish a stable communication link over tens of thousands of kilometers, a laser communication system should have a high-precision beam control ability. Its beam divergence angle must be of a microradian magnitude and its image quality close to the diffraction limit level (Jiang et al., 2015).

In order to reduce the effect of the in-orbit space environment on the primary mirror of the laser communication system, an integrated primary mirror structure made of high-volume-fraction SiC/Al is proposed. As the result of an optimized design, the Peak-to-Valley (PV) value of the surface-shape error of this primary mirror is better than $\lambda/16$ and the Root Mean Square (RMS) value of its surface-shape error is better than $\lambda/67$, when the temperature of the working environment changes by $20 \pm 5°C$. The PV value is better than $\lambda/16$ and the RMS value is better than $\lambda/63$ in the scenario where gravity (9.8 m/s²) and temperature are coupled. The base frequency of the integrated primary mirror is 436 Hz, which can meet the requirements of a space-borne laser communication system.

2 OPTIMIZATION DESIGN OF THE PRIMARY MIRROR

2.1 *Material choice*

Given the impact and vibration of a satellite launch process, and the working environment in orbit, the specific stiffness and thermal stability of the primary mirror material are the most important factors. The primary mirror of a laser communication system working in orbit should be formed of a material with a high specific stiffness and with excellent thermal stability. Generally, the product of these two coefficients is used to evaluate the comprehensive performance of the material; this is called the comprehensive quality factor. Material with a high comprehensive quality factor will meet the requirements for low weight, high rigidity, good thermal conductivity and strong environmental adaptability.

The current materials used for making primary mirrors for optical systems in both domestic and foreign markets are shown in Table 1. All the parameter values of SiC are highly advantageous. However, given its more negative characteristics of having a long manufacturing cycle, being of high cost and being difficult to make light in weight, it has some limitations when in use. High-volume-fraction SiC/Al doesn't share these disadvantages, and has very similar material properties to SiC. It could satisfy the requirements of this project; therefore, high-volume-fraction SiC/Al has been chosen as the material for the primary mirror of our laser communication system.

2.2 *Primary mirror configuration*

Depending on different uses, the ratio of the diameter and thickness of non-metallic mirrors is usually 5–10 (Yoder, 1993). The primary mirror diameter in our study is 150 mm. After carefully considering the other restrictive conditions of the laser communication system, the primary mirror thickness is finally determined to be 18 mm. The support structure of the primary mirror of the optical system generally includes a center support, a 3-point back support, a 6-point back support, a 9-point back support and an 18-point support. The design principle of the support structure is to achieve an effective support function for the mirror with minimal support points (Friedman, 2003). Due to the small diameter of the primary mirror, the influence of gravity on its surface shape is very small. A center support can be used to reduce the influence of the support structure on the primary mirror. According to the lightweight requirements of the laser communication system, the diameter of the primary mirror support ring should be minimized, as shown in Figure 1.

2.3 *Integrated support design*

The primary mirror support is generally made from material which has a thermal expansion coefficient similar to that of the primary mirror (Shao et al., 2015). However, there are still

Table 1. Performance comparison of primary mirror materials.

Material name	SiC	SiC/Al	AlBeMet 162	Zerodur	Fused quartz
Elastic modulus (GPa)	310	235	193	90.6	73
Density (g/cm^3)	2.92	3.03	2.1	2.53	2.20
Specific stiffness	106	78	92	36	33
Thermal conductivity (W/(m*K))	158	230	210	1.64	1.37
Coefficient of thermal expansion (C^{-1})	2.64	7.8	13.9	0.05	0.58
Thermal stability	60	29	15	33	2
Comprehensive quality factor	6354	2287	1388	1175	78

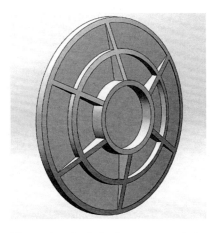

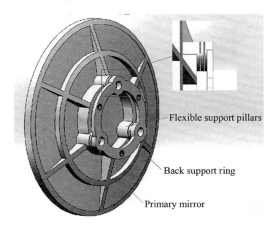

Flexible support pillars

Back support ring

Primary mirror

Figure 1. Optimized structure of the primary mirror.

Figure 2. Diagram of the flexible integration-supporting structure of the primary mirror.

slight differences in the thermal expansion coefficients of the two materials. Therefore, when the ambient temperature changes, the contact surface of the two elements will undergo stress concentrations that will affect the surface shape of the primary mirror. There is a serious effect on laser communication systems close to the diffraction limit level for image quality. In this paper, an integrated support structure for the primary mirror is proposed, and a detailed simulation analysis is carried out. The mirror is connected to the back support ring by three support pillars. For increased structural flexibility, the support pillars are cut with flexible slots (see inset of Figure 2). This design allows the primary mirror to release the restraint stress from the support structure when the ambient temperature changes. Therefore, the primary mirror can be maintained in a relatively free mechanical condition, effectively reducing the influence of the support structure on the primary mirror surface shape. In addition, the design of the integrated structure is more lightweight than the traditional design. The specific structure is shown in Figure 2.

3 SIMULATION EXPERIMENT

Because of the small diameter of the primary mirror, the influence of gravity on the primary mirror surface shape is small. Therefore, the influence of ambient temperature change on the primary mirror surface shape should be the principal subject of analysis. A finite element model was established for this design, which has 107,066 nodes and 64,688 elements, as shown in Figure 3.

The surface-shape error of the primary mirror is analyzed when the ambient temperature rises by 5°C. As the results showed that a temperature rise of 5°C and a drop of 5°C had a similar effect on the primary mirror surface shape, this temperature rise is taken as an example. The results are shown in Figure 4.

The simulation results are shown in Table 2. The results show that when the ambient temperature rises by 5°C, the primary mirror surface shape becomes as shown in Figure 4. The PV value of the primary mirror surface-shape error is $\lambda/16$ and the RMS value is $\lambda/67$. The primary mirror surface shape for the coupled gravity and temperature scenario is shown in Figure 5, and the PV value of the primary mirror surface-shape error is $\lambda/16$ and the RMS value is $\lambda/63$. The results show that gravity has little effect on the primary mirror surface shape.

The mode of the primary mirror was analyzed, and the results are shown in Table 3. The results show that the fundamental frequency of the primary mirror is 436 Hz, and the first-order mode image is shown in Figure 6. It can be seen from the analysis above that the

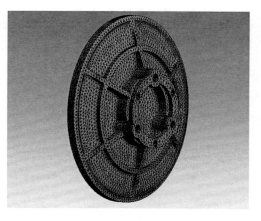

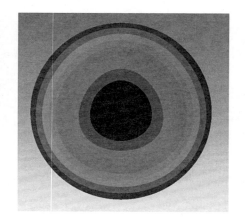

Figure 3. The finite element model of the integrated primary mirror.

Figure 4. Simulation analysis result when the ambient temperature rises by 5°C.

Table 2. The results of the simulation analysis.

Form of load	PV (nm)	RMS (nm)
Coupling of gravity and temperature	39.5	10.1
The ambient temperature rises by 5°C	39.5	9.4
The ambient temperature drops by 5°C	39.6	9.5

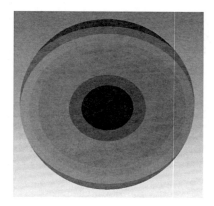

Figure 5. Simulation analysis result in the coupled gravity and temperature scenario.

Table 3. The mode analysis results of the primary mirror.

Number of the mode order	Frequency (Hz)
1	436.33
2	436.68
3	991.63
4	1237.93
5	1847.48
6	1847.78

Figure 6.   Primary mirror deformation in the first-order mode.

integrated support structure not only effectively reduces the surface-shape distortion of the primary mirror caused by temperature change, but also satisfies the rigidity requirement for space-borne laser communication systems.

4   CONCLUSIONS

In this paper, high-volume-fraction SiC/Al was chosen as the primary mirror material for a space-borne laser communication system. The optimized design of the primary mirror was realized via a large amount of simulation analysis. An integrated support structure for the primary mirror is proposed, and the thermal deformation of the primary mirror is reduced by adding flexible slots in the support. The weight of the primary mirror is 0.429 kg, which is about 23% lighter than a traditional primary mirror support design.

Finally, the primary mirror structure was simulated and analyzed. The results show that the PV value of the surface-shape error is $\lambda/16$ and the RMS value of the surface-shape error is $\lambda/67$, when the temperature of the working environment changes by $20 \pm 5°C$. In the scenario of coupled gravity and temperature, the PV value of the surface-shape error is $\lambda/16$ and the RMS value of the surface-shape error is $\lambda/63$. The mode analysis shows that the fundamental frequency of the primary mirror and its integrated support structure is 436 Hz. In summary, optimization of the primary mirror and its integrated support structure can satisfy the project requirements, and gives a lighter weight and more stable performance.

REFERENCES

Friedman, E. (2003). *Photonics rules of thumb*. New York, NY: McGraw-Hill.
Jiang, H.-L., Lun, J., Song, Y.-S., Meng, L.-X., Fu, Q., Hu, Y., … Yu, X.-N. (2015). Research of optical and APT technology in one-point to multi-point simultaneous space laser communication system. *Chinese Journal of Lasers*, *42*(4), 0405008. doi:10.3788/CJL201542.0405008.
Shao, L., Wu, X.-X., Chen, B.-G., Li, J.-F. & Ming, M. (2015). Passive support system of light-weighted SiC primary mirror. *Optics and Precision Engineering*, *23*(5), 1380–1386.
Yoder, P.R. (1993). *Opto-mechanical systems design* (2nd ed.). New York, NY: Marcel Dekker.

Frontier Research and Innovation in Optoelectronics Technology and Industry – Habib & Lewis (Eds)
© 2019 Taylor & Francis Group, London, ISBN 978-1-138-33178-5

A novel few mode fiber with low loss and low crosstalk

Xin Li, Xiao Wang, Hongjun Zheng & Chenglin Bai
Shandong Provincial Key Laboratory of Optical Communication Science and Technology, School of Physics Science and Information Technology, Liaocheng University, Liaocheng Shandong, China

Weisheng Hu
State Key Laboratory of Advanced Optical Communication Systems and Networks, School of Electronic Information and Electrical Engineering, Shanghai Jiao Tong University, Shanghai, China

ABSTRACT: A novel six-cores few-mode fiber with low loss and low crosstalk is proposed in this paper. The fiber characteristics based on the full-vector Finite Element Method (FEM) show that two supermodes with low loss, low crosstalk and low Differential Mode Group Delay (DMGD) are achieved. The proposed fiber performs DMGD flattened profile at a large wavelength range.

1 INTRODUCTION

With a rapid increase of the internet traffic and data center, the space division multiplexing (SDM) has attracted much attention to greatly provide transmission capacity increase for optical fiber communication systems in recent years (Takayuki Mizuno et al. 2016, Werner Klaus et al. 2017, F. J. V. Caballero et al. 2017). The SDM generally employs multiple channels separated in space and implemented using multi-core fiber, few mode fiber and their combination. The researches on multi-core fiber and few-mode fiber have become the frontier of optical communications (Pan Xiaolong et al. 2016, X. L. Pan et al. 2016, Liu Qianqian et al. 2017, He Wen et al. 2017, Hongjun Zheng et al. 2018, Cen Xia et al. 2011, Cen Xia et al. 2013, Li Guifang et al. 2014, Yu Ruyuan et al. 2018). Recently, multi-core supermode fiber, in which there is strong coupling among cores, also becomes the focus of attention (Cen Xia et al. 2011, Cen Xia et al. 2013, Li Guifang et al. 2014, Yu Ruyuan et al. 2018). Compared with step-index single-core fiber, some immediate advantages of using the supermode structure, which the step-index single-core fiber cannot realize, are as follows (Yu Ruyuan et al. 2018). Firstly, the supermodes can perform high mode density and allow more design freedom to obtain the desired propagation characteristics. Secondly, coupling from the isolated cores to the supermode fiber is lossless and therefore must be without mode-dependent loss because the transformation matrix from the uncoupled modes to the supermodes is unitary. Thirdly, the effective areas of these supermodes are larger than that of the corresponding step-index single core few mode fiber for the same bending loss. Finally, the parameters of supermode fiber such as the effective index difference, differential mode group delay (DMGD) and effective area can be easily tuned by changing the geometrical parameters and the indices of the core and cladding.

The pure silica core, which can effectively reduce the attenuation and the fusion loss, is mostly used in single mode fiber (T. Hasegawa et al. 2016, S. Ten. 2016, Y. Tamura. 2018).

Few mode fiber with graded-index distribution can achieve low DMGD and identical mode width of far and near field, and the effective refractive index difference (ERID) in different modes is larger than 0.5×10^{-3} to avoid mode coupling (Pierre Sillard. 2013, Roland Ryf. 2017).

2 RESEARCH METHODS

We propose a novel six-core few mode fiber with pure silica core and graded-index distribution in this paper. The proposed fiber can effectively reduce fiber loss and DMGD using pure silica core and graded-index distribution. Low crosstalk is obtained by employing large ERID ($>0.5 \times 10^{-3}$). The proposed fiber achieves two supermodes operation with high mode density, low loss, low crosstalk and low DMGD, which the step-index single-core fiber cannot realize. Furthermore, the proposed fiber performs DMGD flattened profile at a large wavelength range. The proposed fiber is suitable to apply in high power pump due to its good performance.

Figure 1 shows a cross section of six-cores few-mode fiber. The fiber is consist of six cores (Drak grey part) with the larger refractive index, the trench (French grey part) and pure silica cladding (white part). The radiuses of six cores are all R = R1 = 3 μm, the inner radius of the trench is R2 = 20 μm, the outer radius of the trench is R3 = 30 μm, the cladding radius is R4 = 62.5 μm. The center coordinates of the six cores are (6 μm, 0), (3 μm, $3\sqrt{3}$ μm), (−3 μm, $3\sqrt{3}$ μm), (−6 μm, 0), (−3 μm, −$3\sqrt{3}$ μm), (3 μm, −$3\sqrt{3}$ μm), respectively.

The refractive indices of six cores, trench and pure silica cladding are n1 = 1.444024, n2 = 1.433524, n3 = 1.437024, respectively. The characteristics of electromagnetic fields could be changed by adjusting the cores sizes and refractive indices distribution in the optical fiber.

By solving the electromagnetic field equation and using the full-vector FEM, we obtain the electric field vector and the complex effective index. Then, the dispersion and DMGD, which are important parameters of the fiber, are calculated by our designed matlab program. The chromatic dispersion, which is an important parameter of the fiber, includes the material dispersion and waveguide dispersion. The material (fused silica) dispersion can be calculated from the Sellmeier equation (Zheng Hongjun et al. 2018, Yu Ruyuan et al. 2018).

$$n^2 = 1 + \sum_{j=1}^{m} \frac{B_j \cdot \lambda^2}{\lambda^2 - \lambda_j^2}, \tag{1}$$

and group-velocity dispersion parameter equation

$$D = -\frac{\lambda}{C} \frac{d^2 n}{d\lambda^2}, \tag{2}$$

where n is the refractive index of the transmission medium, λ is the wavelength of incident light, $m = 3$, $B_1 = 0.6961663$, $B_2 = 0.4079426$, $B_3 = 0.8974794$, $\lambda_1 = 0.0684043$, $\lambda_2 = 0.1162414$,

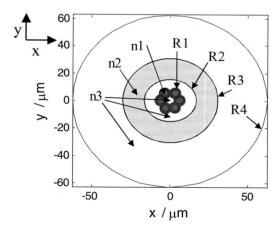

Figure 1.   Cross section of the six-cores few-mode fiber.

$\lambda_3 = 9.896161$, $\lambda_j = 2\pi c / \omega_j$, ω_j is the medium resonance radian frequency, C is the speed of light in vacuum. D, which is the group-velocity dispersion parameter, is useful in practice. The waveguide dispersion could be obtained from Eq. (2) and variation of the real part of effective refractive index n_{eff} with the wavelength.

The DMGD equation is given by (Ruyuan Yu et al. 2018)

$$DMGD = \frac{n_{eff_{nm}} - n_{eff_{01}}}{c} - \frac{\lambda}{c}\left(\frac{\partial n_{eff_{nm}}}{\partial \lambda} - \frac{\partial n_{eff_{01}}}{\partial \lambda}\right). \qquad (3)$$

3 RESEARCH RESULTS

3.1 The electric field distribution

Figure 2 shows the electric field distribution of two supermodes LP_{01} and LP_{11a} at input wavelength of 1.55 μm. If we consider polarization and spatial degeneration, the proposed fiber can support 6 supermodes such as LP_{01} X, LP_{01} Y, LP_{11a} X, LP_{11a} Y, LP_{11b} X, and LP_{11b} Y. The supermodes with strong core-to-core coupling between the cores of a six-cores few mode fiber are obtained, in which the core-to-core distance is much shorter than that in conventional multi-core fiber. Field distribution of supermodes can be seen as a superposition and strong restructuring of isolated modes. So, the six-cores few mode fiber can support larger effective areas and higher mode densities than the conventional fiber, which can effectively reduce the nonlinear effect for fiber optics and optical communications. The characteristics of two supermodes are investigated and those of their degenerate modes are not discussed in this paper because the effective refractive index (ERI), dispersion, effective area and nonlinear coefficient of the degenerate modes are similar.

3.2 Variation of effective indices of two supermodes

Figure 3 shows that effective indices of two supermodes vary with input wavelength. Solid lines with squares and circles are respectively for LP_{01} and LP_{11} supermodes. The effective indices of two supermodes are all decreased with the increase of wavelength. The decrease of LP_{01} supermode with wavelength is faster than that of LP_{11} mode. For a given input wavelength, effective index of LP_{01} mode is larger than that of LP_{11} mode.

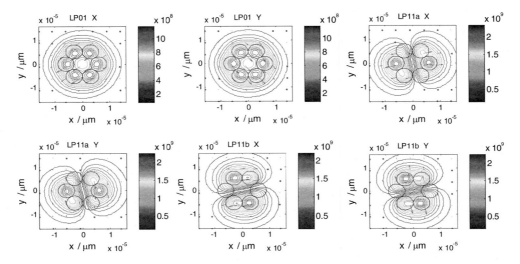

Figure 2. The electric field distribution of supermodes.

3.3 *Variation of dispersion of two supermodes*

Variations of dispersion of the six-cores few-mode fiber with input wavelength are shown in Figure 4. The dotted line, dot-dashed line, and solid line respectively denote material dispersion, waveguide dispersion and total dispersion of LP_{01} (a) and LP_{11} (b) modes. From Figure 4 (a), total dispersion of LP_{01} mode, which is increased with the increase of input wavelength, is less than that of material dispersion from 1.4 µm to 1.7 µm because waveguide dispersion is flattened and lower than –6 ps/(km·nm). From Figure 4 (b), the waveguide dispersions of LP_{11} modes are lower than that of LP_{01} mode from 1.4 µm to 1.7 µm. Total dispersion of LP_{11} mode is obviously less than that of material dispersion at the range of 300 nm. So, variation of total dispersions of two supermodes, which are increased with the increase of input wavelength, are obviously less than those of material dispersions from 1.4 µm to 1.7 µm. Total dispersions of LP_{01} and LP_{11} modes at 1.53 µm are respectively 15.46 and 14.01 ps/(km·nm), which are obviously less than those of standard single mode fiber. Total dispersions of LP_{01} and LP_{11} modes at 1.55 µm are respectively 16.93 and 15.16 ps/(km·nm), which are obviously less than those in the reference (Ruyuan Yu et al. 2018). We will combine transparent performance of the doped silica and the refractive index profile to further optimize the dispersion of the few mode fiber in the future.

3.4 *Variation of DMGD*

Figure 5 shows DMGD vary with input wavelength. Variation of DMGD is slowly decreased and flattened from the wavelength of 1.4 µm to 1.6 µm. For a given input wavelength, DMGD

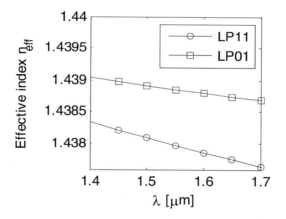

Figure 3.   Variation of the ERI with the wavelength.

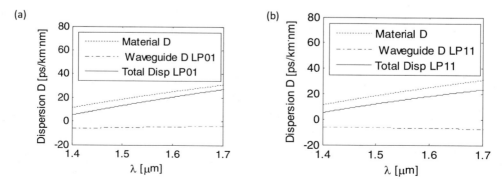

Figure 4.   Dispersion of LP_{01} (a), LP_{11} modes (b).

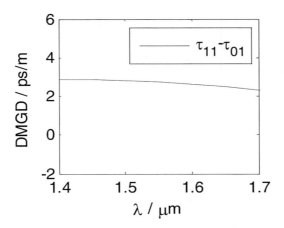

Figure 5. Variation of DMGD with input wavelength.

of LP_{11} is less than 2.85 ps/m. At wavelength of 1.55 μm, DMGD of LP_{11}, which is 2.74 ps/m, is consistent with the reference(Hongjun Zheng et al. 2018).

4 CONCLUSION

The results show that two supermodes with low loss, low crosstalk and low DMGD are achieved. The proposed fiber performs DMGD flattened profile at a large wavelength range. Due to limit of our experimental condition, the results of this paper are obtained from numerical calculation. Although our method is numerical calculation, the idea of our proposed fiber is novel and the fiber characteristics are very good for potential applications. We hope that one can experimentally verify and apply our proposed fiber in the future.

ACKNOWLEDGEMENTS

The work is supported in part by the National Natural Science Foundation of China (Grant No. 61671227 and 61431009), the Shandong Provincial Natural Science Foundation (ZR2011FM015), Taishan Scholar Research Fund of Shandong Province.

REFERENCES

Caballero, F.J.V. et al. 2017. Novel Equalization Techniques for Space Division Multiplexing Based on Stokes Space Update Rule [J]. Photonics, 4(1):12.

Cen Xia et al. 2011. Supermodes for optical transmission [J]. Optics Express, 19(17): 16653–16664.

Cen Xia et al. 2013. Optical Fiber Communication Conference and Exposition and the National Fiber Optic Engineers Conference 1.

Hasegawa, T. et al. 2016. Advances in ultra-low loss silica fibers [J]. Frontiers in Optics, paper FTu2B.2.

He Wen et al. 2017. Few-mode fibre-optic microwave photonic links [J]. Light: Science and Applications, 6, 8.

Hongjun Zheng et al. 2018, Transmission of chirped pulse in optical fibers. Beijing: Science Press 1(2018). (in Chinese).

Li Guifang et al. 2014. Space-division multiplexing: the next frontier in optical communication [J]. Advances in Optics & Photonics, 6(4): 413–487.

Liu Qianqian et al. 2017. A mode division multiplexer with large mode area and graded-index distribution [J]. Journal of optoelectronics laser, 28(11): 1180–1185.

Pan Xiaolong et al. 2016. Low complexity MIMO method based on matrix transformation for few-mode multi-core optical transmission system [J]. Optics Communications, 371(15): 238–242.

Pierre Sillard et al. 2013. Few-Mode Fibers for Space-Division Multiplexed Transmissions [J], European Conference & Exhibition on Optical Communication, 03 (A1):1–3.

Roland Ryf. 2017. Switching and Multiplexing Technologies for Mode-Division Multiplexed Networks [D], Optical Fiber Communication Conference & Exposition, Tu2c.

Takayuki Mizuno et al. 2016. Dense Space Division Multiplexed Transmission Over Multicore and Multimode Fiber for Long-haul Transport Systems [J]. Journal of Lightwave Technology, 34(6): 1484–1493.

Ten, S. 2016. Ultra Low-loss Optical Fiber Technology [J]. Optical Fiber Communication Conference, paper Th4E.5.

Werner Klaus et al. 2017. Advanced Space Division Multiplexing Technologies for Optical Networks [J]. Journal of Optical Communications & Networking, 9(4):C1-C11.

Yoshiaki Tamura. 2018. Ultra-low loss silica core fiber for long haul transmission [J]. Optical Fiber Communication Conference, paper M4B.1.

Yu Ruyuan et al. 2018. A novel three-ring-core few-mode fiber with large effective area and low nonlinear coefficient [J], Optoelectronics Letters. 14 (1):30–35.

Transient Kerr effect and its application in all-optical switching

Chao Li, Zhe-Ming Xu, Shu-Ya Wu & Jun-Fang Wu
School of Physics and Optoelectronic Technology, South China University of Technology, Guangzhou, Guangdong, China

ABSTRACT: We present an analytical model to investigate the dynamic features of the transient Kerr effect in a nonlinear photonic crystal microcavity, and the theoretical predictions agree perfectly with the proposed experimental results. Based on this analytical model, we further derive the critical pump-power condition for the control light in bistable switching based on a nonlinear SCWR system. These results are useful for actual design of optical switching.

1 INTRODUCTION

With the development of integrated photonic systems, high-Q and small-volume nanocavities in two-dimensional (2D) photonic crystals (PCs) have attracted many attentions in recent years (Akahane Y. et al. 2003; Notomi M. et al. 2005; Yang X. D. et al. 2007). These resonant nanocavities are small enough to integrate and can be applied in extremely low-power nonlinear photonic devices because of their strong light–matter interactions. In such a nanocavity, several nonlinear effects, such as Kerr effect, two-photon absorption (TPA), and free-carrier absorption (FCA) could be significantly strengthened, because of their strong light–matter interactions. So far, the dynamic features of Kerr effect, TPA and FCA processes under the pump of ultrashort pulses have been well investigated (Harding P. J. et al. 2007; Barclay P. E. et al. 2005). However, another dynamic process, say, the dynamic interaction between the continuous waves (CWs) and the nonlinear PC cavity, was seldom investigated (Harding P. J. et al. 2007; Yanik M. F. et al. 2003). Actually, under the pump of the CW light, the steady light field coupled in the cavity is oscillating at a frequency that is the same as the pump one. Considering that the response of nonresonant Kerr effect to optical pump is instantaneous (Joannopoulos J. D. et al. 1995; Soljačić M. et al. 2002), the Kerr-effect-induced change of the refractive index should also be very fast and periodical (Yüce E. et al. 2013), so we call this effect transient Kerr effect (TKE). As a result, the resonant frequency will oscillate harmonically between its initial position and a final position that is dependent on the pump intensity. This view has been testified by some reported experiments (Yüce E. et al. 2013), but the related theoretical study is still deficient. Obviously, the transient shift of the resonant frequency will fundamentally affect the transmission and reflection features of a nonlinear PC cavity. Therefore, the investigation of the transient process under the pump of a CW light becomes very important and necessary.

In addition, we will also apply TKE to optical switching, which plays a key role in all-optical communications (Ogusu K. et al. 2008; Nozaki K. et al. 2012; Colman P. et al. 2016). As known to all, to trigger a bistable switching to high-transmission state, a control light should be employed to superpose with the incident signal light. In actual applications, a crucial question is, what is the critical pump conditions of the control light for a bistable switching (i.e., the minimum values of the pump power and pump time of the control light to trigger a bistable switching to high-transmission state)? Unfortunately, this question is rarely studied. Of course, one might obtain these critical values by simulation method after many times of attempts. However, it is time consuming and the achieved results can not be generalized.

Therefore, a theoretical investigation on the critical pump conditions becomes very important in PC switching design work. In this paper, we will present an analytical model and numerical experiments to investigate TKE and its application in all-optical switching.

2 ANALYTICAL MODEL

As is well known that in a linear case (the intensity of the CW pump light is ultralow), the peak transmission occurs only when the pump frequency is equal to the resonance one; once the pump frequency is off resonance, the transmission will drop sharply, and the transmission spectrum (i.e., the excited cavity mode) exhibits a Lorentzian lineshape (Joannopoulos J. D. et al. 1995). However, with the increase of the pump intensity, under the influence of TKE, the resonant frequency will oscillate swiftly between its initial position and a final position that is dependent on the pump intensity. This means the previous "off-resonance" components of the transmission spectrum will now have opportunities to experience the transient state of "on resonance" and therefore be strengthened. Thus, the linewidth of the excited cavity mode is broadened. Obviously, a bigger shift of the resonant frequency will offer more opportunities to those off-resonance components to experience "on-resonance" states, resulting in a further broadening of the cavity mode in frequency domain. We define this TKE-induced broadening quantity as γ_{TKE}, which is approximately proportional to the maximum shift of the resonant frequency $\Delta\omega_0$. Thus, we have

$$\gamma_{\text{TKE}} = s_0 \Delta\omega_0, \tag{1}$$

where s_0 is a dimensionless coefficient and we call it *TKE-induced broadening parameter*, which is a quasi constant and is almost independent of the power or frequency of the pump light. Moreover, s_0 facilitates system design since a single simulation is enough to determine it. Known from Ref. (Soljačić M. et al. 2002), $\Delta\omega_0 = \gamma_L p_{\text{out}}/p_0$, where γ_L is the half of the linewidth of the cavity mode in linear case, p_{out} is the transmitted power, and p_0 is the characteristic power (Soljačić M. et al. 2002). Accordingly, γ_{TKE} can be rewritten as $\gamma_{\text{TKE}} = s_0 \gamma_L p_{\text{out}}/p_0$, and the half of the linewidth of the resonant mode should be corrected as $\gamma = \gamma_L + \gamma_{\text{TKE}} = \gamma_L(1 + s_0 p_{\text{out}}/p_0)$.

Substituting it into the coupled mode theory (Haus H. A. 1984), we obtain the stable transmission and reflection coefficients for the case of a waveguide (WG) directly coupled to a cavity:

$$T = \frac{1}{\left(1 + s_0 p_{\text{out}}/p_0\right)^2} \cdot \frac{\eta}{\left(\delta - p_{\text{out}}/p_0\right)^2 + 1}, \tag{2}$$

$$R = \frac{[(\gamma_1 - \gamma_2 - \gamma_0)/\gamma_L - s_0 p_{\text{out}}/p_0]^2/(1 + s_0 p_{\text{out}}/p_0)^2 + (\delta - p_{\text{out}}/p_0)^2}{(\delta - p_{\text{out}}/p_0)^2 + 1}, \tag{3}$$

where δ is the frequency detuning of the incident CW with respect to the cavity mode, η is the linear peak transmission coefficient; γ_1 and γ_2 are the decay rates of the cavity mode amplitude into the input WG and output WG, respectively, and γ_0 is the intrinsic decay rate of the cavity. From Eqs. (2) and (3), it is found with the increase of the pump power, the red shifts of the transmission/reflection spectra become more and more significant, while the spectral widths become more and more broad, that results in the drop of the "height" for each transmission peak or the "depth" for each reflection valley, as shown in Fig. 1. Obviously, this phenomenon induced by TKE is quite different from that caused by traditional Kerr effect, as will be further shown below.

The contribution of TKE to nonlinear transmission and reflection can also be reflected by several proposed experimental results. For example, in Ref.(Kim M. K. et al. 2007), an all-optical bistable switching in PC resonators with Kerr nonlinearity is investigated experimentally and theoretically. One can see that the measured transmission spectra for different pump

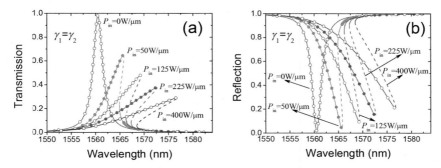

Figure 1. Influence of TKE on (a) the transmission spectra and (b) reflection spectra for different pump powers. The theoretically calculated results are depicted as solid curves, and the FDTD simulation results are also plotted as open circles or dots for comparison purpose.

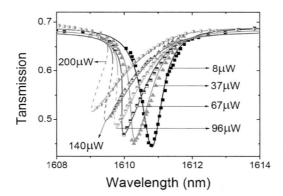

Figure 2. Fitting of the experimental data based on the analytical model of TKE. The theoretical curves calculated by TKE theory are plotted as solid lines, and the experimental data extracted from Fig. 3(a) in Ref. (Kim M. K. et al. 2007) are plotted as discrete dots, for comparison purpose.

powers, i.e., Fig. 3(a) in Ref. (Kim M. K. et al. 2007), is very similar to Fig. 1(b) calculated by the analytical model in our work, but is not in good accord with the transmission spectra calculated by their own model (see Fig. 3(b) in Ref. (Kim M. K. et al. 2007), where only conventional Kerr effect is considered): the analytical model with conventional Kerr effect can not explain why the "valley" of the transmission spectrum will rise with the increase of the pump power (their theoretically-calculated "valley" keeps invariant). However, if we take the contribution of TKE into account, the above-mentioned discrepancy can be well explained: the rise of the "valley" of the transmission spectrum is just one of the features of TKE, as has been discussed before.

Now, let us directly testify the analytical model of TKE by using the experimental data. According to the data extracted from the measured transmission spectra in Fig. 3(a) in Ref. (Kim M. K. et al. 2007), s_0 and p_0 are calculated to be "−0.043" and "−3.65 μW", respectively (note $n_2 < 0$ in this case); $\gamma_1 = \gamma_2 = 1.08 \times 10^{-5}$, and $\gamma_0 = 4.33 \times 10^{-5}$. In Fig. 2, the theoretical curves calculated by TKE theory are plotted as solid lines, and the experimental data extracted from Fig. 3(a) in Ref. (Kim M. K. et al. 2007) are plotted as discrete dots, for comparison purpose. One can see that the experimental results are precisely predicted by TKE theory. This further shows that the analytical model of TKE is effect and can be naturally extended to real 3D cases with Kerr nonlinearity.

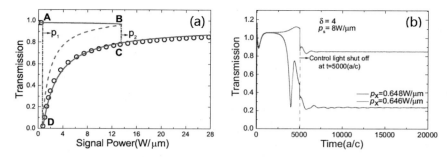

Figure 3. (a) Dependence of nonlinear transmission vs signal power for $\delta = 4$, $s_0 = 0.0036$, and $p_0 = 0.93$ W/μm. calculated to be and The solid lines stand for the theoretically-calculated stable transmission states, and the dashed line represents unstable ones. The scatted dots are the corresponding FDTD simulation results. (b) Time evolutions of bistable switching under different pump powers of the control light. The upper curve: $p_x = 0.646$ W/μm; The lower curve: $p_x = 0.648$ W/μm.

3   APPLICATION OF TKE IN OPTICAL SWITCHING BASED ON SIDE-COUPLED WAVEGUIDE-RESONANT SYSTEM

Side-coupled waveguide-resonator (SCWR) system is fundamental in photonic functional devices, and has been widely used in ultrafast switching, all-optical diodes, logic gates and memories, etc (Nozaki K. 2010; Xia F. N. et al. 2006; Li C. et al. 2017). In the past decade, bistable switching has attracted many attentions for its great potential in all-optical signal processing. To trigger the bistability, the operation powers of signal light and control light should simultaneously satisfy some critical conditions. So far, the critical condition for signal light has been well studied, but that for control light is still deficient. In this paper, we will study the critical pump condition for control light in a bistable switching based on SCWR system.

For simplicity, we temporally do not consider any nonlinear absorptions in the resonator, and only electrical Kerr nonlinearity is taken into account, which provides the highest possible speed owing to its inherently instantaneous response nature. With the increase in the pump intensity, under the influence of TKE, the linewidth of the excited cavity mode will be broadened, and the broadening quantity is proportional to the light intensity inside the resonator. In this case, the half of the linewidth of the cavity mode should be corrected as

$$\gamma = \gamma_L (1 + s_0 p_R / p_0), \tag{4}$$

where p_R is the reflected power.

By using the theory presented in section 2, we obtain

$$T = \frac{(s_0 p_R / p_0)^2 / (1 + s_0 p_R / p_0)^2 + (\delta - p_R / p_0)^2}{(\delta - p_R / p_0)^2 + 1}, \tag{5}$$

Eq. (5) is the nonlinear transmission coefficient for the SCWR system. As an example, in Fig. 3(a) we plot the dependence of nonlinear transmission vs signal power for $\delta = 4$ (the other parameters can be found in the caption). p_1 and p_2 are the signal powers correspond to the left/right borders of the bistable region marked by ABCD.

To trigger the bistability, one can employ high-peak-power Gaussian pulses as the control light. However, it is very difficult to obtain an analytic solution by this way. Therefore, we will employ a CW source with suitable launching time to act as control light. We have shown this method is more efficient for the theoretical study on the critical pump conditions, and the achieved results can be readily generalized to the case of pulsed control light (Li C. et al. 2010).

For simplicity, we suppose the control light has the same frequency and initial phase with the signal light. Accordingly, when the signal light superposed with the control light is incident, we can treat them as a whole, namely, an "equivalent signal light" with incident power of $\left(\sqrt{p_s} + \sqrt{p_x}\right)^2$, where p_s and p_x are the powers of the signal light and control light, respectively. Therefore, to make the switching jump indirectly from the upper branch to the low one, the power of the "equivalent signal light" should be no less than p_2. Thus we have $\left(\sqrt{p_s} + \sqrt{p_x}\right)^2 \geq p_2$, i.e.,

$$p_x \geq (\sqrt{p_2} - \sqrt{p_s})^2, \tag{6}$$

where p_2 is the signal power corresponds to the right border of the bistable region, and can be calculated from Eq. (5). Eq. (6) is the critical pump-power condition for the control light to trigger the bistability when the signal power is p_s.

To testify the theoretical predictions, FDTD simulations are performed on a photonic-crystal structure consists of a two-dimensional square lattice of silicon rods in air. The rods have radii of $r = 0.2a$, and $a = 585$ nm is the lattice constant. By using Eqs. (4) and (5), s_0 and p_0 are calculated to be 0.0036 and 0.93 W/μm, respectively. For $\delta = 4$, p_1 and p_2 are calculated to be 0.55 W/μm and 13.2 W/μm, respectively. Thus, when the incident signal light is $p_s = 8$ W/μm, the critical pump power for the control light is calculated to be $p_x = 0.647$ W/μm. To examine the precision of the critical value predicted by theory, we begin simulations by using $p_x = 0.648$ W/μm and $p_x = 0.646$ W/μm as the power of the CW control light, respectively, for comparison. Fig. 3(b) shows the FDTD results of the temporal transmission curves. One can see when $p_x = 0.648$ W/μm, the bistable switching successfully jumps from the upper state to the lower one; while when p_x is slightly less than the calculated critical value, the switching action will not be triggered. Therefore, the theory presented in this paper is very precise, and we can use this derived critical pump condition for control light to design bistable switchings.

4 CONCLUSIONS

In summary, we have established a physical model to investigate the impacts of TKE on the characteristics of a nonlinear PC nanocavity, and the theoretical predictions agree with the numerical experiment results quite well. It is found with the increase of the pump power, the red shifts of the transmission/reflection spectra become more and more significant, while the spectral widths become more and more broad, and the broadening of the cavity mode is proportional to the shift of the resonant frequency. Using this analytical model, we further derived the critical pump-power condition for the control light in bistable switching based on a nonlinear SCWR system. These results are useful for actual designs of optical switchings.

ACKNOWLEDGMENT

This work is supported in part by the National Natural Science Foundation of China (11774098, 11304099) and Guangdong Natural Science Foundation (2017A030313016, 2013040015639).

REFERENCES

Akahane, Y., Asano, T., Song, B.S. & Noda, S. 2003. High-Q photonic nanocavity in a two-dimensional photonic crystal. *Nature* 425 (6961): 944–947.

Barclay, P.E., Srinivasan, K. & Painter, O. 2005. Nonlinear response of silicon photonic crystal microresonators excited via an integrated waveguide and fiber taper. *Opt. Expres* 13(3): 801–820.

Colman, P., Lunnemann, P., Yu Y. & Mørk, J. 2016. Ultrafast coherent dynamics of a photonic crystal all-optical switch, *Phys. Rev. Lett.* 117(23): 233901.

Harding, P.J., Euser, T.G., Nowicki-Bringuier, Y.R., Gerard, J.M. & Vos. W.L. 2007. Dynamical ultrafast all-optical switching of planar GaAs/AIAs photonic microcavities. *Appl. Phys. Lett.* 91(11):111103.

Haus, H.A. 1984. *Waves and Fields in Optoelectronics*. Englewood Cliffs, NJ, USA: Prentice-Hall.

Joannopoulos, J.D., Meade, R.D. & Winn, J.N. 1995. *Photonic Crystals: Molding the Flow of Light*. Princeton, NJ, USA: Princeton Univ. Press.

Kim, M.K., Hwang, I.K., Kim, S.H., Chang, H.J. & Lee, Y.H. 2007. All-optical bistable switching in curved microfiber-coupled photonic crystal resonators. *Appl. Phys. Lett.* 90(16): 161118.

Li, C., Wang, H., Wu, J.-F. & Xu, W.- C. 2010. Critical conditions of control light for a silicon-based photonic crystal bistable switching. *Opt. Express* 18(16): 17313–17321.

Li, C., Wang, M. & Wu J.- F. 2017. Broad-bandwidth, reversible, and high-nonreciprocal-transmission ratio silicon optical diode. *Opt. Lett.* 42(2): 334–337.

Notomi, M., Shinya, A., Mitsugi, S., Kira, G., Kuramochi, E. & Tanabe, T. 2005. Optical bistable switching action of Si high-Q photonic-crystal nanocavities. *Opt. Express* 13 (7): 2678–2687.

Nozaki, K., Shinya, A., Matsuo, S., Suzaki, Y., Segawa, T., Sato, T., Kawaguchi, Y., Takahashi, R. & Notomi, M. 2012. Ultralow-power all-optical RAM based on nanocavities. *Nature Photon.* 6(4): 248–252.

Nozaki, K., Tanabe, T., Shinya, A., Matsuo, S., Sato, T., Taniyama, H. & Notomi, M. 2010. Sub-femtojoule all-optical switching using a photonic-crystal nanocavity," *Nature. Photon.* 4(7): 477–483.

Ogusu, K. & Takayama, K. 2008. Optical bistability in photonic crystal microrings with nonlinear dielectric materials. *Opt. Express* 16(10): 7525–7539.

Soljačić, M., Ibanescu, M., Johnson, S.G., Fink, Y. & Joannopoulos, J.D. 2002. Optimal bistable switching in nonlinear photonic crystals. *Phys. Rev. E (R)* 66(5): 055601.

Xia, F.N., Sekaric, L. & Vlasov, Y. 2006. Ultracompact optical buffers on a silicon chip. *Nature. Photon.* 1(1): 65–71.

Yang, X.D., Husko, C. & Wong, C.W. 2007. Digital resonance tuning of high-Q/V-m silicon photonic crystal nanocavities by atomic layer deposition. *Appl. Phys. Lett.* 91(16): 051113.

Yanik, M.F., Fan, S.H. & Soljacic, M. 2003. High-contrast all-optical bistable switching in photonic crystal microcavities. *Appl. Phys. Lett.* 83(14): 2739–2741.

Yüce, E., Ctistis, G., Claudon, J., Dupuy, E., Buijs, R.D., Ronde, B. de, Mosk, A.P., Gérard, J.-M. & Vos, W.L. 2013. All-optical switching of a microcavity repeated at terahertz rates. *Opt. Lett.* 38(3): 374–376.

Frontier Research and Innovation in Optoelectronics Technology and Industry – Habib & Lewis (Eds)
© 2019 Taylor & Francis Group, London, ISBN 978-1-138-33178-5

A foresighted measure for RWA algorithms in all-optical mesh networks

Xuhong Li
School of Science, Zhongyuan University of Technology, Zhengzhou, China

Junling Yuan
School of Computer and Communication Engineering, Zhengzhou University of Light Industry, Zhengzhou, China

Lihua Yang
School of Science, Zhongyuan University of Technology, Zhengzhou, China

ABSTRACT: The problem of routing and wavelength assignment is one of the key issues in all-optical mesh networks, and which has been proved to be a NP-hard problem. Most of the existing algorithms to solve the problem are greed-based. Although the greedy idea can get the shortest light-path for the current connection request, it would also worsen the performance of the whole network for a long term. In this paper, we introduce a foresighted measure into greed-based algorithms. The measure can make the algorithms steer by links with few idle wavelengths, and these bypassed links would be left to the connection requests which have to use them. Simulation results show that the foresighted measure can obviously reduce the blocking rate of the whole network, just by a little expense of average end-to-end delay.

1 INTRODUCTION

Nowadays, all-optical mesh networks, based on the dense wavelength division multiplexing (DWDM) technology and optical cross connections (OXCs), are becoming the main structure of the core transport networks. Wavelength routing is the main method to transfer traffics between nodes and which lead the routing and wavelength assignment (RWA) problem to be a key issue in all-optical mesh networks.

The RWA problem ought to be considered in two differential scenes (H. Zang et al. 2000): online scene and online scene. In the online scene, the connection requirements are given by the traffic matrix in advance; we ought to establish route and assign wavelength for each requirement, by minimizing the total number of needed wavelengths. In the online scene, the connection requests arrive over time and each connection has its corresponding holding time; we need to choose route and assign wavelength when a connection request comes and release the occupied wavelengths while a connection is finished, by minimizing the total number of blocked connection requests. Whether in the offline or online scene, the assigned wavelength of all links on the source-destination path would be the same (which is called the constraint of wavelength continuity.) However, if wavelength converters are used at nodes, the constraint of wavelength continuity would be relaxed.

The RWA problem is modeled as an integer linear programming (ILP) formulation by Banerjee and Mukherjee in 1996 and its NP-hardness is further proved (D. Banerjee & B. Mukherjee, 1996). It means that we cannot design an algorithm to obtain the optimal solution of the problem in polynomial time unless $P = NP$. To simplify the design of algorithm, the RWA problem is generally divided into two sub-problems: routing sub-problem and wavelength assignment sub-problem. In routing sub-problem, a suitable source-destination path is found for a connection request, and the choice of wavelength would be done in the wavelength assignment sub-problem. Although design of algorithms in the sub-problems does be simpler than solving the RWA problem as a whole, unfortunately, both of the two sub-problems are *NP*-hard (K. Christodoulopoulos et al. 2010).

For both routing sub-problem and wavelength assignment sub-problem, several heuristic algorithms are proposed in the past years. Zang et al. have given a review of RWA algorithms (H. Zang et al. 2000) in 2000, in which the algorithms for routing sub-problem and wavelength assignment sub-problem are introduced in detail.

- For the routing sub-problem, there are three ways to found source-destination path(s): fixed routing, alternate-fixed routing and adaptive routing. In the fixed routing way, the shortest path between source and destination is found by using the Dijkstra algorithm or Bellman-Ford algorithm. In the alternate-fixed routing way, besides the shortest path, an alternate path (link-disjoint with the shortest one) is also found. The adaptive routing way is designed for networks with wavelength converters, in which the costs of links are associated with the wavelength conversion capability of nodes.
- For the wavelength assignment sub-problem, several wavelength assignment strategies are proposed, such as random, first-fit, least-used, most-used, least-loaded, max-sum, etc. Each strategy has its advantages and can get well performance in specific network and traffic environment.

In recent years, several new methods for solving the RWA problem are proposed. In 2003, Ozdaglar and Bertsekas propose a new integer linear programming formulation for the RWA problem (A. E. Ozdaglar & D. P. Bertsekas, 2003), and the new model can be addressed with highly efficient linear programming methods.

In 2010, Randhawa and Soha proposed several static and dynamic RWA algorithms for all-optical networks (R. Randhawa & J. S. Sohal, 2010), such as maximum empty channel routing, combined path algorithm, etc. The algorithms use the fixed-alternate way to calculate route for a call first and choose wavelength by using their new methods.

In 2011, Han et al. proposed an dynamic algorithm named Bottleneck Links Avoidance with Load Balancing (X. Han et al. 2011), in the algorithm the load on each link is calculated and the links have obviously higher loads than others is defined as bottleneck links. The algorithm calculates the k-shortest paths for connection requests and then chooses the shortest one from the paths without bottleneck links.

Almost all the existing routing algorithms are greed- based. The greedy idea always tries to choose the shortest source-destination path for a connection request; however it would exhaust the wavelengths of some popular links rapidly and result in a bad performance of the whole network in a long term. In this paper, we introduce a foresighted measure into the routing algorithm (J. Yuan et al. 2010). The foresighted measure would steer by links with less idle wavelengths and set apart the links for the connection requests which have to use them. The measure would reduce the blocking rate of the whole network in a long term.

The remainder of this paper is organized as follows. In Section 2 we introduce the foresighted measure in detail. In Section 3 performances of the algorithms with and without foresighted measure are compared by simulations. Finally, brief conclusions are drawn in Section 4.

2   FORESIGHTED MEASURE

An optical network can be expressed by a weighted graph $G(V, E)$. In which, V is the set of nodes and each node represents an optical cross connection or an optical add/drop multiplexer (OADM); E is the set of links and each link represents an optical fiber jointing two nodes. A link always has a weight, and the weight may mean different things, such as the length of the link, the propagation delay, or the remainder capacity, etc.

Most of the existing RWA algorithms are greed-based, and they always try to find a source-destination path with minimum total weight of links. The greed-based algorithm would find the shortest path for the current connection request; however it would also exhaust the idle wavelengths of some bottleneck links. That would result in blocking of the latter arrived connection requests, even though many other links are idle yet.

To avoid that, we introduce a foresighted measure into the greed-based algorithms. The measure would adjust a link's weight according to the number of busy wavelengths on it. For the sake of distinction, we name the adjusted weight as virtual weight and the weight before adjusting as original cost. For the same link, more busy wavelengths it has, bigger its virtual cost is. The foresighted measure results in that the routing algorithm would tend to steer away from the links with bigger virtual cost (i.e. links with less idle wavelengths), unless it has no other choices.

Supposing that the number of wavelengths in each link is W_{total}, the number of occupying wavelengths of a link e is $W_{busy}(e)$, and the original cost of link e is $c_{original}(e)$, then the virtual cost of link e can be calculated by the following formula:

$$c_{virtual}(e) = c_{virtual}(e) \times a^{\frac{W_{busy}(e)}{W_{total}}}$$

In which, the base number a is a constant greater than one. By the formula we can see that, if all the wavelengths of a link are idle, its virtual cost equals its original one; along with the increase of the number of busy wavelengths, the virtual cost would exponentially increase.

Besides, we can adjust the impact of the foresighted measure by changing the value of base number a. We show the variation tendency of virtual cost by an example. Supposing that the original cost of a link e is 3 and the number of total wavelengths on it is 8, the correspondence between the virtual cost and the number of occupying wavelengths is shown in Table 1. From the table we can see that, the virtual cost increases exponentially with the number of busy wavelengths, and the value of the base number affects the virtual cost greatly.

Since the virtual cost of a link is not fixed but varying with the change of link's state, we must update the virtual costs of links in time. The virtual costs of links would be updated at two occasions:

1. When a connection is successfully established, it would occupy some a wavelength on the chosen source-destination path. At this time, the number of occupied wavelengths of the links on the path increases by one, thus the virtual costs of these links would increase.
2. When a connection is completed, the occupied wavelength on the source-destination path would be released. At this time, the number of occupied wavelengths of the links on the path decreases by one, thus the virtual costs of these links would decrease.

We call the proposed algorithm with foresighted measure the modified algorithm, and the algorithm without foresighted measure the initial algorithm. In the modified algorithm, we would calculate the shortest path according to the virtual costs rather than the original costs of links. After a source-destination path is found, we assign it a wavelength by the random and wrap-around rule. By the rule, we randomly choose a wavelength, say w_{start}, as the starting wavelength and then orderly check the states of wavelengths. The serial number of wavelength checked in the i-th round is $(w_{start}+i)\% W_{total}$, in which % is the modulus operator. The wavelengths would be checked one by one until some a wavelength is jointly idle in all the links of the path; if all the wavelengths are checked to be unavailable, the connection request would be blocked.

Table 1. An example to show the correspondence between virtual cost and number of busy wavelengths.

$W_{busy}(e)$	0	1	2	3	4	5	6	7	8
$a = 2$	3.00	3.27	3.57	3.89	4.24	4.63	5.05	5.50	6.00
$a = 3$	3.00	3.44	3.95	4.53	5.20	5.96	6.84	7.85	9.00
$a = 4$	3.00	3.57	4.24	5.05	6.00	7.14	8.49	10.09	12.00

3 SIMULATIONS

In this section, we demonstrate the effect of the foresighted measure by comparing the performances of routing algorithms with and without using the proposed foresighted measure. We do simulations on the NSF-Net, which is the mostly used topology for simulation of all-optical networks. The topology of the NSF-Net is shown in Fig. 1. In which, each cycle represents a node and each line denotes a link; the numbers in the cycles are the nodes' serial numbers, and the numbers on the lines are the corresponding original cost of each link. In our simulations, all the links are duplex and 16 wavelengths are included in each direction.

In all the simulations, traffics are evenly distributed among all the nodes. At each node, the arrival of connection requests is following the Poisson distribution and the holding time of each connection is under the exponential rule. The traffic density at each node equals the product of the mean arrival rate and the average holding time; the traffic density of the whole network is the sum of traffic densities at all the nodes.

We consider two indexes of the network: the blocking rate and the average end-to-end delay. The blocking rate equals the number of succeeded connections divided by the total number of connection requests. The average end-to-end delay equals the sum propagation delay of all the succeeded connections divided by the total number of succeeded connections.

3.1 Chosen of adjusting parameter

We first find the value of base number a that which minimizes the blocking rate of the network. We do simulations by choosing a as a series of values and observe the blocking rates of the network under different traffic densities. The results are shown in Table 2. From the table we can see that, the value of a affects the blocking rate remarkably at low traffic density cases, while only has a slight effect at high traffic density cases. However, at each case, the blocking rate appears its minimal value when a is chosen as 2.5 or 3.0. Hence, in the following simulations, we would fix the value of a as 3.0.

3.2 Comparison of performances

We compare two indexes of the algorithms: blocking rate and average end-to-end delay. The blocking rate is the main index in optical networks, which indicates an algorithm's service ability.

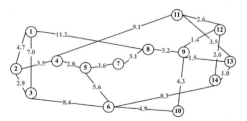

Figure 1.   Topology of the NSF-Net.

Table 2.   Blocking rate (%) of the modified algorithm under different values of a.

Traffic density	100	200	300	400	500	600	700	800	900
a = 1.0	1.13	16.15	30.13	39.07	46.28	51.65	55.61	58.98	61.80
a = 1.5	0.59	14.71	29.63	39.02	46.09	51.53	55.35	58.94	61.68
a = 2.0	0.41	14.34	29.27	38.95	46.02	51.29	55.44	58.96	61.49
a = 2.5	0.34	14.16	29.06	39.01	45.82	**51.26**	**55.27**	**58.77**	**61.44**
a = 3.0	**0.17**	**13.20**	**28.88**	**38.81**	**45.79**	51.54	55.46	58.92	61.65
a = 3.5	0.22	13.34	28.98	38.97	45.83	51.44	55.77	58.83	61.79
a = 4.0	0.26	14.33	29.33	39.21	46.25	51.48	55.32	59.16	61.63

The average end-to-end delay shows the quality of service of an algorithm, and it just need to be not too large. Since the indexes under different traffic densities have a quite large variation, we would separately compare the performances of the algorithms at low and high traffic cases.

Figures 2 and 3 show the blocking rates of the algorithms under low traffic case and high traffic case, respectively. We can see that, the blocking rate increases with the rise of traffic density.

- At low traffic case (less than 90 Erlang), the blocking rate of the modified algorithm is always near to zero, while the blocking rate of the initial algorithm increases rapidly when traffic density is greater than 60 Erlang. It means that the foresighted measure can obviously decrease the blocking rate under low traffic case.
- At high traffic case (greater than 100 Erlang), the blocking rates of both the modified and initial algorithms increase slowly, however the blocking rate of the modified algorithm is always less than that of the initial one. It means that the foresighted measure can also do certain effect under high traffic case.

Hence, the modified algorithm can improve the performance of blocking rate greatly under low traffic case and slightly under high traffic case.

Fig. 4 and Fig. 5 show the average end-to-end delay of the algorithms, under low traffic case and high traffic case, respectively.

- At low traffic case, the average end-to-end delay of the initial algorithm keeps in a fixed level while that of the modified algorithm increases slowly. It is because of that, the initial algorithm would choose the shortest path between source and destination, hence the average end-to-end delay is affected little by the traffic density; the modified algorithm would steer by busy links and which may result in a farther path.

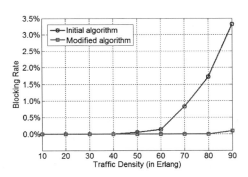

Figure 2. Blocking rate of the initial and modified algorithms under low traffic case.

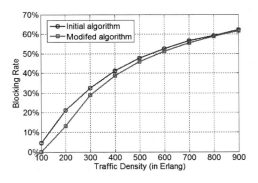

Figure 3. Blocking rate of the initial and modified algorithms under high traffic case.

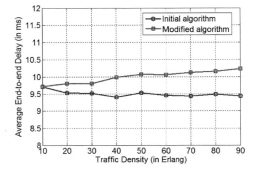

Figure 4. Average end-to-end delay of the initial and modified algorithms under low traffic case.

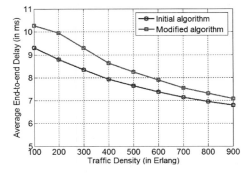

Figure 5. Average end-to-end delay of the initial and modified algorithms under high traffic case.

- At high traffic case, the average end-to-end delay of both the initial algorithm and the modified algorithm decrease slowly. It is because of that, with the rise of traffic density, the connections via small number of links would be more likely been blocked than the connections via many number of links. Since the average end-to-end delay is the mean of successful connections, it would decrease with the rise of traffic density.

Whether under the low traffic case or high traffic case, the average end-to-end delay of the modified algorithm is only slightly bigger than that of the initial algorithm.

Generally, the modified algorithm with the foresighted measure can reduces the blocking rate of the network under each traffic density, just by increasing the average end-to-end delay slightly.

4 CONCLUSIONS

In this paper, we modified greed-based RWA algorithms in all-optical networks by introducing a foresighted measure. Comparing to the initial algorithm, the modified algorithm can obviously reduce the blocking rate of the network, at a little expense of average end-to-end delay. Since blocking rate is the main index for all-optical networks, the foresighted measure is valid. However, as the foresighted measure just improves the performance of greed-based routing algorithms, it is not an optimal measure yet. Hence, many works would be done on the RWA problem in all-optical networks yet.

ACKNOWLEDGEMENT

This work is supported by Natural Science Foundation of China (No.61501406 and 61774277), Higher Educational Scientific Research Program of Henan Province (No. 15A510015 and 15A520032).

REFERENCES

Banerjee D. and B. Mukherjee, A practical approach for routing and wavelength assignment in large wavelength-routed optical networks, IEEE Journal on Selected Areas in Communications, 14, 903 (1996).

Christodoulopoulos K., K. Manousakis, and E. Varvarigos, Online Routing and Wavelength Assignment in Transparent WDM Networks, IEEE/ACM Transactions on Networking, 18, 1557 (2010).

Han X., H. Shi, and Q. Yang, A dynamic routing and wavelength assignment algorithm for WDM networks based on bottleneck link avoidance, 2011 IEEE International Conference on Signal Processing, Communications and Computing (ICSPCC), 1–5 (2011).

Ozdaglar A.E. and D.P. Bertsekas, Routing and wavelength assignment in optical networks, IEEE/ACM Transactions on Networking, 11, 259 (2003).

Randhawa R. and J.S. Sohal, Static and dynamic routing and wavelength assignment algorithms for future transport networks, Optik, 121, 702 (2010).

Yuan J., X. Zhou, J. Wang, X. Yu, and X. Miao, A foresighted strategy for greed-based multicasting algorithms in all-optical mesh networks, Photonic Network Communications, 20, 278 (2010).

Zang H., J.P. Jue, and B. Mukherjee, A review of routing and wavelength assignment approaches for wavelength-routed optical WDM networks, Optical Networks Magazine, 1, 47 (2000).

Frontier Research and Innovation in Optoelectronics Technology and Industry – Habib & Lewis (Eds)
© 2019 Taylor & Francis Group, London, ISBN 978-1-138-33178-5

Intensity and phase of ring Airy Gaussian vortex beam in atmospheric turbulence

Y.Q. Li, L.G. Wang & L. Gong
School of Optoelectronic Engineering, Xi'an Technological University, Xi'an, China

ABSTRACT: In this study, the characteristics of intensity and phase for the ring Airy Gaussian vortex (RAiGV) beam propagation in atmospheric turbulence were numerically simulated, based on the Split-Step Fourier Method (SSFM) and Power Spectrum Inversion Method. Also, the propagation characteristics of the RAiGV beam in the free space were studied. The results obtained showed that at the edge of the phase screen, the interference fringes exist, the phase is distorted and the equiphase line turns into an arc due to the influence of the atmospheric turbulence. Also, it is observed that the smaller the distribution factors is, the greater the change in the intensity and phase wave-front will be, but the fringes interference at the edge of the phase screen is weaker. The results will provide theoretical and technical references for atmospheric optical communication using vortex beams.

1 INTRODUCTION

In recent years, numerous scholars have investigated the generation of Airy beams, their propagation characteristics, and applications. Gu investigated the scintillation of an Airy array beam in the atmosphere, which indicated that the array can greatly decrease scintillation with a value close to the theoretical minimum obtained (Gu, Y. & Gbur, G. 2010). Chu studied the evolution of Airy beams through atmospheric turbulence, and found that the centroid and self-bending of Airy beams were not influenced by atmospheric turbulence (Chu, X. X. 2011). Chen proposed the non-diffraction, auto-acceleration, and self-repairing characteristics of Airy beams propagating through free space (Chen, R.-P. 2011). Deng studied the energy flows and angular momentum density of non-paraxial Airy beams propagation in free space (Deng, D. 2013). Ji investigated the propagation of Airy beams in the atmosphere (but without turbulence) and considered the influence of thermal blooming on their propagation characteristics (Ji, X. 2013). In addition, Tao investigated the far-field average diffusion of Airy beams through non-Kolmogorov atmospheric turbulence (Tao, R. 2013). Chen revealed that both mode-locked and -unlocked radial Airy array beams exhibit strong self-focusing characteristics after the propagation through atmospheric turbulence (Chen, C. Y. 2014). Moreover, Cai revealed the propagation characteristics of Airy beams by using Gaussian beams propagating through atmospheric turbulence (Wen, W. 2015). Chen investigated the propagation properties of the sharply autofocused ring Airy Gaussian vortex beams numerically and some numerical experiments are performed (Chen, B. 2015). Jiang investigated the propagation dynamics of the Circular Airy Beams (CAB) with optical vortices by numerical calculation (Jiang, Y. 2012).

Scholars have investigated the propagation of Airy beams with the focus on their non-linear (auto-accelerating and self-bending) characteristics in vacuum. At present, most research on the propagation of Airy beams through free space or atmospheric turbulence is aimed at the self-repairing and non-diffraction characteristics. However, when beams propagate through atmospheric turbulence, the light field at $z = L$ cannot be expressed because of random fluctuations of the atmospheric refractive index. Therefore, the light-field propagation characteristics of beams can be analyzed by numerical simulation in the absence of experimental conditions. In this work, the split-step Fourier method (SSFM) was applied to simulate RAiGV beam propagation through free space or atmospheric turbulence.

2 THEORETICAL FORMULA

The electric fields of the initial RAiG beams superimposed by a spiral phase in cylindrical coordinate can be expressed as (Chen, B. 2015):

$$U(r,\varphi,0) = A_0 Ai\left(\frac{r_0 - r}{bw}\right) \exp\left(a\frac{r_0 - r}{bw}\right) \exp\left(-\frac{(r_0 - r)^2}{w^2}\right)(r^m e^{im\varphi}) \tag{1}$$

where A_0 is the constant amplitude of the electric field, $A_i(\cdot)$ corresponds to the Airy function, r is the radial coordinate, r_0 represents the radius of the primary Airy ring, and w is a scaling factor, $0 \leq a < 1$ is the exponential truncation factor which determines the propagation distance, and m is the topological charge of the optical vortex. Here, b is a distribution factor parameter. We first introduce the factor b into the RAiGV beams, which can adjust the scale between the ring Airy factor and the hollow Gaussian factor, and it describes in a more realistic way to adjust the initial beams.

The Split-Step Fourier Method (SSFM) is used to simulate the RAiGV beam propagation through the atmospheric turbulence (Yura, H. T. 1972, Martin, J. M. & Flatte, S. M. 1988, Martin, J. M & Flatte, S. M. 1990, Flatte, S. M. 1993, Flatte, S. M. & Gerber, J. S. 2000).

Fig. 1 shows the SSFM is an iterative process. The two-dimensional Fourier transform (FT) is involved in SSFM, so it is proper to solve the algorithm by using a numerical FT algorithm. The two-dimensional FT and inverse FT can both be calculated by fast Fourier transform (FFT) during subsequent iterations. Generation of the random phase screen by the spectrum inversion method is an important step (Flatte, S. M. 1994). Square phase screen is used in this paper, and the mesh widths in x and y directions are the same. The main parameters involved in the simulation of laser propagation include the mesh width $\Delta x, \Delta y$ that conforms to $\Delta x < \pi / \kappa_{max}$ ($\kappa_{max} = \left[0.045 r_0^{-5/3} + 0.025 r_0^{-5/3}(k/\Delta z)^{1/6}\right]^{1/2}$), phase screen width D that satisfies $D > \lambda\Delta z / \Delta x$ (Nelson, D. H. 2000) and phase screen spacing Δz (Flatte, S. M. 1994), which are restricted by the variables of turbulence intensity, wavelength, beam width, and have a certain range of values. Due to the beam propagation along a slant path, the turbulence intensity decreases with the increase of the height, the phase screen spacing is nonuniform.

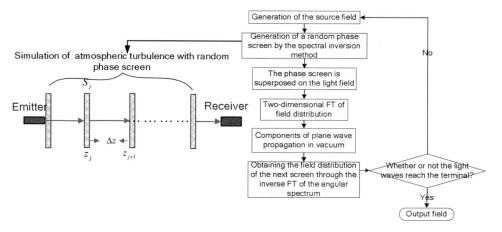

Figure 1.   Flow-process diagram of the split-step Fourier method.

3 RESULTS OF THE NUMERICAL SIMULATION

The SSFM is used to simulate the propagations of the RAiGV beams (w_0 = 3 cm) through atmospheric turbulence.

3.1 *Intensities and phases in free space*

As shown in Figure 2, the radius of the Gaussian hollow is constant, and the effect of the parameter b on the scale between the Airy ring and the Gaussian hollow is great when the distribution factor b varies and the other parameters are fixed. The smaller the value of b is, the smaller the scale between the ring and the Gaussian hollow, the smaller the width of the ring, that is, the more the number of rings in the range of intensity distribution. In contrast, The scale between the ring and Gaussian hollow and the width of the ring become larger, and the number of rings in the range of intensity distribution decreases.

Figure 3 shows the effects of the radius of the primary Airy ring r_0 on the intensities and phases of the RAiGV beam at the source plane. The radius of the Gaussian hollow increases with the increase of r_0. And the width and intensity of the first ring near Gaussian hollow is great affected by r_0, other Airy rings is little affected by it.

Figure 4(a) shows the effect of different truncation parameters on the intensity distribution. The width of the Airy ring is not be affected by the size of the truncation parameter, but the intensities of each ring are affected by it and the first Airy ring near the Gaussian hollow is greatest affected by it. Figure 4(b) represents the effect of different primary radii r_0 on the intensity distribution. The beam intensity increases with the increase of r_0, other conclusions are shown in Figure 3. Figure 4(c) shows the effect of different distribution factors r_0 on the intensity distribution. The radius of Gaussian hollow is not only affected by the parameter b, but also the width of the ring, the scales between the ring and Gaussian hollow (the number of Airy ring in the range of intensity distribution) are affected by it. The conclusions are found to be similar to that shown in Figure 2.

Table 1. Parameter values in the simulation of the beam propagation through atmospheric turbulence.

Laser wavelength λ	Structure constant C_n^2	Phase screen width D	Grid width Δx	Grid numbers N	phase screen spacing Δz
1.06 μm	$1.7 \times 10^{-14}\,\mathrm{m}^{-2/3}$	0.2 m	1.2 mm	512	50 m

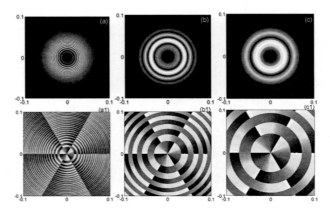

Figure 2. Intensity distributions of the RAiGV beam with topological charge $m = 3$ ($a = 0.05$, $r_0 = 0.01$) at the source plane in free space (a),(b),(c) and the corresponding phase distributions (a1),(b1),(c1), (a) $b = 0.1$; (b) $b = 0.3$; (c) $b = 0.5$.

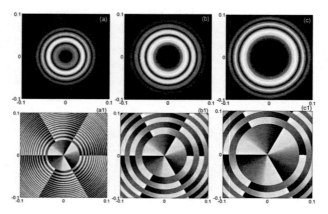

Figure 3. Intensity distributions of the RAiGV beam with topological charge $m = 3$ ($a = 0.05$, $b = 0.3$) at the source plane in free space (a),(b),(c) and the corresponding phase distributions (a1),(b1),(c1), (a) $r_0 = 0.01$; (b) $r_0 = 0.03$; (c) $r_0 = 0.05$.

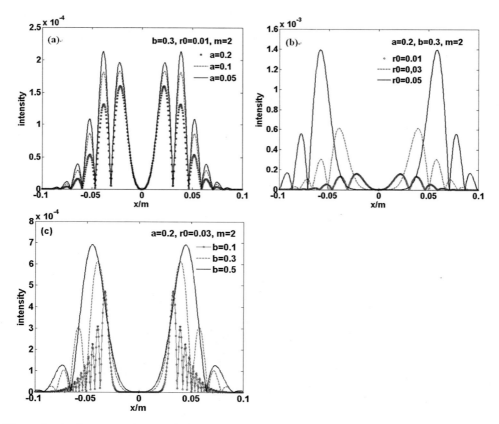

Figure 4. At the source plane, variations of the intensity on the x axis for the RAiGV beam with (a) the truncation parameter a, (b) the radius of the primary Airy ring r_0, and (c) distribution factor b.

3.2 Intensities and phases in the atmospheric turbulence

The parameters are taken as follows: $a = 0.05$, $r_0 = 0.01$, $m = 2$.

It can be seen from Figure 5, the variations of the intensities and phases for the RAiGV beam propagation in atmospheric turbulence at the distance of 300 m and 500 m. As the

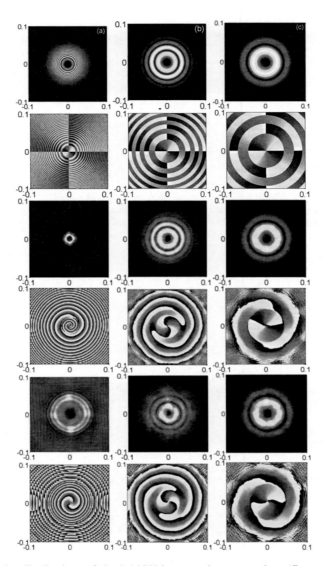

Figure 5. Intensity distributions of the RAiGV beam at the source plane (first row) and the corresponding phase distributions (second row), intensity distributions of the RAiGV beam in atmospheric turbulence at the distance of 300 m (third row) and the corresponding phase distributions (fourth row), intensity distributions of the RAiGV beam in atmospheric turbulence at the distance of 500 m (fifth row) and the corresponding phase distributions (sixth row). (a) $b = 0.1$, (b) $b = 0.3$, (c) $b = 0.5$.

propagation distance increases, the phase is distorted. A phase spin occurs due to the existence of orbital angular momentum. The intensity of the beam results in a random fluctuation, due to the influence of the atmospheric turbulence. At the edge of the phase screen, the interference fringes exist, and the equiphase line turns into an arc. Also, it is observed that the smaller the distribution factors b is, the greater the change in the intensity and phase wavefront will be, but the fringes interference at the edge of the phase screen is weaker.

Therefore, based on the analyses of Figure 5, a Split-Step Fourier algorithm is used to simulate the beam propagation in the atmospheric turbulence. The intensity and phase distributions at the source plane are determined to be the same as those in the free space. The vortices of the RAiGV beams surrounds the beam center to form a circle, and the sum of the phase gradients is $2m\pi$. This is found to be in accordance with the distribution law of the vortex phase factor $\exp(-im\varphi)$. Therefore, the numerical simulation method which is adopted in this study is confirmed to be correct and feasible.

4 CONCLUSIONS

Because of the idealization of theoretical research methods and the complex conditions of experimental research, the numerical simulation method used to study the beam propagation through the atmospheric turbulence is not only used to verify the theory, but also used to compare with the experimental results. In this study, the intensities and phases of the RAiGV beams were numerically simulated, and some basic characteristics of the vortex beam propagation in atmospheric turbulence were obtained. Vortex beams are widely applied in engineering and space optical communication. The results will provide theoretical and technical references for atmospheric optical communication, as well as future information transmissions using vortex beams.

ACKNOWLEDGMENTS

This paper is supported by National Natural Science Foundation of China (Grant No. 61805190, 11504286, 61271110).

REFERENCES

Chen, B. 2015. Propagation of sharply autofocused ring Airy Gaussian vortex beams. *Optics Express* 23(15): 19288–19298.

Chen, C.Y. 2014. Propagation of radial Airy array beams through atmospheric turbulence. *Opt. Lasers Eng.* 52: 106–114.

Chen, R.-P. 2011. Wigner distribution function of an Airy beam. *J. Opt. Soc. Am. A* 28(6): 1307–1311.

Chu, X.X. 2011. Evolution of an Airy beam in turbulence. *Opt. Lett.* 36(14): 2701–2703.

Deng, D. 2013. Energy flow and angular momentum density of nonparaxial Airy beams. *Opt. Commun.* 289: 6–9.

Flatte, S.M. & Gerber, J.S. 2000. Irradiance-variance behavior by numerical simulation for plane-wave and spherical-wave optical propagation through strong turbulence. *J. Opt. Soc. Am* 17: 1092–1097.

Flatte, S.M. 1993. Irradiance variance of optical waves through atmospheric turbulence by numerical simulation and comparison with experiment. *J. Opt. Soc. Am. A* 10: 2363–2370.

Flatte, S.M. 1994. Probability-density functions of irradiance for waves in atmospheric turbulence calculated by numerical simulation. *J. Opt. Soc. Am. A*. 11(7): 2080–2092.

Gu, Y. & Gbur, G. 2010. Scintillation of Airy beam arrays in atmospheric turbulence. *Opt. Lett.* 35(20): 3456–3458.

Ji, X. 2013. Propagation of an Airy beam through the atmosphere. *Opt. Express* 21(2): 2154–2164.

Jiang, Y. 2012. Propagation dynamics of abruptly autofocusing Airy beams with optical vortices. *Optics Express* 20(17): 18579–18584.

Martin, J.M & Flatte, S.M. 1990. Simulation of point-source scintillation through three-dimensional random media. *J. Opt. Soc. Am. A* 7: 838–847.

Martin, J.M. & Flatte, S.M. 1988. Intensity images and statistics from numerical simulation of wave propagation in 3-D random media. *Appl. Opt.* 27: 2111–2126.

Nelson, D.H. 2000. Wave optics simulation of atmospheric turbulence and reflective speckle effects in CO_2 lidar. *Appl. Opt.* 39(12): 1857–1871.

Tao, R. 2013. Average spreading of finite energy Airy beams in non-Kolmogorov turbulence. *Opt. Lasers Eng.* 51: 488–492.

Wen, W. 2015. The propagation of a combining Airy beam in turbulence. *Opt. Commun.* 336: 326–329.

Yura, H.T. 1972. Mutual Coherence Function of a Finite Cross Section Optical Beam Propagating in a Turbulent Medium. *Appl. Opt.* 11: 1399–1406.

Frontier Research and Innovation in Optoelectronics Technology and Industry – Habib & Lewis (Eds)
© 2019 Taylor & Francis Group, London, ISBN 978-1-138-33178-5

Effect of atmospheric refraction on timing deviation of earth-to-satellite time transfer based on dual wavelength

H. Liu, L. Lu, B.F. Zhang, C.X. Wu & J.Y. Wang
Institute of Communication Engineering, Army Engineering University of PLA, Nanjing, Jiangsu, China

ABSTRACT: The distribution of atmospheric refraction introduces timing deviation to earth-to-satellite time transfer based on dual wavelength. In this paper, real-world meteorological data is used to construct a model that can describe the influence of atmospheric refraction distribution on the time transfer. The variation in asymmetric delay deviation and transceiver position deviation with region, month and zenith angle is studied by simulation. Results indicate that a decrease in atmospheric pressure and an increase in atmospheric temperature cause a decrease in the time deviation. The timing deviation caused by the inhomogeneous distribution of atmospheric density and atmospheric dispersion reaches the order of nanoseconds and can be reduced to ten picoseconds through further bidirectional precision alignment. During satellite covering, asymmetric delay deviation and transceiver position deviation vary with zenith angle and geographical position of ground terminals. Therefore, there is a need for real-time correction of the time deviation.

1 INTRODUCTION

The high-accuracy time transfer by earth-to-satellite laser link is essential for the precision, effectiveness and integrity of the time-frequency transfer system. In this context, it has drawn a lot of attention worldwide (Vrancken et al. 2014, Fridelance et al. 1997, Giorgetta et al. 2013, Djerroud et al. 2013). Time transfer in the space can be implemented via an earth-to-satellite laser communication link. The satellite-ground communication system mainly consists of satellite and ground terminals. The satellite terminals usually adopt the transceiver design while the antennae mostly adopt the Cassegrain telescope (Ma et al. 2015) such as the Laser Communications Equipment (LCE) and Laser Utilizing Communications Equipment (LUCE) of Japan, and the OPALE and PASTEL from Europe. In these cases, the uplink and downlink are usually based on optical signal with different wavelengths. The methods for space laser time transfer are similar to these for time transfer in the fiber, and examples include the round tripapproach (Marra et al. 2011) and the two-way comparison approach (Kodet et al. 2016). The timing deviation is a major measure of the time transfer system's performance (Vrancken 2008, Belmonte et al. 2017, Samain et al. 2002). Due to non-uniform atmospheric density distribution and atmospheric dispersion, the use of optical signal with different wavelengths for uplinks and downlinks causes timing deviations, i.e. the asymmetric delay deviation generated during transmission of two-way signal and the time delay deviation that arises from the transceiver position deviation. Therefore, the actual transmission path and time delay of the dual-wavelength optical signal should be studied by modelling the distribution of atmospheric refraction, in order to calibrate the two-way link asymmetric delay deviation and the transceiver position deviation.

In this context, we focus on the delay deviation of time transfer by earth-to-satellite laser link that arises from the dual-wavelength optical signal. The real-world atmosphere meteorological data collected by the ground stations is used to construct a model of the timing deviation. Based on this model, the asymmetric delay deviation and the transceiver position deviation that arise from the non-uniform atmospheric density distribution and atmospheric dispersion are analyzed by simulation.

2 MODEL OF THE INFLUENCE OF ATMOSPHERIC REFRACTIVITY DISTRIBUTION ON THE TIMING DEVIATION

The light tracking method is used to establish the model for the timing deviation. Assume that atmospheric layers are concentric and follow a uniform distribution around the Earth. Light traversing the atmosphere is only refracted at the boundary of two neighboring layers (Yuan 2011). The atmosphere beyond the distance of 32 km is considered to be empty space. The group refractivity of laser in the air can be computed as (Rüeger 2002):

$$N_s = (n_s - 1) \times 10^6 = 287.6155 + \frac{4.88660}{\lambda^2} + \frac{0.06800}{\lambda^4} \tag{1}$$

$$(n-1) \times 10^6 = \left(\frac{273.15}{1013.25} \cdot \frac{P}{T} \cdot N_s \right) - 11.27 \cdot \frac{e}{T} \tag{2}$$

where λ = light wavelength; P = pressure; T = temperature; e = vapor pressure; n = atmosphere group refractive index; N_s = standard atmosphere group refractivity (given an atmosphere temperature of 273.15 K, vapor pressure of 0 hPa, atmospheric pressure of 1013.25 hPa, and carbon dioxide content of 0.0375%); n_s = standard atmosphere group refractive index.

From (1) and (2), it can be seen that the atmospheric refractivity is a function of wavelength, temperature, atmospheric pressure and vapor pressure. Therefore, it can be computed using atmosphere meteorological data. The China Meteorological Data Service Center has released the monthly values of atmosphere observations at specific layers in individual years (China Integrated Meteorological Information Service System), including the average atmospheric pressure, temperature, dew-point temperature gradient of ground surface and constant-pressure surface where the atmosphere meteorological stations are located. Hence, these meteorological data can be used with the model of layer-specific atmosphere refractive index. To satisfy the variables in (1) and (2), the average dew-point temperature gradient needs to be converted to a vapor pressure using (3).

$$e = fE(T_d) \tag{3}$$

where T_d = dew-point temperature; f = moist-air enhancement factor (which can be approximated to 1 for calculation of atmosphere refractivity (Ciddor 1996); E = saturated vapor pressure. The Goff-Gratch formula in (World Meteorological Organization 1988) is adopted to compute the saturated vapor pressure. The geopotential height of constant-pressure surface needs to be converted to an elevation using (4).

$$z = \frac{9.8 \times H_w}{g(z)} = \frac{9.8 \times H_w}{9.784 \cdot (1 - 0.0026 \cdot \cos 2\varphi - 2.8 \times 10^{-7} z)} \tag{4}$$

where g(z) = gravity acceleration; φ = latitude; z = elevation; H_w = geopotential height. Here, a layer-specific model of the atmospheric refractive index can be constructed to determine the refractive index and the distance between the bottom of each layer and the geocenter.

Figure 1 shows the geometrical model for calculation of the timing deviation when using two wavelengths, where O denotes the geocenter, X denotes the ground station, and A denotes the satellite. After the satellite terminal sends the optical signal to the ground terminal along the path *AB*, the ground terminal sends the optical signal to the satellite terminal at the received zenith angle ψ. Because the light with different wavelength has different atmosphere refractivity, the uplink is the path *XB*. Due to the reversibility of the optical paths, the asymmetric delay deviation between the two-way links is the time delay deviation between paths

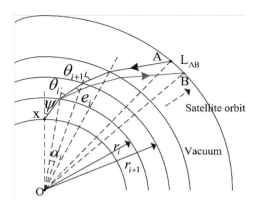

Figure 1. Illustration of a dual-wavelength time transfer by laser link.

XA and XB in Fig. 1. It is equal to the fact that the transmission of optical signal with different wavelengths at the same position and the same zenith angle. Therefore, given the received zenith angle ψ for the ground station, the asymmetric delay deviation between two-way links can be obtained from (5) and (6).

$$t = \begin{cases} \sum \dfrac{n_i r_i \sin(\alpha_i)}{\sin(e_i)C} & \psi \neq 0 \quad i = 1,2,...,N \\[3mm] \sum \dfrac{n_i(r_{i+1}-r_i)}{C} & \psi = 0 \quad i = 1,2,...,N \end{cases} \tag{5}$$

$$\tau_a = \frac{1}{2}(t_{XB} - t_{XA}) \tag{6}$$

where t = time delay; C = light speed in the vacuum; α_i = angle between the incident and exit points in the i-th atmospheric layer with respect to the geocenter; e_i = exit angle of the light at the top of the i-th atmospheric layer; θ_i = incidence angle of the light at the bottom of the i-th atmospheric layer. From Figure. 1, it can be deduced that $\alpha_i = \theta_i - e_i$. According to Snell's law and the sine theorem, we have the following optical path (7).

$$n_1 r_1 \sin \theta_1 = n_i r_i \sin \theta_i = m \quad i = 1,2,...,N \tag{7}$$

where m is a constant; r_1 = the altitude of the ground station (the distance between the bottom of the lowest atmospheric layer and the geocenter); n_1 = atmospheric refractive index corresponding to r_1; θ_1 = received zenith angle ψ of the ground station. Hence, the parameters θ_i, e_i and α_1 corresponding to each atmospheric layer can be obtained from the known angle ψ. Substituting them into Equations (5) and (6) provides the two-way link asymmetric delay deviation.

The transceiver position deviation is equal to the $\overset{\frown}{AB}$ of in Fig. 1(a). The angle between OB and OA is so small that $\overset{\frown}{AB}$ can be approximately equal to the length of AB. Hence, the timing deviation τ_s that arises from the transceiver position deviation can be calculated via Equations (8) and (9).

$$L_{AB} = (r_e + H_s) \cdot \left(\sum \alpha_{ui} - \sum \alpha_{di} \right) \tag{8}$$

$$\tau_s = n_f L/C \tag{9}$$

where r_e = earth radius; H_s = satellite orbit altitude; and α_{ui} and α_{di} denote the angle formed by the incident and exit points of the uplink and downlink in the i-th atmospheric layer with respect to the geocenter. In equation (9), the optical fiber refractivity $n_f = 1.47$.

3 COMPUTATION RESULT AND ANALYSIS

In our analysis we assume that the satellite orbit altitude is 1000 km, the typical 800 nm waveband is selected, the uplink wavelength is 815 nm, and the downlink wavelength is 847 nm (Toyoshima et al. 2004, Yamakawa 2010). Meteorological data collected by stations in Harbin, Haikou, Wuhan and Jiuquan is used. The data relates to temperature, pressure, vapor pressure and geopotential height (China Integrated Meteorological Information Service System). An atmospheric refractive index distribution model is constructed using Equations (1)-(4). The received zenith angle ψ of the ground station is in the range of 0°– 60°. Analysis of the asymmetric delay deviation and transceiver position deviation is performed by simulation and using this model.

3.1 Simulation analysis of the asymmetric delay deviation and position deviation

Figure 2a and Figure 2b show the variation of the two-way link asymmetric delay deviation and position deviation as a function of zenith angle in the case of the Harbin, Haikou, Wuhan, and Jiuquan stations. From Figure 2a, it can be seen that the asymmetric delay deviation in each station increases with the zenith angle. The difference is in the range of several nanoseconds with a maximum of about 3.36 ns. From Figure 2b, it can be seen that the position deviation of each station increases with the zenith angle. The deviation is of the order of hundreds of centimeters with a maximum of about 267.62 cm which equates to approximately 13.11 ns of timing deviation.

Figures 3a and Figure 3b show the monthly variation of asymmetric delay deviation and position deviation over each station given a zenith angle of 60°. It can be seen that for a ground station, the asymmetric delay deviation and position deviation decrease with increasing temperature. This can be attributed to the fact that the increase in temperature reduces air density and alleviates atmospheric refraction. It is especially true for the Harbin station, where the maximum asymmetric delay deviation and position deviation are 0.55 ns and 43.68 cm corresponding to a time deviation of approximately 2.14 ns respectively. This is mainly due to drastic temperature difference between months. The Jiuquan station has the least asymmetric delay deviation and position deviation, upper-bounded to about 0.59 ns and 47.14 cm corresponding to a time deviation of approximately 2.31 ns respectively. This is because of the high latitude of the Jiuquan station, where the low-density air alleviates atmospheric refraction.

3.2 Analysis of the asymmetric delay deviation and position deviation under precision alignment

For the design of an acquisition and tracking link, the two-way link is expected to be aligned accurately so that the bit error rate can fulfill the requirements of satellite-ground laser

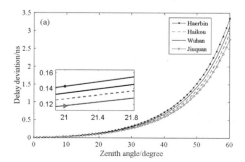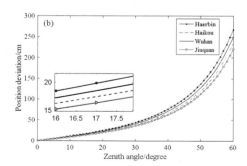

Figure 2. Timing deviation variation as a function of zenith angle for each station in January under the ground terminal acquisition mode. a. Asymmetric delay deviation b. Transceiver position deviation.

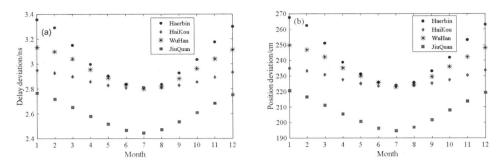

Figure 3. Time deviation for each station in different months given a zenith angle of 60°. a. Asymmetric delay deviation; b. Transceiver position deviation.

Table 1. Minimal position deviation and asymmetric delay deviation obtained after correcting the acquisition andtracking link.

Orbit height (km)	Alignment precision (urad)	Asymmetric delay deviation (ps)	Position deviation (cm)
1000	1	69.1224	4.1861
	0.1	18.7001	0.1456
	0.01	17.0194	0.0109
	0.001	16.9073	0.0109
9000	1	70.3414	8.9328
	0.1	20.3125	0.5730
	0.01	17.8111	0.1550
	0.001	16.9356	0.0087
36000	1	89.8106	33.5910
	0.1	30.3822	6.2205
	0.01	17.0182	0.0621
	0.001	17.0182	0.0621

communication applications. Therefore, the position deviation that arises from atmospheric dispersion has been partly calibrated by the acquisition and tracking system in the engineering applications.

The meteorological data from the Wuhan station in March is used. The orbital height is set to different levels. The reception angle of ground station is set to 60° corresponding to the maximal time deviation. The uplink transmission angle is incremented in the range 59°–60° with a step size of 1 urad, 0.1 urad, 0.01 urad and 0.001 urad. In this way, we can obtain the minimal position deviation after the correction of tracking link and the asymmetric delay deviation corresponding to the minimal position deviation, as shown in Table 1. For a given orbital height, different alignment precision corresponds to different asymmetric delay deviation. The higher the alignment precision, the smaller the asymmetric delay deviation. Given an alignment precision, different orbital heights correspond to different asymmetric delay deviation. Currently, the utmost limit of alignment precision is 0.1 urad (Ma et al. 2015), which corresponds to an asymmetric delay deviation of 18.7001 ps for the low-orbit satellite and an asymmetric delay difference of 30.3822 ps for the synchronous-orbit satellite. If the alignment precision is better than 0.01 urad, the asymmetric delay deviation of each orbit height converges to the same value. From Table 1 we can deduce that in addition to inhibiting the position deviation, further two-way precision alignment can considerably reduce the asymmetric delay deviation to the level of about ten picoseconds. For a dual-wavelength, high-precision space time transfer system, even if the laser link is aligned accurately, it is necessary to calibrate in real-time the asymmetric delay deviation which arises from atmospheric refraction and different link wavelengths.

4 CONCLUSIONS

For a space laser time transfer based on a satellite-ground communication link, the distribution of the atmospheric refractivity and the difference in uplink and downlink wavelengths usually cause an asymmetric delay deviation between the two-way links and a position deviation between the transceivers. In this context, the asymmetric delay deviation and the position deviation were studied by simulation. A case study where the satellite orbital height is 1000 km, the uplink wavelength is 815 nm and the downlink wavelength is 847 nm was presented. Our major conclusions are summarized as follows.

Given a ground station in different months, its asymmetric delay deviation and position deviation decrease with increasing air temperature. For different ground stations at different altitudes, the asymmetric delay deviation and position deviation decrease with decreasing air pressure. Moreover, the asymmetric delay deviation and position deviation increase with zenith angle. It is also found that the asymmetric delay deviation of different ground stations varies by several nanoseconds (maximum of 3.36 ns) and that the position error varies by hundreds of centimeters corresponding to a timing deviation of approximately 13.11 ns. Further two-way precision alignment can reduce the asymmetric delay deviation to the order of ten picoseconds. Therefore, in the case of high-accuracy time transfer based on dual wavelength, it is feasible to reduce the timing deviation to the order of picoseconds by calibrating the asymmetric delay deviation and position deviation in real-time.

ACKNOWLEDGMENTS

This work is partly supported by National Nature Science Foundation of China under (61673393), (61371121) and (61475193). The authors would like to thank the anonymous reviewers for their careful reading and helpful comment.

REFERENCES

Belmonte, A., Taylor, M.T., Hollberg, L., & Kahn, J.M. 2017. Effect of atmospheric anisoplanatism on earth-to-satellite time transfer over laser communication links. *Optics Express* 25(14), 15676.

Ciddor, P.E. 1996. Refractive index of air: new equations for the visible and near infrared. *Applied Optics* 35(9):1566.

Ciddor, P.E., & Hill, R.J. 1999. Refractive index of air. 2. group index. *Appl Opt* 38(9), 1663-7.

Davis, R.S. 1992. Equation for the determination of the density of moist air (1981/91). *Metrologia* 29(1), 67–70.

Djerroud, K., Acef, O., Clairon, A., Lemonde, P., Man, C.N., & Samain, E., et al. 2013. Coherent optical link through the turbulent atmosphere. *Optics Letters* 35(9):1479–81.

Fridelance, P., Samain, E., & Veillet, C. 1997. T2l2 – time transfer by laser link: a new optical time transfer generation. *Experimental Astronomy* 7(3), 191–207.

Giorgetta, F.R., Swann, W.C., Sinclair, L.C., Baumann, E., Coddington, I., & Newbury, N.R. 2013. Optical two-way time and frequency transfer over free space. *Nature Photonics* 7(6), 434–438.

http://data.cma.cn/data/detail/dataCode/B.0021.0002.html.

Kodet, J., Pánek, P., & Procházka, I. 2016. Two-way time transfer via optical fiber providing subpicosecond precision and high temperature stability. *Metrologia* 53(1), 18.

Lovellsmith, J. 2007. An expression for the uncertainty in the water vapour pressure enhancement factor for moist air. *Metrologia* 44(6):L49.

Ma, J., Tan, L.Y., & Yu, S.Y. (eds). 2015. *Satellite Optical Communication*: 229–230. Beijing: National Defense Industry Press.

Mandal, Kumar, Sharma, D.C, & Kumar. 2013. Comparative analysis of different air density equations. *Mapan* 28(1), 51–62.

Marra, G., Slavík, R., Margolis, H.S., Lea, S.N., Petropoulos, P., & Richardson, D.J., et al. 2011. High-resolution microwave frequency transfer over an 86-km-long optical fiber network using a mode-locked laser. *Optics Letters* 36(4), 511–3.

Rüeger, J.M. 2002. Refractive indices of light, infrared and radio waves in the atmosphere.

Samain, E., Dalla, R., & Prochazka, I. 2002. Time walk compensation of an avalanche photo-diode with a linear photo-detection.

Samain, E., Vrancken, P., Guillemot, P., Fridelance, P., & Exertier, P. 2014. Time transfer by laser link (t2l2): characterization and calibration of the flight instrument. *Metrologia* 51(5), 503.

Shiro Yamakawa. 2010. Expanded laser communications demonstrations with OICETS and ground stations. *Proceedings of SPIE—The International Society for Optical Engineering* 7587(6), 75870D-75870D-8.

Toyoshima, M., Reyes, M., Alonso, A., & Sodnik, Z. 2004. Ground-to-satellite optical link tests between japanese laser communications terminal and european geostationary satellite artemis. *Proc Spie* 5338, 1–15.

Vrancken, P. 2008. Characterization of t2l2 (time transfer by laser link) on the jason 2 ocean altimetry satellite and micrometric laser ranging. *Bibliogr.*

WNO. No.49 General meteorological standards and recommended practices. 1988. *Geneva: World Meteorological Organization.*

Yuan, H.M. 2011. Research on Atmospheric Refraction Correction Algorithm and Model for Satellite Laser Range-Finding. *Acta Optica Sinica* 31(4):0401004.

Frontier Research and Innovation in Optoelectronics Technology and Industry – Habib & Lewis (Eds)
© 2019 Taylor & Francis Group, London, ISBN 978-1-138-33178-5

Secrecy performance analysis of NLOS ultraviolet communications over turbulence channels

Weifeng Mou, Tao Pu, Weiwei Yang & Yeteng Tan
Communications Engineering College, Army Engineering University of PLA, Nanjing, China

ABSTRACT: This paper investigates the secure communication over Non-Line-of-Sight (NLOS) Ultraviolet (UV) links suffering from turbulence-induced fading, where the transmitter and the receiver or eavesdropper cone axes lie in the same plane or different planes and can be pointed in arbitrary directions. The closed-form expressions of the Average Secrecy Capacity (ASC), the Secrecy Outage Probability (SOP) and the Probability of Positive Secrecy Capacity (PPSC) are derived to evaluate the secrecy performance. Finally, simulation results are provided to verify the accuracy of our derivations and show the effect of transmission power and elevation angle on the ASC, SOP and PPSC.

Keywords: Physical layer security, ultraviolet communication, log-normal turbulence

1 INTRODUCTION

Recently, Optical wireless communication (OWC) has gained increasing interests because of its high data rate capacity, free spectrum resources and good security (D. Kedar & S. Arnonn, 2004). However, traditional OWC is line-of-sight (LOS) which need perfect alignment between the transmitter and the receiver. Non-line-of-sight (NLOS) ultraviolet (UV) communication can be used as an alternative link or in combination with existing optical and radio frequency (RF) links which relax or eliminate pointing, acquisition and tracking requirements (Z. Xu & B.M. Sadler, 2008). NLOS UV communication is strongly governed by photon interaction with the atmosphere, including scattering, absorption and turbulence. The effects of scattering and absorption have been studied and modeled in (L. Wang et al. 2011, H. Xiao et al. 2011) which ignored the atmospheric turbulence under the assumption of relatively short link distance and clear weather conditions. However, as the link range increases, atmospheric turbulence channel induced fading potentially becomes another degrading factor and should be considered by practical systems (Y. Zuo et al. 2013, M.H. Ardakani et al. 2017).

Compared with RF communication (X. Xu et al. 2016, W. Yang et al. 2016), the OWC is traditionally regarded as an information secure technology with high directionality, but it still suffers from the security risks when an eavesdropper is within the laser beam of the intended receiver (F.J. Lopez-Martinez et al. 2015, M.J. Saber & S.M.S. Sadough 2017). For example, the eavesdropper intentionally blocks the beam to cut the line-of-sight (LOS) of the intended receiver (F.J. Lopez-Martinez et al. 2015), or the footprint of the divergent beam at the receiver end is so large to cover both the intended receiver and the eavesdropper (M.J. Saber & S.M.S. Sadough, 2017). Furthermore, due to atmospheric scattering, confidential information in NLOS UV communications can be eavesdropped easily within the effective communication radius (D. Zou & Z. Xu 2016). Thus, the security threat in NLOS UV communication may be more serious. Moreover, irradiance fluctuation and scintillation attenuation caused by atmospheric turbulence may further degrade the reliability and security performance of OWC link. To the best of our knowledge, the secrecy performance of NLOS UV systems has not yet been studied.

Based on turbulence channel model, we first give a theoretical characterization for secrecy performance in NLOS UV communication systems and aim to provide direct guidelines for system designs. More precisely, we derive the closed-form expressions of the average secrecy capacity (ASC), secrecy outage probability (SOP) and the probability of positive secrecy capacity (PPSC) for intensity modulation/direct detection (IM/DD) technique over log-normal turbulence channels. Simulation and analysis results show that small elevation angle and large transmission power will lead to better secrecy performance. Furthermore, the PPSC is not related to the transmission power.

2 SYSTEM MODEL

We consider a non-coplanar NLOS UV communication system with IM/DD technique in which the transmitter (S) communicates with the legitimate receiver (D) in the presence of a potential eavesdropper (E) who may intercept the information. At any time instant, S transmits a symbol x_s to D with the optical transmission power P_s. Fig. 1 illustrates the non-coplanar NLOS UV link in the atmosphere turbulence channel model (Y. Zuo et al. 2013), where S emits a beam with divergence ϕ_s, elevation angle θ_s and off-axis angle β_{sr} and the receiver $R \in \{D,E\}$ has a field of view (FOV) of ϕ_r, elevation angel θ_r, and off-axis angle β_r, $r \in \{d,e\}$. Here, the transmitter beam divergence is considered small enough that the common scattering volume V_{sr} can be approximated as a frustum (Z. Xu & B.M. Sadler 2008, L. Wang et al. 2011). Moreover, D and E try to optimize pointing direction to maximize the effective scattering volume.

In this case, the common scattering volume can be expressed as $V_{sr} = (\pi/3)\ (D_{sr,1}^2 h_{sr,1} - D_{sr,2}^2 h_{sr,2})$, where $h_{sr,1} = l_{sr} + L_{sr}\phi_r/2$ and $D_{sr,1} = h_{sr,1}\phi_s/2$ are the height and radius of the bottom surface of the larger cone, respectively. Similarly, we have $h_{sr,2} = l_{sr} - L_{sr}\phi_r/2$ and $D_{sr,2} = h_{sr,2}\phi_s/2$ for the smaller cone where l_{sr} and L_{sr} denote the distance between S to V_{sr} and between V_{sr} to D or E, respectively, which can be expressed as

$$l_{sr} = d_{sr}/\left(\cos\theta_s\cos\beta_{sr} + \sin\theta_s\cos\beta_r/\tan\theta_r\right) \tag{1}$$

$$L_{sr} = \sqrt{d_{sr}^2 + l_{sr}^2 - 2d_{sr}l_{sr}\cos\theta_s\cos\beta_{sr}} \tag{2}$$

We assume that the channel state information is known to D and the irradiance of channel remains constant over a symbol duration. Thus, the instantaneous electrical received signal-to-noise-ratio (SNR) for link S-D and S-E can be written as

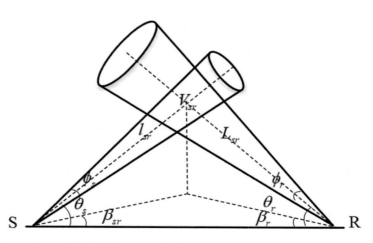

Figure 1.   Non-coplanar NLOS UV communication path in atmosphere turbulence.

$$\gamma_{sr} = \frac{2\left(P_r e\eta/h\nu_s\right)G^2}{\sigma_{\text{thermal,sr}}^2 + \sigma_{\text{shot,sr}}^2} = \frac{2\left(P_r e\eta/h\nu_s\right)^2 G^2}{2G^2 e\left(P_r + i_d\right)\Delta\nu + \left(4k_B T_e \Delta\nu/R\right)} \tag{3}$$

where P_r is the received signal power in D or E, $\sigma_{\text{thermal,sr}}^2$ and $\sigma_{\text{shot,sr}}^2$ are the thermal noise and shot noise, respectively, e is the electronic charge, ν_s is the optical signal transmission rate, η and G represent the quantum efficiency and internal gain of the photomultiplier tube (PMT), respectively. i_d is the gain normalized dark current, $\Delta\nu$ is the filter bandwidth, k_B is the Boltzmann constant. R and T_e represent the output load and temperature, respectively.

It is because that the current gain ($G \approx 10^6$) of PMT is very large, thus shot noise is far less than thermal noise, i.e., $\sigma_{\text{shot,sr}}^2 \leq \sigma_{\text{thermal,sr}}^2$ in (3) (A. Yariv & P. Yeh 2006). Neglecting $\sigma_{\text{shot,sr}}^2$ and assuming the dark current is far more than the received signal power, i.e., $i_d \geq P_r$, the SNR in (3) can be approximately expressed as

$$\gamma_{sr} \approx \frac{\left(P_r e\eta/h\nu_s\right)^2}{e i_d \Delta\nu} \tag{4}$$

Due to the atmosphere turbulence along the NLOS link, the link gain cannot be considered as the constant. As we can know, the log-normal model is well accepted and widely applied to model the stochastic characteristic of channel fading from weak to moderate turbulence strength (Y. Zuo et al. 2013, M.H. Ardakani et al. 2017). Let P_{sr} and P_r denote the optical power arriving at the common volume V_{sr} and the receiver R, respectively. The marginal probability density function (PDF) of P_{sr} and P_r, the conditional PDF of P_r can be expressed as

$$f\left(P_{sr}\right) = \frac{1}{\sqrt{2\pi}\sigma_{sr}P_{sr}}\exp\left\{\frac{\left[\ln\left(P_{sr}/P_{r0}\right) - \langle\ln\left(P_{sr}/P_{r0}\right)\rangle\right]^2}{2\sigma_{sr}^2}\right\} \tag{5}$$

$$f\left(P_r|P_{sr}\right) = \frac{1}{\sqrt{2\pi}\sigma_r P_r}\exp\left\{\frac{\left[\ln\left(P_r/P_{r1}\right) - \langle\ln\left(P_r/P_{r1}\right)\rangle\right]^2}{2\sigma_r^2}\right\} \tag{6}$$

$$f\left(P_r\right) = \int_0^\infty f\left(P_r|P_{sr}\right)f\left(P_{sr}\right)dP_{sr} \tag{7}$$

respectively. where $\langle\cdot\rangle$ is the expectation operation, P_{r0} and P_{r1} are the average optical power arriving at V_{sr} and the receiver R when ignoring the atmosphere turbulence, respectively, $\sigma_{sr}^2 = 1.23C_n^2\left(2\pi/\lambda\right)^{7/6}l_{sr}^{11/6}$ and $\sigma_r^2 = 1.23C_n^2\left(2\pi/\lambda\right)^{7/6}L_{sr}^{11/6}$ are Rytov variance [5], λ is wavelength, C_n^2 is refractive index structure coefficient.

According to (Y. Zuo et al. 2013), P_{r0} and P_{r1} can be expressed as

$$P_{r0} = P_s k_s p\left(\varphi_{sr}\right)\exp\left(-k_e l_{sr}\right)V_{sr}/\left(\Omega_{sr}l_{sr}^2\right) \tag{8}$$

$$P_{r1} = P_{sr}A_R\cos\xi_{sr}\exp\left(-k_e L_{sr}\right)/\left(4\pi L_{sr}^2\right) \tag{9}$$

where $k_s = k_s^R + k_s^M$ is the atmospheric scattering coefficient, k_s^R and k_s^M are the Rayleigh and Mie scattering coefficients, respectively, k_e is the extinction coefficient, A_R is the receiver area, $\Omega_{sr} = 2\pi\left[1 - \cos\left(\varphi_{sr}/2\right)\right]$ is the solid cone angle of transmitter S and φ_{sr} is the scattering angle, $p(\varphi_{sr})$ is the single-scatter phase function (Y. Zuo et al. 2013) and ξ_{sy} is the incident angle of a beam on the transmitter surface. According to (Y. Zuo et al. 2012), $\cos\varphi_{sr}$ and $\cos\xi_{sr}$ can be expressed as

$$\cos\varphi_{sr} = \left(d_{sr}\cos\theta_s\cos\beta_s - l_{sr}\right)/L_{sr} \tag{10}$$

$$\cos\xi_{sr} = l_{sr}\left[\sin\theta_r\sin\theta_s - \cos\theta_r\cos\theta_s\cos(\beta_s+\beta_r)\right]\big/L_{sr} + \left(d_{sr}\cos\theta_r\cos\beta_r\right)\big/L_{sr} \qquad (11)$$

Replacing (5), (6) in (7) and assuming $a_r = \ln P_{r0} + \langle\ln(P_{sr}/P_{r0})\rangle$, $b_r = \ln(P_{r1}/P_{sr}) + \langle\ln(P_r/P_{r1})\rangle$ (Y. Zuo et al. 2013), (7) can be rewritten as

$$f(P_r) = \Psi\exp\left(\frac{\Upsilon^2}{2\sigma^2}\right)\int_0^\infty \frac{1}{\sqrt{2\pi}P_{sr}}\exp\left\{-\frac{\left(\ln P_{sr}+\Upsilon\right)^2}{2\sigma^2}\right\}dP_{sr} \qquad (12)$$

where $\Psi = \left(\sqrt{2\pi}\sigma_r\sigma_{sr}P_r\right)^{-1}\exp\left(-a_r^2\left(2\sigma_{sr}^2\right)^{-1} - \left(\ln P_r - b_r\right)^2\left(2\sigma_r^2\right)^{-1}\right)$, $\sigma = \sigma_{sr}\sigma_r\sqrt{\sigma_{sr}^2 + \sigma_r^2}$, $\Upsilon = 2\sigma_r^2\zeta$, and $\zeta = a_r/\left(2\sigma_{sr}^2\right) + \left(\ln P_r - b_r\right)/\left(2\sigma_r^2\right)$.

Through some integral, the PDF of P_r can be expressed as

$$f(P_r) = \frac{1}{\sqrt{2\pi}P_r\sigma_{\mathrm{NLOS,sr}}}\exp\left(-\frac{\left(\ln P_r - \mu_{\mathrm{NLOS,sr}}\right)^2}{2\sigma_{\mathrm{NLOS,sr}}^2}\right) \qquad (13)$$

where $\sigma_{\mathrm{NLOS,sr}} = \sqrt{\sigma_{sr}^2 + \sigma_r^2}$ and $\mu_{\mathrm{NLOS,sr}} = a_r + b_r$.

For the IM/DD detection technique case, the PDF and CDF of the SNR can be derived as

$$f_{\gamma_{sr}}(\gamma) = \frac{1}{\sqrt{2\pi}\gamma\sigma_{\gamma_{sr}}}\exp\left(-\left(\ln\gamma - \mu_{\gamma_{sr}}\right)^2\big/\left(2\sigma_{\gamma_{sr}}^2\right)\right) \qquad (14)$$

$$F_{\gamma_{sr}}(\gamma) = Q\left(-\left(\ln\gamma - \mu_{\gamma_{sr}}\right)\big/\sigma_{\gamma_{sr}}\right) \qquad (15)$$

where $Q(x) = \left(2\pi\right)^{-1}\int_x^\infty\exp(-t^2/2)dt$ is the Gaussian Q function, $\sigma_{\gamma_{sr}} = 2\sigma_{\mathrm{NLOS,sr}}$ and $\mu_{\gamma_{sr}} = -2\ln\left(\eta h\nu_s\eta^{-1}\sqrt{i_d\Delta\nu/e}\right) + 2\mu_{\mathrm{NLOS,sr}}$.

3 SECRECY PERFORMANCE ANALYSIS

3.1 *Average secrecy capacity*

The secrecy capacity, which is denoted by C_s, is the maximum transmission rate from the transmitter to the intended receiver that maintains perfect secrecy, which can be expressed as

$$C_s = \left[\log_2\left(1+\gamma_{sd}\right) - \log_2\left(1+\gamma_{se}\right)\right]^+ \qquad (16)$$

where $[x]^+ = \max\{0,x\}\cdot\log_2\left(1+\gamma_{sd}\right)$ and $\log_2\left(1+\gamma_{se}\right)$ are the capacity of the main and eavesdropper channels, respectively. Then the average secrecy capacity (ASC) can be given by

$$\bar{C}_s = \int_0^\infty\int_0^\infty C_s f\left(\gamma_{sd}\right)f\left(\gamma_{se}\right)d\gamma_{sd}d\gamma_{se} \qquad (17)$$

Substituting (14) and (16) into (17), we have

$$\begin{aligned}
\bar{C}_s = &\int_0^\infty\log_2\left(1+\gamma_{sd}\right)Q\left(-\frac{\ln\gamma_{sd}-\mu_{\gamma_{se}}}{\sigma_{\gamma_{se}}}\right)f\left(\gamma_{sd}\right)d\gamma_{sd} \\
&-\int_0^\infty\log_2\left(1+\gamma_{se}\right)Q\left(\frac{\ln\gamma_{se}-\mu_{\gamma_{sd}}}{\sigma_{\gamma_{sd}}}\right)f\left(\gamma_{se}\right)d\gamma_{se}
\end{aligned} \qquad (18)$$

It has been proved that log-normal behavior is very difficult to deal with in (G. Pan, 2012). In this paper, we adopt the efficient tool proposed by Holtzmanin (J.M. Holtzman 1992) to deal with (18). According to (J.M. Holtzman 1992), the following lemma can be derived.

Lemma: If $\phi(x)$ is a real function of x, a logarithmic normally distributed variable with logarithmic mean μ and logarithmic variance σ^2, namely, $\ln x \sim N(\mu, \sigma^2)$. Then, the expectation of $\phi(x)$ can be approximated in terms of three points located at e^μ, $e^{\mu + \sqrt{3}\sigma}$ and $e^{\mu - \sqrt{3}\sigma}$, as follows

$$E[\phi(x)] = \frac{2}{3}\phi(e^\mu) + \frac{1}{6}\phi(e^{\mu + \sqrt{3}\sigma}) + \frac{1}{6}\phi(e^{\mu - \sqrt{3}\sigma}) \tag{19}$$

Thus, making use of (19), the ASC can be approximately obtained as follows

$$\bar{C}_s = \frac{2}{3}\left[\phi_1\left(e^{\mu_{\gamma sd}}\right) - \phi_2\left(e^{\mu_{\gamma se}}\right)\right] + \frac{1}{6}\left[\phi_1\left(e^{\mu_{\gamma sd} + \sqrt{3}\sigma_{\gamma sd}}\right) - \phi_2\left(e^{\mu_{\gamma se} + \sqrt{3}\sigma_{\gamma se}}\right)\right]$$
$$+ \frac{1}{6}\left[\phi_1\left(e^{\mu_{\gamma sd} - \sqrt{3}\sigma_{\gamma sd}}\right) - \phi_2\left(e^{\mu_{\gamma se} - \sqrt{3}\sigma_{\gamma se}}\right)\right] \tag{20}$$

where $\phi_1(x) = \log_2(1+x)Q(-(\ln x - \mu_{\gamma se})\sigma_{\gamma se}^{-1})$ and $\phi_2(x) = \log_2(1+x)Q((\ln x - \mu_{\gamma sd})\sigma_{\gamma sd}^{-1})$.

We find that $Q(x)$ is strictly decreasing with x. Thus, $\phi_1(x)$ is strictly increasing and $\phi_2(x)$ is strictly decreasing with x, respectively. Since (20) is the sum of the difference between $\phi_1(x)$ and $\phi_2(x)$, therefore, (20) is an increasing function of x which is related to P_s, θ_s, d_{sr} and so on. (20) indicates the statistical characteristics of the secrecy capacity.

3.2 Secrecy outage probability

The SOP is defined as the probability that the instantaneous secrecy capacity falls below a target rate which can be expressed as

$$p_{so} = \Pr\{C_s < R_s\} \tag{21}$$

Substituting (14), (15) and (16) into (21), we can derive as

$$p_{so} = \int_0^\infty Q\left(-\frac{\ln\left(2^{R_s}(1+y) - 1\right) - \mu_{\gamma sd}}{\sigma_{\gamma sd}}\right)\frac{1}{\sqrt{2\pi}\sigma_{\gamma se}y}\exp\left(-\frac{\left(\ln y - \mu_{\gamma se}\right)^2}{2\sigma_{\gamma se}^2}\right)dy \tag{22}$$

Similarly, making use of (19), we can obtain the approximation for p_{so} as follows

$$p_{so} \approx 1 - \left[\frac{2}{3}Q\left(\psi\left(e^{\mu_{\gamma se}}\right)\right) + \frac{1}{6}Q\left(\psi\left(e^{\mu_{\gamma se} + \sqrt{3}\sigma_{\gamma se}}\right)\right) + \frac{1}{6}Q\left(\psi\left(e^{\mu_{\gamma se} - \sqrt{3}\sigma_{\gamma se}}\right)\right)\right] \tag{23}$$

where $\psi(x) = -\left(\ln\left(2^{R_s}(1+x) - 1\right) - \mu_{\gamma sd}\right)\sigma_{\gamma sd}^{-1}$ which is strictly decreasing with x. Thus, (23) is a strictly decreasing function of x which indicates that the larger x is, the better channel link security is.

3.3 The probability of positive secrecy capacity

The probability of positive secrecy capacity (PPSC) is used to emphasize the existence of the secrecy capacity which can be expressed as

$$p_{psc} = \Pr\{C_s > 0\} = \Pr\{\gamma_{sd} > \gamma_{se}\} \tag{24}$$

Substituting (14) and (15) into (24), we can derive as

$$p_{psc} = \int_0^\infty Q\left(\frac{\ln y - \mu_{\gamma sd}}{\sigma_{\gamma sd}}\right)\frac{1}{\sqrt{2\pi}y\sigma_{\gamma sd}}\exp\left(-\frac{\left(\ln y - \mu_{\gamma se}\right)^2}{2\sigma_{\gamma se}^2}\right)dy \tag{25}$$

Through some calculation, (25) can be rewritten as

$$p_{psc} = \frac{1}{2} - \frac{1}{2} \int_{-\infty}^{\infty} erf\left(\frac{t + \mu_{NLOS,se} - \mu_{NLOS,sd}}{\sqrt{2}\sigma_{NLOS,sd}} \right) \frac{1}{\sqrt{2\pi}\sigma_{NLOS,se}} \exp\left(-\frac{t^2}{2\sigma_{NLOS,se}^2} \right) dt \quad (26)$$

Utilizing (8.259.1) of (I. Gradshteyn & I. Ryzhik 2007), the PPSC can be derived as

$$p_{psc} = \frac{1}{2} - \frac{1}{2} erf\left(\frac{\mu_{NLOS,se} - \mu_{NLOS,sd}}{\sqrt{2\left(\sigma_{NLOS,se}^2 + \sigma_{NLOS,sd}^2 \right)}} \right) \quad (27)$$

From (27), it is obvious that PPSC is not related to the optical transmission power. Moreover, $erf(x)$ is strictly increasing with x, thus, (27) is strictly decreasing function of x which is related to P_s, θ_s, d_{sr} and so on.

4 NUMERICAL RESULTS

In this section, we present numerical results to confirm the accuracy of our derivations and investigate the effect of P_s and θ_s on secrecy performance. The main parameters used in simulations and analysis are set as $k_s^R = 2.4 \times 10^{-4}/m^2$, $k_s^M = 2.5 \times 10^{-4}/m^2$, $AR = 1.77\ cm^2$, $\lambda = 260\ nm$ (similar as H. Xiao et al. 2011), $i_d = 10^{-5}\ A$, $\eta = 0.3$, $G = 10^6$, $\Delta v = 1\ Hz$, $v_s = 10^{12}\ m/s$, $C_n^2 = 10^{-16}\ m^{-2/3}$, $R_s = 10\ bits/sec/Hz$, $d_{sd} = 1\ km$ and $d_{se} = 2\ km$. Furthermore, we set the elevation angle $\theta_d = \pi/3$, $\theta_e = \pi/6$, the axis angle and the divergence angle $\phi_s = \pi/90$, $\phi_d = \phi_d = \pi/4$, respectively.

Fig. 2 illustrates the joint effect of P_s and θ_s on the ASC. We can observe that ASC first increases then approaches to a performance floor as P_s increases. This is because that increasing P_s not only enhances the receive power of legitimate link but also enlarge that of the eavesdropper. What's more, we find that with a small P_s, the ASC is a decreasing function of θ_s. However, with a large P_s, the ASC first decreases to the minimum value then increases as θ_s increases. This is because that the NLOS UV transmission distance keep increasing as θ_s increases which lead to more path loss. Thus, θ_s plays a dominant role for secure transmission when P_s is small.

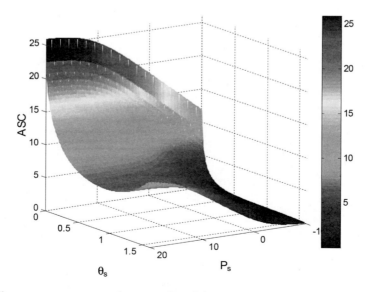

Figure 2. The average secrecy capacity versus P_s and θ_s.

Fig. 3 illustrates the effect of θ_s on SOP when $P_s = 0, 10, 20$ dBW. We can observe that the approximation curve is in general agreement with the simulation and analysis curves which verifies the correctness of our analysis. The SOP keeps decreasing and then approaches to a performance floor as P_s increases. Furthermore, with a fixed P_s, the SOP is a concave function of θ_s which follows the contrary trend of ASC. The reason for these phenomenonis consistent with that in Fig. 2.

Fig. 4 illustrates the joint effect of P_s and θ_s on the PPSC. As we can see, The PPSC has nothing to do with P_s which is consistent with the theoretical result. Moreover, the PPSC is a decreasing function of θ_s. It implies that the smaller θ_s is, the better the link secure connectivity is.

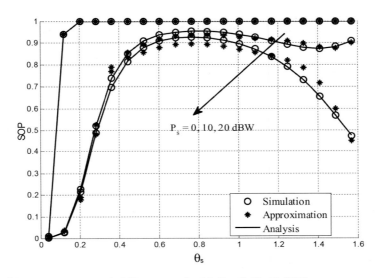

Figure 3.   The secrecy outage probability versus θ_s with $P_s = 0, 10, 20$ dBW.

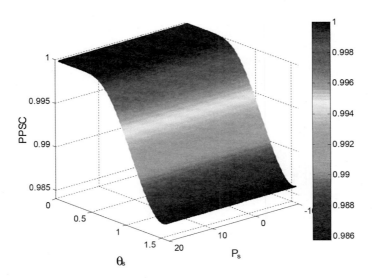

Figure 4.   The probability of positive secrecy capacity versus P_s and θ_s.

333

5 CONCLUSIONS

In this paper, we have investigated the secrecy performance of NLOS UV communication systems with atmospheric turbulence. We derived the closed-form expression of the ASC, SOP and PPSC. Simulation results verified that the large P_s and small θ_s can enhance secrecy performance and the PPSC is not related to P_s.

REFERENCES

Ardakani M.H., A.R. Heidarpour, and M. Uysal, 2017, Performance analysis of relay-assisted NLOS ultraviolet communications over turbulence channels, J. Opt. Commun. Netw., 9(1): 109–118.

Gradshteyn I. and I. Ryzhik, 2007, Table of Integrals, Series and Products, 7thed. San Diego, CA, USA: Academic.

Holtzman J.M., 1992, A simple, accurate method to calculate spread multiple-access error probabilities, IEEE Trans. Commun., 40(3): 461–464.

Kedar D. and S. Arnonn, 2004, Urban optical wireless communication networks: The main challenges and possible solutions, IEEE Commun. Mag., 42(5): s2-s7.

Lopez-Martinez F.J., G. Gomez, and J.M. Garrido-Balsells, 2015, Physical layer security in free-space optical communications, IEEE Photon. J., 7(2): 1–14.

Pan G., E. Eylem and Q. Feng, 2012, Capacity analysis of log-normal channel under various adaptive transmission schemes, IEEE Commun. Lett., 16(3): 346–348.

Saber M.J. and S.M.S. Sadough, 2017, On secure free-space optical communications over M' alaga turbulence channels, IEEE Wireless Commun. Lett., 6(2): 274–277.

Wang L., Z. Xu and B.M. Sadler, 2011, An approximate closed-form link loss model for non-line-of-sight ultraviolet communication in noncoplanar geometry, Opt. Lett., 36(7): 1224–1226.

Xiao H., Y. Zuo, J. Wu, H. Guo and J. Lin, 2011, Non-line-of-sight ultraviolet single-scatter propagation model, Opt. Exp., 9(18): 17864–17875.

Xu X., W. Yang, Y. Cai, and S. Jin, 2016, On the secure spectral-energy efficiency tradeoff in random cognitive radio networks, IEEE J. Sel. Areas Commun., 34(10): 2706–2722.

Xu Z. and B.M. Sadler, 2008, Ultraviolet communication: potential and state-of-the-art, IEEE Commun. Mag., 46(5): 67–73.

Yang W., W. Mou, X. Xu, W. Yang and Y. Cai, 2016, Energy efficiency analysis and enhancement for secure transmission in SWIPT systems exploiting full duplex techniques, IET Commun., 10(14): 1712–1720.

Yariv A. and P. Yeh, 2006, Photonics: Optical Electronics in Modern Communications, 6th ed. New York, NY, USA, Oxford University Press.

Zou D. and Z. Xu, 2016, Information security risks outside the laser beam in terrestrial free-space optical communication, IEEE Photon. J., 8(5): 1–9.

Zuo Y., H. Xiao, J. Wu, Y. Li, and J. Lin, 2012, A single-scatter path loss model for non-line-of-sight ultraviolet channels, Opt. Exp., 20(9): 10359–10369.

Zuo Y., J. Wu, H. Xiao and J. Lin, 2013, Non-line-of-sight ultraviolet communication performance in atmospheric turbulence, China Commun., 10(11): 52–57.

Frontier Research and Innovation in Optoelectronics Technology and Industry – Habib & Lewis (Eds)
© 2019 Taylor & Francis Group, London, ISBN 978-1-138-33178-5

Reconfigurable and tunable microwave photonic filter using a multi-wavelength hybrid-gain-assisted fiber ring laser

Mengmeng Peng
Institute of Science, Chongqing University of Technology, Chongqing, China

Fei Wang & Lun Shi
Institute of Electrical and Electronic Engineering, Chongqing University of Technology, Chongqing, China

ABSTRACT: A reconfigurable and tunable Microwave Photonic Filter (MPF) using a Multi-wavelength Hybrid-gain-assisted Fiber Ring Laser (MHFRL) and a dispersive medium is proposed and experimentally demonstrated. In the scheme, the MHFRL is mainly configured by a High Birefringence Fiber Loop Mirror (Hi-Bi FLM) and a hybrid gain medium. Stable multi-wavelength output can be generated from the MHFRL and serves as multiple taps for the MPF. By adjusting the bias of SOA, the number of multiple taps can be varied, making the MPF's bandwidth reconfigurable. In addition, by changing the length of Polarization Maintaining Fiber (PMF) in the Hi-Bi FLM, the wavelength spacing of multiple taps can be varied, making the MPF's Free Spectral Range (FSR) tunable. The proposed MPF has the merits of good flexibility, stability, and great potential of applications.

Keywords: Microwave photonic filter, multi-wavelength hybrid-gain-assisted fiber ring laser, microwave signal processing

1 INTRODUCTION

MPF is one of the most important microwave photonic processing subsystems, it has been intensively studied in the past few decades (Yao, 2015, Minasian et al., 2013, Capmany et al., 2013). Compared with traditional microwave filter implemented in electrical domain, MPF has many unique characteristics, such as low loss, large bandwidth, seamless tunability, good reconfigurability and immunity to electromagnetic interference (EMI). In recent years, a number of MPFs have been implemented by using multi-wavelength optical source and dispersive medium (Sun et al., 2016, Abreu-Afonso et al., 2012, Xu et al., 2015). The frequency response of MPF can be controlled by altering the dispersion of the multiple taps, or changing the relative amplitude as well as the wavelength spacing of the multiple taps. In general, single mode fiber (SMF), dispersion compensating fiber (DCF) or chirped fiber Bragg grating (CFBG) are used as dispersive medium. As for multi-wavelength optical source, using an array of optical tunable laser diodes (TLDs) is a traditional method to obtain multiple taps, although the control of multiple taps can be easily realized in this way, an array of TLDs leads to a high cost in practical application (Capmany et al., 1999). Hence, designing simple and low-cost configurations to obtain multiple taps for MPFs is highly significant. In (Zhou et al., 2011), the authors used a windowed F-P filter-based multi-wavelength tunable laser to generate multiple taps, a tunable multi-tap bandpass MPF was achieved, the center frequency of the MPF can be tuned by adjusting the F-P filter. However, the tuning process was limited by the step of the F-P filter, making the tunability discontinuous. In (Wu et al., 2017), multiple taps were obtained by slicing the broadband source using reflective and cascaded fiber Mach-Zehnder interferometers (MZIs). A MPF with tunable and selectable multiple passbands was realized, but the stability of MZI was hard to be ensured at room temperature. Beyond these methods, four-wave mixing (FWM) is an effective way to increase the number of optical taps for MPFs, the relative amplitude or the wavelength spacing of multiple taps can be easily controlled in this way. A reconfigurable MPF based on FWM was realized, by adjusting the gain of an optical amplifier and the state of polarization, the relative amplitude of the multiple taps can be adjusted, thus, the shape of MPF can be controlled (Vidal et al., 2012). Another approach to realizing a widely tunable MPF based on FWM was also proposed, by changing the wavelength spacing of multiple taps, FSR of the MPF can be continuously tuned, but the shape of MPF was changed obviously while tuning the FSR (Cao et al., 2017).

In this paper, a reconfigurable and tunable MPF using a MHFRL and a dispersive medium is proposed and experimentally demonstrated. In the MHFRL, the Hi-Bi FLM is used as a comb filter for multi-wavelength operation. A hybrid gain is provided by a semiconductor optical amplifier (SOA) and an erbium-doped optical fiber amplifier (EDFA). Multi-wavelength oscillations at each peak of the Hi-Bi FLM transmission spectrum will be strengthened, leading to the generation of multi-wavelength output. In the proposed scheme, the stable multi-wavelength output from the MHFRL serves as the MPF's multiple taps. By adjusting the bias of SOA, the number of multiple taps can be varied, making the MPF's bandwidth reconfigurable. Additionally, by changing the length of PMF in the Hi-Bi FLM, the wavelength spacing of multiple taps can be varied, making the MPF's FSR tunable. Furthermore, the passband centered at dc frequency is removed due to the use of phase modulation.

2 EXPERIMENTAL SETUP AND PRINCIPLE

The schematic diagram of the proposed MPF is shown in Figure 1. The MHFRL contains a Hi-Bi FLM and a hybrid gain medium. The Hi-Bi FLM is configured by a 50:50 optical coupler (OC, OC1), a polarization controller (PC, PC1) and a length of PMF, it serves as a comb filter for multi-wavelength operation. An SOA and an EDFA1 are used to provide a hybrid gain. The amplified spontaneous emission (ASE) generated from SOA and EDFA1 is amplified and propagates through the Hi-Bi FLM. Multi-wavelength oscillations at each peak of the Hi-Bi FLM transmission spectrum will be strengthened, leading to the generation of multi-wavelength output. Because SOA is a inhomogeneous broadening gain medium at room temperature, it can be used as a model competition reducer. Model competition among each wavelength can be effectively suppressed in the SOA, making the multi-wavelength output operation quite stable (Luo et al., 2009, Ahmad et al., 2009, Ahmad et al., 2008). Then the multi-wavelength output is pow-amplified by an EDFA2. The output of 10% from EDFA2 is injected into an optical spectrum analyzer (OSA) for optical spectrum monitoring, and 90% is used as multiple taps for the MPF. Then the multiple taps are phase-modulated by an input RF signal emitted from a vector network analyzer (VNA). The phase-modulated signal is then launched into a coil of DCF,

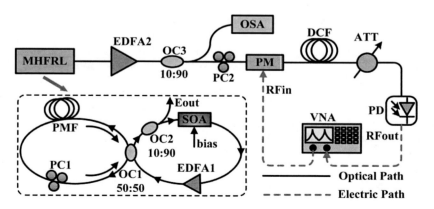

Figure 1.   The schematic diagram of the proposed MPF.

which acts as a wideband dispersive medium. PC2 is used before the phase modulator (PM) to minimize the polarization-dependent loss. The microwave signals are recovered by a high speeding photo-detector (PD). The corresponding frequency response of the MPF can be obtained as S_{21} parameter in the VNA.

The frequency response $H_{RF}(f)$ of the MPF is periodic, it can be indicated as:

$$H_{RF}(f) = R\cos\left(\frac{\beta f^2}{2}\right)\left|\sum_{k=1}^{N} P_k e^{-j[2\pi f(k-2)\Delta\tau]}\right| \tag{1}$$

where R is the responsivity of PD, β is the dispersion parameter, f is the modulation frequency, P_k is weight of the kth optical tap, N is the number of multiple taps, $\Delta\tau$ is the time delay introduced by the dispersive medium. As indicated in Equation 1, the MPF can be reconfigured by changing the number of multiple taps. In this scheme, by adjusting the bias of SOA, the number of multiple taps can be controlled, making the MPF's bandwidth reconfigurable.

FSR of the MPF can be expressed as:

$$FSR = \frac{1}{D \cdot L \cdot \Delta\lambda} \tag{2}$$

where D and L are the dispersion coefficient and the length of DCF, respectively. $\Delta\lambda$ is the wavelength spacing of multiple taps. From Equation 2, for a given length of dispersive medium, FSR of the MPF can be tuned by adjusting the wavelength spacing of multiple taps. Since the Hi-Bi FLM serves as a comb filter for multi-wavelength output operation in this scheme, the wavelength spacing of multiple taps is equal to that of the Hi-Bi FLM transmission spectrum, which can be expressed as:

$$\Delta\lambda = \frac{\lambda^2}{\Delta n \cdot L_{eff}} \tag{3}$$

where Δn is the effective birefringence between two orthogonal polarization modes and L_{eff} is the effective length of PMF. As can be seen in Equation 3, the wavelength spacing is inversely proportional to the length of PMF. Thus, the wavelength spacing of multiple taps can be varied by changing the length of PMF, making the MPF's FSR tunable.

2 RESULTS AND DISCUSSIONS

The parameters of the key devices used in the experiment are summarized as follows: the PM (Photline MPZ-LN-20) has a 3-dB bandwidth of 40 GHz. The DCF (DCF-G.652 C/250) has a length of 2.05 km and a chromatic dispersion of 332.07 ps/nm. The SOA (SOA1117) can be driven at maximum current of 500 mA. The birefringence ratio of the PMF (PMP-1550-1-PM1550-FU-FA) is 0.0004. The resolution of OSA (Anritsu MS9740A) is 0.03 nm. The high speeding PD (u2t XPDV2120R) has a bandwidth of 40 GHz. The microwave signal emitted from the VNA (CETC, AV3629D) is sweeping from 45 MHz to 20 GHz.

Since the number of multiple taps can be varied by adjusting the bias of SOA, the MPF's bandwidth can be easily reconfigured. Figures 2a–2b show the optical spectrum of multiple taps when the bias of SOA is set as 450 mA and 344 mA, respectively. The full span of the spectrum is 40 nm. Because the length of PMF is fixed at 6 m, both of the wavelength spacings in Figures 2a–2b are equal to 0.97 nm, which shows good agreement with the theoretical value (1.001 nm) calculated according to Equation 3. As can be seen, when the bias of SOA is decreased, the number of multiple taps is decreased apparently. The reason is that the SOA provides less gain spectrum with lower bias (Sulaiman et al., 2015). The measured corresponding frequency responses are shown in Figures 2c–2d, respectively. The frequency response of the MPF is periodic, the value of FSR is equal to the center frequency of the first passband (Cao et al., 2017). It can be seen that the 3-dB bandwidth of the passbands is broadened from 247 MHz from 449

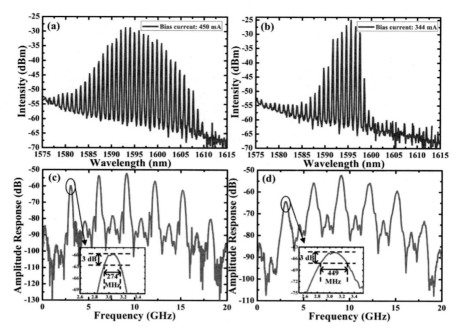

Figure 2. (a–b) Measured optical spectrums of multiple taps when the bias of SOA is set as 450 mA and 344 mA; (c–d) The corresponding frequency responses of the MPF; Inset: the zoom-in view of the first passband.

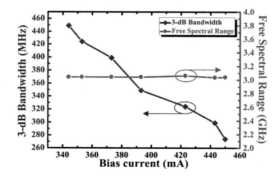

Figure 3. 3-dB bandwidth and FSR of the MPF with different bias of SOA.

MHz when the bias of SOA is adjusted from 450 mA to 344 mA. Note that amplitude fading of the experimental passbands is existing, it is caused by the dispersion slope of the DCF (Abreu-Afonso et al., 2012).

The 3-dB bandwidth of the passbands can be plotted as a function of the bias of SOA, as shown in Figure 3. As the blue curve indicates, when the bias of SOA increases, 3-dB bandwidth of the passbands decreases. Furthermore, the 3-dB bandwidth shows inversely proportional linear dependence with the bias of SOA. The red curve shows that although the bias of SOA increases, FSR of the MPF remains particularly constant, which is fixed around 3.042 GHz. Hence, in the proposed MPF, bandwidth of passbands features flexible reconfigurability.

Next, the tunability of the MPF is discussed. The bias of SOA is set as 450 mA. The wavelength spacing of multiple taps can be varied by changing the length of PMF in the Hi-Bi FLM. Figures 4a–4b show the optical spectrum of multiple taps with different length of PMF. As can be seen, when the length of PMF is decreased from 5 m to 3 m, the wavelength spacing of multiple taps is increased from 1.16 nm to 1.93 nm. The measured corresponding frequency responses are shown in Figures 4c–4d. As can be seen, FSR of the MPF are 2.531 GHz and 1.546 GHz, respectively, which show good agreements with the theoretical values (2.596 GHz and 1.560 GHz) calculated according to Equation 2. When the wavelength spacing of multiple taps is altered, the shape of the MPF also alters slightly. This is mainly because as the wavelength spacing is altered, the number of multiple taps varies slightly due to different power distribution (Abreu-Afonso et al., 2012).

The wavelength spacing of multiple taps and FSR of the MPF can be plotted as functions of the length of PMF, as shown in Figure 5. The blue curve represents the relationship between the wavelength spacing of multiple taps and the length of PMF. The red curve represents the relationship between the FSR and the length of PMF. As can be seen in Figure 5, for a given length of DCF, as the length of PMF increases, the wavelength spacing of multiple taps decreases, and the corresponding FSR of the MPF increases, in accordance with

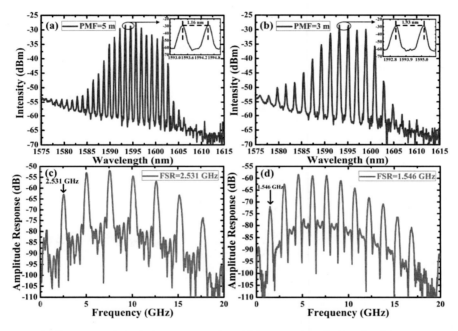

Figure 4. (a–b) Measured Optical spectrums of multiple taps when the length of PMF are 5 m and 3 m; (c–d) The corresponding frequency responses of the MPF; Inset: the zoom-in view of two arbitrary adjacent optical taps.

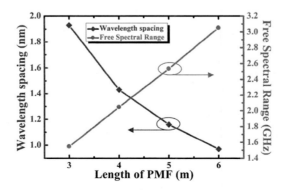

Figure 5. Wavelength spacing of the multiple taps and FSR of the MPF with different length of PMF.

Equation 2 and Equation 3. Furthermore, continuously tuning of FSR can be realized if a length-variable PMF is used. The tunability can also be realized by varying the accumulated dispersion of the DCF.

3 CONCLUSION

We proposed and experimentally demonstrated a reconfigurable and tunable MPF based on a MHFRL and a dispersive medium. In the proposed scheme, stable multi-wavelength output from the MHFRL serves as the MPF's multiple taps. By adjusting the bias of SOA, the number of multiple taps can be varied, making the MPF's bandwidth reconfigurable. Additionally, by changing the length of PMF in the Hi-Bi FLM, the wavelength spacing of multiple taps can be varied, making the MPF's FSR tunable. Thus, by properly controlling the multiple taps, a reconfigurable and tunable MPF can be achieved. The scheme features good characteristics, such as stable, flexible and low-cost, it has potential applications in Radio-over-Fiber (RoF) systems.

ACKNOWLEDGEMENTS

This work was supported by the National Natural Science Foundation of China (61575034), the recruitment program of global experts (WQ20165000357), the Key Project of the Natural Science Foundation Project of Chongqing (cstc2013jjB40002).

REFERENCES

Abreu-Afonso, J., Diez, A., Cruz, J.L. & Andres, M.V. (2012) Continuously Tunable Microwave Photonic Filter Using a Multiwavelength Fiber Laser. *IEEE Photonics Technology Letters*, 24, 2129–2131.

Ahmad, H., Sulaiman, A.H., Shahi, S. & Harun, S.W. (2009) SOA-based multi-wavelength laser using fiber Bragg gratings. *Laser Physics*, 19, 1002–1005.

Ahmad, H., Thambiratnam, K., Sulaiman, A.H., Tamchek, N. & Harun, S.W. (2008) SOA-based quad-wavelength ring laser. *Laser Physics Letters*, 5, 726–729.

Cao, Y., Xu, D., Chen, L., Tong, Z. & Yang, J. (2017) Widely tunable microwave photonic filter based on four-wave mixing. *Optical Engineering*, 56, 026110.

Capmany, J., Mora, J., Gasulla, I., Sancho, J., Lloret, J. & Sales, S. (2013) Microwave Photonic Signal Processing. *Lightwave Technology, Journal of*, 31, 571–586.

Capmany, J., Pastor, D. & Ortega, B. (1999) New and flexible fiber-optic delay-line filters using chirped Bragg gratings and laser arrays. *IEEE Transactions on Microwave Theory & Techniques*, 47, 1321–1326.

Luo, Z., Zhong, W.D., Cai, Z., Ye, C. & Wen, Y.J. (2009) High-performance SOA-based multiwavelength fiber lasers incorporating a novel double-pass waveguide-based MZI. *Applied Physics B*, 96, 29–38.

Minasian, R.A., Yi, X., Chan, E.H.W. & Huang, T.X.H. (2013) Microwave photonic signal processing. *Optics Express*, 21, 22918–22936.

Sun, W., Wang, S., Zhong, X., Liu, J., Wang, W., Tong, Y., Chen, W., Yuan, H., Yu, L. & Zhu, N. (2016) Integrated wideband optical frequency combs with high stability and their application in microwave photonic filters. *Optics Communications*, 373, 59–64.

Vidal, B., Palaci, J. & Capmany, J. (2012) Reconfigurable Photonic Microwave Filter Based on Four-Wave Mixing. *IEEE Photonics Journal*, 4, 759–764.

Wu, R., Chen, H., Zhang, S., Fu, H., Luo, Z., Zhang, L., Zhao, M., Xu, H. & Cai, Z. (2017) Tunable and Selectable Multipassband Microwave Photonic Filter Utilizing Reflective and Cascaded Fiber Mach– Zehnder Interferometers. *Journal of Lightwave Technology*, 35, 2660–2668.

Xu, R., Zhang, X., Hu, J. & Xia, L. (2015) Single-passband microwave photonic filter based on a self-seeded multiwavelength Brillouin-erbium fiber laser. *Optics Communications*, 339, 74–77.

Yao, J. (2015) Photonics to the Rescue: A Fresh Look at Microwave Photonic Filters. *IEEE Microwave Magazine*, 16, 46–60.

Zhou, J., Fu, S., Luan, F., Jia, H.W., Aditya, S., Shum, P.P. & Lee, K.E.K. (2011) Tunable Multi-Tap Bandpass Microwave Photonic Filter Using a Windowed Fabry-Pérot Filter-Based Multi-Wavelength Tunable Laser. *Journal of Lightwave Technology*, 29, 3381–3386.

Frontier Research and Innovation in Optoelectronics Technology and Industry – Habib & Lewis (Eds)
© 2019 Taylor & Francis Group, London, ISBN 978-1-138-33178-5

Research on a space microsatellite laser communication method based on modulating retroreflector technology

J.Y. Ren, H.Y. Sun & L.X. Zhang
Space Engineering University, Beijing, China

ABSTRACT: Satellite miniaturization is a trend for future development as conventional laser communication equipment is large in size, heavy in weight and has a high power consumption. This paper describes a novel concept for laser communication between spacecraft platforms using an optical interrogator on a large host spacecraft and a Modulating Retroreflector (MRR) on a small spacecraft. The weight and power consumption of the MRR is reduced, making it suitable for Size, Weight and Power (SWaP) constrained applications on a small spacecraft platform. We present the design of a laser communication microsatellite based on MRR. A simulation analysis was performed and a close-range experimental verification of the MRR array was conducted.

1 INTRODUCTION

With the development of space activities, the amount of information obtained has grown exponentially, and Radio Frequency (RF) transmission cannot meet the demand for large amounts of data and real-time transmission. Free Space Optical (FSO) links can enable high bandwidth communication that does not require frequency allocation, and which is resistant to interception and jamming (Zeng & Liu, 2017; William & Rabinovich, 2015; Rabinovich & Arnon, 2012). A conventional, direct, FSO link has an actively pointed laser terminal on both ends of the ling, but the size and power consumption of the FSO terminals can sometimes be too large for a small platform. In these cases, an FSO link using a Modulating Retroreflector (MRR) may be advantageous. These links require an active terminal on only one end of the link. On the other end, they use a small retroreflecting terminal, which due to not needing an laser source, reduces the weight and power consumption, not requiring precise pointing, which makes it more suitable for microsatellite platform applications (Quintana et al., 2014).

This paper describes a novel concept for laser communication between spacecraft platforms using an optical interrogator on a large host spacecraft and a MRR on a small spacecraft. The feasibility of array-based MRR communication was analyzed using the returned optical power equation, and simple experimental verification was performed.

2 MRR LINK CONSIDERATION

As shown in Figure 1, modulating retroreflector communications is a promising technology to implement a data link between an active platform and a small platform. The interrogator mainly includes a tracking module, a laser emitting and receiving module. The tracking module is divided into a coarse tracking subsystem and a fine tracking subsystem. The MRR is mainly composed of a modulator and a cat's eye lens, having an original return effect. A continuous wave beam from an optical interrogator has the same structure as a conventional FSO terminal. The MRR reflects light back to the active platform and modulates a CW beam launched from the active platform, as long as the interrogator is within the field

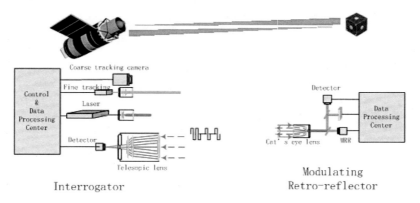

Figure 1.   Diagram of a modulating retroreflector link.

of view of the passive retroreflector. Data can be encoded using either Liquid Crystal (LC) optical shutter or Multiple Quantum Well (MQW) modulators.

The microsatellite receives a signal from the interrogator and demodulates the signal to obtain the command. The laser from the interrogator to the microsatellite is a one-way transmission process. The microsatellite receiving signal power model can adopt the FSO power model. The power received by the MRR terminal, scales as:

$$P_{mrr} = \frac{P_0 D_{MOD}^2 \tau_{mrr} \tau_{atm}}{L^2 \theta_t^2} \tag{1}$$

The microsatellite modulates the incident laser and returns it to the interrogator along the original path. The laser received by the interrogator passes through two-way transmission. For a diffraction-limited system the optical power, retroreflected from the microsatellite to the active platform (Creamer & Gilbreath, 2001), scales as:

$$P_{rec} = \frac{P_{laser} D_{rec}^2 D_{MOD}^4 \tau_{rec} \tau_{mr}^2 \tau_{atm}^2}{L^4 \theta_t^2} \tag{2}$$

where P_{laser} is the power of the laser transmitter on the large platform, D_{rec} is the diameter of the receiving telescope on the large platform, D_{MOD} is the diameter of the modulating retroreflector on the small platform, τ_{rec} is the attenuation through the lens of the receiving telescope on the large platform, τ_{mr} is the attenuation through the lens of the retroreflector on the small platform, τ_{atm} is the loss due to transmission through the atmosphere, L is the range between the two platforms, and θ is its divergence of the transmit beam. It is not influenced by the atmosphere in the space, so $\tau_{atm} = 1$.

From above equation, the returned power in retroreflecting links has the strongest dependence on the range and the retroreflector diameter, both of which scale as fourth powers. The returned optical power falls off as $\frac{1}{\theta_t^2}$. An increase in the size of the MMR diameter and a decrease of the divergence angle can increase the received power, but a decrease of the divergence angle will increase the difficulty of the interrogator tracking.

2.1 Modulating retroreflector

The retroreflect modulator is an important factor affecting the laser communication link. MRR mainly includes LC modulator, a coner-cube modulator and MQW modulator. MQW has faster data rates, lower drive power, lighter weight, better robustness and no polarization sensitivity compared with LC and coner-cube modulators (Rabinovich & Gilbreath, 2002).

The modulation rate of the MQW is determined by the modulator resistance and size (Zhang & Sun, 2013):

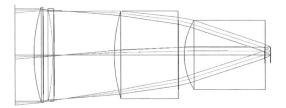

Figure 2.   The structure of a cat's eye lens.

$$F \propto \frac{1}{R_{MOD} D^2_{MOD}} \tag{3}$$

where R_{MOD} is resistance of modulator, D_{MOD} is the diameter of the modulator. From the above equation, the modulation rate of the MQW modulator is inversely proportional to the area size. The larger the MQW area, the lower the modulation rate.

The modulation efficiency of the MQW is expressed as:

$$M = e^{-\alpha on} - e^{-\alpha off} = e^{-\alpha off}\left[C_{MQW} - 1\right] \tag{4}$$

$$C_{MQW} = \frac{e^{-\alpha o}}{e^{-\alpha off}} \tag{5}$$

where $e^{-\alpha on}$ and $e^{-\alpha off}$ indicate the absorption index of MQW in the on and off states. C_{MQW} is modulating contrast of the MQW. Modulation efficiency and contrast are important parameters of the MQW.

The MQW power consumption calculation formula is:

$$W = D^2_{MOD} V^2 F \tag{6}$$

where is drive voltage of the MQW, is the modulation rate of the MQW.

2.2   Cat's eye optical lens

A cat's eye MRR uses a retroreflector formed by a system of lenses and a mirror. The modulator can be placed in the focal plane of the device, partially decoupling the size of the modulator from the size of the optical aperture. The modulator size is now coupled to the focal length and the field of view of the optic (Rabinovich & Goetz, 2003; Peter & William, 2010).

As shown in Figure 2, a cat's eye optical lens is a system that can return parallel incident light in exactly the same direction as the incident angle. The cat's eye optical system approaches the diffraction limit within an effective field of view angle. Assume that the angle between the incident beam and the optical axis of the optical system is φ, also called incident angle. The focal length of the optical system is f, the diameter of the modulator is D_{MOD}, the cat's eye optical system view angle is defined as:

$$\theta_t = 2\arctan\left(\frac{D_{MOD}/2}{f}\right) \tag{7}$$

3   NUMERICAL SIMULATION AND TEXT

From Equation 2, the returned power increases with an increase in the size of the modulator aperture D_{mod}. However, according to Equation 3, when the modulator aperture increases, the modulation rate of the modulator decreases, and the power consumption increases. In order to balance the relationship between MRR apertures, modulation rate and power consumption, this paper proposes a small-aperture array retroreflector modulation design scheme. As Figure 3 shows, there are four small MRRs, and the center lens is an optical imaging lens. In this design, the cat's eye system is only 5 mm in diameter, with a field of view of 30 degrees and a focal length of 0.5 cm.

According to Equations 1 and 2, simulations are performed on the received signal power at the interrogator and MRR terminal. The results are shown in Figure 4. In the simulation, a 1064 nm laser source with a power of 5 W was used. The aperture of the interrogator optical lens is 30 mm, the divergence angle is $1\mu rad$, and the received aperture is 150 mm. Assume that the minimum detectable power of the detector is −80 dBm. As shown in Figure 4, the communication distance of an array of four 5 mm MMRs can reach 50 km. If a single photon detector is used at the interrogator terminal, the communication distance can reach over 100 km.

In order to verify the performance of the MRRs array, a 500 m experiment was conducted in the atmosphere. Different numbers of MRR communication experiments were conducted respectively. In the experiment, three identical MRRs were evenly arranged on a circle with a radius of 150 mm, as Figure 5(a) shows. The three MRRs were numbered as A, B and C, respectively. In the process, using the method of shielding the lens, the return light signal was measured when A was working alone, A and B were working simultaneously and A, B and C were working at the same time. The received signals are shown in Figure 6, Figure 7 and Figure 8.

It can be seen from the experimental results that as the number of MRRs increases, the level of the received signal increases, indicating that the increase in the number of MRRs can increase the signal power at the interrogator terminal. However, the received signal level is not linearly superposed with the increase of the number of MRRs. There are two influencing factors, one is that the MRR array is not completely synchronized, resulting in non-linear

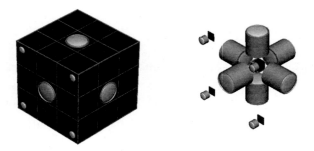

Figure 3. Concept design diagram of the array of a MRR laser communication microsatellite.

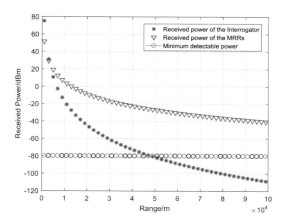

Figure 4. Received power of the interrogator and MRR terminal varies with range.

Figure 5.   (a) MRR array; (b) interrogator.

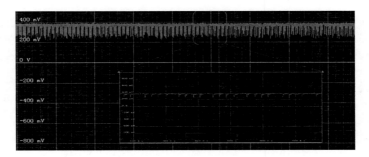

Figure 6.   The return signal when the MRR A works alone.

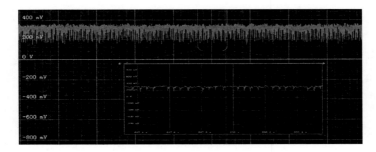

Figure 7.   The return signal when the MRR A and B work.

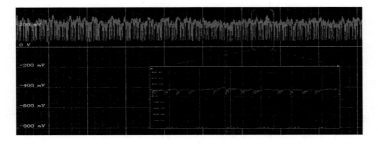

Figure 8.   The return signal when the MRR A, B and C work.

superposition of signals, and the second is the effect of atmospheric turbulence. From the results, it can be seen that with the increase of the number of MRRs, the fluctuation of the return signal is obviously weakened, indicating that the MRR array has a good inhibitory effect on atmospheric turbulence.

4 CONCLUSIONS AND FURTHER RESEARCH EFFORTS

This paper describes a novel concept for laser communication between a large host spacecraft and a MRR on a microsatellite. An array MMR scheme is proposed, and the feasibility is analyzed and simple verification conducted. This research is important, making MRR technologies more attractive in the future.

MRR laser communication technology needs to solve many key technical problems in space applications.

4.1 *Interrogator high precision capture alignment and tracking technology*

Capture alignment and tracking are prerequisites for the establishment of inter-satellite laser communication links. The MRR laser communication is subject to greater shock due to vibration in the two-way transmission and is prone to loss of lock. Therefore, higher requirements are imposed on the capture and alignment technology. The fine tracking module requires a position sensing unit, whose accuracy is critical to reduce the beam spot diameter on the MRR terminal and thus, minimize the geometrical losses. Due to the small divergence angle of the tracked beam, which is greatly affected by the platform vibration, relative motion and so on, a high frame rate camera and sub-pixel subdivision detection technology can be used to realize fine tracking. Intelligent control technology is adopted to improve the tracking system's robustness.

4.2 *High performance MRR technologies*

The retroreflect modulator is a key component of the MRR laser communication link and directly determines the link communication rate. Currently, the modulation methods of the FLC, MEMS and MQW are all intensity modulation, and the modulation method is single, which restricts the application of MRR laser communication. It is necessary to study the retroreflect modulator with high reflectivity and high modulation rate in the frequency modulation.

4.3 *High gain and receive common optical antenna & high sensitivity detection technology*

High sensitivity detection technologies include many key technologies such as single photon detection, automatic gain control, floating threshold selection, polarization reception, coherent reception, phase retention and matching.

REFERENCES

Creamer N.G., G.C. Gilbreath, (2001). Multiple quantum well retromodulators for spacecraft-to-spacecraft laser interrogation, communication, and navigation, In T.J. Meehan & M.F. Stell (Eds.), *15th Annual AIAA/USU Conference on small Satellites*, SSC01-VI-6.
Goetz P.G. and W.S. Rabinovich, (2010). Modulating retro-reflector lasercom systems at the naval research laboratory. In R. Mahon, *An IEEE Conference*, 2302–2307.
Quintana C., G. Erry, A. Gomez, (2014). Novel non-mechanical fine tracking module for retroreflective free space optics. In Y. Thueux & G. Faulkner (Eds.), *SPIE*, 92480S. D. O'Brien.
Rabinovich W.S. and G.C. Gilbreath, (2002). InGaAs multiple quantum well modulating retro-reflector for free-space optical communication. In P.G. Goetz (Ed.), *SPIE*, 4489, 190–201.
Rabinovich W.S. and P.G. Goetz, (2003). Cat's eye quantum well modulating retro-reflectors for free-space optical communications. In R. Mahon & E. Waluschka (Eds.), *SPIE*, 4975.
Rabinovich W.S. and R. Mahon, (2015). Modulating retro-reflectors links in high turbulence: Challenges and solutions. In M.S. Ferraro and J.L. Murphy, *Imaging and Applied Optics*.
Rabinovich W.S. and S. Arnon, (2012). Optical modulating retro-reflectors in advanced wireless optical communication systems. In J.R. Barry & G.K. Karagiannidis, Cambridge University.
Zeng ZH.L. and X. Liu, (2017). Latest developments of space laser communications and some development suggestions. In H. Sun & Z.H.X. Tan (Eds.), *Optical Communication Technology*, 6.
Zhang L.X. and H.Y. Sun, (2013). Progress in free space optical communication technology based on cat-eye modulating retro-reflector. In G.H. Fan & Y.H. Zheng, *Chinese Optics*, 6(5), 681–691.

Frontier Research and Innovation in Optoelectronics Technology and Industry – Habib & Lewis (Eds)
© 2019 Taylor & Francis Group, London, ISBN 978-1-138-33178-5

Performance analysis of a multichannel WDM hybrid optical communication system for long haul communication

D. Shanmuga Sundar
University of Chile, Santiago, Chile

V. Nidhyavijay
Velammal College of Engineering and Technology, Madurai, India

T. Sridarshini
Madras Institute of Technology, Chennai, Tamilnadu, India

A. Sivanantha Raja
Alagappa Chettiar College of Engineering and Technology, Karaikudi, India

ABSTRACT: Hybrid optical communication is an attractive candidate for the wireless network in the near future which utilizes both wired and wireless communication. Optical communication meant for long haul communication has been intended for an eight channel Wavelength Division Multiplexing (WDM) system followed by a Free Space Optics (FSO) called wireless technology. In this paper, hybrid optical communication for an eight channel WDM is performed and different modulation format is analysed for the same link. Dispersion is the major effect present in the optical fiber which causes broadening of pulses and leads to Inter Symbol Interference (ISI). In this paper, dispersion is also compensated by utilizing Optical Phase Conjugation (OPC) which utilizes Four Wave Mixing (FWM) as the non-linear degrading effect. It has been found that the apposite modulation format for the transmitter part of the optical system is Modified Duo-binary Return-to-Zero (MDRZ) which is good for both linear as well as non-linear effects when compared it with Return-to-Zero (RZ), Non-Return-to-Zero (NRZ), Carrier Suppressed Return-to-Zero (CSRZ) and Duo-binary modulation format. The feat of the intended hybrid optical system is scrutinized in terms of Eye Opening, Bit Error Rate (BER), and Q-value etc. Better BER is achieved.

1 INTRODUCTION

Optical communication is a form of telecommunication that uses light and optical fiber as carrier. In last two decades the fiber optic communication has seen a unique growth and technological evolution. Being developed in 1970s, the fiber optic communication system has revolutionized the telecommunication industry and has played a major role in the dawn of the Information Age. Due to the proficient consumption of power and low bit error rate, modified duo-binary return-to-zero (MDRZ) is special for the transmitter part of the system (B. Patnaik et al. 2012), (AnuSheetal et al. 2010). In the direction of exterminating pulse broadening, Optical Phase Conjugation (OPC) is taken into consideration which utilizes Four Wave Mixing (FWM) (C. Lorattanasane et al. 2009).

Four Wave Mixing confines the performance of Dense Wavelength Division Multiplexing (DWDM) systems and dogged that MDRZ format is the best one for transmission system afar 1500 km albeit the transmitter and receiver configuration is intricate by using symmetrical dispersion scheme in the year 2008 (AnuSheetal et al. 2010), (S. Pazi et al. 2009), (S. Singh et al. 2007). K. S. Cheng along with J. Conradi confirmed that for 40 Gb/s system both alternating phase RZ (AP-RZ) format and MDRZ format have less nonlinear pulse interaction by examining the pulse evolution of the optical signals in the first span of Standard Single Mode Fiber (SSMF) in 2002 and also showed that the pulses caused by intra channel FWM between discrete tones can be repressed by the lack of tones in the signal spectrum (K. S. Cheng et al. 2007). Xiaoming Zhu and Joseph M. Kahn together depicted the use of maximum-likelihood (ML) detection in spatial diversity reception to lessen the diversity gain penalty caused by correlation among the fading at different receivers. They also showed that in the dual-receiver case, ML diversity reception surpasses the conventional equal-gain combining (EGC) method. Spatial diversity reception can also help to alleviate turbulence-induced-fading (X. Zhu et al. 2002). This paper demonstrates hybrid optical communication system in which the dispersion effects are compensated using optical phase conjugation technique which utilizes four wave mixing and also best modulation format is analyzed for the same.

2 HYBRID SYSTEM DESIGN

Hybrid optical communication is designed by utilizing OPC technique for the fiber optic link setup and FSO for the wireless part which makes the system hybrid. Figure 1 shows the block diagram for the hybrid optical system. The main consideration for the WDM system is the channel spacing and here an eight channel WDM system is considered. Eight channel WDM channels are multiplexed with the unique modulation format and then it is fed into the OPC module which comprises of DCF fiber and FWM. Finally the demultiplexed output is obtained at the receiver side in which the BER, Q-factor are analyzed. The same link is established for different modulation format and the result of each is compared.

Modified duo binary return-to-zero (MDRZ) is a perfect modulation format for a long haul, high speed and WDM transmission links which has narrow spectral width, low susceptibility to fiber nonlinearity, large dispersion tolerance and good transmission performance (S. Sarkar et al. 2007).

The given input data from PRBS is modulated using optical modulator such as Machzehnder modulator. Figure 2 shows the subsystem of the MDRZ modulation format which

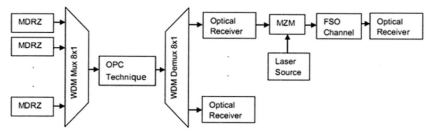

Figure 1.   Block diagram of hybrid optical system.

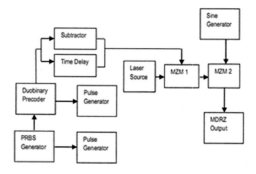

Figure 2.   Subsystem for MDRZ modulation format.

348

is generated by utilizing a delay and subtractor circuit. Then the modulated signal of each modulation format is checked by passing through an optical fiber, the length and bit rate are maintained as constant. After passing through an OPC module, the signal is continued by the Free Space Optics which makes the system as a hybrid. In this system, FSO of Line of Sight is used because of excellent transmission capacity. It is well suited for Local Area Networks (LAN), Metropolitan Area Networks (MAN) and for Point-to-point applications. Hence it is preferred for the short haul communication of the hybrid system

3   RESULTS AND DISCUSSIONS

Table 1, shows the performance analysis of different modulation format and it is found that MDRZ is better for an effective hybrid optical communication when compared to the other modulation format. The Eye diagram obtained for RZ and NRZ modulation formats depicts much dispersion which is shown in Figure 3 and Figure 4.

Table 1.   Performance analysis of the optical system.

Modulation format	Max Q-factor	Min BER	Eye height
RZ	9.77	6.47e-023	1.69e-004
NRZ	10.92	3.80e-028	1.29e-004
CSRZ	9.88	2.06e-023	1.38e-004
Duo binary	8.48	9.84e-018	7.85e-005
MDRZ	10.61	1.078e-026	7.38e-005

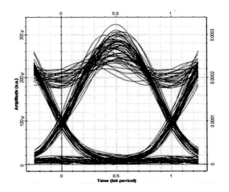

Figure 3.   Eye pattern of the fiber optic link using RZ.

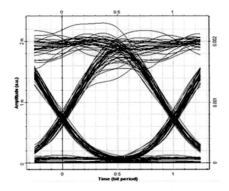

Figure 4.   Eye pattern of the fiber optic link using NRZ.

349

From the annotations it is found that, RZ modulation format has broader optical spectrum which results in decreased tolerance and spectral efficiency. The pulse shape enables robustness to fiber nonlinear effects and also to the polarization mode dispersion.

In contrast to RZ modulation format, NRZ, which has been used in early days since it is insensitive to laser phase noise and also requires low electrical bandwidth for transmitter and receiver has narrow optical spectrum. Hence it has enhanced dispersion tolerance. But the main hitch in this case is, it has the effect of Inter Symbol Interference (ISI) which is not apposite for high bit rates and long distance (X. Wang et al. 1999).

Figure 5 and Figure 6 Shows the Eye diagram of CSRZ and Duobinary modulation format respectively. Analogous to RZ, CSRZ provides better robustness and far less sensitive to fiber nonlinear effects (D. Dahan et al. 2002), (G. Bosco et al. 1994). From the observations it is examined that Duobinary modulation format offers high spectral efficiency and chromatic dispersion tolerance. Albeit it has better dispersion tolerance and suppression of SBS (Stimulated Brillouin Scattering), the main inconvenience in this case is, it has a great impact of fiber non linearities which is the main limiting factor for the utmost transmission length and doable transmission quality. Finally, MDRZ is taken into account.

Figure 7 illustrates the output spectrum of MDRZ modulation format for the power level of 1 mW and for a bit rate of 40 Gb/s. The carrier of the signal is suppressed which results in a narrow spectrum. Then the multiplexed signal is passed through the DCF having zero dispersion and dispersion slope of '0.075 ps/nm/km' along with the pump signal. Figure 8 depicts the eye pattern of the fiber optic link using MDRZ modulation format with distance

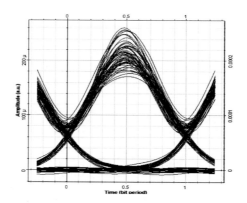

Figure 5.   Eye pattern of the fiber optic link using CSRZ.

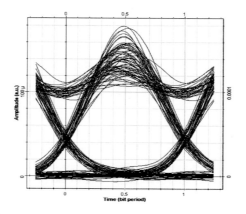

Figure 6.   Eye pattern of the fiber optic link using Duobinary modulation format.

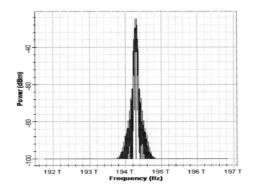

Figure 7.   Optical spectrum of MDRZ format.

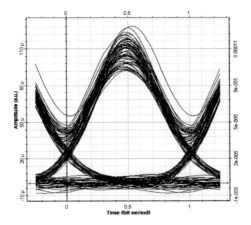

Figure 8.   Eye pattern of the fiber optic link using MDRZ.

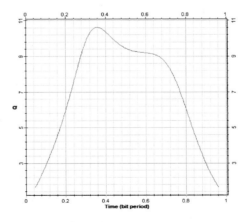

Figure 9.   BER Analyzer of the fiber optic link.

coverage of 200.1 km. The achieved Q-factor is 10.61 and the BER of 1.078×10^{-26}. Thus the Q-factor is maximized with respect to the DCF length (D. C. Kilper et al. 2004). From the Figures 3, 4, 5, 6 and 8, it is confirmed that the MDRZ modulation format is the effective modulation technique for an Optical communication system.

Figure 9 shows the BER Analyzer of the optic link whose Eye height stood at 7.388×10^{-5} which indicates the amount of data that can be sent without dispersion using our proposed system.

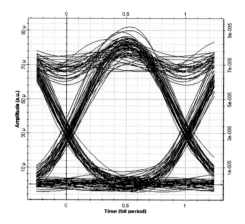

Figure 10.   Eye pattern of the hybrid optic link using RZ.

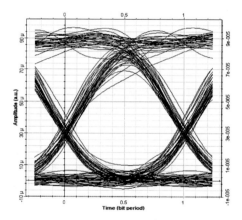

Figure 11.   Eye pattern of the hybrid optic link using NRZ.

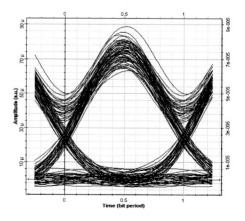

Figure 12.   Eye pattern of the hybrid optic link using CSRZ.

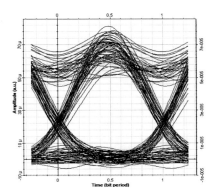

Figure 13. Eye pattern of the hybrid optic link using Duobinary modulation format.

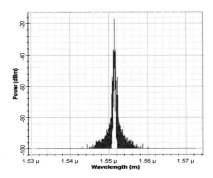

Figure 14. Optical spectrum of the hybrid optic link using MDRZ.

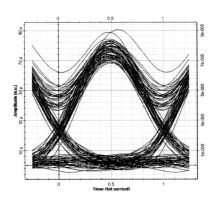

Figure 15. Eye pattern of the hybrid optic link using MDRZ.

Table 2. Performance analysis of the hybrid optical system.

Modulation format	Max Q-factor	Min BER	Eye height
RZ	11.16	3.11e-029	5.56e-005
NRZ	10.25	5.12e-025	5.68e-005
CSRZ	9.22	1.30e-020	4.96e-005
Duobinary Signal	7.68	6.74e-015	3.83e-005
MDRZ	10.39	1.26e-025	5.30e-005

Figure 14 illustrates the optical spectrum of the hybrid optic link whose FSO channel is maintained for a range of 1 km and attenuation of 25 dB/km. From the Figures 10, 11, 12, 13 and 15, once again it is confirmed that the MDRZ modulation technique is the effective technique for an hybrid optical communication system whose Q-factor stood at 10.39 and the BER is achieved as 1.26×10^{-25}.

Table 2 demonstrates the performance analysis of the hybrid optical system which once again proves that MDRZ is the best modulation format for the hybrid optical system.

4 CONCLUSION

In summary, a hybrid optical communication for an eight channel WDM system is proposed and analyzed. The BER of MDRZ modulation format for both optical and hybrid optical communication system using Optical phase conjugation obtained are 1.078e-026 and 1.26e-025 respectively. From the above work it is proved that the dispersion, a major effect can be reduced by using an OPC and also it is confirmed that MDRZ modulation format is the best suited modulation format for the optical links when compared to all other modulation format.

REFERENCES

AnuSheetal, Ajay K. Sharma, R.S. Kaler. 2010. "Simulation of high capacity 40 Gb/s long haul DWDM system using different modulation formats and dispersion compensation schemes in the presence of Kerr's effect", Optik 121, pp 739–749.

BijayanandaPatnaik, P.K. Sahu. 2012. "Optimized ultra-high bit rate hybrid optical communication system design and simulation", Optik-Int. J. Light Electron Opt.

Bosco, G., A. Carena, V. Curri, R. Gaudino, P. Poggiolini. 1994. "The use of NRZ, RZ and CSRZ modulation at 40 Gb/s with narrow DWDM channel spacing," J. Lightwave Technol. 20 (9).

Cheng, K.S., J. Conradi. 2002. "Reduction of pulse-pulse interaction using alternative RZ formats in 40 Gb/s systems", IEEE Photonics Technol. Lett. 14 (1) 98.

Dahan, G. Eisenstein. 2002. "Numerical comparison between distributed and discrete amplification in a pont-to-point 40 Gb/s 40- WDM-based transmission system with three different modulation formats", J. Lightwave Technol., 20.

Kilper, D.C., W. Weingartner, S. Hunsche, A. Azarov. 2004. "Q-factor monitoring using FEC for fault management applications", J. Opt. Netw. 3, pp 651–663.

Lorattanasane, C., K. Kikuchi. 2009. "Design theory of long-distance optical transmission system using midway optical phase conjugation", IEEE J. Lightwave Technol., 15, June 1997.

Pazi, S., C. Chatwin, R. Young, P. Birch, W. Wang. 2009, "Performance of Tanzanian optical DWDM", Eur. J. Sci. Res., 36(4), pp 606–626.

Sarkar, S., S. Dixit, B. Mukherjee. 2007. "Hybrid wireless-optical broadband-access network (WOBAN): a review of relevant challenges", J. Lightwave Technol. 25 (11), pp 3329–3340.

Singh, S., R.S. Kaler. 2007. "Simulation of DWDM signals using optimum span scheme with cascaded optimized semiconductor optical amplifiers", Optik 118, pp 74–82.

Wang, X., K. Kikuchi, Y. Takushima. 1999. "Analysis of dispersion managed optical fiber transmission system using non-return-to-zero pulse format and performance restriction from third-order dispersion", IEICE Trans. Electron. E82-C, pp 1407–1413.

Zhu, X., J.M. Kahn. 2002. "Free-space optical communication through atmospheric turbulence channels", IEEE Trans. Commun. 50, pp 1293–1300.

Frontier Research and Innovation in Optoelectronics Technology and Industry – Habib & Lewis (Eds)
© *2019 Taylor & Francis Group, London, ISBN 978-1-138-33178-5*

Secure performance analysis of optical CDMA systems based on secrecy capacity

Yeteng Tan, Tao Pu, Jilin Zheng, Peng Xiang, Weifeng Mou & Huatao Zhu
Communications Engineering College, Army Engineering University of PLA, Nanjing, China

ABSTRACT: Optical CDMA transmission systems can not only increase system capacity, but also achieve physical-layer security against fiber tapping attacks. In this paper, we examine the information-theoretic security of optical CDMA systems by evaluating the trade off between the achieved information capacity and the confidentiality. In particular, we established a wiretap channel model of optical CDMA systems, based on which we have evaluated the impacts of key system parameters, such as the type of code words, the number of users M, the code lengths N and the input optical power P, on the system security. Our results indicate that key system parameters play an important role and choosing suitable values of these system parameters will be helpful to improve the secure performance of optical CDMA systems to some degree. For example, choosing the code words whose cross-correlation λ values are smaller, increasing the code lengths N or the input optical power P and reducing the number of users M properly can all be used to improve the secure performance of optical CDMA systems.

Keywords: Optical Code Division Multiple Access (OCDMA), wiretap channel model, secrecy capacity, interception probability

1   INTRODUCTION

The conventional fiber-optic communication systems are vulnerable to physical-layer attacks, for example, an eavesdropper can extract the transmission signal by bending the fiber (K. Shaneman and S. Gray, 2004; M. Medard et al., 2011). This type of eavesdropping is relatively easy to implement and could remain largely unnoticed. Optical CDMA systems can address this issue by using optical encoding and decoding technologies. So, enhanced security has often been put forward as an important benefit of optical CDMA technology. In the previous papers, the types and degree of security provided by optical CDMA systems has been examined (T.H. Shake, 2005). But, the wiretap channel models proposed by Thomas H. Shake both make "decoding probability of code words" as quantitative scale to examine physical-layer security. The safety analysis of Shake must be under certain assumptions to estimate the decoding probability of the code words. The decoding of the code words is unable to achieve when the legitimate users take security enhancement strategy (e.g. code words reconstruction), and then the decoding probability of code words can't be analyzed quantitative. So, we need to look for a new theoretical basis and method to evaluate the secure performance of optical CDMA systems when the decoding probability of code words can't be analyzed quantitative.

In this paper, the secure performance of optical CDMA systems is evaluated from information theory viewpoint. Information-theoretic security is a widely accepted and fundamentally provable notion of secrecy in a classical system. When operating in information-theoretic secrecy, the eavesdropper is not better off than randomly guessing the transmitted data, even if he or she can observe the wiretap channel output. The secrecy capacity then quantifies the maximum achievable transmission capacity that can be transmitted to a legitimate user; however, the eavesdropper can't receive any useful information (A.D. Wyner, 1975).

The rest of the paper is organized as follows. First, Section II establishes the wiretap channel model of optical CDMA systems and provides a description of this model. Then, Section III analyzes the secrecy capacity of optical CDMA systems in term of interception probability. The main results and their implications are discussed in detail in Section IV. Finally, Section V concludes the paper.

2 WIRETAP CHANNEL MODEL OF OPTICAL CDMA SYSTEMS

The wiretap channel model is illustrated in Fig. 1. In the model, a legitimate user named Alice wants to send messages D to another user named Bob, but there is an eavesdropper named Eve who can extract the transmitted signals. The message D sent by Alice is firstly encoded by optical encoder, and then the coded signals from different users are coupled into the fiber channel by the coupler and transmit. At the legitimate user's receiving end, the coded signals from all users come into the decoder d, where optical signal D from user d is recovered, meanwhile part of optical signals I from other users will mix into it and form multi-access interference (MAI) whose statistical character is determined by the cross-correlation of the code words. Then optical signal D and I come into the photodiode and produce photocurrents which consist of two parts: signal current I_d and interference current N_I, here N_I has added shot noise. Because photodiode is a square-law device, optical signals from different sources will produce cross terms when converting into electronic signals. Cross terms between optical signals D and I are called prime beat noise (PB) and cross terms between optical signals I from different users are called secondary beat noise (SB). There will also be thermal noise T induced by photodiode in the system at the same time (T. Pu et al., 2006). So, we mainly consider user signal (D), multi-access interference (MAI), prime beat noise (PB), secondary beat noise (SB), and thermal noise (T) etc. in the wiretap channel models.

According to Kerckhoffs's principle, eavesdropper knows all the parameters of encryption algorithm except the secret key itself that the legitimate users use in the data encryption systems. The same as physical-layer encryption systems, eavesdropper knows the data rate, the type of physical-layer encryption and the algorithm type, but he or she doesn't know the specific parameter of the encryption algorithm. So, we assume that eavesdropper knows the encoding type, the coding rate and the structure of code words, but he or she doesn't know the specific code words that the legitimate users use in the optical CDMA systems (C.E. Shannon, 1949). The eavesdropper can only choose some codes from random sequences with the period $n = 2^r-1$ as decoder's code word, then the decoder is unmatched in the wiretap channel. Therefore, the signal currents in the legitimate and wiretap channel are shown as Table 1 below.

Where λ_i is the cross-correlation of decoder d and encoder i, $\lambda_{i'}$ are the cross-correlation of eavesdropper's decoder and encoder i, R is the response of PD and P is the input optical power.

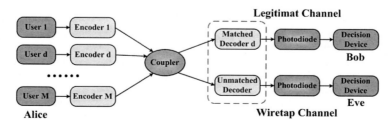

Figure 1. Wiretap channel model of optical CDMA systems.

Table 1. The signal currents.

Signal	Legitimate channel	Wiretap channel
D	$RP\lambda_d^2$	$RP\lambda_{d'}^2$
MAI	$RP\sum_{i=1,i\neq d}^{m}\lambda_i^2$	$RP\sum_{i=1,i\neq d}^{m}\lambda_{i'}^2$
PB	$2RP\sum_{i=1,i\neq d}^{m}\lambda_d\lambda_i$	$2RP\sum_{i=1,i\neq d}^{m}\lambda_{d'}\lambda_{i'}$
SB	$2RP\sum_{i=1}^{m}\sum_{j=i+1}^{m}\lambda_i\lambda_j$	$2RP\sum_{i=1}^{m}\sum_{j=i+1}^{m}\lambda_{i'}\lambda_{j'}$
T	$\sqrt{4k_BTB_e/R_L}$.	

3 SECRECY CAPACITY OF OPTICAL CDMA SYSTEMS

This section characterizes the secrecy capacity of optical CDMA systems in terms of interception probability. We start by deriving secrecy capacity of optical CDMA systems for each choice of code words. According to Shannon's theorem (C.E. Shannon, 1948), the secrecy capacity (J. Barros and M.R.D. Rodrigues, 2006) of Optical CDMA systems can be represented as:

$$C_S = C_M - C_W \qquad (1)$$

where

$$C_M = \log_2(1+ SNR_M)$$
$$= \log_2\left(1+ \frac{\langle D_M^2\rangle}{\langle T^2\rangle+\langle MAI_M^2\rangle+\langle PB_M^2\rangle+\langle SB_M^2\rangle}\right) \qquad (2)$$

is the capacity of the legitimate channel and

$$C_W = \log_2(1+ SNR_W)$$
$$= \log_2\left(1+ \frac{\langle D_W^2\rangle}{\langle T^2\rangle+\langle MAI_W^2\rangle+\langle PB_W^2\rangle+\langle SB_W^2\rangle}\right) \qquad (3)$$

denotes the capacity of the wiretap channel.

The legitimate and wiretap channels are both random because the users and the eavesdropper choose code words randomly. So, the capacities of the legitimate and wiretap channels and the secrecy capacities of optical CDMA systems are also both random. We can't predict which of the code words will be chosen. As a result, the secrecy capacity of Optical CDMA systems is randomly distributed. We are now ready to characterize the interception probability (K. Guan et al., 2015) as follow:

$$p_{\text{int}}(R_S) = P(C_S < R_S) \qquad (4)$$

i.e. the probability that the secrecy capacities corresponding to random choice of code words is less than a target secrecy rate $R_S > 0$. When setting the secrecy rate R_S, Alice is proposing that the capacity of the wiretap channel is given by $C_W' = C_M - R_S$. If $R_S < C_S$, the actual capacity of the wiretap channel is worse than Alice's estimate, i.e. $C_W < C_W'$, and so the transmission in perfect secrecy can be achievable. Otherwise, if $R_S > C_S$ then $C_W > C_W'$ and information-theoretic security is compromised.

4 RESULTS AND DISCUSSION

In this section, we take optical CDMA systems based on bipolar Gold codes and 2-D PCPC codes for example to evaluate the impact of several system parameters, such as the number of users M, the code length L and the input optical power P, etc. by using numerical simulation approach. Firstly, for a given set of parameters, we run a simulation that generates 10^5 random choices of code words. For each choice of code words, we calculate the corresponding C_S by using Eq. (1). Based on the statistics generated by these random choices, we then calculate the interception probability (for a given communication rate R_s) or the outage (for a certain interception probability p_{int}) secrecy capacity by using Eq. (4) or (5), respectively. The simulation parameters are set as follow: extinction ratio $ext = 10\ dB$, responsivity of PD $R = 0.65$, bandwidth of PD $Be = 2 \times 10^7\ Hz$, equivalent load impedance $Re = 1000\Omega$, absolute temperature $T = 290K$.

4.1 Impact of the input optical power P

In Fig. 2 we plot the outage secrecy capacity as functions of the input optical power P for the Gold codes whose code length is 127 and the PCPC codes whose code length is 121, respectively, under the case of $M = 8$. We observe that the outage secrecy capacity will increase with increase of the input optical power P, and the larger interception probability p_{int} corresponds to the larger outage secrecy capacity. For Gold codes, when the input power P is small (<300 nW), the outage secrecy capacity increases quickly, then as P continues to increase, the outage secrecy capacity increases slower, and finally it stabilizes to a constant value with the input optical power P increasing. For PCPC codes, the increased tendency is similar to Gold codes, but the input optical power P corresponding to the stable point is larger. This is because the user signals and noises (MAI, PB and SB) in the channel will both increase as the power P gets large, but the augmenter of the user signals is larger relative to the noises. So, the SNRs and capacities of the legitimate and wiretap channels will both increase. And because the augmenter of the legitimate channel capacity is larger than that of the wiretap channel, the secrecy capacity of optical CDMA systems will get larger. When the input optical power P increases to some value, the systems will be saturated and the secrecy capacity won't increase. Therefore, increasing the input optical power P properly is a method of enhancing the secrecy capacities of optical CDMA systems.

4.2 Impact of the code length L

Fig. 3 show that whether for the Gold codes (a) or for the PCPC codes (b), the outage secrecy capacity increases along with the increase of the code length under the case of $M = 8$, $P = 500$ nW, and the larger interception probability p_{int} corresponds to the bigger outage secrecy capacity for the same code length. This is because the normalized cross-correlations

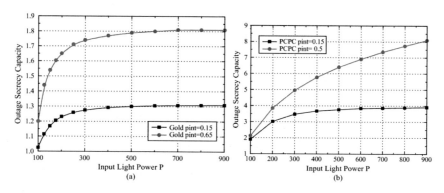

Figure 2.   Outage secrecy capacities vs. input optical power for Gold codes (a) and PCPC codes (b).

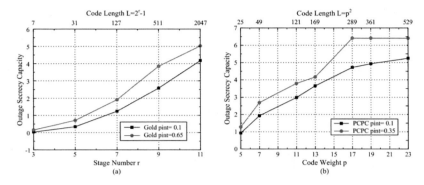

Figure 3.   Outage secrecy capacities vs. code length for Gold codes (a) and PCPC codes (b).

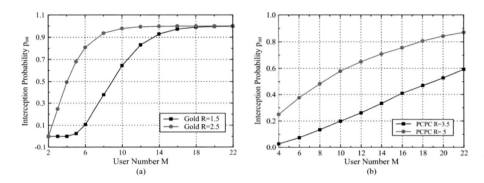

Figure 4.   Interception probability vs. user number for Gold codes (a) and PCPC codes (b).

of codes in the legitimate channel words decrease with the increase of the number of the code length L, which will lead to the decrease of the noises (MAI, PB and SB) and the increase of SNR of the legitimate channel. So, the capacity of the legitimate channel will also increase. But, for the wiretap channel, the number of the code length L has no influence on the user signals and noises, and then the SNR and capacity of the wiretap channel is independent of the number of the code length L. Then, the secrecy capacities of optical CDMA systems will increase with the increase of the number of the code length L. Therefore, we can enhance the secure performance of optical CDMA systems by increasing the number of the code length L.

4.3   Impact of the number of users M

In Fig. 4, we plot the interception probabilities p_{int} as functions of the number of users M for the Gold codes whose code length is 127 and the PCPC codes whose code length is 121, respectively, under the case of $P = 500$ nW. We observe that the interception probabilities p_{int} increase along with the increase of the number of users M, and final it reaches to 1. And for the same number of uses M, the larger the secrecy rate R_S is, the bigger the interception probability p_{int} is. This is because the noises (MAI, PB and SB) in the channels are getting larger as the number of users M increases, and then the SNRs and channel capacities of the legitimate and wiretap channels will be smaller. The secrecy capacity will reduce due to the larger reduction of the legitimate channel relative to the wiretap channel. So, when the secrecy rate R_S is a constant value, the interception probability p_{int} will increase with the increase of M. And the increase of the number of users M will go against the enhancement of system security performance.

5 CONCLUSION

In conclusion, we evaluated the secure performance of optical CDMA systems quantitatively by using interception probability and outage secrecy capacity from information theory viewpoint. And, we assessed the impacts of key system parameters on the secure performance of optical CDMA systems. In particular, we established a wiretap channel model and studied the secrecy capacity of optical CDMA systems. At the same time, we also assessed the impacts of key system parameters, such as the type of the code words, the number of users M, the number of the code length L and the input optical power P, on the interception probability and the secrecy capacity. Our results show that the system parameters play an important role in improving the secure performance of optical CDMA systems. Choosing the code words whose cross-correlations are smaller and the proper values of the system parameters will help to enhance the secrecy capacity of the systems and decrease the interception probability of secure information at the same communication rate. This will provide theoretical guidance for further improving the secure performance of optical CDMA systems.

This work was partly supported by the National Natural Science Foundation of China (NSFC) under grants with nos. 61174199, 61475193, and 61177065; and the Jiangsu Natural Science foundation under grants BK20120058 and BK20140069.

REFERENCE:

Barros, J., M.R.D. Rodrigues, 2006, Secrecy Capacity of Wireless Channels, IEEE International Symposium on Information Theory, 356–360.

Guan, K., A.M. Tulino, P.J. Winzer, and E. Soljanin, 2015, Secrecy Capacities in Space-Division Multiplexed Fiber Optic Communication Systems, IEEE Transactions on Information Forensics & Security, 10(7): 1325–1335.

Medard, M., D. Marquis, R.A. Barry, and S.G. Finn, 2011, Security issues in all-optical networks, Srii Global Conference, 11(3): 42–48.

Pu, T., H. Zhang, Y. Guo, M. Xu, and Y. Li, 2006, Evaluation of beat noise in OCDMA system with non-Gaussian approximated method, Journal of Lightwave Technology, 24(10): 3574–3582.

Shake, T.H., 2005, Confidentiality performance of spectral-phase-encoded optical CDMA, Journal of Lightwave Technology, 23(4): 1652–1663.

Shake, T.H., 2005, Security performance of optical CDMA against eavesdropping, Journal of Lightwave Technology, 23(2): 655–670.

Shaneman, K., and S. Gray, 2004, Optical network security: Technical analysis of fiber tapping mechanisms and methods for detection & prevention, IEEE Military Communications Conference, 2(2): 711–716.

Shannon, C.E., 1948, A mathematical theory of communication, The Bell System Technical Journal, 27(3): 3–55.

Shannon, C.E., 1949, Communication Theory of Secrecy Systems, Bell System Technical Journal, 28(4): 656–715.

Wyner, A.D., 1975, The wire-tap channel, The Bell System Technical Journal, 54(8): 1355–1387.

Frontier Research and Innovation in Optoelectronics Technology and Industry – Habib & Lewis (Eds)
© 2019 Taylor & Francis Group, London, ISBN 978-1-138-33178-5

A 3-D high-precision indoor positioning strategy using the ant colony optimization algorithm based on visible light communication

P.F. Wang
School of Materials Science and Engineering, South China University of Technology, Guangzhou, China

W.P. Guan
School of Automation Science and Engineering, South China University of Technology, Guangzhou, China

S.S. Wen, Mengyuan Sun & Qi Peng
School of Materials Science and Engineering, South China University of Technology, Guangzhou, China

Y.X. Wu
School of Automation Science and Engineering, South China University of Technology, Guangzhou, China

ABSTRACT: Recently, Visible Light Communication (VLC) has gradually become a research hotspot in indoor environments, but unfortunately the results of most existing 3-D positioning strategies are not satisfactory. This paper proposes a 3-D high-precision indoor positioning strategy using the Ant Colony Optimization algorithm based on visible light communication. The Ant Colony Optimization algorithm is a powerful population-based heuristic bionic evolutionary algorithm that can be used to solve global optimization problems, and the 3-D indoor positioning strategy can be transformed into an optimal solution problem. Therefore, for 3-D indoor positioning, the optimal receiver coordinate can be obtained by the Ant Colony Optimization algorithm. Our simulation results show that the average positioning error is 0.44 cm. The extended experiment in motion scene positioning also show that 96.04% of positioning errors are below 1.006 cm. Our experiment result proves that the mentioned positioning strategy satisfies the requirement of cm-level indoor positioning. Therefore, this strategy may be considered as one of the competitive indoor positioning candidates in the future.

1   INTRODUCTION

Indoor positioning has been a topic of growing interest for the last few years as the demand for accurate Location-Based Services (LBS). Due to the wide application scenarios such as large shopping malls and underground parking lots, the demand for high-accuracy indoor positioning is becoming larger and larger. However, traditional indoor positioning systems, such as Wireless Local Area Networks (WLANs), Zigbee, Ultra-Wideband (UWB) and Bluetooth, deliver positioning accuracies from tens of centimeters to a few meters (Yang et al., 2012). Therefore, a high accuracy LED-based indoor positioning system using Visible Light Communication (VLC) technology is put forward. To date, Visible Light Positioning (VLP) based on PD has been deeply explored and there are some measurements to calculate the location of the terminal such as Time of Arrival (TOA) (Alavi & Pahlavan, 2006; Shi et al., 2015), Time Difference of Arrival (TDOA) (Wann et al., 2006), Angle of Arrival (AOA) (Dakkak et al., 2011) and Received Signal Strength (RSS) (Won et al., 2013; Wang & Zhu, 2008). To take the cost, difficulty and accuracy of indoor positioning into consideration, RSS-based positioning is preferred due to its low cost and high accuracy (Zhang & Kavehrad, 2012).

Hence, the distance from the LED to the terminal can be determined by the gain differences. At the same time, multiple LED transmitters or optical receivers are often used to achieve the gain difference in indoor localization. However, when using multiple transmitters, the Inter-Cell Interference (ICI) may be a serious problem in VLP. Therefore, in our prior works (Guan et al., 2016; Guan et al., 2017), to reduce inter-cell interference caused by the presence of multiple reference points in the positioning system, the visible light from LED base stations installed on the ceiling was modulated in Code Division Multiple Access (CDMA) at different spreading code, which realizes the code division multiplexing to separate the overlapping signals in the time domain and frequency domain. The effectiveness and performance of the CDMA in VLP is well documented, so there will not be an in-depth description in this paper. For those readers who are interested in the CDMA, please refer to our previous reports.

Many previous works about 3-D indoor visible light positioning have been done. However, those 3-D localization methods still have many limitations. For example, in Yasir et al. (2014), Yasir et al. used the light sensor and accelerometer of the smart phone to measure the light intensity and locate the smart phone combined with the low complexity algorithm, and an error less than 25 cm was achieved. This method requires some additional devices, which increases the cost. In Yang et al. (2014), Yang et al. used a single transmitter and multiple tilted optical receivers to realize 3-D indoor positioning based on gain difference, which is a function of the AOA and the RSS. It is therefore complex in calculation. In Lim (2015), Lim proposed a maximum likelihood approach for the positioning system, which enhanced the positioning performance by employing an iterative maximum likelihood approach and adopting the least square solution as an initial guess. However, the accuracy was not high.

In order to solve these above-mentioned problems, we propose a 3-D high-precision indoor positioning strategy using the Ant Colony Optimization algorithm for 3-D visible light indoor positioning. The Ant Colony Optimization algorithm is a powerful population-based heuristic bionic evolutionary algorithm that can be used to solve global optimization problems, and the 3-D indoor positioning can be transformed into an optimal solution problem. Therefore, the Ant Colony Optimization algorithm can get the best positioning coordinates in 3-D indoor positioning, and this is the first time the Ant Colony Optimization algorithm has been applied to visible light positioning. With Ant Colony Optimization, the search individuals can converge to the optimal solution of the three-dimensional localization problem in a short path in three-dimensional space. This article combines the global search with the local search strategy to avoid the search process into a local optimization. Compared with existing 3-D positioning systems, it has higher accuracy, improves the search efficiency and does not need additional devices. The results show that while using the Ant Colony Optimization algorithm for 3-D indoor positioning, the maximum error is 6.31 cm and the mean error is 0.44 cm.

2 SYSTEM PRINCIPLE AND POSITIONING ALGORITHM

2.1 *Indoor wireless optical channel*

The indoor visible light positioning model is shown in Figure 1. All LEDs are installed on the ceiling to satisfy the requirement of lighting, and the receiver is located at a certain height in the room. The CDMA modulated signals are transmitted from LEDs to the receiver that is located at a certain height in the room by using the VLC technique. In this article, the radiant intensity of an LED can be assumed to follow a Lambertian radiation pattern, due to its large beam divergence. So, the Line-of-Sight (LOS) channel gain of this visible light indoor positioning system can be given by Guan et al. (2017):

$$H(0) = \begin{cases} \dfrac{(m+1)A}{2\pi d^2} \cos^m(\theta)\cos(\phi)T_s(\phi)G(\phi), 0 \leq \phi \leq \phi_c \\ 0, \phi \geq \phi_c \end{cases} \tag{1}$$

where the parameters are as follows. d is the distance between LED anchor nodes and the positioning terminal. A is the area of an optical detector. θ is the irradiant angle relative to

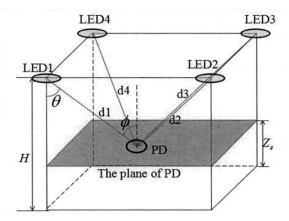

Figure 1.   Indoor visible light positioning system.

the vertical axis of the LED. ϕ is the incident angle relative to the receiving axis. ϕ_c is the field-of-view of the receiver. $T_s(\phi)$ is the gain of the optical filter, and $G(\phi)$ is the gain of the optical concentrator. m is the Lambertian order and is defined as:

$$m = -\frac{\ln 2}{\ln(\cos(\theta_{1/2}))},$$

where $\theta_{1/2}$ is the half-power angles of the transmitter (LED).

When the emitted optical power P_t is received by the receiver, the received optical power P_r can be represented as:

$$P_r = H(0) \cdot P_t = \frac{(m+1)A}{2\pi d^2} \cos^m(\theta)\cos(\phi)T_s(\phi)G(\phi)P_t \tag{3}$$

Since it is assumed that the plane at which the receiver is located is parallel to the ceiling, the incident angle ϕ is equal to the irradiant angle θ. We can also simplify Equation 3 to:

$$P_r = \frac{C}{d^2}\cos^{m+1}(\theta)P_t \tag{4}$$

where

$$C = \frac{(m+1)A}{2\pi}T_s(\phi)G(\phi) \tag{5}$$

The distance between the LED and the receiver can be represented as d and can eventually be represented as:

$$d = \sqrt{\frac{C}{H(0)}\cos^{m+1}(\theta)} \tag{6}$$

2.2   3-D positioning algorithm based on ant colony optimization

First, we are going to introduce the Ant Colony Optimization algorithm. The basic idea of the Ant Colony Optimization algorithm comes from the principle of the shortest path in the foraging of ants in nature. It is a simulation optimization algorithm that simulates the foraging behavior of ants. In the application of 3-D indoor positioning, four LEDs installed in the ceiling of the room send localized ID information; different ID information is modulated using CDMA technology and is then received by the PD. When the received optical power is detected by the PD, the best positioning coordinate can be obtained by using the Ant Colony Optimization algorithm. The solution space is positioning coordinates which is in the room. By applying the route optimization of the ant colony algorithm, the search individuals can converge to the optimal solution of the three-dimensional localization problem in a short path in three-dimensional space, which can greatly improve the computational efficiency.

2.2.1 *Define the search space range and initialize the algorithm parameters*
Set the indoor space to a range of (X, Y, Z). After setting the range, the algorithm searches for the optimal anchor point within the specified space. In the ant colony algorithm, the initial input of the algorithm is the population size of ant colony M, the maximum number of iterations Max, the transition probability constant P, the pheromone volatilization factor ρ, and the algorithm termination conditions. Changing the algorithm initial input conditions will affect the algorithm convergence speed and positioning accuracy.

2.2.2 *Create pheromone concentration function*
From Equation 5 we learn that the distance between the LED and the receiver can be expressed as:

$$d = \sqrt{\frac{C}{H(0)} \cos^{m+1}(\theta)} \tag{7}$$

To relax the complexity of the system, suppose that the transmitting and receiving plane are parallel, so that $\theta = cos^{-1}\frac{h}{d}$, where h is the vertical distance between the LED and the receiver. Therefore, the distance between the LED and the receiver can be represented as:

$$d^{(i)} = \sqrt[4]{Ch^2 \frac{P_t}{P_r^{(i)}}} = \sqrt[4]{Ch^2 \frac{1}{H^{(i)}(0)}} \qquad (i = 1,2,3,4) \tag{8}$$

The coordinate of LED(i) is $X_{LED(i)} = (x^{(i)}, y^{(i)}, z_t), (i = 1,2,3,4)$. The coordinate of ant individual is $X_{ANT(n)} = (x_n, y_n, z_n), (1 \le n \le populationsize)$. The distance between each ant's individual location and the LED(i) can be expressed as:

$$d_r^{(i)} = \sqrt{\left[\left(x_n - x^{(i)} \right)^2 + \left(y_n - y^{(i)} \right)^2 + \left(z_n - z_t \right)^2 \right]} \tag{9}$$

Calculate the deviation of each ant's distance from the transmitter and the actual distance, and then use it as a pheromone concentration function. In this localization problem, if an ant gets closer to the best anchor point, the pheromone concentration should be lower. However, in the ant colony algorithm, ant individuals move in the direction of high pheromone concentration.

So we construct the pheromone concentration function:

$$\tau = -\sqrt{\sum_{i=1}^{4} \left(d_r^{(i)} - d^{(i)} \right)^2} \tag{10}$$

2.2.3 *Determine the search strategy*
There are two search strategies in the ant colony algorithm: global search and local search. In the problem of the optimal locating solution, the greater the pheromone concentration is, the smaller is the deviation from the optimal solution, indicating that the probability of the existence of the optimal solution in the vicinity of the ant is greater. Therefore, we should adopt the local search strategy. Otherwise, it is considered that there is a small probability of the optimal solution being near the place, and a global search strategy should be adopted.

Define the transition probability function of ants as:

$$P_i = \frac{\tau_i - \tau_{\min}}{\tau_i} \tag{11}$$

where τ_i is the pheromone concentration of the ith ant, $\tau_{\min}$ is the lowest pheromone concentration of all ants. The initial input of the algorithm is given the transition probability constant P. If $P_i \le P$, then the local search strategy should be adopted; on the contrary, if $P_i > P$, the global search strategy should be adopted.

In the local search, in order to improve the search efficiency, define the step factor for a single search:

$$\lambda = \frac{1}{T} \tag{12}$$

where T is the number of searches, and λ is the moving step in the local search of the ant. It can be seen that the moving step is relatively large at the beginning of the algorithm, which helps to enlarge the search range. As the number of searches increases, the moving step size becomes smaller, which helps improve convergence accuracy.

In the global search, just move randomly within the specified three-dimensional space.

2.2.4 *Calculate pheromone concentration values*

The pheromone concentration value should be recalculated after each search and it can be expressed as:

$$\tau_{i,k} = \rho\tau_{i,k-1} + \tau_{newplace} \tag{13}$$

The meaning of this function is that the pheromone concentration value of the old position is attenuated proportionally, with the addition of the pheromone concentration value of the new position, where k is the number of iterative calculations; i represents the ant's sequence number; ρ is the pheromone volatile constant, which is the initial input value of the algorithm.

2.2.5 *Select the elimination mechanism*

We use the roulette wheel selection to select individuals in the ant population. In this localization problem, the pheromone concentration function is used to measure the individual's distance from the optimal solution. When the value of pheromone concentration function is larger, the closer it is to the optimal solution, and the ant individual should be reserved with a higher probability. Conversely, if the pheromone concentration function is small, the ant individual should be reserved with a low probability. The probability that an individual is eliminated can be expressed as:

$$P_{e\,\text{limination}}^{(i)} = \frac{\tau_i}{\sum\limits_{x=1}^{M} \tau_x} \tag{14}$$

where τ_i is the pheromone concentration value at the location of the ant individual; M is the population size. In order to maintain the size of the ant population unchanged, ant individuals that are eliminated are replaced with the current best (pheromone concentration maximum) ant individuals.

2.2.6 *Set the algorithm termination conditions*

Generally speaking, two types of termination conditions need to be set. When the calculated value of pheromone concentration of a single individual is less than a given threshold, it means that the algorithm can be exited due to the calculation accuracy being reached. When the number of iterations exceeds the set threshold, indicating that convergence is too slow, the positioning algorithm should be restarted.

3 SIMULATION AND ANALYSIS

3.1 *Design of indoor 3-D positioning simulation model*

In order to test the performance of the Ant Colony Optimization algorithm in 3-D positioning, modeling and simulation of the established algorithm model are carried out based on MATLAB software. In this section, an indoor visible light positioning system with a 3 m × 3 m × 4 m environment is established, as shown in Figure 1. Each of these four LED bulbs will transmit localized ID information, and receive by the positioning terminal whose estimated position can be calculated by the Ant Colony Optimization algorithm. The main parameters of the model are shown in Table 1.

3.2 *Result and discussions*

In order to estimate the performance of the Ant Colony Optimization algorithm in 3-D indoor positioning, in this section a multipoint positioning test is adopted and the positioning is performed at different heights. The resolution of the test positions is 0.5 m, from 0.2 m to 2.0 m in the room, and 250 positions are included. In this test, positioning tests are performed at ten different heights of the plane. The height was set to 0.2 m, 0.4 m, 0.6 m, 0.8 m, 1 m, 1.2 m, 1.4 m, 1.6 m, 1.8 m, and 2 m, respectively. For space reasons, only the positioning results with a height of 0.2 m and 2.0 m are listed here, as shown in Figure 2. There are 25 real tested position points in every plane, where the symbol "Δ" represents the estimated position and the symbol "×"represents the real position.

We can tell from Figure 2 that the proposed positioning system performs very well in the whole room and the real position is very close to the position estimated by the Ant Colony Optimization algorithm. To assess the performance of a positioning system more accurately and directly, we conduct further quantitative analysis.

The result in Figure 3 shows that the effect of the Ant Colony Optimization algorithm for 3-D positioning is good. Moreover, analysis of the above data shows that we can know

Table 1. Parameters of the 3-D positioning system.

Parameter	Value
Room size (L × W × H)/m	3 m × 3 m × 4 m
Position of each LED (x,y,z)/m	LED1 (0,0,4)
	LED2 (3,0,4)
	LED3 (3,3,4)
	LED4 (0,3,4)
LED power $[P_t]$ /W	5
The ϕ_c of PD /deg	90
The effective area of PD $[A]$ / cm^2	1
The gain of optical filter $[T_s(\phi)]$	1
The gain of optical concentrator $[G(\phi)]$	1
The half-power angles of LED $[\theta_{1/2}]$ /deg	60
The order of Lambert's luminous intensity $[m]$	1
Population size $[m]$	150
Maximum number of iteration $[Max]$	200
Transition probability $[P]$	0.2
Pheromone volatilization factor $[\rho]$	0.8
Pheromone concentration threshold	−0.0001

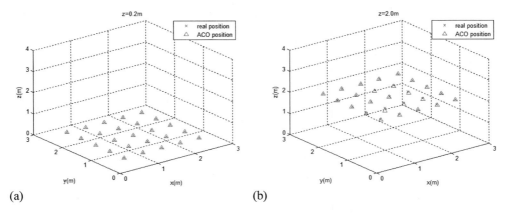

(a)　　　　　　　　　　　　　　　　　　(b)

Figure 2.　Distribution of the real position point and its ACO estimated position point, respectively: (a) The positioning result of 0.2 m; (b) The positioning result of 2 m.

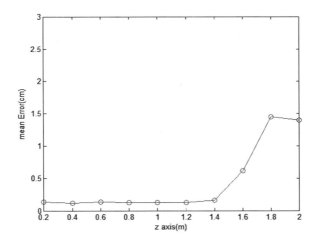

Figure 3.　Mean error of different height.

when the height is low, and that the errors of different heights are basically the same; when the height rises to a certain height, the error significantly increases with the increase of the height. It is because the transmitting angle of the LED light source (θ) enlarges with the increase of the height of the receiving plane, and when they increase to a certain extent, the luminous intensity of the LED is weakened, so that the attenuation of the optical signal received by the receiver increases. Therefore, the error of the optical power received by the receiver increases, which leads to the reduction of the positioning accuracy. Figure 3 is the line chart for the average error between each of 25 tested points and their respective estimated points at different heights. It can be seen that the error will increase significantly when the height is higher.

Figure 4(a) is the histogram of error between each of 250 tested positioning points and their respective estimated positioning points in a multipoint positioning test. The maximum error is 6.31 cm, the average error is 0.44 cm, and most of the positioning errors are below 1 cm. Figure 4(b) is the Cumulative Distribution Function (CDF) curves of the position errors. From Figure 4 we can see that 94.8% of the positioning errors are below 1.6 cm. The results show that the effect of the Ant Colony Optimization algorithm for 3-D positioning is good, and that the average error can reach mm level.

367

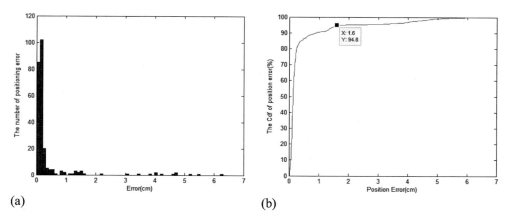

(a)   (b)

Figure 4.   Histogram of position errors and the Cumulative Distribution Function (CDF) curves of position errors.

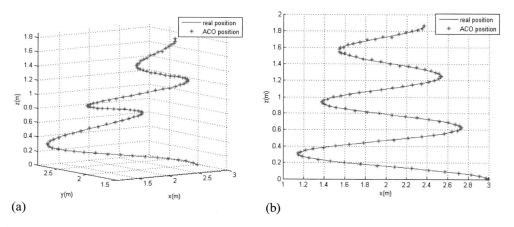

(a)   (b)

Figure 5.   3-D position in motion scene and the horizontal view of positioning result.

3.3   *Extended experiment and result analysis*

In order to evaluate the performance of the Ant Colony Optimization algorithm in motion scene positioning, an extended experiment of trajectory tracking is carried out. In our simulation, a random path is given by assuming a moving target in the room, and there are 96 samples in total.

Figure 5(a) shows that the algorithm performs well in the motion scene. In this Figure, the blue line is the path and the red spots "*" are the estimated positions that track the path. In order to better present the results, Figure 5(b) shows the horizontal view of the positioning results. The results show that with the proposed algorithm, the estimated results follow the path very well, as we can see.

To present the 3-D positioning result directly, the histogram of the positioning error and the CDF curves of the position error in the motion scene are shown in Figure 6. As is shown in Figure 6(a), most of the positioning errors in the motion scene are below 0.4 cm, and with the increase in the height, the positioning error is larger, which is consistent with the analysis in section 3.2. From Figure 6(b) we can see that 96.04% of the positioning errors in the motion scene are below 1.006 cm. It can be concluded from the above results that the Ant Colony Optimization algorithm performs very well in the motion positioning scene.

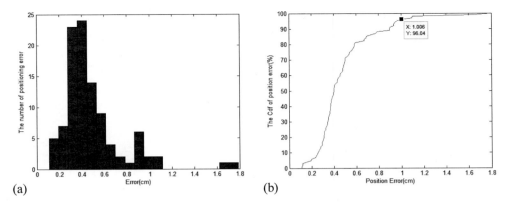

(a)   (b)

Figure 6. Histogram of position error in the motion scene and the Cumulative Distribution Function (CDF) curves of position error in the motion scene.

4 CONCLUSION

In this paper, we proposed a 3-D high-precision indoor positioning strategy using the Ant Colony Optimization algorithm. We transform the VLC positioning problem into a global optimization problem and use the aforementioned algorithm to solve it, for the reason that the Ant Colony Optimization algorithm is a powerful population-based heuristic bionic evolutionary algorithm that can be used to solve global optimization problems. It is the first time that the Ant Colony Optimization algorithm has been applied to visible light positioning, which is innovative. In this positioning strategy, each LED transmits a unique signal to the terminal and the algorithm estimates the position according to the messages received by the terminal. Our simulation result shows that the average error is about 0.44 cm. Most of the positioning errors are less than 1 cm. Among them, 94.8% of the positioning errors are below 1.6 cm. Then, in order to evaluate the performance of the Ant Colony Optimization algorithm in the motion scene positioning, an extended experiment of trajectory tracking is carried out. The data shows that the vast majority of the positioning errors are below 1.1 cm, and 96.04% of the positioning errors are below 1.006 cm. The experiment results verify that our mentioned strategy can achieve cm-level indoor positioning. Thus, the 3-D indoor positioning based on the Ant Colony Optimization algorithm has potential application value in various indoor positioning scenes.

REFERENCES

Alavi, B. & Pahlavan, K. (2006). Modeling of the TOA-based distance measurement error using UWB indoor radio measurements. *IEEE Communications Letters*, *10*(4), 275–277.

Dakkak, M., Nakib, A., Daachi, B., Siarry, P. & Lemoine, J. (2011). Indoor localization method based on RTT and AOA using coordinates clustering. *Computer Networks the International Journal of Computer & Telecommunications Networking*, *55*(8), 1794–1803.

Guan, W., Wu, Y., Wen, S., Chen, Y. & Chen, H. (2016). Indoor positioning technology of visible light communication based on CDMA modulation. *Acta Optica Sinica*, *36*(11), 1–9.

Guan, W., Wu, Y., Xie, C., Chen, H., Cai, Y. & Chen, Y. (2017b). High-precision approach to localization scheme of visible light communication based on artificial neural networks and modified genetic algorithms. *Optical Engineering, 56*(10), 106103.

Guan, W., Wu, Y., Wen, S., Chen, H., Yang, C., Chen, Y. & Zhang, Z. (2017a). A novel three-dimensional indoor positioning algorithm design based on visible light communication. *Optics Communications*, *392*, 282–293.

Lim, J. (2015). Ubiquitous 3D positioning systems by led-based visible light communications. *IEEE Wireless Communications*, *22*(2), 80–85.

Shi, A., Duan, J., Zou, Y. & Shi, A. (2015). Impact of multipath effects on theoretical accuracy of TOA-based indoor VLC positioning system. *Photonics Research*, 3(6), 296.

Wang, W.D. & Zhu, Q.X. (2008). RSS-based Monte Carlo localisation for mobile sensor networks. *IET Communications*, 2(5), 673–681.

Wann, C.D., Yeh, Y.J. & Hsueh, C.S. (2006). Hybrid TDOA/AOA indoor positioning and tracking using extended Kalman filters. *Vehicular Technology Conference. VTC 2006-Spring. IEEE* (pp. 1058–1062). IEEE.

Wen, S., Guan, W., Wu, Y., Yang, C., Chen, H., Zhang, Z. & Chen, Y. (2016). High precision three-dimensional iterative indoor localization algorithm using code division multiple access modulation based on visible light communication. *Optical Engineering*, 55(10), 106105.

Won, Y.Y., Yang, S.H., Kim, D.H. & Han, S.K. (2013). Three-dimensional optical wireless indoor positioning system using location code map based on power distribution of visible light emitting diode. *IET Optoelectronics*, 7(3), 77–83.

Yang, S.H., Kim, D.R., Kim, H.S., Son, Y.H. & Han, S.K. (2012). Indoor positioning system based on visible light using location code. *International Conference on Communications & Electronics* (pp. 360–363). IEEE.

Yang, S.H., Kim, H.S., Son, Y.H. & Han, S.K. (2014). Three-dimensional visible light indoor localization using AOA and RSS with multiple optical receivers. *Lightwave Technology*, 32(14), 2480–2485.

Yasir, M., Ho, S.W. & Vellambi, B.N. (2014). Indoor positioning system using visible light and accelerometer. *Journal of Lightwave Technology*, 32(19), 3306–3316.

Zhang, W. & Kavehrad, M. (2012). A 2-D indoor localization system based on visible light LED. *Photonics Society Summer Topical Meeting Series* (pp. 80–81). IEEE.

Frontier Research and Innovation in Optoelectronics Technology and Industry – Habib & Lewis (Eds)
© *2019 Taylor & Francis Group, London, ISBN 978-1-138-33178-5*

Free-space-based multiple-access frequency transfer based on optical frequency comb

Danian Zhang, Jiyuan Chen & Yimei Li
Time and Frequency Research Center, School of Automation Engineering, University of Electronic Science and Technology of China, Chengdu, Sichuan, China

Dong Hou
Time and Frequency Research Center, School of Automation Engineering, University of Electronic Science and Technology of China, Chengdu, Sichuan, China
School of Electronics Engineering and Computer Science, Peking University, Beijing, China

ABSTRACT: We demonstrate a free-space-based multiple-access frequency transfer with an optical frequency comb by using a passive phase conjunction correction technique. Timing fluctuations and Allan deviations are both measured to characterize the excess frequency instability incurred during the frequency transfer process. By reproducing a 2 GHz radiofrequency signal at a middle point over a 60-m long free-space link in 5000s, the total Root-Mean-Square (RMS) timing fluctuation was measured to be about 224 fs with a fractional frequency instability yon the order of 8×10^{-14} at 1 s and 1×10^{-16} at 1000s. This free-space-based multiple-access frequency transfer with passive phase conjunction correction can be used to disseminate a stable frequency signal at arbitrary point in a free-space link.

1 INTRODUCTION

Transfer of timing and frequency signal are important to precision scientific and engineering applications such as metrology, optical-microwave frequency standards, optical communication, radar, and navigation (Levine, J. 2008; Li, B. *et al.* 2006; Wang, W. Q. *et al.* 2008). Recently, optical free-space links have become an attractive option for dissemination of timing and frequency signals as it can provide higher flexibility than fiber links (Foreman, S. M. *et al.* 2007). In the past decade, there are some important works to demonstrate the transfer of frequency signals via optical free-space links (Sprenger, B. *et al.* 2009; Gollapalli, R. P. and Duan, L. 2010; Giorgetta. F. R. *et al.* 2013; Kang, J. *et al.* 2014; Chen, S. *et al.* 2017; Sun, F. *et al.* 2017). In some of these frequency transfer experiments, atmospheric transfers of radio-frequency (RF), optical frequency, and optical frequency comb (OFC) signals with ultra-low timing deviation have been achieved, where kinds of phase compensation schemes ranging from two-way time-frequency transfer (TWTFT) to active optical/microwave phase delay line, were proposed to suppress the turbulence-affected timing fluctuation over free-space optical link.

In these before mentioned works of free-space frequency transfer, the transferred frequency signal with active phase compensation have high stability. However, we can only recover the disseminated frequency signal between specific sites using the currently-used free-space frequency transfer schemes, while with fiber link, we can reproduce the signals at arbitrary point in the fiber link. In the past few years, there are some prior experiments to demonstrate the multiple-access frequency disseminations via fiber links, in which ultra-low noise extractions of RF signal (Gao, C. *et al.* 2012; Li, H. *et al.* 2016), optical frequency signal (Bai, Y. *et al.* 2013; Bercy, A. *et al.* 2014), time signal (Sliwczynski, L. and Krehlik, P. 2015; Yuan, Y. *et al.* 2017), and OFC (Zhang, S. *et al.* 2017) at arbitrary point of fiber links have been achieved. This kind of multiple-access frequency transfer can provide a flexibility of reproduction of ultra-sable timing-frequency signal at arbitrary section of a fiber link without design of any independent signal dissemination, while for free-space link, it is still a problem that multiple-access frequency transfer cannot be realized with currently-used atmospheric frequency transfer techniques.

In this paper, we demonstrate a free-space-based multiple-access frequency transfer with a frequency comb using passive phase conjunction correction. In our experiment, we first deployed a 60-m long free-space OFC transmission link with our before-used frequency transfer setup (Sun, F. *et al.* 2017), and then reproduced a stable radio-frequency signal at the middle point in the entire free-space pathway, where the middle point was chosen arbitrarily. The experiment result shows that the root-mean-square (RMS) timing fluctuation of the reproduced 2 GHz microwave signal at the middle point of the 60 m free-space link was ~224 fs within 5000s in a normal environment, and the relative fractional frequency instability of the transmission link in a normal outdoor environment is on the order of 8×10^{-14} at 1 s and 1×10^{-16} at 1000s.

2 EXPERIMENTAL SETUP FOR FREE-SPACE-BASED MULTIPLE-ACCESS FREQUENCY TRANSFER

2.1 *Schematic of the free-space-based multiple-access frequency transfer*

In a multiple-access transfer of timing and frequency signal with an ultra-low timing fluctuation at a point in the free-space link, the biggest problem is that the air turbulence introduces excess phase noise or timing jitter into the reproduced frequency signal at an arbitrary point in the frequency transmission link (Sinclair, L. C. *et al.* 2014; Robert, C. *et al.* 2016), which indicates the stability of the direct reproduced frequency signal at any point of the free-space link is deteriorated. Therefore, to reduce the stability deterioration in a multiple-access atmospheric frequency transfer, the timing fluctuation affected by turbulence should be suppressed. In the before presented multiple-access frequency transfer via fiber link, a phase correction technique were proposed with bidirectional phase compensation, in which the timing fluctuations of the reproduced signals at arbitrary point of fiber links have been suppressed. Therefore, based on the phase correction idea, we proposed a free-space-based multiple-access frequency transfer scheme with a fem to second OFC using a passive phase correction technique.

Figure 1 shows the experimental setup of our free-space-based multiple-access frequency transfer with an OFC. In our previous comb-based frequency transfer experiments, an Er: fiber mode-locked laser (MLL) was used as the OFC source, and disseminated from the transmitter to the receiver with a passive phase conjunction correction technique (Sun, F. *et al.* 2017). Here, the comb-based frequency transmission link is rebuilt, where the frequency signal is transferred

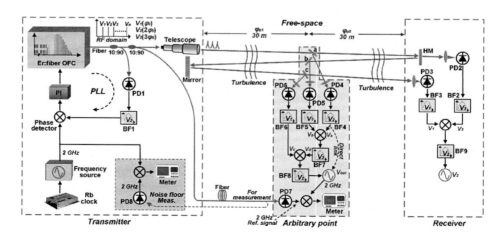

Figure 1. Experimental setup of free-space-based multiple-access frequency transfer with OFC by using passive phase conjunction correction. PD: photodiode, HM: half-reflecting mirror, PI: proportional-integral controller, BPF: band-pass filter. $V_1 \dots V_n$, are harmonics of the OFC.

from transmitter and receiver with the phase conjunction correction. In this frequency transmission link, an Er: fiber MLL with repetition frequency of 100 MHz and center wavelength of 1550 nm, is tightly locked to an Rb clock referenced RF source (Agilent, E4421B) with a phase-locked loop (PLL) at 2 GHz. A laser pulse beam with optical power of 60 mW generated from the laser is directly launched into free-space. With the help of a half-reflecting mirror (HM) on the receiver and a mirror on the transmitter, the laser light travels three times over the optical free-space link between transmitter and receiver (see Fig. 1). On the receiver, part of the laser beam which travels the free-space link once and the other part of the beam which travels the free-space link three times are both detected by two high-speed photodiodes (PD2 and PD3) respectively. By mixing the third harmonic of the microwave detected by PD1 and fundamental signal of the microwave detected by PD2, a clean and stable microwave signal at the lower sideband is achieved because of the natural elimination of timing fluctuation with the down-converting function of the mixer. This process realizes a stable frequency transfer from transmitter to receiver, which has been demonstrated in our previous report (Sun, F. *et al.* 2017). After this, to achieve a multiple-access frequency transfer in this paper, we arbitrarily chose a point of the optical free-space link to reproduce the comb frequency signal. In the arbitrary section of the rebuilt free-space frequency transmission link, we coupled three beams on the laser travel path at three nodes **a**, **b** and **c** via three partially reflecting mirrors (10:90). Three high-speed photodiodes (PD4, PD5, and PD6) and three band-pass filters (BF4, BF5, and BF6) were used to extract the three microwave signals V_a, V_b, and V_c, respectively. Here, V_a is the extracted second harmonic signal of MLL, and V_b and V_c are the extracted fundamental frequency signals of MLL. After the extraction, V_a is mixed with V_b via a RF mixer, and its higher sideband is extracted via a band-pass filter (BF7) to produce a new intermediate signal V_d. Next, by mixing the new intermediate signal V_d and V_c, and extracting its lower sideband signal, a clean and stable microwave signal at the lower sideband is achieved because of the natural elimination of timing fluctuation with the up-converting and down-converting function of the mixers. Note that, for the recovered microwave signal at this arbitrary point in the free-space link, the timing fluctuation affected by turbulence is eliminated naturally by the passive phase conjunction correction scheme. The mechanism of the elimination of the timing fluctuation in the multiple-access frequency transfer will be explained in detail below.

2.2 *Principle and mechanism of timing fluctuation suppression*

As shown in Fig. 1, we assume the frequency-stabilized OFC has a fundamental frequency V_1, second harmonic frequency V_2, and third harmonic frequency V_3 in RF domain; the initial phase of the fundamental frequency signal is φ_0. Accordingly, the initial phases of the second harmonic are $2\varphi_0$. We also assume the air turbulence introduces a phase fluctuation φ_p to the transmitted signal between the transmitter and receiver, a phase fluctuation φ_{p1} between the transmitter and arbitrary point, and a phase fluctuation φ_{p2} between the receiver and arbitrary point, at the fundamental frequency over a one-trip free-space link. Naturally we have $\varphi_p = \varphi_{p1} + \varphi_{p2}$. In this case, on the arbitrary point, the phase delay of the second harmonic signal V_a coupled from *a* is given by $\varphi_a = 2\varphi_0 + 2\varphi_{p1}$, the phase delay of fundamental frequency signal V_b coupled from *b* is given by $\varphi_b = \varphi_0 + \varphi_p + \varphi_{p2}$, and the phase delay of fundamental frequency signal V_c coupled from *c* is given by $\varphi_c = \varphi_0 + 2\varphi_p + \varphi_{p1}$, respectively. With mixing V_a and V_b, and extracting its higher sideband, a new intermediate signal V_d is produced, and its phase delay, therefore, is given by $\varphi_d = \varphi_a + \varphi_b = 3\varphi_0 + 2\varphi_p + \varphi_{p1}$. Next, by mixing V_d and V_c, and extracting its lower sideband, a final microwave signal V_{out} is produced, and its phase delay, therefore, is given by $\varphi_{out} = \varphi_d - \varphi_c = 2\varphi_0$. Note that, for the recovered microwave signal V_{out} at this arbitrary point, the phase fluctuation affected by turbulence is eliminated naturally by the passive phase conjunction correction scheme. To verify this multiple-access frequency transfer technique with passive phase conjunction correction, an actual OFC-based multiple-access frequency transfer experiment has been conducted.

2.3 *Experimental setup*

Our frequency transmission link was located at the top floor of the engineering building of our university. The distance between the transmitter and receiver was 60 meters, and the middle point of the transmission link was arbitrarily chosen as the multiple-access frequency transfer point, in which a stable RF frequency has been extracted. In this multiple-access frequency transfer experiment, we chose 1 GHz as the fundamental frequency V_1 (10th harmonic of the OFC), and accordingly the frequency of V_2 is 2 GHz (20th harmonic of the OFC). With the passive phase correction technique described above, a clean and stable 2 GHz RF signal should be extracted on the arbitrary point. To estimate the timing fluctuation and stability of the resulting 2 GHz RF signal, we mixed it with a frequency reference signal which was coupled from the transmitter via a 30-m long fiber link, to produce a DC error output and sent it to a digital voltage meter (Keysight, 34461 A), for data recording and analysis. To compare the qualities of the transmitted frequency signals with and without timing fluctuation suppression, a direct link was also designed (Fig. 1). In this direct link setup, a 2 GHz microwave was extracted from PD4 and BF4, and directly mixed with the reference signal.

3 EXPERIMENTAL RESULTS AND DISCUSSION

Our multiple-access frequency transfer experiment was conducted in a normal night. We believe the quality of the frequency extracted at the arbitrary point should be improved significantly when compared to the uncorrected direct link. Because the passive phase conjunction correction can suppress the additional timing fluctuations caused by turbulence, Here, we measured the timing drifts and relative frequency stabilities of the reproduced OFC signals with and without phase conjunction correction. The timing drift results are shown in Fig. 2. Here, Curve (i) and (ii) are the measured timing drift of the transferred 2 GHz RF signal without and with timing fluctuation suppression. It can be calculated that the RMS timing fluctuation without and with timing fluctuation suppression are 1.3 ps and 224 fs within 5000s, respectively. In addition, the 30-m short fiber could introduce a few extra phase noise. Therefore, we tested the timing fluctuation of the reference signal coupled from the fiber, by comparing it to the RF reference on the transmitter (see Fig. 1). The measured residual timing drift can be treated as the measurement floor of our multiple-access frequency transfer experiment. Curve (iii) shows the residual timing drift of the reference signal coupled form the short fiber, and its RMS timing drift is approximately 45 fs within 5000s.

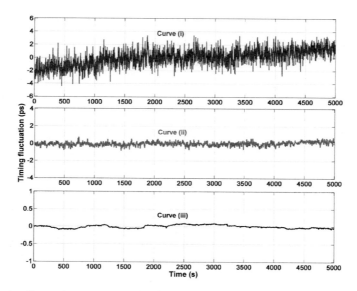

Figure 2. Timing fluctuation results for free-space-based multiple-access frequency transfer. Curve (i): Without timing fluctuation suppression. Curve (ii): With timing fluctuation suppression. Curve (iii): The result for the 30-m long short fiber link as a measurement floor. Sample rate is 1 point/second for all curves.

374

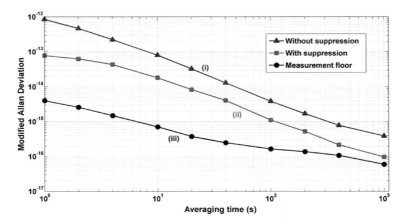

Figure 3. Instability results for atmospheric frequency transfer, (i) Relative Allan deviation between transferred microwave and reference signal without timing fluctuation suppression; (ii) Relative Allan deviation with timing fluctuation suppression; (iii) Allan deviation for a short link as measurement floor.

To evaluate the stability of the multiple-access frequency transfer, we calculated the Allan deviation of the transferred 2 GHz frequency signal at the arbitrary point. Figure 3 demonstrates the instability results of the transferred RF signal. Curve (i) and (ii) shown in Fig. 3 are the relative Allan Deviations without and with timing fluctuation suppression. These curves show the instability of the multiple-access frequency transmission link without timing fluctuation suppression is 8×10^{-13} at 1 s and 4×10^{-16} at 1000s, and the instability of the transmission link with timing fluctuation suppression is 8×10^{-14} at 1 s and 1×10^{-16} at 1000s, respectively. We also demonstrates the instability measurement floor as Curve (iii) that is obtained directly from the short fiber link. Note that, curve (iii) is merely the lower bound of instability incurred during the multiple-access frequency transfer experiment. This is because it was measured with the short fiber link, and most of the turbulence and vibration effects were cancelled out. This Allan deviation measurement floor is limited in our case by the stability of the frequency source and the noise introduced by the fiber link. By the comparison of curve (i) and curve (ii), we find that the instability of the transferred signal at the arbitrary point with phase correction is reduced by more than half order of magnitude at 1000s due to the timing fluctuation suppression. This proves that the phase conjunction correction can effectively eliminate the residual timing fluctuation of the multiple-access frequency transmission link. When comparing the instability of our transfer result and that of a commercial Cs clock (Microsemi, "DS-5071a", 2014) or H-master clock (Symmetricom, "MHM-2010", 2011), we find that the instability of the multiple-access frequency transmission link is lower than those of the clocks. Therefore, we believe that disseminating a Cs or H-maser clock signal at arbitrary point over a short free-space link is feasible with the atmospheric comb-based multiple-access frequency transfer technique proposed in this paper.

4 CONCLUSIONS

We have demonstrated a free-space-based multiple-access frequency transfer with OFC by using passive phase conjunction correction. The RMS timing fluctuation for a 2 GHz frequency signal extracted at middle point over a 60-m long free-space link was measured to be approximately 224 fs within 5000s with a fractional frequency instability on the order of 8×10^{-14} at 1 s and of 1×10^{-16} at 1000s. The achieved instability demonstrates that the proposed multiple-access frequency transfer setup with passive phase conjunction correction promises a high flexibility of timing and frequency transfer at arbitrary point of a free-space transmission link. For instance, the proposed setup can be used to transfer a Cs or H-master clock signal at arbitrary point over a free-space transmission link. In the future, we will attempt to build a free-space-based multiple-access frequency transmission link with a lower short-time timing fluctuation and a longer distance by using a higher power OFC and higher frequency harmonic.

ACKNOWLEDGMENT

This work was supported by the National Natural Science Foundation of China (No. 61601084), and the State Key Lab. of Advanced Optical Communication Systems and Networks, China.

REFERENCES

Bai, Y., Wang, B., Zhu, X., Gao, C., Miao, J., and Wang, L. 2013. Fiber-based multiple-access optical frequency dissemination. *Opt. Lett.* 38(17), 3333–3335.

Bercy, A., Guellati-Khelifa, S., Stefani, F., Santarelli, G., Chardonnet, C., Pottie, P., Lopez, O., and Amy-Klein, A. 2014. In-line extraction of an ultrastable frequency signal over an optical fiber link. *J. Opt. Soc. Am.* B 31(4), 678–685.

Chen, S., Sun, F., Bai, Q., Chen, D., Chen, Q., and Hou, D. 2017. Sub-picosecond timing fluctuation suppression in laser-based atmospheric transfer of microwave signal using electronic phase compensation. *Opt. Commun.* 401(15), 18–22.

Foreman, S.M., Holman, K.W., Hudson, D.D., Jones, D.J., and Ye, J. 2007. Remote transfer of ultrastable frequency references via fiber networks. *Rev. Sci. Inst.* 78(2): 021101.

Gao, C., Wang, B., Chen, W., Bai, Y., Miao, J., Zhu, X., Li, T., and Wang, L. 2012. Fiber-based multiple-access ultrastable frequency dissemination. *Opt. Lett.* 37(22), 4690–5692.

Giorgetta. F.R., Swann, W.C., Sinclair, L.C., Baumann, E., Coddington, I., and Newbury, N.R. 2013. Optical two-way time and frequency transfer over free space. *Nat. Photonics* 7(6), 435–439.

Gollapalli, R.P. and Duan, L. 2010. Atmospheric timing transfer using a femtosecond frequency comb, *IEEE Photon. J.* 2(6): 904–910.

Kang, J., Shin, J., Kim, C., Jung, K., Park, S., and Kim, J. 2014. Few-femtosecond-resolution characterization and suppression of excess timing jitter and drift in indoor atmospheric frequency comb transfer. *Opt. Express* 22(21), 26023–26031.

Levine, J. 2008. A review of time and frequency transfer methods. *Metrologia* 45(6): 162–174.

Li, B.H., Rizos, C., Lee, H.K., and Lee, H.K. 2006. A GPS-slaved time synchronization system for hybrid navigation. *GPS Solut.* 10(3): 207–217.

Li, H., Wu, G., Zhang, J., Shen, J., and Chen, J. 2016. Multi-access fiber-optic radio frequency transfer with passive phase noise compensation. *Opt. Lett.* 41(24), 5672–5675.

Microsemi, "DS-5071a", http://www.microsemi.com/products/timing-synchronizationsystems/time-frequency-references/ cesium-frequency-standards/5071a, (2014).

Robert, C., Conan, J.M., and Wolf, P. 2016.Impact of turbulence on high-precision ground-satellite frequency transfer with two-way coherent optical links. *Phys. Rev. A.* 93(3), 033860.

Sinclair, L.C., Giorgetta, F.R., Swann, W.C., Baumann, E., Coddington, I., and Newbury, N.R. 2014. Optical phase noise from atmospheric fluctuations and its impact on optical time-frequency transfer. *Phys. Rev. A.* 89(2), 023805.

Sliwczynski, L., and Krehlik, P. 2015. Multipoint Joint Time and Frequency Dissemination in Delay-Stabilized Fiber Optic Links. *IEEE Trans. Ultrason. Ferroelectr. Freq. Control* 62(3), 412–420.

Sprenger, B., Zhang, J., Lu, Z., and Wang, L. 2009. Atmospheric transfer of optical and radio frequency clock signals. *Opt. Lett.* 34(7): 965–967.

Sun, F., Hou, D., Zhang, D., Tian, J., Hu, J., Huang, X., and Chen, S. 2017. Femtosecond-level timing fluctuation suppression in atmospheric frequency transfer with passive phase conjection correction. *Opt. Express* 25(18), 21312–21320.

Symmetricom, "MHM-2010", http://www.symmetricom.com/products/frequency-references/active-hydrogen-maser/MHM-2010/, (2011).

Wang, W.Q., Ding, C.B., and Liang, X.D. 2008. Time and phase synchronisation via direct-path signal for bistatic synthetic aperture radar systems. *IET Radar Sonar.* Nav. 2(1): 1–11.

Yuan, Y., Wang, B., Gao, C., and Wang, L. 2017. Fiber-based multiple access timing signal synchronization technique. *Chin. Phys.* B26(4), 040601.

Zhang, S. and Zhao, J. 2015. Frequency comb-based multiple-access ultrastable frequency dissemination with $7 \times 10(-17)$ instability. *Opt. Lett.* 40(1), 37–40.

A cat's eye modulating retroreflector based on focal plane misalignment

L.X. Zhang, H.Y. Sun, J.Y. Ren & T.Q. Zhang
Space Engineering University, Beijing, P.R. China

ABSTRACT: Free space optical communication based on modulating retroreflector uses a modulating retroreflector terminal instead of a traditional communication terminal to build a link. Compared with classical free space optical communication systems, a system based on a cat's eye modulating retroreflector has great advantages. It can build communication links more rapidly, and its passive terminal is smaller and lighter with a lower power consumption. In this paper, a new cat's eye modulating retroreflector based on focal plane misalignment is developed. When the focal plane of a cat's eye lens is misaligned/aligned, the reflection direction of the beam will change. The change will cause the intensity of retro-reflected beam change to achieve intensity modulation. An experimental link was built using a cat's eye modulating retroreflector which uses a Micro-Electro-Mechanical Systems (MEMS) array with high reflection film placed on the focal plane. The link range is about 1 km, with the highest rate of 10 kbps. The results show that the cat's eye modulating retroreflector based on focal plane misalignment is available. It could realize amplitude modulation, and the data rate is decided by the switching rate of focal plane.

1   INTRODUCTION

A Modulating Retroreflector (MRR) couples a passive optical retroreflector with electro-optic shutters to allow free space optical communication with a laser and an Acquisition/Tracking/Pointing (ATP) system required on only one end of the link. In operation, the active end emits a Continuous Wave (CW) laser beam to illuminate the MRR on the other end. The MRR imposes a modulation on the interrogating beam and passively retroreflects the modulated beam back to the active end, exactly in the incidence direction. For this type of system reduced a laser and an ATP system, it is attractive for asymmetric communication links for which there are strict limits of volume, weight and power consuming on one end of the link. The MRRs demonstrated to date have used a large modulator placed in front of the aperture, or as one of the faces of a corner cube retroreflector (W. S. 2006; Johan et al., 2007), and a relatively smaller modulator placed on the focal plane of a cat's eye modulating retroreflector (W. S., 2007; Peter et al., 2012). The cat's eye modulating retroreflector is based on the cat's eye effect, which is affected by the focal plane misalignment. Some work on the effect of the influence of focal plane misalignment on the cat-eye effect has been done, including using numerical simulation (Zhang et al., 2009; Zhao et al., 2010) and experimental simulation. As different misalignment of focal plane can cause a power retroreflected change, a cat's-eye modulating retroreflector can be created using a Micro-Electromechanical Systems (MEMS) mirror which is placed at the focal plane of a cat's eye lens. Compared with other CEMRR using a spatial light modulator such as multiple quantum well, liquid crystal, EOM and so on, focal plane misalignment devices could use MEMS, Giant Magnetostrictive Material (GMM) and other cheap material, which will reduce the cost of CEMRR greatly. What is more, by coating different high reflecting film on MEMS or GMM placed on the focal plane, CEMRR based on focal plane misalignment can easily meet different communication link wavelength needs, which is more convenient.

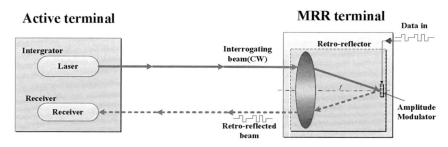

Figure 1. Free space communication link based on a cat's-eye modulating retroreflector.

In this paper, we use the theory of geometric optics to analyze the influence of focal plane misalignment to cat-eye effect, and we present analytical solution of influence of focal plane misalignment to active area and retroreflected light intensity. We use a MEMS mirror as focal plane to build a modulating retroreflect to build a communication link, and we prove the possibility of Cat's Eye Modulating Retroreflector (CEMRR) based on focal plane misalignment use the link.

2 FREE SPACE OPTICAL COMMUNICATION BASED ON A CAT'S EYE MODULATING RETROREFLECTOR

Figure 1 shows a typical link of free space optical communication based on CEMRR. The link includes two terminals, which are an active terminal and the MRR terminal. The active terminal is a classical optical communication terminal, which controls the interrogator to emit a CW laser beam to interrogate the MRR terminal. In the MRR terminal, the cat's eye retroreflector focuses the interrogating laser beam onto the modulator placed in the focal plane, which is driven according to the input data to impose a modulation on the amplitude of the interrogating beam. Then, the cat's eye retroreflector reflects the modulated beam exactly to the active terminal. The active terminal controls the receiver to receive the retroreflected beam and demodulates the signals to close the link. As a laser communication link, the link depends on laser power, beam divergence, pointing accuracy, link lose, receiver sensitivity and so on. However, for a MRR link, the key point of the link is the cat's eye retroreflector, which determines whether the retroreflected beam could be reflected to the active terminal and whether the reflected laser intensity could be modulated.

3 INFLUENCE OF FOCAL PLANE MISALIGNMENT ON THE CAT'S EYE EFFECT

The key part of a cat's-eye modulating retroreflector free space optical communication system is the CEMRR, which is mainly made up of a cat's eye lens and a spatial light modulator placed on the focal plane. For the link range is far enough, when the laser beam reaches the cat's eye the laser wavefront can be approximated as plane wave. A perfect cat's eye lens can be presented as Figure 2(a), which is a 4f system, where the radius is r and the focal length is f. When the incidence beam reaches the cat's eye lens, the lens focuses the beam onto the focal plane. When the focal plane is aligned, the focused beam will get through a certain area prescribed as the effective aperture (shown as Figure 2(b)) of the lens, again according to the incidence angle of the beam with the direction exactly opposite the incidence angle. For the beam in cat's eye lens keeps straight and the beam is seen as plane wave, the retroreflected beam intensity is proportional to effective aperture.

However, when the focal plane is misaligned, the cat's eye lens model and the effective aperture are equivalent, as shown Figure 3, according to geometric optics. When the

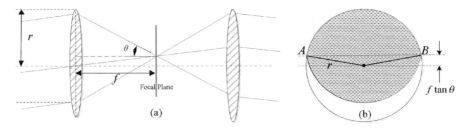

Figure 2.   Optics model (a) and effective aperture with incidence angle θ (b) of a perfect cat's eye lens.

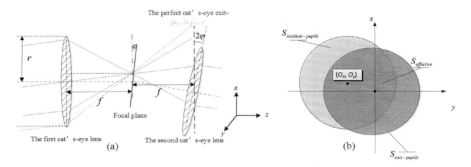

Figure 3.   Optics model (a) and the effective aperture (b) of a misaligned focal plane cat's eye lens.

misalignment is φ, the exit-pupils' normal direction will change compared to the perfect cat's eye lens, which will change the effective aperture of the cat's eye greatly. Suppose misalignment occurs on the XY plane, incidence angle of beam is (θ_x, θ_y), according to the geometric optics, the retroreflected beam angle (θ'_x, θ'_y) can be presented as Equation 1.

$$\begin{cases} \theta'_x = \theta_x + 2\varphi \\ \theta'_y = \theta_y \end{cases} \tag{1}$$

(O_x, O_y) is the center of the incident-pupils, which is presented as Equation 2:

$$\begin{cases} O_x = f\tan\theta'_x + f\tan\theta'_y \\ O_y = 2f\tan\theta'_y \end{cases} \tag{2}$$

According to geometric optics, the incident-pupil of the receiving beam with incidence angle (θ_x, θ_y) of a cat's eye lens with focal plane misalignment φ can be presented as Equation 3:

$$S_{incident-pupils} = \begin{cases} 1 & (x-O_x)^2 + (y-O_y)^2 \le r^2 \\ 0 & other \end{cases} \tag{3}$$

The exit-pupils' area is the projection of the second lens of the cat's-eye lens to the perfect cat's eye exit-pupils' plane, which can be presented as Equation 4:

$$S_{exit-pupils} = \begin{cases} 1 & x^2 + \left(\dfrac{y}{\cos(2\varphi)}\right)^2 \le r^2 \\ 0 & other \end{cases} \tag{4}$$

379

Table 1. Relationship between the misalignment and the retroreflected signal.

Misalignment	Retroreflected beam power	Retroreflected signal
Misaligned	Weak	0
Aligned	Strong	1

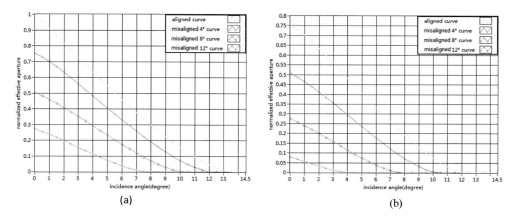

Figure 4. Normalized effective aperture with different misaligned angle vs. incidence angle (0, θ_y) (a) and incidence angle (6°, θ_y) (b).

The effective aperture of a focal plane misaligned cat's eye is the cross area of the incident-pupils and the exit-pupils, so it is presented as Equation 5:

$$S_{effective} = \begin{cases} 1 & x^2 + \left(\dfrac{y}{\cos(2\varphi)} \right)^2 \le r^2, (x - O_x)^2 + (y - O_y)^2 \le r^2 \\ 0 & other \end{cases} \tag{5}$$

When a MEMS mirror is placed at the focal plane of cat's eye lens with its normal axis coincide with optic axis, the MEMS deflects a certain angle called focal plane misalignment according to the data to be transferred, which can cause the reflected laser change direction, and finally equivalent to the effective aperture of the lens changes, which will make the retroreflected beam power change to present different signal. The relationship between the misalignment and the retroreflected signal is shown in Table 1.

According to the relationship between the misaligned and the retroreflected signal, using the MEMS mirror's different statement to present the code '0' and '1', we can modulate the signal to be transferred with On-Off Key (OOK) modulation, amplitude modulation and differential amplitude modulation.

4 NUMERICAL SIMULATION

We use LabVIEW in the Windows operation system to simulate the influence of focal plane misalignment on a cat's eye modulating retroreflector. Suppose the effective view of the angle of the cat's eye lens is 30°, the maximum focal plane misalignment is 12°, the radius of the cat's eye lens is one unit, when incidence angle is (0, θ_y) and (6°, θ_y), we get a normalized effective aperture of the cat's eye modulating retroreflector with the focal plane aligned, misaligned 4°, 8° and 12° as shown in Figure 4(a) and Figure 4(b).

According to Figure 4(a) and Figure 4(b), the effective aperture of the cat's eye varies greatly with focal plane misalignment at different incidence angles, which means that when a cat's eye lens with a focal plane status changed from aligned to misaligned, active terminals with different incidence angles could all receive a variety of retroreflected signals. When the misalignment is greater than 8°, the effective area could reduce to 50%, and when the misalignment is greater than 12°, the effective area could reduce to less than 30%.

When define contrast ratio as retroreflected beam power at aligned status to the power at misaligned status, the simulation results means that the contrast ratio of CEMRR based on focal plane misalignment is greater than two when misalignment is bigger than 8°, and greater than three when misalignment is bigger than 12°, which is big enough for modulating.

5 EXPERIMENT

In order to prove the validity of a CEMRR based on focal plane misalignment, we built an experimental system using a cat's eye retroreflector with a diameter of 20 mm, effective view of angle of 30°, effective focal plane of 8 mm × 8 mm. A DLP9500 Digital Micromirror Device (DMD), built by MEMS and produced by Texas Instruments, was placed on the cat's-eye retroreflector focal plane. The DLP9500supports a maximum switching rate 1bit B/W of 17857 Hz, and its micromirror array is 1920 × 1080, micromirror pitch is 10.8 μm, and active area is 20.7 × 11.7 mm², which can cover the whole effective focal plane area. The DMD has three states of +12°, −12° and flat state, and we controlled its +12° normal axis coincidence with a cat's eye lens optical axis which is shown in Figure 5. Therefore, the retroreflected power would reach the maximum at +12°and reach the minimum at −12°. The cat's-eye modulating retroreflector using a MEMS DMD is shown in Figure 6. The relationship between the retroreflected signal power and DMD state is shown in Table 2.

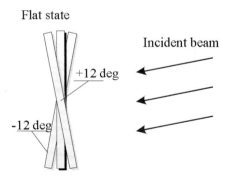

Figure 5.   DMD status.

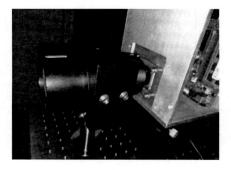

Figure 6.   Cat's-eye modulating retroreflector using DMD as a focal plane.

Table 2.   Relationship between retroreflected signal power and DMD state.

DMD state	Retroreflected beam power
+12°	Weak
−12°	Strong

Figure 7.   Active terminal of CEMRR FSO.

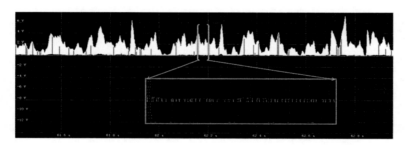

Figure 8.   Classical signal of active terminal received of 10 Kbps.

Using the CEMRR above, together with an active terminal, we built a CEMRR FSO link covering a range of 950 m in December 2017. The active terminal is shown in Figure 7, which consists of a 1550 nm CW laser with a maximum power of 5 W, emitting an angle of beam of 50 urad, a receiving lens with a diameter of 60 mm and a detector with a minimum detective power of 10^{-7} W. When the visibility was 8 Km, and the atmospheric condition was light turbulence, we got a classical retroreflected random signal with data rate of 10 Kbps as shown in Figure 8.

From the results above, we could get a good communication signal using CEMRR based on focal plane misalignment, and the signal was of high SNR, which made the processing and demodulation easily realized.

We stored the received signal using a DAQ board, used an adaptive threshold to process and demodulate the signal and achieved a correct communication signal as shown in Figure 9. The original signal received is shown above, and the processed and demodulated signal is shown below. From the results we could see that when the focal plane was aligned, the retroreflected beam had very strong intensity, while when misaligned, the retroreflected beam was almost equal to zero, which was of great advantage to transmitting a digital signal.

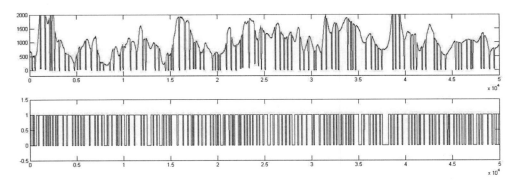

Figure 9.   Original signal received (above) and signal processed and demodulated (below).

6   CONCLUSION

This paper proved the possibility of CEMRR based on focal plane misalignment, which will be cheaper, more convenient and flexible compared with other CEMRR. Based on focal plane misalignment to build a cat's eye modulating retroreflector, the retroreflected beam intensity will change varying with the misalignment of the focal plane to the cat's eye lens optics axis, which will realize the amplitude modulation of the CEMRR free space optical communication link, and the link speed is limited by the focal plane switching rate. Using faster material, a higher speed free space optical link could be built, and the larger view of angle the cat's eye optics is, the larger area the link could cover.

REFERENCES

Johan, O., Fredrik, K., Lars, S., Kun, W., Qin, W., Stephane, J., Susanne, A., Bertrand, N. (2007). A high-speed modulated retro-reflector communication link with a transmissive modulator in a cat's eye optics arrangement. *Proceedings of SPIE, 6736*, 673619.

Rabinovich, W.S., Ferraro, M.S., Suite, M.R., Moore, C.I., Suite, M.R., Ferraro, M.S., Goetz, P.G. (2012). Retro-reflector diversity effects in free-space optical links. *Proceedings of SPIE, 8380.*

Rabinovich, W.S., Goetz, P.G, Mahon, R., Swingen, L., Murphy, J.L., Ferraro, M.S., Burris, H.R., Moore, C.I., Suite, M.R., Gilbreath, G.C., Binari, S.C., Klotzkin, D., (2007). 45-Mbit/s cat's eye modulating retroreflectors. *Optical Engineering, 46*(10), 104001.

Rabinovich, W.S., Mahon, R., Goetz, P.G, Swingen, L., Murphy, J.L., Ferraro, M.S., Ray, B., Suite, M.R., Moore, C.I., Gilbreath, G.C., Binari, S.C. (2006). 45 Mbps cat's eye modulating retro-reflector link over 7 Km. *Proceedings of SPIE, 6304*, 63040Q.

Zhang B., Niu Y.X., Zhang C. (2009). Analysis of influence factors for reflected wave power of "cat-eye" target. *Infrared and Laser Engineering, 38*(3), 419–423.

Zhao Y.Z., Sun H.Y, Fan P.S, Li, W., Zhao, X.L. (2010). Laser reflection characteristics of cat-eye effect under large angle oblique incidence. *High Power Laser and Particle Beams, 22*(7), 1457–1461.

Frontier Research and Innovation in Optoelectronics Technology and Industry – Habib & Lewis (Eds)
© 2019 Taylor & Francis Group, London, ISBN 978-1-138-33178-5

Direct detection of a single-channel 112 Gb/s PAM-4 signal using an 18 GHz directly modulated laser and Maximum-Likelihood Sequence Estimation (MLSE) equalization

J.L. Zhang
School of Science, Beijing University of Posts and Telecommunications, Beijing, China

C. Xu, G.J. Gao, H.D. Liu & Z. Yang
State Key Laboratory of Information Photonics and Optical Communications,
Beijing University of Posts and Telecommunications, Beijing, China

ABSTRACT: In this paper, we experimentally demonstrate single-channel 112 Gb/s PAM-4 signal transmission using a commercial 18 GHz Directly Modulated Laser (DML), achieving a record receiver sensitivity of −9 dBm. Several digital signal processing techniques are used to improve the system performance, including digital duo-binary encoding, digital pre-compensation, receiver-side Least-Mean Square (LMS) algorithm, and Maximum-Likelihood Sequence Estimation (MLSE).

1 INTRODUCTION

As a result of the rapid increase in mobile and cloud services, imperative demand for short-reach optical communication with lower cost and higher capacity has appeared for the vast data centers and metro networks applications associated with them (K. Zhong et al., 2015; Sadot et al., 2015; Cartledge et al., 2014). Intensity Modulation and Direct Detection (IM/DD) is well-suited to providing such low-cost interfaces. Various advanced modulation formats, such as Pulse Amplitude Modulation (PAM), Multi-band Carrierless Amplitude and Phase (Multi-CAP) modulation and Discrete Multi-Tone (DMT) modulation, have been demonstrated with up to 112 Gb/s per channel line rate for such applications. As one of the promising solutions, PAM-4 modulation can provide a line rate above 100 Gb/s for a single channel and various Digital Signal Processing (DSP) techniques have been employed for improving its performance (K. Zhong et al., 2015; Sadot et al., 2015; Cartledge et al., 2014). Recently, Kikuchi et al. (2015) reported the generation of a 102 Gb/s Nyquist PAM-4 signal based on a commercial 1300 nm 28 Gb/s Electro-absorptive Modulated Laser (EML). Suhr et al. (2015) reported the transmission of a 56 Gb/s duo-binary PAM-4 signal over 10 km Single-Mode Fiber (SMF) using a 10 Gb/s Directly Modulated Laser (DML), in which a 56 GSa/s Digital-to-Analog Convertor (DAC) with digital Bessel filter are applied jointly to effectively obtain the seven-level partial-response PAM-4 signal. K. P. Zhong et al. (2015) demonstrated the highest rate for PAM-4 modulation with single-channel 128 Gb/s and four-channel 500 Gb/s based on 25 Gb/s EML Transmitter Optical Sub-Assembly (TOSA) and 25 Gb/s PIN Receiver Optical Sub-Assembly (ROSA), in which digital equalizer, post-filter and Maximum-Likelihood Sequence Estimation (MLSE) are employed in the receiver-end digital signal processing.

In this paper, we experimentally demonstrate single-channel 112 Gb/s PAM-4 signal transmission with −9 dBm sensitivity. To the best of our knowledge, this is the highest sensitivity to which 112 Gb/s PAM-4 transmission has been demonstrated with commercial 18-GHz-only DML. By combining the transmitter-side digital duo-binary encoding of PAM-4 signal, digital pre-compensation, receiver-side Least-Mean Square (LMS) algorithm, MLSE, and duo-binary decoding, we successfully achieved a receiver sensitivity of about −9 dBm using a Vestigial Sideband (VSB) filter with 7% overhead Hard-Decision Forward Error Correction (HD-FEC).

2 PRINCIPLES

For high baud rate signaling, such as 112 Gb/s PAM-4, transmitter- and receiver-side DSP plays a critical role in improving system performance such as sensitivity and Chromatic Dispersion (CD) tolerance, especially when low-cost and low-bandwidth directly modulated lasers are used. The transmitter DSP in this work consisted of three blocks, which are PAM-4 mapping, duo-binary encoding of PAM-4, and channel pre-compensation. The receiver DSP consists of the following parts: 1) seven-level training sequence aided least-mean square algorithm; 2) maximum-likelihood sequence estimation (MLSE); 3) duo-binary decoding; 4) PAM-4 signal de-mapping.

As one of the partial-response functions, the duo-binary encoding is used to narrow the bandwidth of the PAM-4 signal, which is fulfilled by a two-tap Finite Impulse Response (FIR) filter with equal weights (Suhr et al., 2015; Zhong et al., 2015; Chen et al., 2014). Because such operation introduces correlation between adjacent symbols, pre-coding at the transmitter is necessary to avoid error propagation. At the receiver, duo-binary decoding is fulfilled by a simple 'modulus-4' operation (Xie, 2015). Digital channel pre-compensation is another important procedure to overcome the bandwidth limitation from low-cost optics. In this work, a segmented pre-compensation method is used at the transmitter side to achieve the optimum performance.

At the receiver, LMS is applied first to estimate the channel response and mitigate the Inter-Symbol Interference (ISI) with long memory preliminary. Then, the MLSE algorithm is used mainly to mitigate the residual ISI induced by the bandwidth limitation or chromatic dispersion according to a sequence of bits (Olmos et al., 2013). The principle of MLSE is to maximize the probability of $p[y(D)|a(D)]$. We associate the formal power series in the delay operator D (D-transform):

$$a(D) = a_0 + a_1 D + a_2 D^2 + \dots \tag{1}$$

which will itself be referred to as the input sequence, with $y(D)$ referred to as the received symbol sequence. Symbol $s(D)$ represents the state sequence, and the relationship between $s(D)$ and $a(D)$ is one-to-one correspondence. Thus, $p[y(D)|a(D)]$ can be represented by $p[y(D)|s(D)]$. Considering that the noise is independent, the log likelihood, $\ln p[y(D)|a(D)]$, can be broken into a sum of independent increments:

$$\ln p\big[\, y(D)\,|\,s(D)\,\big] = \sum_k \ln f\big[\, y_k - x\big(s_{k-1}, s_k\big)\big] \tag{2}$$

where $f(\cdot)$ is the probability density of noise; y_k and s_k are received sequence and state at time k. is the input sequence corresponding to the transfer from s_{k-1} to s_k.

As shown in Figure 1, $S_{(k,i)}$ represents the i-th (1«i»m^v) state at time k. The symbols m and v are referred to as the level number of signals and tap number of MLSE, respectively. Variables ak and ak' are two possible input sequences (length = v) at time k. At time k, each state stores one most-possible path and the corresponding probability. At the next time, $k+1$, each state chooses and stores the most-possible path from a certain set of states at time k according to the stored weight value and the Euclidean distance between $y(k+1)$ and $a(k+1)$. Then, the symbols before time $k+1$ can be determined by comparing the path stored weight value in each state $S_{(k+1,i)}$.

On the other hand, considering that the bandwidth of DML is still smaller than the encoded PAM-4 signal, further spectrum shaping is inevitable. The bandwidth Compression Ratio (CR) is defined as $(B_{PAM} - B_{LPF})/B_{PAM}$, where B_{PAM} is the baud rate of the PAM-4 signal,

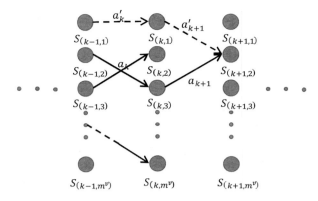

Figure 1.   Survivors of successive recursions of Viterbi algorithm.

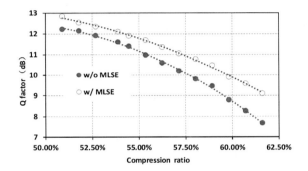

Figure 2.   Q versus CR for 112 Gb/s duo-binary encoded PAM-4 signal.

and B_{LPF} is the 3 dB bandwidth of a Bessel Low-Pass Filter (LPF), which is used for emulating the bandwidth response in numerical simulation.

Simulation is carried out to investigate the impact of CR on Q-factor performance. A fifth-order Bessel filter is adopted to filter the encoded PAM-4 signal. Figure 2 shows the Q factor versus CR for the encoded PAM-4 signal. According to this figure, about 1-dB Q penalty is observed at 55% CR ratio, corresponding to 25 GHz LPF for the 56-Gbaud PAM-4 signal. When CR is around 56%, around 1 dB improvement can be obtained by applying MLSE. When CR is larger than 60%, the Q-factor improvement from MLSE becomes more significant (larger than 1 dB). The spectrum shaping can be carried out at the same time with the S21 pre-compensation by DSP at the transmitter.

3   EXPERIMENTAL SETUPS

Figure 3 shows the experimental setup for the transmission of a single-channel 112 Gb/s PAM-4 signal based on a commercial 18 GHz DML. The insets are the spectra of signals at the transmitter and receiver, respectively. At the transmitter, the offline DSP is applied to encode the PAM-4 sequence with a duo-binary filter, and pre-compensate the bandwidth limited influence with the pre-measured S21 response. Then, the output signal is sent by the Arbitrary Waveform Generator (AWG) running at 56 GSa/s, achieving the 112 Gb/s in one single channel. After amplification by a linear RF driver with 3 dB bandwidth of 25 GHz, the signal is modulated by the DML at 1.53 µm (3 dB bandwidth: 18 GHz). The optical power at the output of DML is set at 6 dBm. An Erbium-Doped Fiber Amplifier (EDFA) is applied to

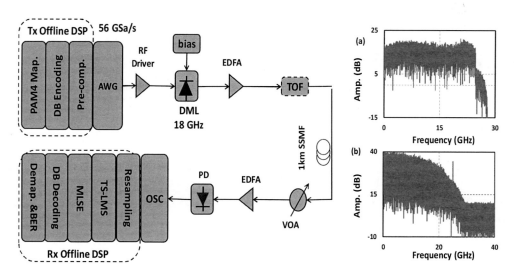

Figure 3. Experimental setup of single-channel 112 Gb/s PAM-4 signal over 1 km SSMF based on commercial 18 Gb/s DML.

adjust the launch power into the fiber. After that, a Tunable Optical Filter (TOF) is employed optionally, which shapes the signals into a VSB signal for chirp management.

After the transmission over a span of Standard SMF (SSMF), the Received Optical Power (ROP) of signal is adjusted by a Variable Optical Attenuator (VOA). Another EDFA is used to reboot the optical power into a 40 GHz PIN-photodetector without Transimpedance Amplifier (TIA). The detected RF output is sampled by a Tektronix real-time scope with 100 GSa/s. In the receiver offline DSP, a 37-taps LMS is first adopted to recover the duo-binary signal with an oversampling rate of two samples per symbol. The first 3% of the received data is used as the training sequence for the LMS. After that, the signal is compensated by MLSE equalization and then decoded by a duo-binary decoder with simple 'modulus-4' operation, followed by the PAM-4 de-mapping and Bit Error Ratio (BER) calculation. Listing and numbering

4 RESULTS AND DISCUSSION

The measured system response is shown in Figure 4, which includes DAC/ADC, RF driver, DML and Photodetector (PD). Among these devices, the key factor limiting the system response performance is the DML, of which the 3 dB bandwidth is only 18 GHz. It can be observed from Figure 4 that the amplitude drops rapidly beyond 21 GHz and the noise dominates the response for frequencies above 25 GHz.

At the transmitter, we pre-compensate the signal in the frequency domain according to the pre-measured system response curve. For the frequency within 18 GHz, direct pre-compensation is applied. Because significant noise is observed for the frequency ranging from 20 to 25 GHz, a smoothing window is applied to the curve for noise suppression between these frequencies. Because the frequency component beyond 25 GHz makes little contribution to the whole system performance, compensation in this frequency range can corrupt the signal quality in the lower frequency range and is therefore omitted, which corresponds to a 55% ratio CR (see Figure 2). The spectra after pre-compensation at the transmitter and at the receiver are shown, respectively, in insets (a) and (b) of Figure 3.

The system performance measured Back-To-Back (BTB) for the 112 Gb/s PAM-4 signal, using only S21 pre-compensation without CR, smoothed S21 pre-compensation with 55%

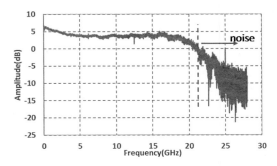

Figure 4.   Measured system response (S21 parameter) including DAC/ADC, RF driver, DML and PD.

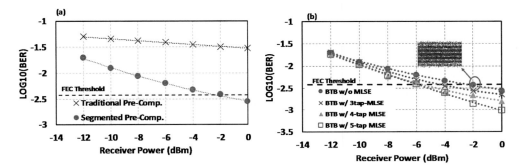

Figure 5.   Receiver sensitivity measured BTB: (a) using only S21 pre-compensation or CR and smoothing; (b) using complete pre-compensation with different-tap MLSE.

CR are shown in Figure 5(a). Without CR and smoothing, the performance is very poor. The optimal receiver sensitivity (BTB) is around −2 dBm by applying both CR and smoothing, for the 7% HD-FEC.

With the exception of the duo-binary coding and transmitter pre-compensation, MLSE is adopted at the receiver to compensate for the bandwidth limitation and non-linearity induced by low-cost optics. The performance of MLSE with different taps is shown in Figure 5b. The inset shows the seven-level signal, restored from LMS. In addition, the optimal receiver sensitivity can be lowered further by employing MLSE. With an increase in the MLSE tap number, the performance keeps improving. To balance the trade-off between system performance and the required DSP resources, a five-tap MLSE is applied and an optimal receiver sensitivity of −6 dBm is obtained.

In the following test, we investigate the influence of chromatic dispersion on the system performance, where the CD and chirp interact with each other in a complicated manner. A five-tap MLSE is again used to explore the system performance after up to 10 km SSMF transmission, as shown in Figure 6. Under the interaction between CD and chirp, the Q penalty varies for different transmission distances. After 7 km SSMF transmission, a minimum Q penalty of 0.5 dB is observed. By filtering the signal into a VSB signal with a TOF at the transmitter, chirp management operation is performed to mitigate the interaction between CD and chirp.

In the practical communication system, Wavelength-Division Multiplexing (WDM) devices can achieve the same effect. The optical spectra with and without chirp management are shown in Figure 7. It is observed with chirp management that the spectrum is obviously narrowed, which mitigates the interaction between CD and chirp. The system performance by chirp management is shown in Figure 8. For the BTB performance, the optimal receiver

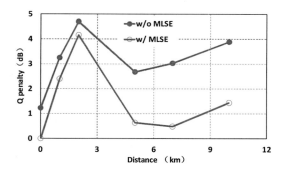

Figure 6. The Q penalty versus the transmission distance (SSMF).

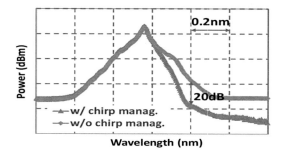

Figure 7. Optical spectra with and without chirp management.

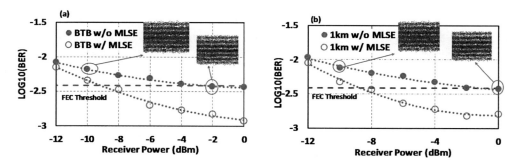

Figure 8. Receiver sensitivity for the 112 Gb/s PAM-4 signal with chirp management measured: (a) BTB; (b) at 1 km SSMF.

sensitivity is still −2 dBm without the MLSE algorithm. By using MLSE, the optimal receiver sensitivity can be optimized down to −9 dBm, providing a significant improvement.

The experimental results for a VSB signal after 1 km SSMF transmission are shown in Figure 8b, representing the system performance with and without VSB-based chirp management. As shown in Figure 8b, after 1 km SSMF transmission, the optimal receiver sensitivity without MLSE is 0 dBm, exhibiting a 2-dB performance penalty compared with the BTB case. However, by applying five-tap MLSE, the optimal receiver sensitivity after 1 km SSMF transmission is not degraded in comparison to the BTB results. As a rough estimate, with an optimal receiver sensitivity of −9 dBm for the BTB case, the 112 Gb/s PAM-4 signal with chirp management can be transmitted over 40 km (SSMF).

5   CONCLUSION

We have experimentally demonstrated the transmission of single-channel 112 Gb/s PAM-4 with about −9 dB sensitivity enabled by 18 GHz DML and direct detection. To the best of our knowledge, this is the highest sensitivity that 112 Gb/s transmission has demonstrated with only 18 GHz 3 dB bandwidth and MLSE. Results show that the adopted scheme can be a potential technology for the next-generation 100 Gb/s short-reach metro network.

ACKNOWLEDGMENTS

This work has been partially supported by the National Natural Science Foundation of China under Grants 61302085 and 62171189, the Fundamental Research Funds for the Central Universities under Grant 2016RCGD20, and the Fund of IPOC under Grant 2017ZT12.

REFERENCES

Cartledge, J.C. & Karar, A.S. (2014). 100 Gb/s intensity modulation and direct detection. *Journal of Lightwave Technology*, *32*(16), 2809–2814.

Chen, S., Xie, C. & Zhang, J. (2014). Comparison of advanced detection techniques for QPSK signals in super-Nyquist WDM systems. *IEEE Photonics Technology Letters*, *27*(1), 105–108.

Fludger, C. (2016). Digital signal processing in optical communications. In *Proceedings of Optical Fiber Communication Conference, Anaheim, CA* (Paper W3G.4).

Forney, G.D. (1972). Maximum-likelihood sequence estimation of digital sequences in the presence of intersymbol interference. *IEEE Transactions on Information Theory*, *18*, 363–378.

Kikuchi, N., Hirai, R. & Fukui, T. (2015). Single-VCSEL 100-Gb/s short-reach system using discrete multi-tone modulation and direct detection. In *Proceedings of Optical Fiber Communication Conference, Los Angeles, CA* (Paper Tu2H.2).

Olmos, J.J.V., Suhr, L.F., Li, B. & Monroy, I.T. (2013). Five-level polybinary signaling for 10 Gbps data transmission systems. *Optics Express*, *21*(17), 20417–20422.

Sadot, D., Dorman, G., Gorshtein, A., Sonkin, E. & Vidal, O. (2015). Single channel 112Gbit/sec PAM4 at 56Gbaud with digital signal processing for data centers applications. *Optics Express*, *23*(20), 991–997.

Suhr, L.F., Olmos, J.J.V., Mao, B., Xu, X., Liu, G.N. & Monroy, I.T. (2015). Direct modulation of 56 Gbps duobinary-4-PAM. In *Proceedings of Optical Fiber Communication Conference, Los Angeles, CA* (Paper TH1E.7).

Xie, C. & Chen, S. (2015). Quadrature duobinary modulation and detection. In *Proceedings of Optical Fiber Communication Conference, Los Angeles, CA* (Paper W4 K.6).

Zhong, K., Zhou, X., Gui, T., Tao, L., Gao, Y., Chen, W. & Lu, C. (2015). Experimental study of PAM-4, CAP-16, and DMT for 100 Gb/s short reach optical transmission systems. *Optics Express*, *23*(3), 1176–1189.

Zhong, K.P., Chen, W., Sui, Q., Man, J.W., Lau, A.P.T., Lu, C. & Zeng, L. (2015). Experimental demonstration of 500Gbit/s short reach transmission employing PAM4 signal and direct detection with 25Gbps device. In *Proceedings of Optical Fiber Communication Conference, Los Angeles, CA* (Paper TH3 A.3).

Frontier Research and Innovation in Optoelectronics Technology and Industry – Habib & Lewis (Eds)
© 2019 Taylor & Francis Group, London, ISBN 978-1-138-33178-5

Broadband near-infrared luminescence of porous silicon heat-treating at high temperature

Sa Chu Rong Gui
College of Engineering, Bohai University, Jinzhou, Liao Ning, China
Institute of Automation, Bohai University, Jinzhou, Liao Ning, China

Wujisiguleng Bao, Ying Fu & Wei Wang
College of New Energy, Bohai University, Jinzhou, Liao Ning, China

ABSTRACT: The influence of the annealing temperature on the Photoluminescence (PL) of hydrofluoric-treated Porous Silicon (PSi) films prepared at the same current densities has been investigated. The PSi was annealed at between 800 and 1,100°C in air and then the samples were re-annealed in N_2 at the higher temperature of 1,300°C. Oxidized PSi shows very broad near-infrared luminescence. The PL intensity and the peak position strongly depend on the annealing temperature. The PL peak first blue shifts from 850°C drastically, and then red shifts gradually from the annealing temperature at 1,000°C. After re-annealing in N_2 at 1,300°C, a broad and long wavelength near-infrared luminescence peak at around 1,130 nm was observed in the PSi-N_2, and the PL peak was red shifted. This is due to the high degree oxidized Si^{4+} state decreasing to three suboxide states, Si^{1+}, Si^{2+}, or Si^{3+}, when annealing at high temperature in the N_2 atmosphere.

1 INTRODUCTION

Porous Silicon (PSi) is easily prepared by anodic electrochemical etching, stain-etching or photo-etching in a hydrofluoric acid-based solution (Shichi et al., 2011; Gelir et al., 2017). The Photoluminescence (PL) property of PSi structures makes it a potential material for the applications in Si-based optoelectronic technology (Wang et al., 2012; Boukherroub et al., 2001). PSi shows efficient visible to near-infrared photoluminescence at room temperature. The photostability can be improved by oxidizing the PSi surface and the oxidation of PSi can also result in the enhancement of the PL quantum efficiency (Sawada et al., 1994). Since PSi is compatible with Si processes, it is known to be the most important material in Si-based photonics.

In past years, the PL properties of PSi and the oxidation of PSi have been widely studied (Kan et al., 2005; Nakamura et al., 2010). Quantum-size effects have been suggested as explanations for visible luminescence in a variety of other Si-based materials, including Si-rich SiO_2 (Kanemitsu et al. 1992) and ultrafine Si particles. The existence of luminescent chemical products such as polysilane (Tsai et al., 1991) or siloxene (Xu et al., 1996) in a PSi layer has also been suggested as being the origin of the luminescence. In recent years, oxidation effects of PSi on PL were reported (Kan et al., 2005; Nakamura et al., 2010). The reports infer that the surface passivation, either by hydrogen or oxygen, is one of the requisite conditions for obtaining strong PL efficiency in PSi. Gelir et al. (2017) concluded that the diffusion of the oxygen to PSi layers takes place in two different stages. In the first stage, the oxygen molecules are adsorbed on the surface of the PSi layer, while they diffuse through into the pores in the second stage. For all cases the diffusion in the first stage is much faster than in the second stage. The oxygen diffusion increases as the pore size and the temperature increase in the first stage. In the second stage, while the diffusion increases with the pore size, it almost does not change with temperature.

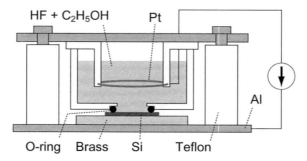

Figure 1. Anodization cell used for the fabrication of PSi thin film.

In this paper, the influence of the annealing temperature and annealing atmosphere on the PL of hydrofluoric-treated PSi has been investigated. The PSi was annealed at between 800 and 1,100°C in air, and after the sample was re-annealed in N_2 at the higher temperature of 1,300°C. It was observed that the PL intensity was steeply quenched after annealing at 900°C and recovered at above 1,100°C. After re-annealing in N_2 at 1,300°C, the PL intensity was increased significantly and the PL peak was red shifted gradually. Perhaps the Si-O bond was decomposed after annealing in the N_2 atmosphere at high temperature. The detailed mechanism was under study.

2 EXPERIMENTAL PROCEDURE

One hundred oriented P-type Si wafers with the resistivity of 20 mΩ·cm are used as initial substrates for the preparation of porous Si. 100 m-thick free-standing porous Si monolayers are prepared by electrochemical anodization in a mixture of Hydrofluoric (HF) (46 wt. % in water) and ethanol (HF:ethanol = 1:1) with the anodic current density of 70 mA/cm². A schematic diagram of the porous Si formation setup is shown in Figure 1. The obtained porous Si film is oxidized at between 800 and 1,100°C in air for one hour (PSi-air). The samples were then annealed at 1,300°C for 30 minutes in N_2 atmosphere (PSi-N_2).

The PL spectra are measured by using a single grating monochromator, equipped with a liquid-nitrogen-cooled InGaAs diode array. The excitation source is a 488 nm line of an argon ion laser. For all the PL measurements, the spectral response of the detection system was corrected by using the reference spectrum of a standard tungsten lamp. All the measurements were carried out at room temperature.

3 RESULT AND DISCUSSIONS

In the first part of the study, the plain-view Transmission Electron Microscope (TEM) image of the as-prepared free-standing PSi was measured. In Figure 2, the dark regions correspond to PSi skeletons, while the bright regions correspond to pores. The pore size is several tens of nanometers. The porosity of the PSi is estimated from the effective refractive index, which is about 58% (Föll et al., 2002; Kanungo et al., 2010).

In Figure 3(a), the PL spectra of the PSi oxidized between 800 and 1,100°C under the excitation at 488 nm are shown. The broad Near-Infrared (NIR) PL is observed in the PSi-air, the peak position being in the range 790 to 900 nm. The PL intensity and the peak position strongly depend on the annealing temperature. From the normalized PL spectra (Figure 3(b)) we can see that the PL peak first blue shifts from 850°C drastically, and then red shifts gradually from the annealing temperature at 1,000°C. For detailed study, the PL intensity depending on the annealing temperature is shown in Figure 4(b). The PL intensity is steeply quenched after annealing at 900°C and recovered at above 1,000°C. When the oxidation

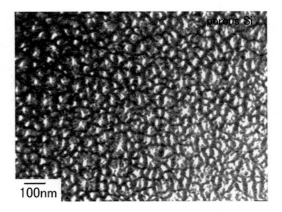

Figure 2.   TEM image of as-prepared porous Si.

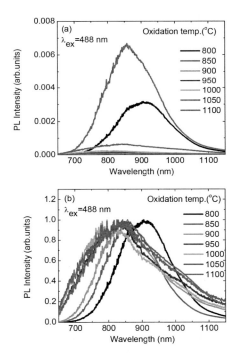

Figure 3.   (a) PL spectra of PSi annealing from 800 to 1,100°C; (b) Normalized PL spectra of the PSi-air.

temperature is below 800°C, PSi films are not fully oxidized and are brownish; the transmittance spectra of oxidized PSi films are shown in the report by Shichi et al. (2011). In our sample, the PSi was maybe fully oxidized at 850°C; thus the PL intensity was significantly increased at 850°C. The mechanism of intensity quenching and recovering of PL depending on oxidation temperature was suggested in the report by Nakamura et al. (2010), indicating that the formation of a thin disordered SiO_2 over layer results in the quenching of the PL intensity. The growth of a high-quality SiO_2 layer was necessary for the recovery of the PL intensity at high temperature. The possible mechanism of the PL red shift can be considered. Gole et al. (1998a, 1998b, 2006) studied the origin of the PL red shift from the green to the orange-red spectral region in PSi. They concluded that the red shift is due to the oxidative reactions of the PSi-surface-bound silanone group and subsequent transformation to the silanone-based silicon oxyhydrides. Our observed PL red shift after thermal oxidation may be caused by the essentially same oxidative reactions as those reported by Gole et al.

For a detailed study of the PL and the origin of the luminescence of annealed PSi, the samples oxidized from 850 to 1,000°C were re-annealed at 1,300°C in N_2 atmosphere (PSi-N_2). The PL of re-annealed samples are shown in Figure 5(a). It is interesting that broad and long wavelength near-infrared luminescence peak at around 1,130 nm was observed in the PSi-N_2, and the PL peak was red shifted gradually. The red shift is possibly due to the SiO_2 bond being decreased to three suboxide states Si^{+1}, Si^{+2}, and Si^{+3}. Compared to the PL intensity of PSi oxidized in air, the PL intensity of PSi-N_2 was enhanced significantly. The enhanced factor reached was as high as 170 times. The possible oxidation states of Si are the three suboxide states Si^{1+}, Si^{2+}, and Si^{3+}, which represent Si atom bonding to 1, 2 and 3 oxygen atoms, respectively; the SiO_2 state was bonding to four oxygen atoms (Nakamura et al., 2010). After annealing in N_2 atmosphere at high temperature, the Si^{4+} state will decrease to the three suboxide states Si^{1+}, Si^{2+} or Si^{3+}. The PL intensity of PSi-N_2 depending on oxidation temperature is shown in Figure 5(b). The PL intensity decreases with increasing oxidation temperature. It is possible that the high degree oxidized PSi is severely decomposed at 1,300°C in N_2 atmosphere. The detailed mechanism was studied by measuring XRD and XPS spectra.

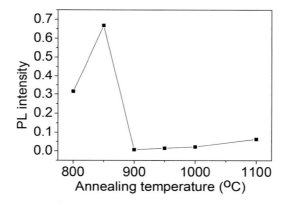

Figure 4.   Annealing temperature dependence of PL intensity of PSi-air.

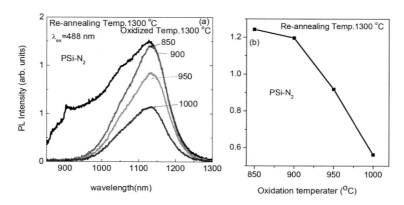

Figure 5.   (a) PL spectra of re-annealed PSi-N_2 at 1,300°C; (b) PL intensity dependence on oxidation temperature.

4 CONCLUSION

This paper presents and discusses the luminescence of PSi by high temperature oxidizing in air and annealing in N_2 atmosphere. Oxidized PSi shows very broad near-infrared luminescence, and the PL intensity and the peak position significantly depend on the annealing temperature. After re-annealing in N_2 at 1,300°C, a broad and long wavelength near-infrared luminescence peak at around 1,130 nm was observed in the PSi-N_2, and the PL peak was red shifted. It may be that the Si-O bond was decomposed after annealing in N_2 atmosphere at high temperature. The detailed mechanism was studied by measuring XPS and XPS spectra.

ACKNOWLEDGMENT

This project is financially supported by the National Natural Science Foundation of China (61503042 and 51704029), Education Science and Technology Research Project of Liaoning (LQ2017008 and 20170540017) and the Bohai University Doctor Project (bsqd201439).

REFERENCES

Boukherroub, R., Wayner, D.D., Sproule, G.I., Lockwood, D.J. & Canham, L.T. (2001). Stability enhancement of partially-oxidized porous silicon nanostructures modified with ethyl undecylenate. *Nano* Letters, *1*(12), 713–717.

Föll, H., Christophersen, M., Carstensen, J. & Hasse, G. (2002). Formation and application of porous silicon. *Materials Science and Engineering: R: Reports*, *39*(4), 93–141.

Gelir, A., Yargi, O. & Yuksel, S.A. (2017). Elucidation of the pore size and temperature dependence of the oxygen diffusion into porous silicon. *Thin Solid Films*, *636*, 602–607.

Gole, J.L. & Dixon, D.A. (1998a). Transformation, green to orange-red, of a porous silicon photoluminescent surface in solution. *J Phys Chem B*, *102*(1), 33–39.

Gole, J.L. & Dixon, D.A. (1998b). Potential role of silanones in the photoluminescence-excitation, visible photoluminescence-emission, and infrared spectra of porous silicon. *Phys Rev B*, *57*(19), 12002–12015.

Gole, J.L., Veje, E., Egeberg, R.G., Ferreira da Silva, A., Pepe, I. & Dixon, D.A. (2006). Optical analysis of the light emission from porous silicon: A hybrid polyatom surface-coupled fluorophor. *J Phys Chem B*, *110*(5), 2064–2073.

Kanemitsu, Y., Sizuki, K., Uto, H. & Masumoto, Y. (1992). On the origin of visible photoluminescence in nanometer-size Ge crystallites. *Appl. Phys. Lett.*, *61*(18), 2187–2189.

Kanungo, J., Maji, S., Mandal, A.K., Sen, S., Bontempi, E., Balamurugan, A.K., ... Basu, S. (2010). Surface treatment of nanoporous silicon with noble metal ions and characterizations. *Appl Surf Sci*, *256*(13), 4231–4240.

Nakamura, T., Ogawa, T., Hosoya, N. & Adachi, S. (2010). Effects of thermal oxidation on the photoluminescence properties of porous silicon. *Journal of Luminescence*, *130*(4), 682–687.

Sawada, S., Hamada, N., & Ookubo, N. (1994). Mechanisms of visible photoluminescence in porous silicon. *Phys. Rev. B*, *49*(8), 5236–5240.

Sharma, S.N., Banerjee, R. & Barua, A.K. (2003). Effect of the oxidation process on the luminescence of HF-treated porous silicon. *Current Applied Physics*, *3*(2–3), 269–274.

Shichi, S., Fuji, M. & Hayashi, S. (2011). Ultraviolet true zero-order wave plate made of birefringent porous silica. *Optics Letters*, *36*(19), 3951–3953.

Tsai, C., Li, K.H., Sarathy, J., Shih, S., Campbell, J.C., Hance, B.K. & White, J.M. (1991). Thermal treatment studies of the photoluminescence intensity of porous silicon. *Appl Phys Lett*, *59*(22), 2814–2816.

Wang, C., Hu, B. & Yi, H.H. (2012). The study of structure and optoelectronic properties of ZnS and ZnO films on porous silicon substrates. *Optik-International Journal for Light and Electron* Optics, *123*(12), 1040–1043.

Xu, Y.K. & Adachi, S. (2009). Properties of green-light-emitting anodic layers formed on Si substrate in HF/MnO_2 mixed solution. *J Appl Phys*, *105*(11), 1046–21.

Optoelectronic devices and integration

Frontier Research and Innovation in Optoelectronics Technology and Industry – Habib & Lewis (Eds)
© *2019 Taylor & Francis Group, London, ISBN 978-1-138-33178-5*

Optimization of amplifier-combination schemes in an ultra-long-span optical-fiber transmission system

R.H. Chi, H.T. Zhao & L.Y. Li
School of Internet of Things and Software Technology, Wuxi Vocational College of Science and Technology, Wuxi, China

ABSTRACT: This paper analyzes amplifier-combination schemes and reports their application in a 325-km, 10×10 Gbit/s ultra-long-span repeater-less transmission system. Four different amplifier-combination schemes, which include an Erbium-Doped Fiber Amplifier (EDFA), a Raman amplifier and a Remote Optically Pumped Amplifier (ROPA), are proposed and studied in detail. An amplifier-combination scheme involving an EDFA, co-pumped Raman amplifier and ROPA achieves a good balance between system performance and cost. Unrepeated link with 77.5-dB power budget and 10-dB optical signal-to-noise ratio are obtained with this optimized amplifier-combination scheme. The research results have clear significance for designers deploying amplifiers in ultra-long-distance transmission systems.

Keywords: EDFA, Raman amplifier, remote optically pumped amplifier, amplifier-combination schemes, ultra-long-span optical-fiber transmission system

1 INTRODUCTION

There is a growing interest in ultra-long transmission technologies for applications in terrestrial transport, island hopping, lake crossings and even bridging long spans. These unrepeatered systems provide a highly effective solution for point-to-point connections over several hundred kilometers without the need for in-line repeaters. Over the years, a lot of methods have been used to achieve as long a distance as possible without any in-line active elements (Bissessur et al., 2012; Gainov et al., 2014; Zhu et al., 2013).

Almost all unrepeatered transmissions use an Erbium-Doped Fiber Amplifier (EDFA) and a Raman fiber amplifier due to their good noise characteristics (Bousselet et al., 2008; Cai et al., 2008; Du et al., 2008; Lucero et al., 2009). A directional Raman amplifier usually includes a Co-pumped Raman Amplifier (CoRA) and a Counter-pumped Raman Amplifier (CtRA). A CoRA may cause nonlinear effects due to the high signal power and the high pump Raman power co-injected into the fiber. On the other hand, a remotely pumped EDFA (ROPA) is an effective way to extend the span, but some customers do not like ROPA schemes because it is difficult to build the Remote Gain Unit (RGU) module at the construction site.

In some practical cases, for example, on a 300~400 km unrepeatered transmission line, the operators have trouble choosing the amplifier configuration because they want a good balance between performance and cost. In this paper, we analyze four amplifier-combination configurations. This work is carried out on the basis of our previous work (Chi et al., 2015), in which we reported two simple amplifier-combination schemes (options A and B) based on a 325-km unrepeatered transmission line. In this work, we continue to study the performance of ROPA/Raman configurations, including two novel cases: ROPA/CtRA (option C) and ROPA/CoRA (option D). This paper also discusses the amplifier configuration and describes the Optical Signal-to-Noise Ratio (OSNR) and the trade-off with nonlinear tolerance. Our experimental results show that an EDFA/CoRA/ROPA scheme affords a significant perform-ance increase in an ultra-long-haul unrepeatered Dense Wavelength-Division Multiplexing (DWDM) system and gives a good balance between cost and performance.

2 EXPERIMENTAL SETUP AND PROPOSED AMPLIFIER-COMBINATION SCHEMES

Figure 1 shows the experimental setup. The transmitters consisted of ten Distributed Feed-back (DFB) lasers at frequencies from 192.5 to 193.5 THz on the 100-GHz-spaced ITU frequency grid. The ten DWDM signals with Enhanced Forward-Error Correction (EFEC, 7% overhead) at 11.2 Gb/s are amplified by a Booster Amplifier (BA) followed by a Variable Optical Attenuator (VOA), and then launched into the 325-km fiber link. The maximum output power of the BA is 24.4 dBm.

The link consisted of two sections using G.652 fiber. The first section of fiber span, from the transmitter to the RGU, was 245 km (span loss 49 dB), and the second section from the RGU to the receiver was 80 km (span loss 16 dB). The total link loss at 1550.12 nm is 65 dB (including connector loss) with a mean chromatic dispersion of 17 ps/nm/km at 1550.12 nm. Most of the 6120 ps/nm chromatic dispersion is compensated in the receiving terminal by fiber Dispersion Compensation Modules (DCMs). Loss in the DCMs is compensated by a Pre-Amplifier (PA). Finally, the ten signals are de-multiplexed and detected by the line termi-nal receivers individually. The transmission system is evaluated by measuring the OSNR with an Optical Spectrum Analyzer (OSA) and measuring the signal Bit Error Rate (BER) with an optical network tester (JDSU ONT-506).

To realize a 325-km transmission line without any in-line active elements, amplifiers such as an EDFA, Raman amplifier or ROPA are indispensable. In the application, the Raman amplifier can either be co-pumped (CoRA) or counter-pumped (CtRA). The ROPA can be pumped though the transmission fiber or pumped by a span of bypass fiber. To avoid Raman transfer from the red band to the blue band, we chose to use a bypass ROPA.

There are several choices when deploying these amplifiers in the transmission line. In this paper, we propose and compare four different amplifier-combination schemes, which are shown in Figure 2. Option A is a bidirectional Raman scheme which uses a bidirectional Raman amplifier and an EDFA (a combination of BA + CoRA + CtRA + PA). Option B is an EDFA/ROPA scheme (BA + ROPA + PA). A ROPA-plus-CtRA scheme (BA + ROPA + CtRA + PA, option C) is also tested. Finally, we tested a CoRA/ROPA scheme (combination of BA + CoRA + ROPA + PA, option D).

The CoRA and CtRA are pumped by three semiconductor diodes at wavelengths of 1425, 1439 and 1457 nm. Their total pump power is nearly 1 W. Thanks to the multiple pump wavelengths, the Raman pump modules can provide flexible gain profiles. In the schemes of options B, C and D, we use a bypass ROPA in which a 15 m-long high gain efficiency Erbium-Doped Fiber (EDF) is placed 80 km away from the receiver side and is remotely pumped by a Remote Pump Unit (RPU). The RPU provides pump power at 900 mW at wavelengths of 1465 and 1475 nm. In the experiment, the CoRA, CtRA, RGU and RPU are

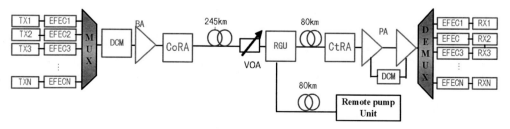

Figure 1.   Schematic diagram of the unrepeatered transmission link.

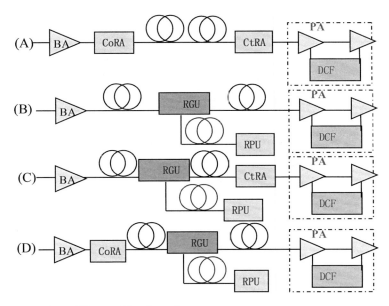

Figure 2.   Proposed amplifier-combination schemes.

placed in line according to Figure 1. Only the RGU is a passive module. The other three are active modules and can be controlled to turn the pump on or off. 'Pump on' indicates their presence, otherwise, 'pump off' means they are not present.

3   EXPERIMENTAL RESULTS

The transmission system was optimized with approximately −510 ps/nm of dispersion pre-compensation (at the transmitter side of the span) and a single launch power of 13.7 dBm/ch. In addition to reducing the OSNR requirement, the EFEC also has an Stimulated Brillouin Scattering (SBS) suppression function (Li, 2017), which can increase the transmitter power to 23.7 dBm without significant SBS phenomena.

In option A, the power of the three co-pumped lasers is 912 mW, which provides about 8 dB distributed Raman on/off gain, and the 794-mW counter-pumped pumps provide 16.5 dB Raman on/off gain. During the test we slightly adjusted the VOA attenuation and got a 67 dB power budget (BER = 10^{-5} before correction).

In option B, the remote pump power is about 794 mW, which transmitted the 80 km bypass fiber with a 0.238 dB/m loss; the pump power reaching the RGU module is about 9.9 mW and the EDFA gain is about 23.46 dB. The power budget in option B is 72 dB.

In option C, there is a CtRA on the receiver side. The Raman pump power is about 900 mW in total. The remote pump power is about 824 mW and the RGU gain is around 24 dB. The system has a 73.5 dB link budget. Although option C has an extra CtRA in comparison to option B, this CtRA has not produced much effect. The final power budget is only 1.5 dB better than option B.

In option D, at the transmitter site, the CoRA pump power is 912 mW, which causes a 7.9 dB Raman gain. At the receiver site, the remote pump power is 794 mW. When the CoRA pump is turned off and on, we find that the OSNR can be increased by 4–5 dB. Therefore, we can further increase the VOA attenuation and finally attained a power budget of 77.5 dB.

Figure 3 shows the measured optical spectra at the transmitter and receiver Q-factors when the link loss is 74 dB. The measured OSNR is shown and the average received OSNR over ten channels at 0.1 nm Resolution Bandwidth (RBW) was 13.10 dB. The BERs of all channels were measured. Figure 3 also shows all channel Q-values. The average Q-factor is 8.84 dB and

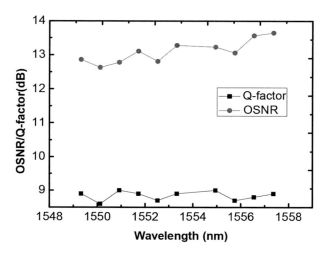

Figure 3. Received OSNR (at 0.1 nm resolution bandwidth) and Q-factors (74 dB link loss).

Table 1. Transmission performance of different amplifier-combination schemes.

Option	Amplifier configurations	Power budget (dB)	OSNR (dB) (1550.12 nm)	Cost ($)
A	BA+CoRA+CtRA+PA	67	9.8	5100
B	BA+ROPA+PA	72	10	3400
C	BA+CtRA+ROPA+PA	73.5	10.2	5400
D	BA+CoRA+ROPA+PA	77.5	10	5400

the worst channel has a Q-factor of 8.6 dB; all channels are therefore above the EFEC limit of 8.5 dB and would yield a BER below 10^{-15} after correction by EFEC.

For option D, the error-free results are obtained on all channels when the link loss is 74 dB. We have do error test on 10×9.95 Gb/s (Synchronous Digital Hierarchy (SDH) signal) and 10×10.31 Gb/s (10GE LAN Ethernet signal) in our experiment. Error-free results are obtained in 24 hours for all channels.

Table 1 summarizes the power budget, OSNR and cost of the four amplifier-combination schemes. We can see that the cost is relatively similar for options A, C and D. However, option D attains the largest power budget and a similar level of OSNR to the other options. Taking performance and cost into account, we think that option D is the best choice. From the transmission distance perspective, option A can only support a transmission line of 300–350 km. The other three cases can support transmission lines of 350–400 km. These factors should also be taken into account when configuring an amplifier scheme.

4 SUMMARY

In this paper, we demonstrated 10×10 Gb/s DWDM channel transmission over a 325-km unrepeatered link with G.652 fiber. The unrepeatered transmission is achieved by using different amplifier-combination schemes. Four kinds of amplifier-configuration structures are tested and the performance is verified in the 325-km transmission system. Experimental results show that different amplifier-combination schemes can be used for the same transmission distance. The combination of an EDFA/CoRA/ROPA scheme provides a good balance between performance and cost. The performance was tested in detail and a 77.5 dB power budget and 10 dB OSNR were obtained for this scheme. The research results of this paper have clear significance for designers deploying amplifiers in ultra-long-distance transmission systems.

REFERENCES

Bissessur, H., Etienne, S., Bousselet, P., Ruggeri, S. & Mongardien, D. (2012). 6 Tb/s unrepeatered transmission of 60 × 100 Gb/s PDM-RZ-QPSK channels with 40 GHz spacing over 437 km. In *Proceedings of European Conference and Exhibition on Optical Communication (ECOC '12)* (paper Mo.1.C.3). Washington, DC: Optical Society of America.

Bousselet, P., Mongardien, D.A., Brindel, P., Bissessur, H., Dutisseuil, E., Brandon, E. & Brylski, I. (2008). 485 km unrepeatered 4 × 43 Gb/s NRZ-DPSK transmission. In *Proceedings of Optical Fiber Communication Conference/National Fiber Optic Engineers Conference* (paper OMQ7). Washington, DC: Optical Society of America.

Cai, J., Bakhshi, B. & Nissov, M. (2008). Transmission of 8 × 40 Gb/s RZ-DQPSK signals over a 401 km unrepeatered link. In *Proceedings of Optical Fiber Communication Conference/National Fiber Optic Engineers Conference* (paper OMQ5). Washington, DC: Optical Society of America.

Chi, R., Li, A. & Sun, X. (2015). Optimized amplification scheme in unrepeated DWDM system over 325 km G.652 fiber. In *Proceedings of 14th International Conference on Optical Communications and Networks*. Piscataway, NJ: Institute of Electrical and Electronics Engineers. doi:10.1109/ICOCN.2015.7203718.

Du, M., Yu, J. & Zhou, X. (2008). Unrepeatered transmission of 107 Gb/s RZ-DQPSK over 300 km NZDSF with bi-directional Raman amplification. In *Proceedings of Optical Fiber Communication Conference/National Fiber Optic Engineers Conference* (paper JthA47). Washington, DC: Optical Society of America.

Gainov, V., Gurkin, N., Lukinih, S., Makovejs, S., Akopov, S., Ten, S., … Sleptsov, M. (2014). Record 500 km unrepeatered 1 Tbit/s (10 × 100 G) transmission over an ultra-low loss fiber. *Optics Express*, *22*(19), 22308–22313.

Li, L. (2017). *Digital band suppression SBS adjustable optical transceiver module*. Chinese patent no. CN104579493B. Retrieved from https://patents.google.com/patent/CN104579493B/.

Lucero, A., Foursa, D.G. & Cai, J. (2009). Long-haul Raman/ROPA-assisted EDFA systems. In *Proceedings of Optical Fiber Communication Conference/National Fiber Optic Engineers Conference* (paper OThC3). Washington, DC: Optical Society of America.

Zhu, B., Borel, P., Carlson, K., Jiang, X., Peckham, D. & Lingle, R. (2013). Unrepeatered transmission of 3.2 Tb/s (32 × 120 Gb/s) over 445 km fiber link with Aeff managed span. In *Proceedings of Optical Fiber Communication Conference/National Fiber Optic Engineers Conference* (paper OTu2B.2). Washington, DC: Optical Society of America.

SiN micro-resonator optical frequency comb with varying FSR spacing

Yu Gao, Yinglu Zhang, Yuedi Ding, Shanlin Zhu, Cheng Zeng, Qingzhong Huang & Jinsong Xia
Wuhan National Laboratory for Optoelectronics, Huazhong University of Science and Technology, Wuhang, Hubei, China

ABSTRACT: In this paper, we have experimentally realized Kerr Optical Frequency Combs (OFCs) with the bandwidth of ~200 nm using general single pump approach in a silicon nitride micro-resonator with a quality factor of 8.2×10^5. In addition, we have observed the OFCs generation with varying free spectral range (FSR) spacing using a feedback loop. By selecting different resonant wavelengths through the tunable filter, OFCs with the frequency spacing varying from 1 to 6-fold FSRs are successfully achieved. The proposed OFC with varying FSR spacing is a promising tool form any potential and practical applications, such as optical communication systems, optical metrology, arbitrary optical waveform generation and optical signal processing.

1 INTRODUCTION

Optical frequency combs (OFCs) (Kippenberg, Holzwarth et al. 2011, Pasquazi, Peccianti et al. 2018) have attracted much attention in the last decade. They are widely used in the fields of high-speed optical communication (Pfeifle, Brasch et al. 2014, Marin-Palomo, Kemal et al. 2017), spectroscopy (Suh, Yang et al. 2016, Yu, Okawachi et al. 2017), optical metrology (Papp, Beha et al. 2014), arbitrary optical waveform generation (Jiang, Huang et al. 2007, Ferdous, Miao et al. 2011), optical signal processing (Hamidi, Leaird et al. 2010, Supradeepa, Long et al. 2012), and sources for quantum entanglement (Reimer, Kues et al. 2016, Kues, Reimer et al. 2017).

Many of the previous OFCs generation technologies are based on the mode-locked lasers, such as Ti: sapphire laser (Gohle, Udem et al. 2005), Er-doped fiber laser (Gambetta, Ramponi et al. 2008), and Yb: fiber laser (Adler, Cossel et al. 2009). However, these commercial technologies are severely limited by their high cost, large volume and strict working environment. Besides, OFCs can be generated with optical modulators by phase modulation (Ozharar, Quinlan et al. 2008) or intensity modulation (Zhou, Zheng et al. 2011). Nevertheless, the bandwidths of the combs are restricted by the achievable modulation depth. Compared to the technologies based on mode-locked lasers and optical modulators, micro-resonator-based OFCs utilizing Kerr nonlinear effect exhibit enormous advantages due to their low cost, compactness, broadband, and ultra-high repetition rate. Their bandwidths can cover multiple optical communication bands, including the C, L and U bands. Since their compact structures, it is able to be on-chip integrated. Micro-resonator-based OFCs have been developed rapidly in recent years and have been demonstrated in a variety of platforms, including calcium fluoride (Savchenkov, Matsko et al. 2008), silicon nitride (Pfeifle, Brasch et al. 2014, Huang, Yang et al. 2015, Marin-Palomo, Kemal et al. 2017), aluminum nitride (Jung, Xiong et al. 2013), high-index doped silica glass (Pasquazi, Caspani et al. 2013) as well as diamond (Hausmann, Bulu et al. 2014).

In this paper, we combine the feedback loop with general single pump approach to realize the OFC with varying free spectral range (FSR) spacing in a silicon nitride (SiN) micro-resonator. By tuning the passband of the tunable filter, the frequency spacing of the OFC can vary from 1 to 6-fold FSRs. We have also achieved an OFC with a bandwidth of ~200 nm in the experiment.

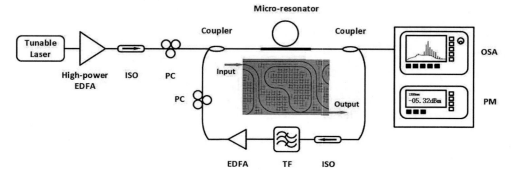

Figure 1. Experimental setup for OFCs generation with feedback loop. EDFA, erbium-doped fiber amplifier; ISO, Isolator; PC, Polarization Controller; TF, Tunable Filter; OSA, Optical Spectrum Analyzer; PM, Power Meter. The inset shows the micro-resonator formed by 1.5 μm × 0.8 μm SiN waveguides.

2 EXPERIMENT

As shown in the inset of Fig. 1, the OFC is generated through a one-bus coupled micro-ring resonator based on SiN waveguides with a cross section of 1.5 μm × 0.8 μm. The measured quality factor and FSR of the micro-ring resonator are 8.2×10^5 and 100 GHz, respectively.

Figure 1 illustrates the experimental setup for combs generation with the feedback loop. The light from the tunable semiconductor laser is amplified by a high power erbium-doped fiber amplifier (EDFA) and as pump light. After going through an isolator and a polarization controller, the pump light is combined with the light in the feedback loop by a 90/10 coupler and then coupled into the chip. The measured coupling loss at each facet is approximately 4.5 dB. The light coupled out of the chip is divided into two parts. 10% of the light is used to characterize the OFCs by monitoring the output optical power with a power meter and recording the optical spectrum with an optical spectrum analyzer. 90% of the light is coupled into the feedback loop, which consists of an isolator, a tunable filter, an EDFA and a polarization controller. The tunable filter in the loop removes the residual pump light but letting one neighboring resonant wavelengths pass. The polarization controller is used to adjust the polarization of the light in the loop.

3 RESULTS AND DISCUSSION

Figure 2 presents the measured spectra when the passband of tunable filter is tuned to different resonant wavelengths of the micro-resonator. The micro-resonator is pumped at 1560.92 nm and 32 dBm (the output power of the high-power EDFA). Meanwhile, we observe that the optical power measured by the power meter shows a significant decrease, indicating that the pump light is resonating with and coupled into the micro-resonator. Then turn on the EDFA in the feedback loop whose output power is set to 30dBm. When the passband of the tunable filter with 0.5-nm bandwidth locates at 1560.14 nm (corresponding to 1-FSR), multiple comb lines are observed as shown in Fig. 2(a). Figures 2(b)–2(f) show the spectra when the passband of tunable filter is moved to 1559.34 nm, 1558.54 nm, 1557.75 nm, 1556.96 nm, and 1556.17 nm, corresponding to 2 to 6-fold FSR of the micro-ring resonator respectively. The feedback loop introduce an additional gain profile which can significantly alter the dynamics of OFC generation. This feedback loop provides an effective way to generate combs with flexible FSR spacing.

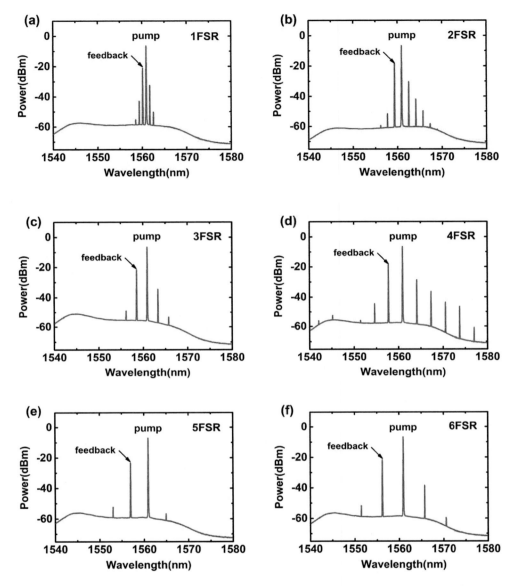

Figure 2. Spectra of the OFCs with flexible FSR spacing. (a) 1-FSR, (b) 2-FSR, (c) 3-FSR, (d) 4-FSR, (e) 5-FSR, (f) 6-FSR.

The measured spectra of the OFCs without and with the feedback loop in the system are shown in Figs. 3(a) and 3(b), respectively. Firstly, we pump the micro-resonator at 1561.55 nm and 36 dBm (the output power of the high-power EDFA), and the measured spectrum with the bandwidth of ~200 nm is shown in Fig. 3(a). Then we employ the feedback loop in the experimental setup and tune the passband of the tunable filter at 1558.37 nm with 0.5-nm bandwidth. The measured OFC spectrum is presented in Fig. 3(b) when the loop EDFA with 27 dBm power is turned on. The output optical power at 1558.37 nm is increased by ~25 dB. The experimental result demonstrates that when the selected comb is covered by the passband of the filter, the output optical power of the selected comb can be improved. This technology can be used to optimize OFCs.

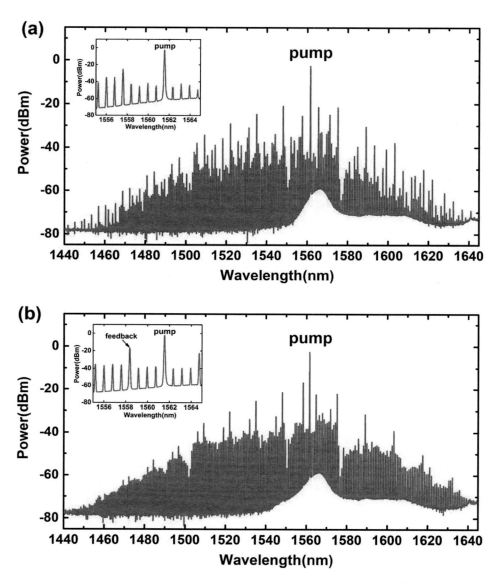

Figure 3.   OFC spectra generated in the SiN micro-resonator. (a) Without feedback loop, (b) With feedback loop. The insets of (a) and (b) show the zoom-ins of the spectra.

4   CONCLUSION

We experimentally demonstrate a OFCs generation scheme in a silicon nitride micro-resonator which has a quality factor of 8.2×10^5. Through combining the feedback loop and tuning the passband of the tunable filter, OFCs with varying FSR spacing from 1- to 6-fold FSRs are successfully achieved in the experiments. In addition, we pump the micro-resonator without and with the feedback loop and realize an OFC with a bandwidth of ~200 nm. These CMOS-compatible and on-chip integrated OFCs with varying FSR spacing will play an important role in many fields and achieve practical solutions for many key applications.

FUNDING

National Natural Science Foundation of China (NSFC) (61335002, 11574102, 61675084, 61775094); National High Technology Research and Development Program of China (2015 AA016904).

REFERENCES

Adler, F., K.C. Cossel, M.J. Thorpe, I. Hartl, M.E. Fermann and J. Ye (2009). "Phase-stabilized, 1.5 W frequency comb at 2.8–4.8 μm." *Optics Letters* **34**(9): 1330–1332.

Ferdous, F., H. Miao, D.E. Leaird, K. Srinivasan, J. Wang, L. Chen, L.T. Varghese and A.M. Weiner (2011). "Spectral line-by-line pulse shaping of on-chip microresonator frequency combs." *Nature Photonics* **5**: 770.

Gambetta, A., R. Ramponi and M. Marangoni (2008). "Mid-infrared optical combs from a compact amplified Er-doped fiber oscillator." *Optics Letters* **33**(22): 2671–2673.

Gohle, C., T. Udem, M. Herrmann, J. Rauschenberger, R. Holzwarth, H.A. Schuessler, F. Krausz and T.W. Hänsch (2005). "A frequency comb in the extreme ultraviolet." *Nature* **436**: 234.

Hamidi, E., D.E. Leaird and A.M. Weiner (2010). "Tunable Programmable Microwave Photonic Filters Based on an Optical Frequency Comb." *IEEE Transactions on Microwave Theory and Techniques* **58**(11): 3269–3278.

Hausmann, B.J.M., I. Bulu, V. Venkataraman, P. Deotare and M. Lončar (2014). "Diamond nonlinear photonics." *Nature Photonics* **8**: 369.

Huang, S.W., J. Yang, J. Lim, H. Zhou, M. Yu, D.L. Kwong and C.W. Wong (2015). "A low-phase-noise 18 GHz Kerr frequency microcomb phase-locked over 65 THz." *Scientific Reports* **5**: 13355.

Jiang, Z., C.-B. Huang, D.E. Leaird and A.M. Weiner (2007). "Optical arbitrary waveform processing of more than 100 spectral comb lines." *Nature Photonics* **1**: 463.

Jung, H., C. Xiong, K.Y. Fong, X. Zhang and H.X. Tang (2013). "Optical frequency comb generation from aluminum nitride microring resonator." *Optics Letters* **38**(15): 2810–2813.

Kippenberg. T.J., R. Holzwarth and S.A. Diddams (2011). "Microresonator-Based Optical Frequency Combs." *Science* **332**(6029): 555.

Kues, M., C. Reimer, P. Roztocki, L.R. Cortés, S. Sciara, B. Wetzel, Y. Zhang, A. Cino, S.T. Chu, B.E. Little, D.J. Moss, L. Caspani, J. Azaña and R. Morandotti (2017). "On-chip generation of high-dimensional entangled quantum states and their coherent control." *Nature* **546**: 622.

Marin-Palomo, P., J.N. Kemal, M. Karpov, A. Kordts, J. Pfeifle, M.H.P. Pfeiffer, P. Trocha, S. Wolf, V. Brasch, M.H. Anderson, R. Rosenberger, K. Vijayan, W. Freude, T.J. Kippenberg and C. Koos (2017). "Microresonator-based solitons for massively parallel coherent optical communications." *Nature* **546**: 274.

Ozharar, S., F. Quinlan, I. Ozdur, S. Gee and P.J. Delfyett (2008). "Ultraflat Optical Comb Generation by Phase-Only Modulation of Continuous-Wave Light." *IEEE Photonics Technology Letters* **20**(1): 36–38.

Papp, S.B., K. Beha, P. Del'Haye, F. Quinlan, H. Lee, K.J. Vahala and S.A. Diddams (2014). "Microresonator frequency comb optical clock." *Optica* **1**(1): 10–14.

Pasquazi, A., L. Caspani, M. Peccianti, M. Clerici, M. Ferrera, L. Razzari, D. Duchesne, B.E. Little, S.T. Chu, D.J. Moss and R. Morandotti (2013). "Self-locked optical parametric oscillation in a CMOS compatible microring resonator: a route to robust optical frequency comb generation on a chip." *Optics Express* **21**(11): 13333–13341.

Pasquazi, A., M. Peccianti, L. Razzari, D.J. Moss, S. Coen, M. Erkintalo, Y.K. Chembo, T. Hansson, S. Wabnitz, P. Del'Haye, X. Xue, A.M. Weiner and R. Morandotti (2018). "Micro-combs: A novel generation of optical sources." *Physics Reports* **729**: 1–81.

Pfeifle, J., V. Brasch, M. Lauermann, Y. Yu, D. Wegner, T. Herr, K. Hartinger, P. Schindler, J. Li, D. Hillerkuss, R. Schmogrow, C. Weimann, R. Holzwarth, W. Freude, J. Leuthold, T.J. Kippenberg and C. Koos (2014). "Coherent terabit communications with microresonator Kerr frequency combs." *Nature Photonics* **8**: 375.

Reimer, C., M. Kues, P. Roztocki, B. Wetzel, F. Grazioso, B.E. Little, S.T. Chu, T. Johnston, Y. Bromberg, L. Caspani, D.J. Moss and R. Morandotti (2016). "Generation of multiphoton entangled quantum states by means of integrated frequency combs." *Science* **351**(6278): 1176.

Savchenkov, A.A., A.B. Matsko, V.S. Ilchenko, I. Solomatine, D. Seidel and L. Maleki (2008). "Tunable Optical Frequency Comb with a Crystalline Whispering Gallery Mode Resonator." *Physical Review Letters* **101**(9): 093902.

Suh, M.-G., Q.-F. Yang, K.Y. Yang, X. Yi and K.J. Vahala (2016). "Microresonator soliton dual-comb spectroscopy." *Science* **354**(6312): 600.

Supradeepa, V.R., C.M. Long, R. Wu, F. Ferdous, E. Hamidi, D.E. Leaird and A.M. Weiner (2012). "Comb-based radiofrequency photonic filters with rapid tunability and high selectivity." *Nature Photonics* **6**: 186.

Yu, M., Y. Okawachi, A.G. Griffith, M. Lipson and A.L. Gaeta (2017). "Microresonator-based high-resolution gas spectroscopy." *Optics Letters* **42**(21): 4442–4445.

Zhou, X., X. Zheng, H. Wen, H. Zhang, Y. Guo and B. Zhou (2011). "All optical arbitrary waveform generation by optical frequency comb based on cascading intensity modulation." *Optics Communications* **284**(15): 3706–3710.

Frontier Research and Innovation in Optoelectronics Technology and Industry – Habib & Lewis (Eds)
© 2019 Taylor & Francis Group, London, ISBN 978-1-138-33178-5

Development of a dual-channel LSPR biosensing system for detection of melamine using AuNPs-based aptamer

Q.Q. Guo, S. Wang, H. Zhang, D.X. Li, J.J. Shang, H.F. Sun, Y.T. Yang, L.Z. Ma & J.D. Hu
College of Mechanical and Electrical Engineering, Henan Agricultural University, Zhengzhou, Henan Province, China

ABSTRACT: This study proposed a dual-channel Localized Surface Plasmon Resonance (LSPR) biosensing system via optical fibers for colorimetric detection of melamine using gold nanoparticles (AuNPs)-based aptamer. The aptamer which has high affinity and specificity towards melamine is employed to be the recognition element to capture melamine molecules in AuNPs solution. Quantitative analysis of melamine was performed by measuring spectral absorbance ratio of AuNPs-based aptamer. Under optimized conditions, the linear responses with the square values of correlation coefficient of 0.98 and 0.96 are obtained between the absorbance peak ratio A_{640}/A_{520} (absorbance at 640 nm and 520 nm) and melamine concentration in the range of 0 to 1 µM from channel 1 and channel 2, respectively. The difference of absorbance peak ratio A_{640}/A_{520} obtained from two channels is less than 0.03 when the solution containing a certain concentration of melamine was placed into both channels.

1 INTRODUCTION

Adulteration of food and drink has become one of the major problems in food industry (Chi et al. 2010; Bala et al. 2016). For instance, an adulterant, melamine ($C_3H_6N_6$) is a kind of white odorless crystalline powder with a high nitrogen content of 66.6% by mass due to containing three amino groups (Chang et al. 2017). Melamine is illegally added into milk to make it look protein rich (Filazi et al. 2012). Melamine in milk can result in kidney stones, renal failure, and even death in humans. The detection of melamine in food has become more and more important in the field of food safety (Vail et al. 2007; Li et al. 2014; Bala et al. 2015). Currently, the conventional detection methods of melamine involve high performance liquid chromatography (HPLC), enzyme linked immunosorbent assay (ELISA), and gas chromatography-mass spectrometry (GC-MS). Those methods have many advantages such as high selectivity and low limits of detection, however, they need expensive instruments, complicated sample pretreatment and trained operators (Wei et al. 2010; Bala et al. 2016). In this paper, an effective dual-channel LSPR biosensing system via optical fibers was put forward to the colorimetric detection of melamine using gold nanoparticles-based aptamer. The color of AuNPs solution shifts from red to blue due to the aggregation of AuNPs along with the increase of the concentration of melamine. The calibration curves of absorbance ratio (A_{640}/A_{520}) calculated from absorption spectra have been established, and square values of correlation coefficient of 0.98 and 0.96 are obtained from both channels. Based upon the experimental results, the melamine at a low concentration of below 1 µM can be rapidly and sensitively detected by the proposed LSPR biosensing system via optical fibers.

2 EXPERIMENTAL

2.1 *Synthesis of gold nanoparticles*

100 mL of aqueous 0.01% HAuCl$_4$ solution was heated to boiling followed by the rapid addition of 1 mL of 1% sodium citrate solution. The color changes from pale yellow to red,

indicating the formation of AuNPs. The AuNPs solution was further boiled for an additional 10 min and allowed to cool down at room temperature under stirring. Transmission electron microscope (TEM) image of the AuNPs solution was shown in Figure 1 A, and the spectra of AuNPs solution obtained from this preparation were monitored by UV-vis spectrophotometer. It is clearly found that the maximum absorption wavelength of AuNPs solution is about 520 nm (see Figure 1B).

2.2 Reagents

100 µM aptamers (5'-TTTTTTTTTTTTTTTTTTTTTTTTTTTTTTTTTTTTT-3') were custom synthesized by Sangon Biotechnology Co., Ltd. (Shanghai, China). Pure melamine was obtained from Sigma-Aldrich (USA). All solvents and reagents were of analytical grade and were used without further purification. Milli-Q water having a resistivity of 18.2 MΩ cm is used in experiments. The glassware was thoroughly rinsed with aqua regia and dried in air before use.

2.3 Principle of the LSPR-based biosensor

The principle of this method is based on the analysis of LSPR spectrum of noble metal nanoparticles (gold). When a small nanoparticle (AuNP) (R/λ < 0.1, where R is the radius of the AuNPs and λ is the wavelength of incident light) is irradiated by incident light, the electromagnetic field causes the free electrons to oscillate coherently. The LSPR resonance occurs when the frequency of the incident light is consistent with the vibration frequency of the free electrons in noble metal. The extinction can be accurately represented by using the quasistatic approximation equation,

$$E(\lambda) = \frac{24\pi^2 r^3 \varepsilon_m^{3/2} N}{\lambda \ln(10)} \frac{\varepsilon_i}{(\varepsilon_r + \chi \varepsilon_m)^2 + \varepsilon_i^2} \tag{1}$$

where, ε_m is the dielectric constant of the surrounding medium; $\varepsilon = \varepsilon_r + i\varepsilon_i$ is the dielectric constant of the bulk metal; r is the radius of the AuNPs and N is the electron density. The factor χ that appears in Eq. (1) is a magnitude of two for a spherical particle (Kelly et al. 2003).

2.4 AuNPs-based LSPR for the detection of melamine

The synthesized AuNPs without aptamer are stable in aqueous solution due to the repulsion force between citrate anions on the surface of AuNPs, which prevents the AuNPs from aggregating. When aptamer occurs in the AuNPs solution, aptamer was conjugated on the surface of AuNPs

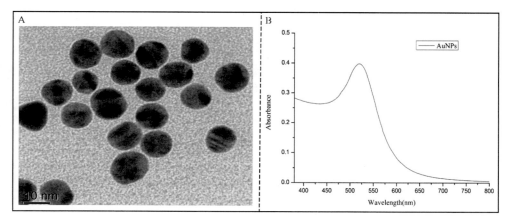

Figure 1.   A) TEM image of the AuNPs solution, (B) UV-vis absorption spectra of AuNPs with the diameter of 13.5 nm.

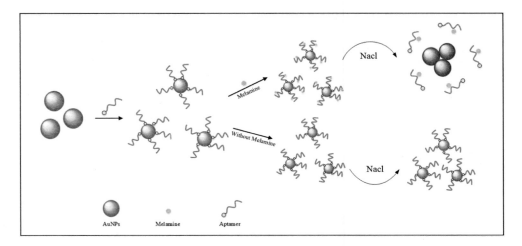

Figure 2. Schematic illustration of the principle of melamine utilizing aptamer-modified gold nanoparticles.

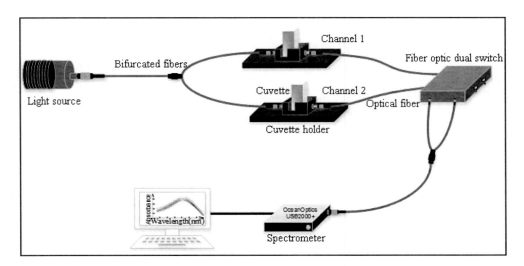

Figure 3. Schematic diagram illustrating the principle of the dual-channel LSPR biosensing system via optical fibers.

based on the binding of a strong Au-N bond. In the absence of melamine, the AuNPs are stable in aqueous solution owing to its random coil structure of the aptamer bases of outward negative charge. Consequently, AuNPs solution without melamine could keep well dispersed in red color (see Figure 2), although the electrolyte NaCl is added. However, in the presence of melamine molecules, the aptamer specifically binds to melamine due to the stronger affinity. As a result, desorption of the aptamer from surface of the AuNPs leads to expose AuNPs to the electrolyte NaCl. The color of AuNPs solution is changed from red to blue during the aggregation phase.

3 MEASUREMENT DESIGN

The schematic diagram of the LSPR biosensing system was shown in Figure 3. This LSPR biosensing system consists of a broadband light source (tungsten halogen lamp) that covers the spectral range from 200 nm to 1700 nm, two premium bifurcated optical fibers with SMA905 connectors, a fiber optic dual switch (FOS) and a miniature spectrometer with a linear charge coupled device (CCD) array. By adjusting the FOS manually, the spectral data from both channels are shown on the computer, graphically.

4   RESULTS

4.1   *Quantitative detection of melamine*

Melamine with concentrations of 0 μM, 0.2 μM, 0.4 μM, 0.8 μM and 1.0 μM was prepared in water, respectively. 40 μL of 1 μM aptamer were mixed with 100 μL of AuNPs solution, diluted with 200 μL water and incubated for 24h at room temperature. Afterwards, 100 μL of different concentrations of melamine were added into the AuNPs solution and the solution was again incubated for 30 min. Finally, 60 μL of NaCl were added into the AuNPs solution by following the measurement principle (see Figure 2).

The experimental results of the absorbance spectrum at different concentrations of melamine from channel 1 were shown in Figure 4. The absorbance peak ratio A_{640}/A_{520} (absorbance at 640 nm and 520 nm) exhibits a linear response to the melamine concentration in the range of 0 to 1 μM. The calibration curve is plotted by using the absorption peak ratio A_{640}/A_{520} as an ordinate. The linear equation of $Y = 0.85C + 0.15$ (C is the concentration of melamine) was obtained with the square of the correlation coefficient of 0.98.

From Figure 5, the absorbance was obtained from channel 2 when the concentration of melamine in the range of 0 to 1 μM was added into the aqueous solution. The absorption peak ratio of A_{640}/A_{520} is also linearly correlated with the concentration of melamine. The linear equation of $Y = 0.86C + 0.12$ (C is the concentration of melamine) was obtained with the square of the correlation coefficient of 0.96.

As described above, the linear equations obtained from channel 1 and channel 2 are expressed with $Y = 0.85C + 0.15$ and $Y = 0.86C + 0.12$, respectively. Obviously, the difference of absorbance peak ratio A_{640}/A_{520} calculated from two channels is less than 0.03 when the AuNPs solution containing a certain melamine concentration in the range of 0 to 1 μM was placed into both channels.

4.2   *Specificity measurement*

Various interfering substances, such as glucose, fructose, lactose and ammeline were applied to verify the selectivity of this LSPR biosensing system in order to put this method into

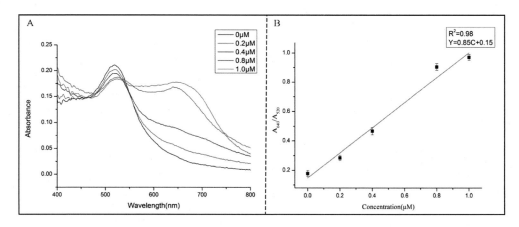

Figure 4.   Absorption spectrum obtained from channel 1 with different concentrations of melamine.

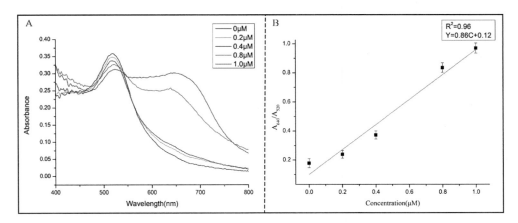

Figure 5. Absorption spectroscopy obtained from the channel 2 with different concentrations of melamine.

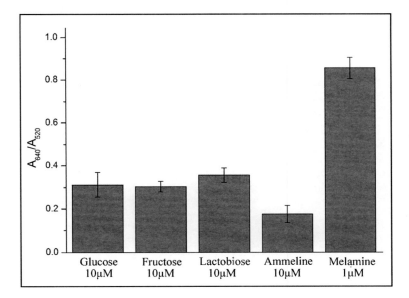

Figure 6. Relative responses obtained from aptamer solutions which were treated with different interfering substances.

the practical detection of milk samples. Selectivity studies were carried out using interfering substances with the concentration of 10 μM comparing to measurement results from the concentration of 1 μM of melamine. The experimental results show that only melamine could induce the aggregation of AuNPs and change the color of AuNPs from red to blue, which demonstrated that the aptamer modified AuNPs has good specificity for the detection of melamine. All other interfering substance were unable to aggregate the AuNPs, and the solution remained in red color (see Figure 6).

5 CONCLUSIONS

In this work, the dual-channel LSPR biosensing system via optical fibers for detection of melamine by using AuNPs-based aptamer is developed and validated. The results from the proposed method revealed that the absorbance peak ratio A_{640}/A_{520} (absorbance at 640 nm and 520 nm) for each channel is linear response to the melamine concentration in the range of 0 to 1 μM. The square values of correlation coefficient of 0.98 and 0.96 between the absorbance peak ratio A_{640}/A_{520} and melamine concentration in the range of 0 to 1 μM are obtained from channel 1 and channel 2, respectively. The proposed dual-channel biosensing system is entirely devoid of the use of complex sample preparation, enzyme, or sophisticated instruments, thus offering its potential for the rapid and efficient colorimetric detection of absorbance in general. Moreover, the very small difference of absorbance was existed between both channels, providing this dual-channel LSPR biosensing system can be used for effective detection of plant hormones, such as, auxins, cytokinins, gibberellins, and abscisic acid (ABA) andochratoxin A (OTA) upon the LSPR biosensing method, besides melamine.

REFERENCES

Bala, R., Kumar, M., Bansal, K., Sharma, R.K., & Wangoo, N. 2016. Ultrasensitive aptamer biosensor for malathion detection based on cationic polymer and gold nanoparticles. *Biosensors & Bioelectronics* 85:445–449.

Bala, R., Sharma, R.K., & Wangoo, N. 2015. Highly sensitive colorimetric detection of ethyl parathion using gold nanoprobes. *Sensors & Actuators B Chemical* 210:425–430.

Bala, R., Sharma, R.K., & Wangoo, N. 2016. Development of gold nanoparticles-based aptasensor for the colorimetric detection of organophosphorus pesticide phorate. *Analytical & Bioanalytical Chemistry* 408(1):333–338.

Chang, K., Wang, S., Zhang, H., Guo, Q., Hu, X., & Lin, Z. 2017. Colorimetric detection of melamine in milk by using gold nanoparticles-based LSPR via optical fibers. *Plos One* 12(5):e0177131.

Filazi, A., Sireli, U.T., Ekici, H., Can, H.Y., & Karagoz, A. 2012. Determination of melamine in milk and dairy products by high performance liquid chromatography. *Journal of Dairy Science* 95(2):602–608.

Hong, C., Liu, B.H., Guan, G.J., Zhang, Z.P., & Han, M.Y. 2010. A simple, reliable and sensitive colorimetric visualization of melamine in milk by unmodified gold nanoparticles. *Analyst* 135(5): 1070.

Kelly, K.L., Coronado, E., Lin, L.Z., & Schatz, G.C. 2003. The optical properties of metal nanoparticles: the influence of size, shape, and dielectric environment. *Cheminform* 34(16):668–677.

Li, Y., Xu, J., & Sun, C. 2014. Cheminform abstract: chemical sensors and biosensors for the detection of melamine. *Rsc Advances* 5(2):1125–1147.

Vail, T., Jones, P.R., & Sparkman, O.D. 2007. Rapid and unambiguous identification of melamine in contaminated pet food based on mass spectrometry with four degrees of confirmation. *Journal of Analytical Toxicology* 31(6):304–312.

Wei, F., Lam, R., Cheng, S., & Lu, S. 2010. Rapid detection of melamine in whole milk mediated by unmodified gold nanoparticles. *Applied Physics Letters* 96(13):133702–133703.

Frontier Research and Innovation in Optoelectronics Technology and Industry – Habib & Lewis (Eds)
© *2019 Taylor & Francis Group, London, ISBN 978-1-138-33178-5*

X-ray absorption gratings fabricated via nanoparticles for differential phase-contrast imaging

J. Li, J. Huang, X. Liu, J. Guo & Y. Lei
Key Laboratory of Optoelectronic Devices and Systems of Ministry of Education and Guangdong Province, College of Optoelectronic Engineering, Shenzhen University, Shenzhen, China

ABSTRACT: The advantages of grating-based X-ray Differential Phase-Contrast Imaging (DPCI) in detecting the composite materials constituted by light elements have been demonstrated by many researchers. However, the fabrication of absorption gratings still remains a great challenge, especially large-area ones. In practice, the high-temperature micro-casting technique hinders the fabrication of such gratings due to the fragility and absorption homogeneity of large-area absorption gratings. Here, we propose and implement a novel fabrication method to avoid the high-temperature process. We use tungsten nanoparticles as an X-ray absorbing material, and insert them into the grating structures with carriers of organic solvent and emulsifying agent and the assistance of a negative-pressure process. In addition, we compare the X-ray projection absorbing contrast between two gratings with the same grating structure (period of 42 µm, depth of 150 µm) but fabricated by micro-casting and nanoparticle filling. Our research shows this method can provide lower-cost large-area absorption gratings, promoting the further development of DPCI.

Keywords: X-ray Imaging, Phase-Contrast, X-ray Grating, nanoparticle filling, micro-casting technique

1 INTRODUCTION

In the early stage of X-ray phase-contrast imaging, synchrotron radiation and micro-focus X-ray sources were indispensable illuminating sources to guarantee the requirement of coherence (Wilkins et al., 1996; David et al., 2002; Momose et al., 2003). However, both of these sources block access to the widespread application of this imaging technique, because of the conflict between convenience and brightness. An appropriate source for X-ray phase-contrast imaging appeared when an absorption grating was introduced downstream of an X-ray tube (Pfeiffer et al., 2006). The absorption grating divides the conventional X-ray tube into an array of individually coherent emitters, but mutually incoherent sources, and each line source satisfies the requirement of spatial coherence for the self-image of the phase grating and has a constructive effect on the imaging process (Pfeiffer et al., 2006; Du et al., 2016). Consequently, potential applications of grating-based X-ray Differential Phase-Contrast Imaging (DPCI) are promising, such as early diagnosis of lesions, characterization of polymers in materials science, non-destructive testing in industry, and guaranteeing of safety (Stampanoni et al., 2011; Miller et al., 2013; Momose et al., 2014; Bachche et al., 2017).

As an important technology for the fabrication of various kinds of microstructures, LIGA (lithography, electroplating, and molding) has conventionally been used for the fabrication of X-ray absorption gratings. In principle, microstructures with large area and High Aspect Ratio (HAR) can be achieved using this technology, which relies, however, mostly on a synchrotron source (Noda et al., 2008; Schröter et al., 2017). In 2007, David et al. presented alternative methods to develop absorption gratings, the processes of which include photolithography, anisotropic wet etching and electroplating. However, the aspect ratio is insufficient to acquire a high-contrast image at high X-ray photon energies. A micro-casting technique has been developed to fabricate HAR and low-cost absorption gratings. It involves a three-step process: formation of the HAR grating structure in a silicon wafer; surface modification enhancing the wettability of the structure surface with the molten material; filling the etched structure with molten metal (Lei et al., 2013, 2016). In 2017, a hot embossing process was implemented using eutectic gold–tin foils (Au 80 Wt% / Sn 20 Wt%) to be filled into Si templates (size 70×70 mm^2) (Romano et al., 2017). Although the metals' melting temperatures in both methods, bismuth and Au–Sn alloy, are low, changes in temperature may, unfortunately, induce a bent grating surface, especially in the fabrication of large-area absorption gratings.

In this paper, we propose an innovative nanoparticle filling method to fabricate absorption gratings without the high-temperature process. To obtain better absorption, we chose tungsten nanoparticles that have large X-ray mass attenuation coefficients as the absorbing materials. They are prepared as a turbid liquid made up of ethanol and an emulsifying agent, OP-10, and induced into the grating structures formed by Deep Reactive Ion Etching (DRIE) in silicon wafers. Compared with micro-casting, this method can avoid the high-temperature process that is generally responsible for the distortion of large-area absorption gratings, and be carried out in a low-cost manner in laboratories. Furthermore, its feasibility has been verified by a comparison between the image contrast curves obtained by X-ray projection to a tungsten-nanoparticle grating and a micro-cast bulk bismuth grating.

2 EXPERIMENTS

Our fabrication procedure consists of three steps. The first step, the absorption grating, was set with a period of 42 μm, a sidewall width of 10.5 μm and a depth of 150 μm in a silicon wafer, which has been designated according to our imaging system (Lei et al., 2013). Because of its large period and low aspect ratio, such an absorption grating structure was easily accomplished by DRIE, including the standard photolithography process used for pattern definition on the silicon wafer, and the Bosch process for trench formation. This process has been described elsewhere (Bui et al., 2015; Genova et al., 2018). A cross-sectional micrograph obtained by a Scanning Electron Microscope (SEM) is shown in Figure 1.

Although the width of the structure grooves is much larger than the diameter of the tungsten nanoparticles (their nominal average diameter is 50 nm), these nanoparticles cannot be filled directly into these grooves in large numbers, even using a centrifugal device. Therefore,

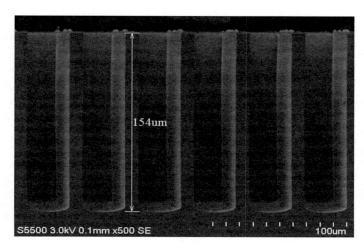

Figure 1. SEM image of the section profile of the absorption grating.

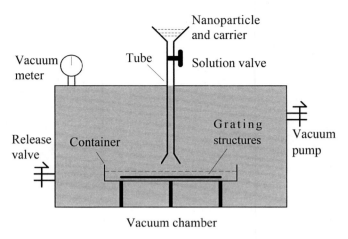

Figure 2.   Diagrammatic sketch of the vacuum chamber used to release air in the grating grooves.

in the second step, special solutions are employed as carriers to induce nanoparticles into the grooves, including into their bases. Surface modification is the key factor affecting the material filling ratio and should be implemented on the grating so that a wetting layer can be formed on the surfaces, including the side walls (Lei et al., 2016). In our experiment, the material of the wetting layer is the same as the nanoparticle carrier, which was composed mainly of ethanol and the emulsifying agent, OP-10. We uniformly mixed ethanol and OP-10 at the ratio of 50:1 by volume using an ultrasonic machine for 7 minutes at a power of 750 W. At the same time, the grating structure was placed right-side up into a container located in a vacuum chamber and the air from the grooves was evacuated for 30 minutes. A diagrammatic sketch of the vacuum chamber is shown in Figure 2. Then the solution flowed into the container through a tube on the top until it covered the grating surface. The structure was taken out of the chamber approximately 15 minutes later and was dried at a temperature of 80°C. Finally, this grating structure was returned to the vacuum chamber, which was evacuated again for 30 minutes, and then covered with the nanoparticle carrier. The carrier had been pre-prepared using the following steps: the tungsten nanoparticles were weighed at the ratio of 1:200 with the carrier solvent, and dispersed uniformly into it using an ultrasonic machine at a power of 750 W for 8 minutes.

For the third step, after the right-side up immersion of the grating structure in the carrier, the container was taken out of the chamber and the carrier was left to stand for 8 hours to allow the nanoparticles to fall slowly into the structure surface in the atmosphere, so that the nanoparticles would fall freely into the grooves in a certain proportion until they reached the bottom.

3   RESULTS AND DISCUSSIONS

Figure 3 shows the image obtained from the SEM, focusing on the filling of tungsten nanoparticles in the absorption gratings. It can be seen that nanoparticles have filled to the bottom of the grooves and no air is observed, which indicates the surface modification process has worked. In addition, we can see the dense arrangement of these nanoparticles, which differs significantly from the state of natural arrangement without the use of any solution, and this is important in an absorption grating, because of blocking more X-rays.

To compare the absorption performance, we also filled the grooves of the same Si substrate as the above-mentioned grating structures with molten bismuth using micro-casting technology (Lei et al., 2016). The outcome of the filling was observed by SEM and its cross-sectional micrograph is shown in Figure 4, where the black and gray parts are the Si substrate and the bismuth filling, respectively.

Figure 3. SEM cross-sectional micrographs of absorption grating filled by tungsten nanoparticles.

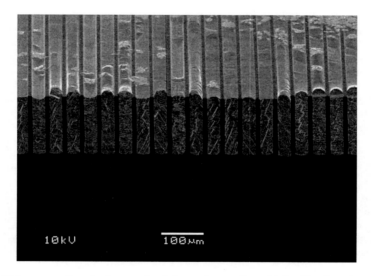

Figure 4. SEM cross-sectional micrograph of absorption grating obtained by the micro-casting technique; black and gray parts are the Si substrate and the bismuth filling, respectively.

To test and analyze quantitatively the effect of the gratings on the absorption of X-rays, we constructed an X-ray imaging system using a micro-focus source (5 μm @ 4 W) and a flat panel detector with pixel size of 75×75 μm based on a geometrical projection. The distance between the source and detector was set to 1 m and the absorption grating was placed downstream of the source at approximately 1.5 cm, which represents an amplification factor of 68. The voltage and current were set to 40 kV and 80 μA, respectively, and the exposure time to 5 s. After the acquisition of absorption images from both absorption gratings, the nominalized contrasts according to the distribution of pixel value were obtained, as shown in Figure 5, where the red and blue lines represent the contrasts obtained from the bulk bismuth and tungsten-nanoparticle absorption gratings, respectively. Obviously, the contrast of the red line (50.0%) is larger than that of the blue one (44.9%). The smaller contrast occurring

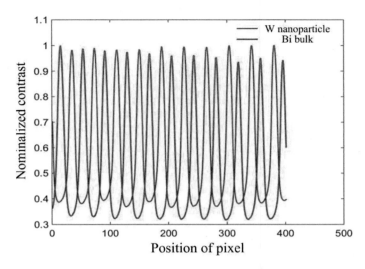

Figure 5. A micro-focus X-ray source is used to illuminate the bulk bismuth and tungsten-nanoparticle absorption gratings, with the contrasts of the resulting images shown as red and blue lines.

in the nanoparticle filling is ascribed to the inevitable gaps among these nanoparticles, even though the X-ray absorption coefficient of bulk tungsten is larger than that of bulk bismuth. We can observe that the grooves are not completely filled by the tungsten nanoparticles but are full of the bulk bismuth, which may be considered as another reason for the smaller contrast. If a complete filling with tungsten nanoparticles could be achieved, there would have been a larger absorption of the X-rays and, perhaps, an equivalent absorption effect to that of the bulk bismuth.

4 CONCLUSIONS

This contribution presents our efforts in improving the technology for fabricating absorption gratings used in DPCI assembly. A novel method for the fabrication of absorption gratings has been developed using metal nanoparticles as the absorbing material. This process uses ethanol and an emulsifying agent as the carrier to induce these nanoparticles into the grooves of the grating structure and cause them to arrange compactly, which is beneficial to improving the effective mass coefficient of the nanoparticles. Negative pressure is used to ensure the acquisition of a complete filling with few air bubbles in the grooves. Thus, an absorption grating with a high filling ratio has been obtained. Furthermore, the results show that the absorption contrast of a tungsten-nanoparticle grating is only slightly less than that of a bulk bismuth grating, demonstrating that this novel fabrication of an absorption grating offers an easy implementation with low cost in laboratories.

ACKNOWLEDGMENTS

We thank Mrs. Xu for her assistance in the fabrication of the gratings used in this work. This research is supported by the National Special Foundation of China for Major Science Instrument (61227802), the National Natural Science Foundation (61405120, 61605119, 61571305, 11674232), the NSF Natural Science Foundation of Shenzhen University (2017017), and the Natural Science Foundation of Shenzhen (JCYJ20150324141711612, JCYJ20170302142617703).

REFERENCES

Bachche, S., Nonoguchi, M., Kato, K., Kageyama, M., Koike, T., Kuribayashi, M. & Momose, A. (2017). Laboratory-based X-ray phase-imaging scanner using Talbot-Lau interferometer for nondestructive testing. *Sci Rep*, 7, 6711.

Bui, T.T., Tu, H.P. & Dang, M.C. (2015). DRIE process optimization to fabricate vertical silicon nanowires using gold nanoparticles as masks. *Adv Nat Sci Nanosci Nanotechnol*, 6, 045016.

David, C., Bruder, J., Rohbeck, T., Grünzweig, C., Kottler, C., Diaz, A., ... Pfeiffer, F. (2007). Fabrication of diffraction gratings for hard X-ray phase contrast imaging. *Microelectron Eng*, 84, 1172–1177.

David, C., Nöhammer, B., Solak, H.H. & Ziegler, E. (2002). X-ray phase contrast imaging using a shearing interferometer. *Appl Phys Lett*, 81, 3287–3289.

Du, Y., Lei, Y., Liu, X., Huang, J., Zhao, Z., Guo, J., ... Niu, H. (2016). A low cost method for hard X-ray grating interferometry. *Phys Med Biol*, 61, 8266–8275.

Genova, V.J., Agyeman-Budu, D.N. & Woll, A.R. (2018). Time multiplexed deep reactive ion etching of germanium and silicon – A comparison of mechanisms and application to X-ray optics. *J Vac Sci Technol B Nanotechnol Microelectron*, 36, 011205.

Lei, Y., Du, Y., Li, J., Huang, J., Zhao, Z., Liu, X., ... Niu, H. (2013). Application of Bi absorption gratings in grating-based X-ray phase contrast imaging. *Appl Phys Express*, 6, 117301.

Lei, Y., Liu, X., Li, J., Guo, J. & Niu, H. (2016). Improvement of filling bismuth for X-ray absorption gratings through enhancement of wettability. *J Micromech Microeng*, 26, 065011.

Miller, E.A., White, T.A., McDonald, B.S. & Seifert, A. (2013). Phase contrast X-ray imaging signatures for security applications. *IEEE Trans Nucl Sci*, 60, 416–422.

Momose, A., Kawamoto, S., Koyama, I., Hamaishi, Y., Takai, K. & Suzuki, Y. (2003). Demonstration of X-ray Talbot interferometry. *Jpn J Appl Phys*, 42, L866–L868.

Momose, A., Yashiro, W., Kido, K., Kiyohara, J., Makifuchi, C., Ito, T., ... Tanaka, J. (2014). X-ray phase imaging: From synchrotron to hospital. *Philos Trans A Math Phys Eng Sci*, 372, 20130023.

Noda, D., Tanaka, M., Shimada, K., Yashiro, W., Momose, A. & Hattori, T. (2008). Fabrication of large area diffraction grating using LIGA process. *Microsyst Technol*, 14, 1311–1315.

Pfeiffer, F., Weitkamp, T., Bunk, O. & David, C. (2006). Phase retrieval and differential phase-contrast imaging with low-brilliance X-ray sources. *Nat Phys*, 2, 258–261.

Romano, L., Vila-Comamala, J., Kagias, M., Vogelsang, K., Schift, H., Stampanoni, M. & Jefimovs, K. (2017). High aspect ratio metal microcasting by hot embossing for X-ray optics fabrication. *Microelectron Eng*, 176, 6–10.

Schröter, T.J., Koch, F.J., Kunka, D., Meyer, P., Tietze, S., Engelhardt, S., ... Mohr, J. (2017). Large-area full field X-ray differential phase-contrast imaging using 2D tiled gratings. *J Phys D: Appl Phys*, 50, 225401.

Stampanoni, M., Wang, Z., Thüring, T., David, C., Roessl, E., Trippel, M., ... Hauser, N. (2011). The first analysis and clinical evaluation of native breast tissue using differential phase-contrast mammography. *Invest Radiol*, 46, 801–806.

Wilkins, S.W., Gureyev, T.E., Gao, D., Pogany, A. & Stevenson, A.W. (1996). Phase-contrast imaging using polychromatic hard X-rays. *Nature*, 384, 335–338.

Frontier Research and Innovation in Optoelectronics Technology and Industry – Habib & Lewis (Eds)
© 2019 Taylor & Francis Group, London, ISBN 978-1-138-33178-5

High-efficiency and folding silicon solar cell modules

Yijie Li, Yaoju Zhang & Chaolong Fang
*College of Physics and Electronic information Engineering, Wenzhou University, Wenzhou,
Zhejiang Province, China*

ABSTRACT: A high efficiency and foldable silicon solar cell module is demonstrated. The experimental module consists of nine silicon solar cells glued at a soft Polydimethylsiloxane (PDMS) polymer substrate and coated by a PDMS wrinkle reflective (AR) film. The large-scale nano-wrinkle AR film is fabricated by a simple and low-cost plasma surface oxidation method. Optical measurement and electrical characterization show the nanowrinkle AR film can reduce the front-side reflectance and improve the photoelectric conversion efficiency of solar cell modules. Interestingly, the silicon solar cell module encapsulated with soft film is foldable, which can be mounted on the curved surface of some outdoor equipment and tools, such as, car roof and illumination tools.

Keywords: Antireflection coating; PDMS; Wrinkle; Foldable; Efficiency; Solar cell module

1 INTRODUCTION

Ever increasing energy demands and the commitments to reduce CO_2 emissions have brought renewable energy sources to the fore. As one of the green energy source with highest potential, solar energy would definitely be a best option for future energy demand since it is superior in terms of availability, cost effectiveness and efficiency compared to other renewable energy sources (Paulo & Geciane 2017, Trinh et al. 2018). An increasing number of researchers have been attracted to explore various solar cells to convert solar radiation into electrical energy, such as crystalline Si (c-Si) (Hara et al. 2015, Ye et al. 2015), amorphous Si (a-Si) (Lin et al. 2016), CdTe (Wolden et al. 2016, Yu et al. 2011), GaAs (Kosten et al. 2013), organic (Chen et al. 2014, Espinosa et al. 2012) and perovskite solar cell (Tavakoli et al. 2015), and so on. Since the crystalline Si (c-Si) solar cells have many advantages of high power conversion efficiencies, long lifespan, reliable performance, and proven manufacturability, they remain in the most widespread use in the photovoltaic industry (Dumont et al. 2017, Kannan & Vakeesan 2016). Due to its inflexibility and fragility, the c-Si solar cell encapsulated by transparent coverglass is generally installed on the rigid flat substrate (Lipomi et al. 2011). The glass-encapsulation might result in heavy mass of solar cell modules. More importantly, rigid solar cell modules are unable to be directly installed on some equipment and tools with a curved surface, such as the roof of a building, vehicle, and lighting tools without specialized facilities and installation condition. So it is necessary to develop foldable and flectional c-Si solar cell modules with high efficiency.

On the other hand, c-Si solar cell modulus generally encapsulated by flat coverglass lead to inefficient utilization of a part of incoming photon. In this regards, various antireflective (AR) schemes have been developed to achieve high-efficiency solar cell devices. Traditionally, a single-layer homogeneous dielectric film was mounted on the PV cell to reduce reflectance loss based on destructive interference, which needs to meet the following condition: $\lambda/4n_c$, (λ is wavelength and $n_c = (n_a n_s)^{1/2}$, n_a and ns are refractive indexes of air and film). Consequently, the film thickness needs to be precisely controlled once the material of the film is decided. Particularly, the reflectance loss is suppressed only at a specific wavelength increased significantly with the incident angle of light deviates from the normal direction. Meanwhile, micro/nanostructures, such as nano/micropyramid (Lu et al. 2016, Papet et al. 2006), micro/nanolens array (Chen & Hong 2016), micro/nanocone (Leem et al. 2015, Tsui et al. 2014), subwavelength grating (Chong et al. 2012, Esteban et al. 2010, Jia et al. 2016, Kanamori et al. 2001, Zi et al. 2017) and nanowire array (Huang et al. 2012, Jung et al. 2010, Kelzenberg et al. 2010) have been discovered with a broadband and omnidirectional light-harvesting capability. However, these structures have been fabricated with costly methods, such as lithographic and wet/dry etching (Aabdin et al. 2017, Mavrokefalos et al. 2012), e-beam direct write (Perl et al. 2014), nanoprint, chemical self-assembly approaches (Askari & Islam 2012, Mayer et al. 2016).

In our work, a novel and well-engineering method to fabricate flexible AR nanowrinkle film using a simple and low-cost plasma surface oxidation method. The flexible AR films can be naturally and closely adhered on a c-Si solar modulus, consisting of nine c-Si solar cells mounted on the flexible substrate. Mechanical test displays the c-Si solar modulus can be curved and folded. Optical measurement shows the flexible nanowrinkle pattern can reduce the front-side reflectance of the modulus from 5.2% to 7.2%. Electrical characteritics demonstrates the flexible AR nanowrinkle film can improve the short-circuit current from 0.409 A to 0.421 A, thus improving the power conversion efficiency of the modulus. The demonstrated Si solar cell modules can be directly installed on some equipment and buildings with a curved surface.

2 EXPERIMENTAL DETAILS

The fabrication of the solar cell with PDMS wrinkle AR coating was divided into two steps. The first step is combination and encapsulation of solar cells. First, c-Si solar cell panels were sequentially put into deionized water, absolute ethanol and acetone and cleaned ultrasonically for 5 min. After surface modification using plasma surface cleaning technique to facilitate adhension between panel and epoxy resin, these panels soldered with the electrode wires using an electric iron. Three cell panels were connected in series and three-in-series cells were connected in parallel. Then, epoxy resin prepolymer and crosslinking agent with mass ratio of 3:1 were mixed and casted on the upper and lower surfaces of the panel degassed and baked at 80°C for 2h. Subsequently, Polydimethylsiloxane (PDMS) monomer (GE RTV 615 component A) and its crosslinking agent (GE RTV 615 component B) with a weight ratio of 10:1 were mixed, and the mixture was degassed in the vacuum drying oven and baked at 80°C for 4h. After cured, the PDMS film was placed on glass and the PDMS can naturally and closely adhere to the glass substrate. Then the degassed premixed PDMS was casted on the PDMS film and paved the surface naturally. Finally, the array cell modulus was put on the liquid premixed PDMS and backed.

The second step is fabrication of PDMS wrinkle AR coating. The fabrication procedures of a PDMS AR coating with an ordered wrinkle were demonstrated in Figure 1. First, a 200

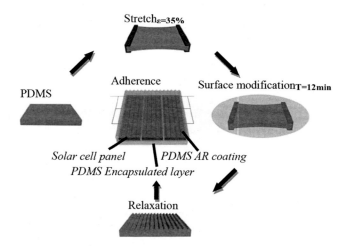

Figure 1. Fabrication procedures of a folding solar cell module with PDMS nanowrinkle AR film.

μm thick rectangle PDMS film with an area of 80 mm × 80 mm was mounted on a strain stage. Two ends of the PDMS film were firmly fixed on two clamps of the strain stage and stretched to 25% in one direction. After surface modification by a plasma etching device for 10 min, the stretching PDMS film gradually recovered to the original unstretched state. Finally, the PDMS film was put on the cell module. The PDMS film can be naturally and closely adhered to the cell module because of the strong Van der Waals interaction between PDMS and flat surface of rigid and flexible materials.

3 RESULTS AND DISCUSSIONS

Figure 2 shows images of a 9-cell-panel folding module with nanowrinkle film under flat and curved state, where the each panel area in the module is 2 cm × 2 cm. Note that the module displays a diffraction effect indicates their excellently ordered distribution of nanowrinkles. Figure 3 shows atomic force microscopy (AFM) and scanning electron microscopy (SEM) nanowrinkle images, which obviously present an ordered periodic nanowrinkle pattern, which indicates a high-quality nanowrinkle AR film has been obtained using plasma surface oxidation method. The detailed optical performance characterization will be discussed later. The preparation mechanism of the nanowrinkle pattern can be explained with the following rationale. Since the PDMS film under stretching state was modified in oxygen environment using a plasma etching device, the PDMS film surface is oxidized to change elastic modulus of the PDMS film surface. When the imposed stretching force is removed, difference between the oxidized layer and PDMS film causes contraction inconsistency, which leads to formation of nanowrinkle. In our experiment, the obtained nanowrinkle period and depth are 933 nm and 280 nm.

To quantitatively characterize the AR effect of PDMS nanowrinkle on the solar cell module, reflectance spectra in a wavelength range from 400 to 1000 nm were obtained using LAMBDA™ 1050 UV/Vis/NIR spectrometer along with 150-mm integrating sphere (PerkinElmer, Inc, Sheldon, CT, USA). Figure 4 shows the reflectance spectra of a plane-epoxy-resin-encapsulated Si device and a device covered with a PDMS nanowrinkle AR film measured with normal incident light. It can be seen that the reflectance of the device with the nanowrinkle AR film is decreased by ca. 2.1% for the given wavelength range, compared to the one without PDMS nanowrinkle film. This reduction of front-side reflectance is primarily because the nanowrinkles enable the light to couple light into the device (Mayer et al. 2016).

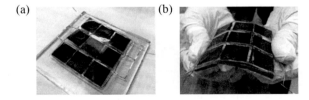

Figure 2. (a) and (b) Physical images of a folding silicon solar cell module under flat and curved state.

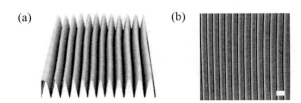

Figure 3. (a) AFM images of nanowrinkles. (b) SEM images of nanowrinkles. Scale bar 1000 nm.

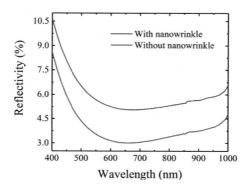

Figure 4. Reflectance measurements of a folding silicon solar cell module with and without PDMS nanowrinkle film.

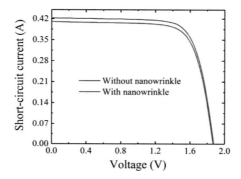

Figure 5. The current–voltage (*I–V*) curve of the device with and without the nanowrinkle AR film.

Table 1. Electrical performance parameters of solar cells.

Device	V_{oc} (V)	I_{sc} (A)	FF (%)	η (%)
w/wrinkle	1.92	0.421	71.3	16.0
w/o wrinkle	1.91	0.409	71.3	15.5

The aim to introduce optical AR effect on solar module is to improve photon utilization efficiency, eventually to enhance device performance. Therefore, electrical characterization of a folding silicon solar cell module is performed systematically with and without attaching the nanowrinkle AR film for the sake of comparison. Figure 5 shows the current–voltage (*I–V*) curves of the device with and without the nanowrinkle AR film, which are acquired with a Newport Oriel PVIV Station using under 1 Sun illumination. Obviously, the nanowrinkle AR film can improve power conversion efficiency (PCE) of the silicon solar cell module. Meanwhile, the measured electrical parameters of the silicon solar cell module with and without attaching the nanowrinkle AR film are summarized in the Table 1. The short-circuit currents of the silicon solar cell module with and without attaching the nanowrinkle AR film are 0.421 A and 0.409 A, respectively, which indicates ca. 2.9% improvement. While the open-circuit voltage and fill factor almost keep constant. Consequently, the nanowrinkle AR film leads to the solar cell PCE increase from 15.5% to 16.0%.

4 CONCLUSION

In conclusion, we present a folding silicon solar cell module consisting of 9 epoxy-resin-packaged solar cell panels sandwiched between a flexible PDMS nanowrinkle AR film and a flat PDMS substrate. Mechanical test demonstrates the c-Si solar modulus can be curved and folded. The flexible PDMS nanowrinkle film has been used as an additional AR layer mounted on a folding silicon solar cell module to improve the power conversion efficiency. The optical property investigations have shown that the nanowrinkle can reduce the module front-side reflectance. The systematic solar cell device electrical characterizations clearly showed that the flexible PDMS nanowrinkle AR film can considerably improve device short-circuit current density, leading to a conspicuous increase of power conversion efficiency. The proposed solar cell modules can be mounted on the curved surface of some outdoor equipment and tools, such as, car roof and illumination tools.

REFERENCES

Aabdin, Z., Xu, X., Sen, S., Anand, U., Kral, P. & Holsteyns, F., et al. 2017. Transient clustering of reaction intermediates during wet etching of silicon nanostructures. *Nano Letters* 17(5): 2953–2958.

Askari, D. & Islam, N. 2012. Improving the Pumping Efficiency of a Micropump Using Hydrophobic Nanocomposite Coating. *ASME 2012 International Mechanical Engineering Congress and Exposition* 2012: 275–279.

Chen, J., Zhou, L., Ou, Q., Li, Y., Shen, S. & Lee, S., et al. 2014. Enhanced light harvesting in organic solar cells featuring a biomimetic active layer and a self-cleaning antireflective coating. *Advanced Energy Materials* 4(9): 1289–1295.

Chen, W.H. & Hong, C.N. 2016. 0.76% absolute efficiency increase for screen-printed multicrystalline silicon solar cells with nanostructures by reactive ion etching. *Solar Energy Materials & Solar Cells* 157: 48–54.

Chong, T.K., Wilson, J., Mokkapati, S. & Catchpole, K.R. 2012. Optimal wavelength scale diffraction gratings for light trapping in solar cells. *Journal of Optics* 14(2): 24012–24020.

Dumont, L., Benzo, P., Cardin, J., Yu, I.S., Labbé, C. & Marie, P., et al. 2017. Down-shifting si-based layer for si solar applications. *Solar Energy Materials & Solar Cells* 169: 132–144.

Espinosa, N., García-Valverde, R., Urbina, A., Lenzmann, F., Manceau, M. & Angmo, D., et al. 2012. Life cycle assessment of ito-free flexible polymer solar cells prepared by roll-to-roll coating and printing. *Solar Energy Materials & Solar Cells* 97(97): 3–13.

Esteban, R., Laroche, M. & Greffet, J.J. 2010. Dielectric gratings for wide-angle, broadband absorption by thin film photovoltaic cells. *Applied Physics Letters* 97(22): 2757.

Hara, K., Jonai, S. & Masuda, A. 2015. Potential-induced degradation in photovoltaic modules based on n-type single crystalline si solar cells. *Solar Energy Materials & Solar Cells* 140: 361–365.

Huang, B.R., Yang, Y.K., Lin, T.C. & Yang, W.L. 2012. A simple and low-cost technique for silicon nanowire arrays based solar cells. *Solar Energy Materials & Solar Cells* 98(98): 357–362.

Jia, Z., Cheng, Q., Song, J., Si, M. & Luo, Z. 2016. Optical properties of a grating-nanorod assembly structure for solar cells. *Optics Communications* 376: 14–20.

Jung, J.Y., Guo, Z., Jee, S.W., Um, H.D., Park, K.T. & Lee, J.H. 2010. A strong antireflective solar cell prepared by tapering silicon nanowires. *Optics Express* 3(19): A286–292.

Kanamori, Y., Hane, K., Sai, H. & Yugami, H. 2001. 100 nm period silicon antireflection structures fabricated using a porous alumina membrane mask. *Applied Physics Letters* 78(2): 142–143.

Kannan, N. & Vakeesan, D. 2016. Solar energy for future world:—a review. *Renewable & Sustainable Energy Reviews* 62: 1092–1105.

Kelzenberg, M.D., Boettcher, S.W., Petykiewicz, J.A., Turner-Evans, D.B., Putnam, M.C. & Warren, E.L., et al. 2010. Enhanced absorption and carrier collection in Si wire arrays for photovoltaic applications. *Nature Materials* 9(3): 239–244.

Kosten, E.D., Atwater, J.H., Parsons, J., Polman, A. & Atwater, H.A. 2013. Highly efficient gaas solar cells by limiting light emission angle. *Light Science & Applications* 2(1): e45.

Leem, J.W., Choi, M. & Yu, J.S. 2015. Multifunctional microstructured polymer films for boosting solar power generation of silicon-based photovoltaic modules. *Acs Applied Materials & Interfaces* 7(4): 2349–2358.

Leem, J.W., Song, Y.M. & Yu, J.S. 2013. Biomimetic artificial Si compound eye surface structures with broadband and wide-angle antireflection properties for Si-based optoelectronic applications. *Nanoscale* 5(21): 10455–10460.

Lin, Y., Xu, Z., Yu, D., Lu, L., Yin, M. & Tavakoli, M.M., et al. 2016. Dual-layer nanostructured flexible thin-film amorphous silicon solar cells with enhanced light harvesting and photoelectric conversion efficiency. *Acs Appl Mater Interfaces* 8(17): 10929–10936.

Lipomi, D.J., Tee, B.C., Vosgueritchian, M. & Bao, Z. 2011. Stretchable organic solar cells. *Advanced Materials* 23(15): 1771–1775.

Lu, C., Ji, Z., Xu, G., Wang, W., Wang, L. & Han, Z., et al. 2016. Progress in flexible organic thin-film transistors and integrated circuits. *Science Bulletin* 61(14): 1081–1096.

Mavrokefalos, A., Han, S.E., Yerci, S., Branham, M.S. & Chen, G. 2012. Efficient light trapping in inverted nanopyramid thin crystalline silicon membranes for solar cell applications. *Nano Letters* 12(6): 2792–2796.

Mayer, J., Gallinet, B., Offermans, T. & Ferrini, R. 2016. Diffractive nanostructures for enhanced light-harvesting in organic photovoltaic devices. *Optics Express* 24(2): A358–373.

Papet, P., Nichiporuk, O., Kaminski, A., Rozier, Y., Kraiem, J. & Lelievre, J.F., et al. 2006. Pyramidal texturing of silicon solar cell with TMAH chemical anisotropic etching. *Solar Energy Materials & Solar Cells* 90(15): 2319–2328.

Paulo, A.F.D. & Geciane, P. 2017. Solar energy technologies and open innovation: a study based on bibliometric and social network analysis. *Energy Policy* 108: 228–238.

Perl, E.E., Mcmahon, W.E., Farrell, R.M., Denbaars, S.P., Speck, J.S. & Bowers, J.E. 2014. Surface structured optical coatings with near-perfect broadband and wide-angle antireflective properties. *Nano Letters* 14(10): 5960–5964.

Tavakoli, M.M., Tsui, K.H., Zhang, Q., He, J., Yao, Y. & Li, D., et al. 2015. Highly efficient flexible perovskite solar cells with antireflection and self-cleaning nanostructures. *Acs Nano* 9(10): 10287–10295.

Trinh, C.T., Preissler, N., Sonntag, P., Muske, M., Jäger, K. & Trahms, M., et al. 2018. Potential of inter-digitated back-contact silicon heterojunction solar cells for liquid phase crystallized silicon on glass with efficiency above 14%. *Solar Energy Materials & Solar Cells* 174: 187–195.

Tsui, K.H., Lin, Q., Chou, H., Zhang, Q., Fu, H. & Qi, P., et al. 2014. Low-cost, flexible, and self-cleaning 3d nanocone anti-reflection films for high-efficiency photovoltaics. *Advanced Materials* 26(18): 2805–2811.

Wolden, C.A., Abbas, A., Li, J., Diercks, D.R., Meysing, D.M. & Ohno, T.R., et al. 2016. The roles of znte buffer layers on cdte solar cell performance. *Solar Energy Materials & Solar Cells* 147: 203–210.

Ye, X., Zou, S., Chen, K., Li, J., Huang, J. & Cao, F., et al. 2015. 18.45%—efficient multi—crystalline silicon solar cells with novel nanoscale pseudo—pyramid texture. *Advanced Functional Materials* 24(42): 6708–6716.

Yu, R., Ching, K.L., Lin, Q., Leung, S.F., Arcrossito, D. & Fan, Z. 2011. Strong light absorption of self-organized 3-d nanospike arrays for photovoltaic applications. *Acs Nano* 5(11): 9291–9298.

Zi, W., Hu, J., Ren, X., Ren, X., Wei, Q.B. & Liu, S. 2017. Modeling of triangular-shaped substrates for light trapping in microcrystalline silicon solar cells. *Optics Communications* 383: 304–309.

Frontier Research and Innovation in Optoelectronics Technology and Industry – Habib & Lewis (Eds)
© *2019 Taylor & Francis Group, London, ISBN 978-1-138-33178-5*

Emulsion synthesis of size-tunable green emitting halide perovskite $CH_3NH_3PbBr_3$ colloidal quantum dots

Y.H. Liu, X. Tong, Q. Xu, S.J. Hao & Y.D. Ren
School of Physics and Electrical Information Engineering, Daqing Normal University, Daqing, China

ABSTRACT: We controllably synthesized $CH_3NH_3PbBr_3$ quantum dots with a tunable spectrum with the emission peaks covering the range from green (523.6 nm), to blue and eventually to deep violet (432.9 nm), which is wider than that of quantum dots obtained without changing the halide component. The mechanism of photoluminescence spectral blueshift of $CH_3NH_3PbBr_3$ quantum dots was investigated. The $CH_3NH_3PbBr_3$ quantum dots with 5–30 μL n-octylamine showed an ideal color-saturated green emission and a narrow full width at a half-maximum of 19–24 nm. The photoluminescence quantum yielded up to 90.2%. All these properties indicate that these quantum dots can provide the effective data support for application to white LEDs, and may potentially be used as single-component multicolor-emitting materials, which can be applied to lighting and display technology.

Keywords: Quantum Dots; Halide Perovskites; White LED

1 INTRODUCTION

Halide perovskites with a typical chemical formula of ABX_3 (where A and B are a monovalent and bivalent cation, respectively, and X is a monovalent halide anion) have been considered as promising materials in various applications, due to their very flexible composition and versatile properties (Protesescu et al., 2015; Li et al., 2016; Sun et al., 2017). Recently, colloidal perovskite Quantum Dots (QDs) were successfully synthesized with enhanced Photoluminescence (PL) properties via simple reprecipitation or hot injection method (Levchuk et al., 2017; Huang et al., 2015). Among them, increasing attention has been devoted to organic-inorganic hybrid perovskites $CH_3NH_3(= MA)PbX_3$, X = Cl, Br, I) colloidal QDs due to the low cost, and their unique PL properties, as well as the superior electrical conductivity, such as high PL Quantum Yield (PLQY), narrow Full Width at Half-Maximum (FWHM) of emission, wide tunable absorption and emission ranges, easy and fast generation of the charge carriers, and high chromatogram purity. These superior properties have attracted extensive exploration in optoelectronic devices, such as LEDs, Electroluminescence (EL) devices, information displays, photodetectors, and solar cells, as well as low-threshold lasers (Xing et al., 2016; Deng et al., 2016). However, at present most reports on perovskite colloidal QDs with high luminescence and a wide range of emission spectrum (between 400 and 800 nm) can be prepared by fine-tuning the molar ratio of the precursors (Deng et al., 2016), the composition of constituted halide ions, or simple halide mixing to vary their compositions or nanostructures.

Recently, Zhong's group synthesized size-tunable green $MAPbBr_3$ colloidal QDs, with PLQYs generally in the range of 80–92%, by varying the amount of a demulsifier, and pointed out the potential for applications in thin-film displays, solid-state lighting, and other solution-processed optoelectronic devices (Huang et al., 2015). Huang et al. synthesized the bandgap-tunable $MAPbBr_3$ nanocrystals. The emission peaks covered the range of 455–516 nm, and through the variation of the precursor and the ligand concentrations (Huang et al., 2017). As aforementioned, it is feasible for $MAPbBr_3$ QDs to accurately regulate the amounts of ligands.

Inspired by the above information, for the first time, we controllably synthesized pure MAPbBr$_3$ colloidal QDs with brightly luminescent, stability and a tunability emission peak from green, bluish-green, blue and eventually to violet. Different amounts of ligands (oleic acid and n-Octylamine (OLAM)) were used and found to play a significant effect on the PL properties of colloidal MAPbBr$_3$ QDs to achieve high stability and wide color ranges. The chromaticity coordinates of MAPbBr$_3$ QDs moved gradually from green toward the white region. Such study would cause further developments in the colloidal perovskite QDs field, and provide a guidance to improve the existing synthetic methods.

2 EXPERIMENT

2.1 Materials and methods

CH$_3$NH$_3$PbBr$_3$ QDs were prepared via the microemulsion method from starting materials of lead bromide (PbBr$_2$) (Aladdin, 99%), methylamine (33 wt% in absolute ethanol) (Aladdin), OLAM (Aladdin, ≥ 99%), hydrobromic acid (HBr) (Aladdin, 48 wt% in water), OA (Alfa Aesar, 90%,), DMF (Beijing Chemical Reagent Co., Ltd., China, analytical grade), n-hexane (C$_6$H$_{14}$) (Beijing Chemical Reagent Co., Ltd., China, analytical grade), acetone (Beijing Chemical Reagent Co., Ltd., China, analytical grade), and tert-butanol (Beijing Chemical Reagent Co., Ltd., China, analytical grade). All reagents were used as received without further purification.

2.2 Synthetic procedures

The Methylammonium Bromide (MABr) powders were synthesized with MA and HBr. Firstly, an excess of HBr (48 wt% in water) was slowly added to MA solution (33 wt% in absolute ethanol) under vigorous stirring (the molar ratio of MA and HBr was less than 1). The solution was kept in an ice bath under stirring for 2 h. Then the reacted solution was rotary evaporated with a pressure of −0.1 MPa at 45°C for 30 min. The residual was washed three times with diethyl ether for precipitation and dried under vacuum (60°C, 5 h) for future use.

0.16 mmol MABr was dissolved in 300 μL DMF and 0.2 mmol PbBr$_2$ powder was dissolved in 500 μL DMF to form precursor solutions. 500 μL OA and a certain amount of OLAM (5–100 μL) as ligands were added dropwise into 10 mL hexane, respectively. Then, precursor solutions were added dropwise into this n-hexane under vigorous stirring. During the titration, emulsion was formed. After that, 8 mL acetone was injected into the solution to initiate a demulsion process. The obtained solutions were centrifugated at 6,000 rpm for three min to obtain precipitates, which contain the as-prepared yellow-green colloidal QDs. Then the precipitates were redissolved into 4 mL n-hexane to extract the colloidal QDs. After another centrifugation at 5,000 rpm for three min, and after precipitates had been discarded, a bright green colloidal solution was obtained. The colloidal solution, which was scaled up ten-fold, can be reprecipitated from the solution by adding tert-butanol and then dried into solid-state powder for further use.

2.3 Characterizations

High-Resolution Transmission Electron Microscopy (HRTEM) images were taken using a JEOL-JEM-2100F TEM electron microscope, operating at 200 kV operating voltage. The PL spectra measurements were recorded with an F-380 fluorescence spectrometer (Tianjin Gangdong Sci. & Tech. Development Co., Ltd., China). The absorption spectra of the QDs dissolved in n-hexane were measured using a UV-6100 UV-vis spectrophotometer (Shanghai Mapada Instruments Co., Ltd., China). The XPS measurements were performed on a ULVAC-PHI machine (PHIQUANTERA-II SXM) with Al Kα as the X-ray source. The absolute PLQY of the diluted QD solutions, defined as the ratio between the photons emitted and absorbed by the sample, were determined using a fluorescence spectrometer with an integrating sphere (C9920-02, Hamamatsu Photonics, Japan), excited at a wavelength of 450 nm using a blue LED light source.

3   RESULTS AND DISCUSSION

The Transmission Electron Microscopy (TEM) and the corresponding High-Resolution TEM micrograph of colloidal MAPbBr$_3$ QDs are shown in Figures 1a–b. These nearly spherical QDs were monodispersed. The lattice fringes, which indicate high crystallinity, were clearly observed on each particle. It is clearly seen that the interplanar distance is 2.9 Å, corresponding to the (200) planes. Moreover, the Fast Fourier Transform (FFT) image is also supplied at the top right corner in Figure 1b. The result illustrates that the single crystalline phase is formed. It also reveals the presence of (200) plane, further confirming the formation of the perovskite structure. Figure 1c presents a chart of the crystallite size distribution for the obtained MAPbBr$_3$ QDs, which was determined directly from HRTEM images. The average size calculated from the analysis of about 200 QDs is around 3.67 nm, with a size deviation of 0.8 nm. High-resolution XPS spectra were obtained for the major photoelectron peaks of all elements to provide chemical state information, which showed the XPS spectra of Br 3d, Pb 4f, and N 1 s spectra for the resulting QDs. As shown in Figure 1d, the Br 3d peak could be fitted into two peaks with binding energies of 67.35 and 68.40 eV, which correspond to the inner and surface Br ions, respectively. Figure 1e shows two symmetric peaks of Pb $^4f_{7/2}$ and Pb $^4f_{5/2}$ at binding energy values 137.48 and 142.34 eV, respectively. The N 1 s spectrum plot (Figure 1f) shows the peak with binding energy at 401.01 eV. This peak is ascribed to methylamine.

We further investigated the optical properties of MAPbBr$_3$ QDs. Figures 2a–b show the absorption (left) and emission (right) spectra of selected MAPbBr$_3$ QDs with varying amounts of OLAM. Here, when OA content was set to 500 µL, the amount of OLAM was adjusted from 5 to 100 µL during the emulsion process. The UV-vis absorption spectrum of the MAPbBr$_3$ QDs (5 µL OLAM) had a band edge at 2.42 eV. The PL spectrum was centered at 523.6 nm (green) with a standard Gaussian profile and the narrow FWHM value of only 19 nm, indicating the superior color saturation. The sample had a relatively small Stokes shift of about 50 meV, which was consistent with a direct exciton recombination process.

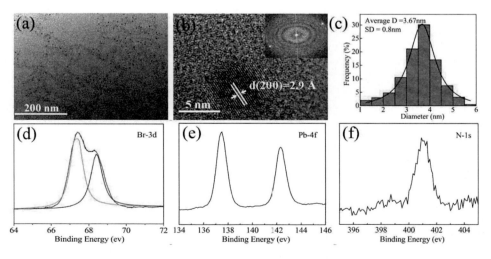

Figure 1.   (a) TEM and (b) HRTEM micrographs of colloidal MAPbBr$_3$ QDs (20 µL OLAM amount); the inset shows the corresponding FFT pattern, which exhibits a series of small diffraction spots; (c) size distribution histogram of colloidal MAPbBr$_3$ QDs; (d–f) XPS spectra corresponding to Br 3d (d), Pb 4f (e), and N 1 s (f) of MAPbBr$_3$, respectively.

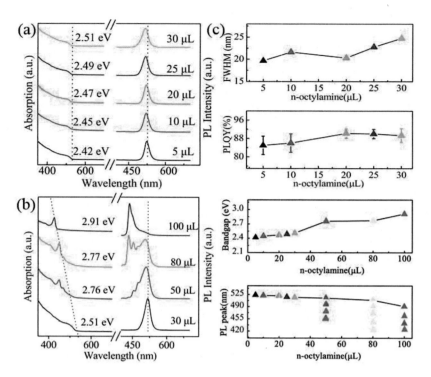

Figure 2. (a) and (b) Optical absorption (left) and the respective PL (right) spectra of size-tunable MAPbBr$_3$ QDs with varied amounts of OLAM; (c) the FWHM, PLQYs, bandgap and PL emission peak as a function of the concentration of OLAM, respectively.

As shown in Figures 2a–c, the absorption spectra exhibited a gradual blueshift with increasing amount of OLAM, due to the increased bandgap from 2.42 eV to 2.91 eV. In addition, Figure 2a and c exhibit symmetric and narrow PL spectra with line widths of 19–24 nm at FWHM, indicating the narrow size distributions (5–30 μL). Corresponding to the absorption spectra, the PL peak was slightly blueshifted to 517 nm (Figure 3a). However, upon further varying amounts of OLAM, an asymmetric emission band and multiple emission peaks were observed, as shown in Figure 2b. When the amount of OLAM was adjusted to 50 μL, the corresponding PL spectrum was more complicated, which showed a main peak at around 514.7 nm (green) with a nonGaussian distribution. Gaussian curve fitting analysis of the spectrum was performed, as shown in Figure 3a, which revealed that indeed there are other PL peaks located at 494 nm (bluish-green), 471.8 nm (blue) and 454.7 nm (blue). When the amount of OLAM was adjusted to 80 μL, the PL spectrum was much more complicated. The resolved PL spectrum showed five individual peaks of 508.7 nm (green), 484.3 nm (sky-blue), 470.9 nm (blue), 453.1 nm (blue) and 432.9 nm (deep violet) (as shown in Figure 3b). When the amount of OLAM was adjusted to 100 μL, the PL spectrum could be divided into four peaks at 490.8 nm (bluish-green), 453.4 nm (blue), 445.8 nm (blue) and 435.3 nm (deep violet) (as shown in Figure 3c). These Gaussian peaks were found to give the best fit to the corresponding spectrum. These peaks also exhibited a gradual blueshift. Another important observation is that the PL emission gradually shifted from green (523.6 nm) to deep violet (432.9 nm), and that with a significant decrease in intensities of the green peak (514.7 nm), the intensities of the blue and the violet emission peaks relatively increase. With the amount of OLAM varied from 5 μL to 100 μL, the color of the colloidal QDs gradually evolves from green to deep violet, corresponding to the PL spectra. It is worth mentioning that the blueshift of the bandgap and emission peak from green to violet and the enhancement of

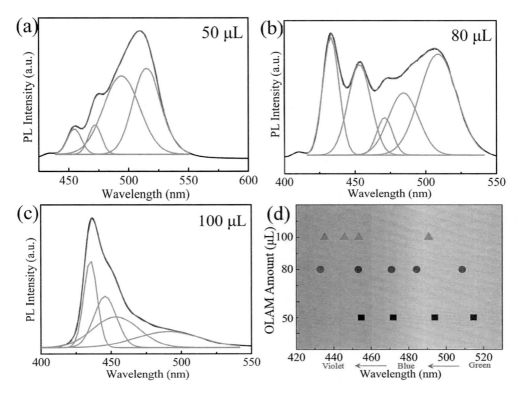

Figure 3. (a) The PL of 50 μL OLAM can be divided into four peaks; (b) the PL of 80 μL OLAM can be divided into five peaks; (c) the PL of 1,000 μL OLAM can be divided into four peaks; (d) emission peak-dependent OLAM content of MAPbBr$_3$ QDs.

PLQY could be attributed to the Pb-Br bond length and bond angles in the inorganic part, due to OLAM concentration or surface trap passivation effect which interacted between the surface of the MAPbBr$_3$ and the amine group of OLAM. The dependence of concentration of OLAM on the absorption and PL properties are displayed in Figure 2c. The PLQYs of these samples were determined using a fluorescence spectrometer equipped with an integrated sphere to collect emission in all angles under the excitation of 365 nm illumination. The PLQY for 5 μL OLAM was obtained as 85 ± 4%, and it was increased to a maximum of 90.2 ± 2% for 20 μL OLAM, but the PLQY is reduced to 89.3% for the 30 μL OLAM. Surprisingly, in this case the PLQY of MAPbBr$_3$ QDs was very low at 0.08% when they were shifted to blue emission. Also note that the brightness of green emission decreases with an increasing amount of OLAM. In particular, the most optimal PLQY of QDs is 20 μL OLAM content.

4 CONCLUSION

In summary, we have successfully prepared the MAPbBr$_3$ QDs with novel high luminescence, stability, and spectrally narrow green emissions via improved microemulsion. XPS results proved that the as-synthesized products have the same perovskite phase. There was a strong dependency of the bandgap on the choice of organic ligand. Here, the OA content was a controlled variable at 500 μL to stabilize the microemulsion, and the amount of OLAM was adjusted from 5 to 100 μL during the emulsion process. Absorption/PL spectra and theoretical calculations evidenced the green luminescence of MAPbBr$_3$ QDs with FWHM values of

19–24 nm and PLQYs of 85–90.2%, which originated from the direct exciton recombination emission of excitons with a large binding energy of 2.42 eV. The PL emission could be altered from 523.6 to 432.9 nm through the modulation of the organic OLAM content. Our findings will provide a new strategy for the fabrication of perovskite-related halide QDs and could further strengthen their competitiveness in the field of lighting or display.

ACKNOWLEDGMENTS

The authors acknowledge the support of the Science Foundation of Heilongjiang Province (Grant No. QC2015066) and Instructional Technology Plan of Daqing City (Grant No. zd-2017-09).

REFERENCES

Deng, W., Xu, X., Zhang, X., Zhang, Y., Jin, X., Wang, L., ... Jie, J. (2016). Organometal halide perovskite quantum dot light-emitting diodes. *Advanced Functional Materials*, 26(26), 4797–4802.

Huang, H., Raith, J., Kershaw, S.V., Kalytchuk, S., Tomanec, O., Jing, L., ... Rogach, A.L. (2017). Growth mechanism of strongly emitting $CH_3NH_3PbBr_3$ perovskite nanocrystals with a tunable band gap. *Nature Communications*, 8(1), 996.

Huang, H., Susha, A.S., Kershaw, S.V., Hung, T.F. & Rogach, A.L. (2015a). Control of emission color of high quantum yield $CH_3NH_3PbBr_3$ perovskite quantum dots by precipitation temperature. *Advanced Science*, 2(9), 1500194.

Huang, H.L., Zhao, F.C., Liu, L.G., Zhang, F., Wu, X.G., Shi, L., ... Zhong, H. (2015b). Emulsion synthesis of size-tunable $CH_3NH_3PbBr_3$ quantum dots: An alternative route toward efficient light-emitting diodes. *ACS Applied Materials & Interfaces*, 7(51), 28128–28133.

Levchuk, I., Osvet, A., Tang, X.F., Brandl, M., Perea, J.D., Hoegl, F., ... Brabec, C.J. (2017). Brightly luminescent and color-tunable formamidinium lead halide perovskite $FAPbX_3$ (X = Cl, Br, I) colloidal nanocrystals. *Nano Letters*, 17(5), 2765–2770.

Li, X.M., Wu, Y., Zhang, S.L., Cai, B., Gu, Y., Song, J., ... Zeng, H. (2016). $CsPbX_3$ quantum dots for lighting and displays: Room-temperature synthesis, photoluminescence superiorities, underlying origins and white light-emitting diodes. *Advanced Functional Materials*, 26(15), 2435–2445.

Protesescu, L., Yakunin, S., Bodnarchuk, M.I., Krieg, F., Caputo, R., Hendon, C.H., ... Kovalenko, M.V. (2015). Nanocrystals of cesium lead halide perovskites ($CsPbX_3$, X = Cl, Br, and I): Novel optoelectronic materials showing bright emission with wide color gamut. *Nano Letters*, 15(6), 3692–3696.

Sun, X.G., Shi, Z.F., Li, Y., Lei, L.Z., Li, S., Wu, D., ... Li, X. (2017). Effect of CH3 NH3I concentration on the physical properties of solution-processed organometal halide perovskite $CH_3NH_3PbI_3$. *Journal of Alloys and Compounds*, 706, 274–279.

Xing, J., Yan, F., Zhao, Y., Chen, S., Yu, H., Zhang, Q., ... Xiong, Q. (2016). High-efficiency light-emitting diodes of organometal halide perovskite amorphous nanoparticles. *ACS Nano*, 10(7), 6623–6630.

Frontier Research and Innovation in Optoelectronics Technology and Industry – Habib & Lewis (Eds)
© 2019 Taylor & Francis Group, London, ISBN 978-1-138-33178-5

All-dielectric asymmetrical metasurfaces based on mesoscale dielectric particles with different optical transmissions in opposite directions through full internal reflection

I.V. Minin & G.V. Shuvalov
FGUP "SNIIM", Novosibirsk, Russia

O.V. Minin
Tomsk State University, Tomsk, Russia

ABSTRACT: Optically asymmetric metasurfaces are structures that pass light in one direction, but block it in another. This paper presents a numerical study of all-dielectric asymmetric metasurfaces based on a photonic nanojet effect from asymmetric dielectric particles, which have a full internal reflection in one direction. It has been shown that for conical dielectric particles, the formation of near-field localization near the shadow side of the particle depends on the direction of particle illumination. The criterion of critical cone angle for full internal reflection is described. The numerical simulations confirm the full internal reflection for the illumination of the conical particle from one side and the formation of the photonic jet from the other side. The asymmetric all-dielectric metastructure described in the paper can be used not only as an optical diode, but also as a transparent electrode and light-trapping structure for thin-film solar cells.

1 INTRODUCTION

Optically asymmetric metasurfaces are structures that pass light in one direction, but block it in another. In addition, an all-optical diode is an essential component for optical computer development, which is based on different optical transmission in opposite directions. These structures are remarkable for practical applications and, for example, can be used as light-trapping layers for thin-film solar cells (Chen et al., 2016).

The use of microlenses to create asymmetric transmission, where the microlenses focus and guide light through small holes, was described in Tvingstedt et al. (2008). An asymmetry effect was obtained, but the structure had a thickness and unit cell lateral size of about 100 μm, or more than 100–200 times wavelength. There have also been numerous reports regarding optical diodes (Greenberg & Orenstein, 2004; Feng et al., 2011; Serebryannikov, 2009; Wang et al., 2011; Lu et al., 2011). An efficient routine for creating optical diodes is via time-reversal symmetry breaking or spatial-inversion symmetry breaking (Serebryannikov, 2009), which could lead to an optical isolation in any device where the forward and backward transmissivity of light is very different.

Unidirectional on-chip optical diodes, based on the directional bandgap mismatch effect of two 2D square-lattice photonic crystals comprising a heterojunction structure and the break of the spatial-inversion symmetry, were analyzed by Wang et al. (2011). A compact optical diode design with a length of 2 wavelengths, based on a photonic crystal waveguide spatial mode converter, was discussed by Liu et al. (2012).

However, these asymmetrical metasurface designs, operating on spatial (rather than polarization) modes, have all been periodic devices, which require many periods to achieve efficient mode conversion. Furthermore, some of these designs use materials that have intrinsic loss.

This paper presents a numerical study of all-dielectric asymmetric metasurfaces based on a photonic nanojet effect from asymmetric dielectric particles, which have a full internal reflection in one direction.

2 ASYMMETRIC METASURFACE PRINCIPLES

To the best of our knowledge, nobody has used arbitrary three-dimensional (3D) mesoscale dielectric particles without transverse symmetry of the particle producing the effect of a photonic jet to create optically asymmetric structures. The application of the nanojet effect can allow substantial miniaturization of the structure and open the door to easier methods of fabrication.

The main idea is as follows: the field localization near the shadow surface of the dielectric mesoscale particles depends not only on the shape of the particle, but also on its orientation in space relative to the direction of light illumination. In addition, all parameters, such as S_{11} and S_{12} (transmission and reflection coefficients), are dependent on the shape and orientation of the particle.

It should be noted that the minimal dimensions of the particle (diameter and length) are about one wavelength (very compact). In addition, we might obtain an optical isolation based on the full internal reflection of light in one direction, which is a sought-after object of fundamental difficulty in integrated photonics.

3 SIMULATION RESULTS

Although analytical solutions of the vector diffraction problem can be obtained for selected objects (sphere, half plane, cylinder) (Minin & Minin, 2015, 2016), the boundary conditions on the electromagnetic field for other dielectric structures make an analytical solution impossible (Minin et al., 2016; Pacheco-Peña et al., 2016; Li et al., 2016; Pham et al., 2017; Yue et al., 2017; Liu et al., 2018; Mahariq et al., 2017).

In order to evaluate the focusing performance of the structure, we used the transient solver of CST Microwave Studio™ commercial software (Dassault Systèmes, Paris, France), together with an extra-fine hexahedral mesh with a minimum mesh size of $\lambda_0/45$. CST Microwave Studio™ is used to apply the Finite Integration Technique (FIT), which in the time domain increases more slowly with the problem size than other commonly employed methods.

It should be noted that the nature of EM radiation scattering on a particle implies the use of a dimensionless size parameter, which is constructed as the ratio of the equivalent radius of the scatterer to the EM wavelength. Therefore, as long as this ratio is preserved and the article's refractive index contrast remains the same, we are free to choose particles of any desired size illuminated by radiation in any desired wavelength range.

Let us consider the example of a non-spheroidal axisymmetric particle as a unit cell. Here, and below in the 3D simulation, the dielectric particle was illuminated vertically (E_y) by a plane-polarized wave. The particle was placed in a vacuum ($n_0 = 1$), and open boundary conditions were used. The dielectric material refractive index was 1.42. The radius of the axisymmetric pyramidal particle base was 0.5 λ_0, its height was 1.0 λ_0, and the incident radiation direction was from the cone point.

Simulations by Minin and Minin (2014, 2016) show that the value of the field intensity at Full Width Half-Maximum (FWHM) on the optical axis at the x-axis was 0.49 λ_0, and at the y-axis was 0.45 λ_0, which is less than the classical diffraction limit; the length of a jet (at half-maximum) was 1.37 λ_0. The field intensity enhancement was about I ≈ 3.96. For the conical axicon-like particle with a base radius 0.5 λ_0 and height 0.22 λ_0, FWHM was 0.75 λ_0 and field intensity enhancement was about I ≈ 1.5.

However, when we increase the base radius of the conical particle twofold, the full internal reflection is observed in one direction (see Figures 1a and 1b).

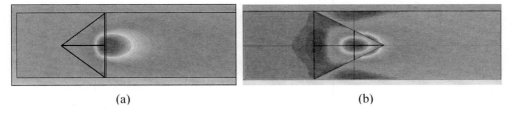

(a)   (b)

Figure 1.   Conical particle with illumination from: (a) an apex; (b) the base side.

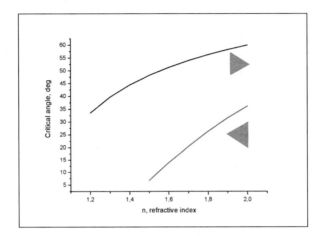

Figure 2.   Critical angle of conical particle vs refractive index contrast.

It has been shown (Minin & Minin, 2016) that the conditions of non-availability of the full reflection effect are $\alpha > \arccos(1/n)$ for illuminating the particle from the base, and $\alpha > arctg\left(\sqrt{n^2 - 1} - 1\right)$ $(n > \sqrt{2})$ (see Figure 2) for illuminating the particle from the apex.

Moreover, the conical (pyramidal) shape of the particles contributes to a small reflection coefficient for the incident radiation from the side of the vertex, and to an increased reflection coefficient when the radiation falls from the side of the base. In the first case, the conical (pyramidal) shape of the particle takes precedence over other forms of the particle to minimize the reflection coefficient (Minin & Minin, 2014, 2018).

4   CONCLUSIONS

A numerical study of all-dielectric asymmetric metasurfaces based on a photonic nanojet effect from asymmetric dielectric particles, which have a full internal reflection in one direction, has been presented and described briefly. The asymmetric metastructure described above can be used not only as an optical diode, but also as a transparent electrode and light-trapping structure for thin-film solar cells.

ACKNOWLEDGMENT

This work was partially supported by the Mendeleev scientific fund of Tomsk State University.

REFERENCES

Chen, H.T., Taylor, A.J. & Yu, N. (2016). A review of metasurfaces: Physics and applications. *Reports on Progress in Physics, 79*, 076401. doi:10.1088/0034-4885/79/7/076401.

Feng L., Ayache, M., Huang, J., Xu, Y.-L., Lu, M.-H., Chen, Y.-F., ... & Scherer, A. (2011). Nonreciprocal light propagation in a silicon photonic circuit. *Science, 333*, 729–733.

Greenberg, M. & Orenstein, M. (2004). Irreversible coupling by use of dissipative optics. *Optics Letters, 29*, 451–453.

Li, Y., Fu, Y., Minin, I.V. & Minin, O.V. (2016). Ultra-sharp nanofocusing of graded index photonic crystals-based lenses perforated with optimized single defect. *Optical Materials Express, 6*(8), 781–792.

Liu, C.-Y., Yen, T.P., Minin, O.V. & Minin, I.V. (2018). Engineering photonic nanojet by a graded-index micro-cuboid. *Physica E, 98*, 105–110.

Liu, V., Miller, D.A.B. & Fan, S. (2012). Ultra-compact photonic crystal waveguide spatial mode converter and its connection to the optical diode effect. *Optics Express, 20*(27), 28388–28397.

Lu, C., Hu, X., Yang, H. & Gong, Q. (2011). Ultrahigh-contrast and wideband nanoscale photonic crystal all-optical diode. *Optics Letters, 36*, 4668–4670.

Mahariq, I., Giden, I.H., Minin, I.V., Minin, O.V. & Kurt, H. (2017). Strong electromagnetic field localization near the surface of hemicylindrical particles. *Optical and Quantum Electronics, 12*(49), 423–427.

Minin, I.V. & Minin, O.V. (2014). Photonics of isolated dielectric particles of arbitrary 3D shape – A new direction of optical information technologies. *Vestnik NSU, 12*(4), 59–70. Retrieved from http://www.nsu.ru/xmlui/handle/nsu/7717.

Minin, I.V. & Minin, O.V. (2015). Photonics of mesoscale nonspherical and non axysimmetrical dielectric particles and application to cuboid-chain with air-gaps waveguide based on periodic terajet-induced modes. In *Proceedings of the 17th International Conference on Transparent Optical Networks, Budapest, 2015* (paper We.D6.6).

Minin, I.V. & Minin, O.V. (2016). Diffractive optics and nanophotonics: Resolution below the diffraction limit. London, UK: Springer.

Minin, I.V. & Minin, O.V. (2018). *Fully optical diode*. Russian patent no. RU178617U1. Retrieved from https://patents.google.com/patent/RU178617U1/en.

Minin, I.V., Minin, O.V. & Nefedov, I.S. (2016). Photonic jets from Babinet's cuboid structures in the reflection mode. *Optics Letters, 41*(3), 7–11.

Pacheco-Peña, V., Minin, I.V., Minin, O.V. & Beruete, M. (2016). Comprehensive analysis of photonic nanojets in 3D dielectric cuboids excited by surface plasmons. *Annalen der Physik, 528*(9), 684–692.

Pham, H.-H.N., Hisatake, S., Minin, O.V., Nagatsuma, T. & Minin, I.V. (2017). Enhancement of spatial resolution of terahertz imaging systems based on terajet generation by dielectric cube. *APL Photonics, 2*, 056106. doi:10.1063/1.4983114.

Serebryannikov, A.E. (2009). One-way diffraction effects in photonic crystal gratings made of isotropic materials. *Physical Review B, 80*, 155117. doi:10.1103/PhysRevB.80.155117.

Tvingstedt, K., Dal Zilio, S., Inganas, O. & Tormen, M. (2008). Trapping light with micro lenses in thin film organic photovoltaic cells. *Optics Express, 16*(26), 21608–21615.

Wang, C., Zhou, C.-Z. & Li, Z.-Y. (2011). On-chip optical diode based on silicon photonic crystal heterojunctions. *Optics Express, 19*, 26948–26955.

Yue, L., Yan, B., Monks, J.N., Wang, Z., Minin, I.V. & Minin, O.V. (2017). Intensity-enhanced apodization effect on an axially illuminated circular-column particle-lens. *Annalen der Physik, 530*(2), 1700384. doi:10.1002/andp.201700384.

Frontier Research and Innovation in Optoelectronics Technology and Industry – Habib & Lewis (Eds)
© 2019 Taylor & Francis Group, London, ISBN 978-1-138-33178-5

Ultra-wide dynamic range fiber-optic SPR sensor based on phase interrogation

Z.L. Song, Y.B. Guo, T.G. Sun, M.T. Liu & Y. Zheng
College of Communication Engineering, Jilin University, Changchun, Jilin Province, China

ABSTRACT: An ultra-wide dynamic range fiber-optic surface plasmon resonance sensor based on phase interrogation is reported. The sensing head is based on a side-polished Polarization-Maintaining optical Fiber (PMF) with the configuration of Coupled Plasmon Waveguide Resonance (CPWR). The phase detection is accomplished by performing interference between the TE mode in sensing channel and TM mode in reference channel. The dynamic range of the sample's refractive index has been extended to 0.444 in terms of Refractive Index Units (RIU), and the resolution on average is 3.43×10^{-7} RIU.

Keywords: SPR sensor, optical fiber, phase interrogation

1 INTRODUCTION

Surface plasmon resonance (SPR) is a collective free electron oscillation which occurs at a metal-dielectric interface when the wave-vector horizontal component of the incident light matches the propagation constant of the surface plasmon wave (Deng et al. 2017, Huang et al. 2012a). In 1997, Salamon et al. reported a SPR sensor based on coupled plasmon waveguide resonance (CPWR) (Salamon et al. 1997). The sensor incorporates a waveguide layer beneath the surface of the conventional SPR sensor in their design, as shown in Figure 1(b). Different from the conventional SPR sensors, the interference of the waveguide layer in the CPWR device causes sharp dips in both the transverse magnetic (TM, p-wave component) and transverse electric (TE, s-wave component) modes (Grotewohl et al. 2016, Chien & Chen 2004).

Several methods have been used to date to monitor the excitation of SPR, including angle, wavelength, intensity and phase interrogation techniques. Compared with amplitude, the phase of SPR reflected light changes much abruptly. Therefore, phase interrogation is the most sensitive excitation monitoring method (Deng et al. 2017). However, the narrow dynamic range has greatly limited the versatility and applicability of phase sensitive SPR

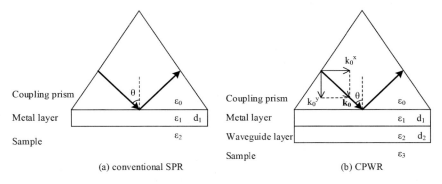

Figure 1. Configurations of conventional SPR and CPWR.

sensors (Huang et al. 2012b). In 2011, Y.H. Huang et al. reported a design combining phase detection and angular interrogation, which had a resolution of 2.2×10^{-7} RIU with a wide dynamic range of over 0.06 RIU (Huang et al. 2011). In 2013, Yonghong Shao et al. proposed a wavelength multiplexing phase sensitive SPR imaging sensor offering a resolution of 2.7×10^{-7} RIU with a dynamic range of 0.0138 RIU (Shao et al. 2013). For the reason that the optical configurations of phase-interrogated SPR sensors which depend on phase-extraction techniques such as optical heterodyne, polarimetry, ellipsometry and interferometry are more complex than those of the other sensor types (Deng et al. 2017), the phase interrogation method has not yet been widely explored in the context of the fiber-optic SPR sensors (Gasior et al. 2018).

In this paper, the combination of interferometry and coupled plasmon waveguide resonance with different incident wavelengths can extremely expand the dynamic range of SPR sensor, which is based on a side polished polarization-maintaining optical fiber.

2 THEORY AND CONFIGURATION

2.1 Theory

In general, SPR sensing is based on the Kretschmann configuration, which consists of a high refractive index prism, thin gold film and the sample, as shown in Figure 1(a).

In the four-layer configuration of CPWR (Salamon et al. 1997), the intensity and the phase of the reflected light are determined using the complex reflection coefficients of the multilayer medium structure which is based on a combination of the Fresnel equations and interference theory (Deng et al. 2017, Huang et al. 2012a, b). And the complex reflection coefficients of the whole multilayer structure are given by:

$$r_{j,3}^{p,s} = \frac{r_{j,j+1}^{p,s} + r_{j+1,3}^{p,s} \exp(2id_{j+1}k_{j+1}^{y})}{1 + r_{j,j+1}^{p,s} r_{j+1,3}^{p,s} \exp(2id_{j+1}k_{j+1}^{y})}, (i = \sqrt{-1}, j = 0,1) \tag{1}$$

where the angle of incidence is θ, the thickness of the metal layer and the waveguide layer are d_1 and d_2, and the dielectric coefficients of the prism, the metal layer, the waveguide layer and the sample are ε_0, ε_1, ε_2 and ε_3, respectively.

The wave-vector vertical components of the incident light in each layer are

$$k_j^y = \frac{2\pi}{\lambda} \sqrt{\varepsilon_j - \varepsilon_0 \sin^2 \theta}, (j = 0,1,2,3) \tag{2}$$

The complex reflection coefficients of adjacent layers are given by:

$$\begin{cases} r_{j,j+1}^{p} = \dfrac{\varepsilon_{j+1}k_j^y - \varepsilon_j k_{j+1}^y}{\varepsilon_{j+1}k_j^y + \varepsilon_j k_{j+1}^y} \\ r_{j,j+1}^{s} = \dfrac{k_{j+1}^y - k_j^y}{k_{j+1}^y + k_j^y} \end{cases}, (j = 0,1,2) \tag{3}$$

The principal feature of the CPWR configuration is the inclusion of the additional waveguide layer, which induces the waveguide resonance mode (Salamon et al. 1997). When a plane wave in TM or TE mode is incident upon the waveguide layer, the light wave reflects and refracts at the interface between the two different media. The action of multiple beams reflection and refraction generates optical path differences between the reflected beam at different positions, which in turn generate interference in the reflectivity spectrum (Grotewohl et al. 2016), as shown in Figure 2. The CPWR structure exhibits dual as well as multiple

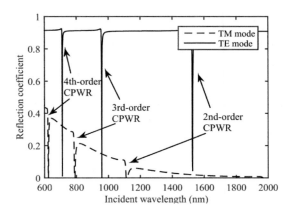

Figure 2. Reflection coefficient of TM and TE modes. (Step-index silica optical fiber with core diameter of $d = 8.5$ μm and NA = 0.125. The length of sensing region is $L = 5$ cm. Fiber core is covered by a 65 nm gold layer coating with 590 nm TiO$_2$ film. The refractive index of sample $n_3 = 1.333$ RIU, and incidence angle is 85.1°.)

resonances depending upon the thickness of the waveguide layer. This structure not only provides improved resolution of the dielectric constant but also protects the plasmon generating metal layer from chemical degradation (Chien & Chen 2004).

2.2 Configuration

Figure 3 presents a schematic configuration of the four-layer fiber-optic SPR sensing head in longitudinal section which consists of a side-polished layout with 65 nm Au film and 590 nm TiO$_2$ layer directly coating onto the fiber core. The length of the sensing region is set to be $L_1 = 5$ cm, and the length of the sample region is L_2. The meridional optical rays in the sensing region undergo a number of reflections given by (Chiu et al. 2005):

$$\begin{cases} N_1(\theta) = \dfrac{L_1 - L_2}{2d \tan\theta}, (air) \\ N_2(\theta) = \dfrac{L_2}{2d \tan\theta}, (sample) \end{cases} \quad (4)$$

where θ is the angle of incident optical ray with respect to the fiber-metal interface in the sensing region, which is greater than or equal to 85.05° at the wavelength of 1550 nm, and d is the diameter of the fiber core. In equation (4), N_1 is the number of reflections in the sensing region without sample, and N_2 is the number of reflections which the meridional optical rays passing through the sensing region of sample undergo. The CPWR trough will deepen when L_2 is enlarged and the phase difference variation between the s-wave and the p-wave polarization components will increase.

The cross section of the polarization-maintaining optical fiber (PMF) whose beat length is less than 5.0 mm at the wavelength of 1550 nm has been shown in Figure 4. The numerical aperture (NA) is 0.125 and the core diameter is 8.5 μm. Note that the two orthogonal solid lines (i.e., the slow and fast axes) represent the principal axes of the PMF (Piliarik et al. 2003). When the s-wave component of the laser beam is aligned with the slow axis of the optical fiber, the polarization state of incident light will be maintained. With the proper wavelength, the s-wave is coupled into the waveguide layer and the phase of TE mode changes abruptly. Meanwhile, the laser beam of the reference channel simply propagates along the length of the fiber (Lo et al. 2011).

Figure 5 shows the schematic setup for the proposed fiber-optic SPR sensing system. The illumination light is provided by a tunable laser source working in S-band (1460 nm ~ 1530 nm),

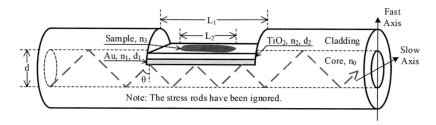

Figure 3. Schematic configuration of side-polished fiber-optic SPR sensing head in longitudinal section.

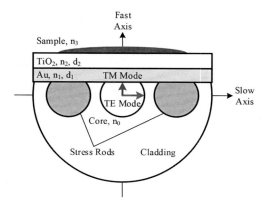

Figure 4. Schematic configuration of side-polished fiber-optic SPR sensing head in cross section.

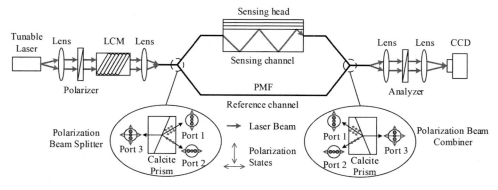

Figure 5. Schematic of fiber-optic SPR sensing system.

C-band (1530 nm ~ 1565 nm) and L-band (1565 nm ~ 1625 nm), respectively. The present system should first locate the CPWR dip by a lamp and a spectrometer and then adjust the wavelength of the tunable laser for phase detecting (Shao et al. 2013). When the refractive index of sample changes in sensing system, the resonant dip will also shift. Then the light is passed through a 45° polarizer and a liquid crystal modulator (LCM) with its slow axis oriented at 0° from the p-polarized beam (Huang et al. 2011). The LCM is able to generate retardation larger than 6λ between the fast and the slow axes, which is essential for accurate phase retrieval. The light is then coupled into the port 3 of a polarization beam splitter with s-wave oriented at 0° from the slow axis. The port 1 is connected to the sensing channel and port 2 is linked to the reference channel. After passing through the sensing head, the s-component undergoes phase change on account of CPWR while the p-component transmitting in reference channel remains unaffected. The two polarization components are finally recombined by a polarization beam combiner and produce an interference signal through the use of an analyzer with an azimuth angle β relative to the principal axis of the PMF (Li et al. 2008, Ng et al. 2010). When β is adjusted until the intensity of s-wave and p-wave are equal, the interference signal is strongest. Then the charge coupled device (CCD) captures the intensity response of the interference signal while the LCM modulates the polarization retardation continuously (Kashif et al. 2014). The CPWR phase of each sensor site (i.e., a small collection of pixels) is obtained by analyzing a large number of image frames using an image correlation algorithm based on the technique reported in (Shao et al. 2013).

The complex reflection coefficients of the s-wave passing through sensing head and p-wave passing through reference channel at a given incident angle are (Hlubina et al. 2015):

$$\begin{cases} r_s(\theta) = \left| r_{air}(\theta)^{N_1(\theta)} r_{sample}(\theta)^{N_2(\theta)} \right| \exp\{i[N_1(\theta)\Phi_{air}(\theta) + N_2(\theta)\Phi_{sample}(\theta)]\} \\ r_p(\theta) = \left| r_p(\theta)^{[N_1(\theta)+N_2(\theta)]} \right| \exp\{i[N_1(\theta) + N_2(\theta)]\Phi_p(\theta)\} \approx 1 \end{cases} \quad (5)$$

Total phase difference variation $\Delta\Phi(\theta)$ resulting from all the reflections inside the sensing region at a given incident angle between the s-wave and the p-wave polarization components is given by:

$$\Delta\Phi(\theta) = \Phi_s(\theta) - \Phi_p(\theta) \quad (6)$$

It is useful to define the sensitivity of the sensing structure to the refractive index n of the surrounding medium, which can be regarded as the basic parameter quantifying the performance of the sensor. The sensitivity can be defined as (Su et al. 2005):

$$S_{n,\Phi} = \frac{\Delta\Phi(\theta)}{\Delta n} \quad (7)$$

Limit of detection (LOD) characterizes the minimal measurable variation of refractive index of the sample is given by:

$$LOD = \frac{\sigma_\Phi}{S_{n,\Phi}} \quad (8)$$

where σ_Φ is the resolution of phase detection in typically which approximate to 0.01° (Markowicz et al. 2007).

3 RESULT AND DISCUSSION

A step-index polarization-maintaining silica optical fiber with core diameter of $d = 8.5$ μm and NA = 0.125 is selected in this work. The sensing head consists of a side-polished layout with 65 nm Au film and 590 nm TiO$_2$ layer directly coating onto the fiber core. The length of sensing region is $L_1 = 5$ cm. Figure 2 shows the reflection coefficient of TE and TM modes in sensing channel with the incident angle $\theta = 85.1°$. The refractive index of sample $n_3 = 1.333$ RIU with length of the region $L_2 = 5$ cm. It is obviously that higher-order CPWR dips are generated in both TM and TE modes when the thickness of waveguide layer increases.

The phase difference $\Delta\Phi(\theta)$ between TE mode transmitting in sensing channel and TM mode in reference channel is shown in Figure 6. The incident angle θ is 85.1° and the length of sample region $L_2 = 5$ cm. The wavelength range is 1467 nm ~ 1605 nm and the wavelength interval is 2 nm. It can be observed that for each wavelength change, phase undergoes an abrupt change only for a small refractive index (RI) range, and phase changes for each wavelength are much moderate. Thus the sensor output can be defined as the maximum phase change from all the wavelengths during the RI changing.

Figure 7 shows the relationship between sensitivity and RI of sample in the fiber-optic SPR sensing system. The mean of sensitivity in the range of RI is $2.92 \times 10^{4 \circ}$/RIU.

Figure 8 shows the relationship between LOD and RI of sample in the fiber-optic SPR sensing system. The mean of LOD in the range of RI is 3.43×10^{-7} RIU.

In Table 1, the performance parameters of fiber-optic SPR sensing system, including wavelength, range of RI, dynamic range, mean of sensitivity and mean of LOD, have been listed out. Through simulation analyses, samples with refractive index ranging from 1.000 RIU ~ 1.444 RIU have been calculated and the phase response of sensor accumulate up to 13,000° at an ultra wide dynamic range of 0.444 RIU without showing any saturation effect.

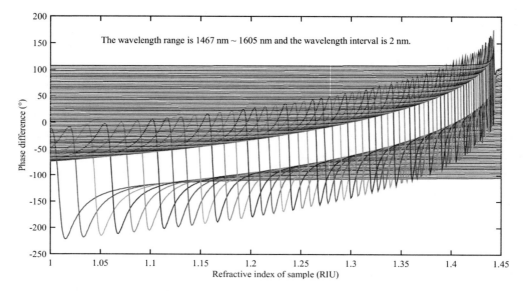

Figure 6. Phase difference of TE mode transmitting in sensing channel and TM mode in reference channel.

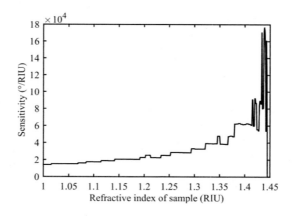

Figure 7. Sensitivity of SPR sensing system.

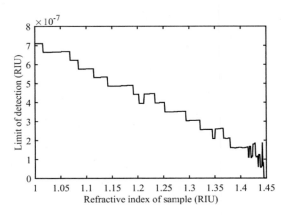

Figure 8. LOD of SPR sensing system.

Table 1. The performance parameters of fiber-optic SPR sensing system on average.

Num.	Wavelength	Range of RI	Dynamic range	Mean of sensitivity	Mean of LOD
	nm	RIU	RIU	°/RIU	RIU
1	1467 ~ 605	1.000 ~ 1.444	0.444	2.92×10^4	3.43×10^{-7}

4 CONCLUSION

In this paper, a fiber-optic SPR sensor based on phase interrogation is proposed. On account of the combination of phase interrogation and coupled plasmon waveguide resonance with different incident wavelengths, the sensing system has ultra wide dynamic range and high sensitivity. The dynamic range of the sample's refractive index has been extended to 0.444 RIU, and the resolution (i.e. LOD) on average is 3.43×10^{-7} RIU.

ACKNOWLEDGMENT

This work was supported by the Natural Science Foundation of Jilin Province under Grant No. 20160101245 JC.

REFERENCES

Chien, F.C. & Chen, S.J. 2004. A sensitivity comparison of optical biosensors based on four different surface plasmon resonance modes. *Biosensors & Bioelectronics* 20(3):633–642.
Chiu, M.H., Wang, S.F. & Chang, R.S. 2005. D-type fiber biosensor based on surface-plasmon resonance technology and heterodyne interferometry. *Optics Letters* 30(3):233–235.
Deng, S.J., Wang, P. & Yu, X.L. 2017. Phase-sensitive surface plasmon resonance sensors: recent progress and future prospects. *Sensors* 17(12):2819–2832.
Gasior, K., Martynkien, T., Mergo, P., et al. 2018. Fiber-optic surface plasmon resonance sensor based on spectral phase shift interferometric measurements. *Sensors & Actuators B Chemical* 257:602–608.
Grotewohl, H., Hake, B. & Deutsch, M. 2016. Intensity and phase sensitivities in metal/dielectric thin film systems exhibiting the coupling of surface plasmon and waveguide modes. *Applied Optics* 55(30):8564–8570.
Hlubina, P., Duliakova, M., Kadulova, M., et al. 2015. Spectral interferometry-based surface plasmon resonance sensor. *Optics Communications* 354(11):240–245.

Huang, Y.H., Ho, H.P., Wu, S.Y., et al. 2011. Phase sensitive SPR sensor for wide dynamic range detection. *Optics Letters* 36(20):4092–4094.

Huang, Y.H., Ho, H.P., Wu, S.Y., et al.. 2012a. Detecting phase shifts in surface plasmon resonance: a review. *Advances in Optical Technologies* 2012:1–12.

Huang, Y.H., Ho, H.P., Kong, S.K., et al. 2012b. Phase sensitive surface plasmon resonance biosensors: methodology, instrumentation and applications. *Annalen Der Physik* 524(11):637–662.

Kashif, M., Bakar, A.A.A., Arsad, N., et al. 2014. Development of phase detection schemes based on surface plasmon resonance using interferometry. *Sensors* 14(9):15914–15938.

Li, Y.C., Chang, Y.F., Su, L.C., et al. 2008. Differential-phase surface plasmon resonance biosensor. *Analytical Chemistry* 80(14):5590–5595.

Lo, Y.L., Chuang, C.H. & Lin, Z.W. 2011. Ultrahigh sensitivity polarimetric strain sensor based upon D-shaped optical fiber and surface plasmon resonance technology. *Optics Letters* 36(13):2489–2491.

Markowicz, P.P., Law, W.C., Baev, A., et al. 2007. Phase-sensitive time-modulated surface plasmon resonance polarimetry for wide dynamic range biosensing. *Optics Express* 15(4):1745–1754.

Ng, S.P., Wu, C.M., Wu, S.Y., et al. 2010. Differential spectral phase interferometry for wide dynamic range surface plasmon resonance biosensing. *Biosensors & Bioelectronics* 26(4):1593–1598.

Piliarik, M., Homola, J., Maníková, Z., et al. 2003. Surface plasmon resonance sensor based on a single-mode polarization-maintaining optical fiber. *Sensors & Actuators B Chemical* 90:236–242.

Salamon, Z., Macleod, H.A. & Tollin G. 1997. Coupled plasmon-waveguide resonators: a new spectroscopic tool for probing proteolipid film structure and properties. *Biophysical Journal* 73(5):2791–2797.

Shao, Y.H., Li, Y., Gu D.Y., et al. 2013. Wavelength-multiplexing phase-sensitive surface plasmon imaging sensor. *Optics Letters* 38(9):1370–1372.

Su, Y.D., Chen, S.J. & Yeh, T.L. 2005. Common-path phase-shift interferometry surface plasmon resonance imaging system. *Optics Letters* 30(12):1488–1490.

Frontier Research and Innovation in Optoelectronics Technology and Industry – Habib & Lewis (Eds)
© 2019 Taylor & Francis Group, London, ISBN 978-1-138-33178-5

Microwave photonic phase shifter in a polarization-maintaining fiber Bragg grating immune to wavelength drift

M.M. Sun, R.J. Zhu, Z.F. Fang & X. Zhou
No.724 Research Institute of China Shipbuilding Industry, Nanjing, Jiangsu, China

ABSTRACT: In this paper, an optical phase shifter based on a Polarization-Maintaining Fiber Bragg Grating (PMFBG) with continuously tunable phase shift independent of the optical wavelength drifting is proposed, which has been demonstrated by using the nonlinear birefringence effects in the PMFBG. A 360° phase shifter at 10 GHz is demonstrated by tuning the input optical power or the polarization angle of the polarization controller.

1   INTRODUCTION

The Microwave Photonic (MWP) phase shifter makes it possible to generate, transfer, and process the microwave and millimeter-wave signals directly in the optical domain (Capmany & Novak, 2007; Lim & Li, 2013), which is one of the key components in conventional RF engineering applications, such as microwave filters and reconfigurable front-ends. Tunable phase shifters also find applications in phased array antennas (VanBlaricum, 1994) for modern radar wing to their advantages over electronic steering systems such as the immunity to electromagnetic interference, the squint-free wide instantaneous bandwidth, and the light weight and small volume desirable for airborne applications.

Different phase shift techniques have been proposed, including the dispersion technique (Blais & Yao, 2009), the frequency mixing technique (Lee & Udupa, 1999), the vector sum technique (Bui et al., 2005) and the nonlinear optical technique (Chen et al., 2009). Recently, dispersion technology has been widely used due to its high reliability and low cost, such as dispersive fiber, fiber grating (Barmenkov et al., 2010), photonic crystal fiber (Wei et al., 2009), and so on. The phase shift dispersion technique referred to above can obviously be controlled by wavelength tuning (Blais & Yao, 2009), but it also means the phase shift is susceptible to wavelength drift as a result of environmental factors, which is unacceptable. The fiber Bragg grating has also been reported utilizing static and moving strain perturbation (Caucheteur et al., 2010) and discrete wavelength scanning. The devices proposed in these papers involve mechanical moving parts and deformation, and do not allow for the continuous operation desirable for high beam-steering accuracy, requiring an expensive tunable light source.

In this paper, we propose and demonstrate a simple photonic-assisted microwave phase shifter based on a Polarization-Maintaining Fiber Bragg Grating (PMFBG) that combines traditional dispersion and nonlinear techniques to insulate against wavelength drift. The continuous phase shifter is implemented by use of third-order nonlinear birefringence in the PMFBG, which can acquire refractive index difference between Transverse-Electric (TE) and Transverse-Magnetic (TM) modes in the optical domain by tuning the input launch optical power or the polarization angle of the Polarization Controller (PC).

2   STRUCTURE AND PRINCIPLES

The proposed phase shifter system demands a proper input signal, which presents two phase-locked orthogonal polarization carriers in the optical spectrum. These optical carriers obtain different phase shifts in the PMFBG, then beat at a Photodetector (PD) and generate a controllable phase-shifted signal. Figure 1 shows the schematic block diagram of the proposed phase shifter. First, a Continuous-Wave (CW) light wave from a laser source is sent to a Single Sideband Suppressed Carrier Mach–Zehnder Modulator (SSB-SC MZM) to generate two phase-locked carriers. The modulated microwave signal is half of the initial frequency provided by the signal generator. The optical spectrum presents two phase-locked carriers, which are 8.4 GHz apart. The optical field at the output of the modulator is given by $E_{SSB}(t)$:

$$E_{SSB}(t) = \hat{x} \bullet (A_1 e^{j\omega_c t} + A_2 e^{j(\omega_c + \omega_{RF})t}) \qquad (1)$$

where A_1 and A_2 are the amplitudes of the central and sideband carriers, ω_c and ω_{RF} are the frequencies of the optical carrier and the RF signal, respectively, and x and y are the orthogonal polarization directions, respectively.

By tuning the differential phase between the two arms of the Delay Interferometer (DI) after the Erbium-Doped Fiber Amplifier (EDFA) with a gain of G, the optical Double Sideband (DSB) signal is filtered and the two phase-locked carriers are de-multiplexed, and then combined as a linearly polarized light by a polarizing beam combiner which can be expressed as $E_{PBC}(t)$:

$$E_{PBC}(t) = \hat{x} A_1 G e^{j\omega_c t} + \hat{y} A_2 G e^{j(\omega_c + \omega_{RF})t} \qquad (2)$$

The spectral response and dispersion characteristics of the PMFBG are shown as curves "X" (red) and "Y" (blue), respectively, in Figure 2. The spectral response is almost the same in the wavelength of 1550 nm. The phase shift between x and y modes is due to the existing

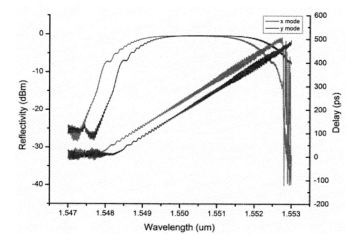

Figure 1.   Optical experimental set-up for the RF phase shifter.

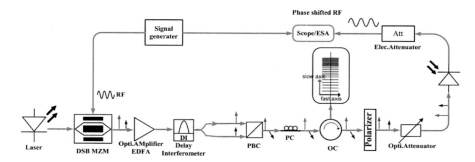

Figure 2.   The spectral response and dispersion of the PMFBG.

difference between the two modes' refractive index, which can be given by (and is defined as the birefringence and the period of the grating) and it is then only a few picometers that barely influence the amplitude of the two modes. Otherwise, the dispersions are approximately a constant value of 43 ps/nm in the broad spectrums.

After being amplified by an EDFA to control the launch power, the optical signal is then sent to a PMFBG via a polarization controller and an Optical Circulator (OC). The optical signal after the PC and OC can be expressed as $E_{pc}(t)$ and $E_{oc}(t)$:

$$E_{PC}(t) = \hat{x}G(A_1 e^{j\omega_c t}\cos\theta - A_2 e^{j(\omega_c + \omega_{RF})t}\sin\theta) \\ + \hat{y}G(A_2 e^{j(\omega_c + \omega_{RF})t}\cos\theta + A_1 e^{j\omega_c t}\sin\theta) \tag{3}$$

$$E_{oc}(t) = \hat{x}G\eta(A_1 e^{j\omega_c t}\cos\theta - A_2 e^{j(\omega_c + \omega_{RF})t}\sin\theta)e^{i\varphi_1} \\ + \hat{y}G\eta(A_1 e^{j\omega_c t}\cos\theta + A_2 e^{j(\omega_c + \omega_{RF})t}\sin\theta)e^{i\varphi_2} \tag{4}$$

where θ is the polarized angle of the PC, φ_1 and φ_2 are the phase shifts induced by the two orthogonal polarization signals in the PMFBG, respectively, and η is the reflectivity of the grating.

Then the phase-shifted orthogonal polarization signal is sent to a $\pi/4$ polarizer and the two orthogonal polarization components interfere. The optical field signal at the output of the polarizer is given by:

$$E_p(t) = \frac{\sqrt{2}}{2}\hat{x}\{G\eta(A_2 e^{j(\omega_c + \omega_{RF})t}\cos\theta - A_1 e^{j\omega_c t}\sin\theta)e^{i\varphi_1} \\ + G\eta(A_1 e^{j\omega_c t}\cos\theta + A_2 e^{j(\omega_c + \omega_{RF})t}\sin\theta)e^{i\varphi_2}\} \tag{5}$$

Finally, the output signal from the polarizer is detected by a Photodetector (PD), and the alternating current part of the output current from the PD is given by:

$$i_{AC}(t) = RG^2\eta^2 A_2(A_1\cos^2\theta - A_1\sin^2\theta)\cos(\omega_{RF}t + \Delta\varphi) \\ = RG^2\eta^2 A_1 A_2\cos 2\theta\cos(\omega_{RF}t + \Delta\varphi) \tag{6}$$

where R is the responsivity of the PD, and $\Delta\varphi = \varphi_1 - \varphi_2$ is the difference of phase shift between the two modes. Therefore, the difference in phase, $\Delta\varphi$, induced by the birefringence effects is directly translated to the phase of the origin RF signal due to the third-order nonlinear effect. A detailed theoretical derivation is given below that shows that the phase shift can be controlled by the gain of the EDFA and the angle of the PC.

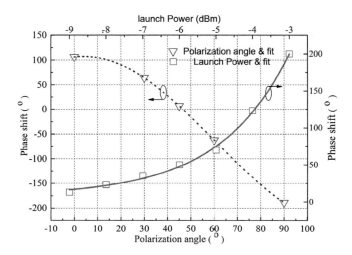

Figure 3.   Phase shift as function of polarization angle.

451

A polarization-maintaining fiber has two principal axes along which the fiber is capable of maintaining the state of linear polarization of the incident light. These axes are called slow and fast axes, according to the speed at which light polarized along them travels inside the fiber. Assuming $n_x > n_y$, n_x and n_y are the effective refractive index along the slow and fast axes, respectively. In addition, because of the nonlinear birefringence effects in the polarization-maintaining fiber, the nonlinear contributions Δn_x and Δn_y are given by:

$$\Delta n_x = n_2 \left(|E_x|^2 + \frac{2}{3}|E_y|^2 \right); \quad \Delta n_y = n_2 \left(|E_y|^2 + \frac{2}{3}|E_x|^2 \right) \tag{7}$$

$$\Delta n = (\Delta n_x - \Delta n_y) = n_2 \frac{1}{3}\left(|E_x|^2 - |E_y|^2 \right) = n_2 \frac{1}{3}\left(|E_{pc}(x)|^2 - |E_{pc}(y)|^2 \right)$$
$$= n_2 \frac{1}{3} G^2 (A_1^2 \cos^2 \theta + A_2^2 \sin^2 \theta - A_2^2 \cos^2 \theta - A_1^2 \sin^2 \theta) \tag{8}$$
$$= n_2 \frac{1}{3} G^2 (A_1^2 - A_2^2)\cos 2\theta$$

where n_2 is the nonlinear parameter.

Furthermore, according to the theory of fiber grating, the phase shift difference in the grating is proportional to the effective refractive index, which can be expressed as:

$$\Delta \varphi = k \bullet n_2 \frac{1}{3} G^2 (A_1^2 - A_2^2)\cos 2\theta \bullet \omega_{RF} \tag{9}$$

where k is the structure parameter of the fiber Bragg grating, which can express the dispersion. So the tunable phase shift of the RF signal can be achieved by changing the input launch optical power or the polarization angle of the PC, which have been validated experimentally, as illustrated in Figure 3.

3 EXPERIMENT AND DISCUSSION

Figure 3 shows that by turning the polarization angle of the PC from 0 to 90° when the input launch optical power is fixed at –4 dBm, a total phase shift of more than 300° at 8.4 GHz is achieved. The fitting curve shows that the experimental data conforms to the cosine function of the theoretical derivation.

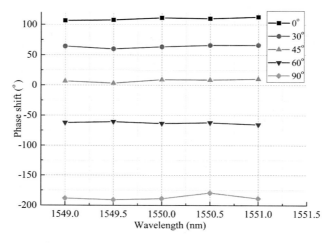

Figure 4. Phase shift at five different polarization angles of PC over a wavelength range from 1549 to 1551 nm.

Figure 3 also shows the measured and fitted phase shifts as a function of launch power when the polarization angle of the PC is fixed at 45°. By increasing the launch power from −9 to −3 dBm, the proposed structure provides a gradual-increment trend phase shifter that can be continuously tuned over a range of 200°.

As can be seen from Figure 4, between five different given polarization angles of the PC, a constant phase shift over a wavelength range from 1549 to 1551 nm is achieved. The phase shift is independent of the optical wavelength in a 2 nm range approximate to the spectral range of the PMFBG.

4 CONCLUSIONS

In this paper, a MWP phase shifter based on a PMFBG with continuously tunable phase shift independent of the optical wavelength drifting is proposed, which has been demonstrated by using the nonlinear birefringence effects in the PMFBG. A total phase shift of more than 300° at 8.4 GHz is demonstrated by tuning the polarization angle of the PC from 0 to 90° and the input launch optical power.

REFERENCES

Barmenkov, Y.O., Cruz, J.L. & Díez, A. (2010). Electrically tunable photonic true-time-delay line. *Optics Express*, *18*(17), 17859–17864.

Blais, S. & Yao, J. (2009). Photonic true-time delay beamforming based on superstructured fiber Bragg gratings with linearly increasing equivalent chirps. *Journal of Lightwave Technology*, *27*(9), 1147–1154.

Bui, L.A., Mitchell, A. & Ghorbani, K. (2005). Wide-band RF photonic second order vector sum phase-shifter. *IEEE Microwave and Wireless Components Letters*, *15*(5), 309–311.

Capmany, J. & Novak, D. (2007). Microwave photonics combines two worlds. *Nature Photonics*, *1*(6), 319–330.

Caucheteur, C., Mussot, A. & Bette, S. (2010). All-fiber tunable optical delay line. *Optics Express*, *18*(3), 3093–3100.

Chen, H., Dong, Y. & He, H. (2009). Photonic radio-frequency phase shifter based on polarization interference. *Optics Letters*, *34*(15), 2375–2377.

Lee, S.S. & Udupa, A.H. (1999). Demonstration of a photonically controlled RF phase shifter. *IEEE Microwave and Guided Wave Letters*, *9*(9), 357–359.

Lim, C. & Li, G. (2013). Microwave photonics: Current challenges towards widespread application. *Optics Express*, *21*(19), 22862–22867.

VanBlaricum, M.L. (1994). Photonic systems for antenna applications. *IEEE Antennas and Propagation Magazine*, *36*(5), 30–38.

Wei, L., Xue, W. & Chen, Y. (2009). Optically fed microwave true-time delay based on a compact liquid-crystal photonic-bandgap-fiber device. *Optics Letters*, *34*(18), 2757–2759.

Frontier Research and Innovation in Optoelectronics Technology and Industry – Habib & Lewis (Eds)
© 2019 Taylor & Francis Group, London, ISBN 978-1-138-33178-5

Field trial of active coexistence unit over a gigabit passive optical network with live broadband subscribers

D. Tarsono, A. Ahmad, K. Khairi & N.A. Ngah
Communication Technology, Telekom Malaysia Research & Development, Cyberjaya, Malaysia

ABSTRACT: The field trial demonstrates the performance of an Active Coexistence (ACEX) unit in a live Gigabit Passive Optical Network (GPON). The ACEX unit, which is installed at the central office together with the optical line terminal system, acts as a booster and pre-amplifier, respectively, for the downstream and upstream optical signals. The field trial showed the successful operation of the ACEX unit and GPON in a single optical distribution network.

1 INTRODUCTION

Gigabit Passive Optical Network (GPON) technology is widely deployed throughout several countries and the number of broadband users has tremendously increased in the past few years. Telekom Malaysia (TM), for example, has rolled out High-Speed Broadband (HSBB) utilizing GPON technology since 2008, serving more than one million subscribers in major Malaysian cities. Consequently, Malaysian internet traffic growth has increased at a Compound Annual Growth Rate (CAGR) of 53.5% between 2012 and 2017, largely contributed to by mobile internet, video streaming, Over-The-Top (OTT) content, cloud computing, and social media (MyIX, 2017).

As the number of subscribers increases, service providers are considering plans for their networks to accommodate larger numbers of subscribers, which will involve more branching or longer distances, utilizing the same infrastructure. Subsequently, the link budget of the networks increased compared to the legacy which requires Passive Optical Network (PON) extender to increase system margin as a solution.

Recent works demonstrate the development of PON extenders for legacy GPON and high-speed 10G PON technologies (Dalla Santa et al., 2016; Le Guyader et al., 2012; Ok & Seok, 2016; Tarsono et al., 2017). Le Guyader et al. (2012) developed a dual-reach extender based on Semiconductor Optical Amplifiers (SOAs) for the coexistence of GPON and 10G PON with an extended power budget at 31 dB (30-km Single Mode Fiber (SMF)). The extender used two SOA units installed in-line in the Optical Distribution Network (ODN) at 10 km after the Optical Line Terminals (OLTs). The long-reach extender was then improved (Ok & Seok, 2016), whereby the design was implemented over 100 km of SMF. The design was demonstrated over a 10G Ethernet PON (EPON) with a maximum splitting at 128. However, both designs are installed in-line in the ODN, which requires a power feed resulting in a complex Fiber-To-The-Home (FTTH) network architecture. In contrast, Dalla Santa et al. (2016) designed a reach extender-based Raman amplifier installed at the Central Office (CO) where the OLT was installed. The extender supported over 50 km of SMF with a splitting at 64 Optical Network Units (ONUs). However, when using high amplification such as Raman the optical power in the transmission fiber needs to be controlled to avoid a non-linearity effect such as Stimulated Brillouin Scattering (SBS), which distorts system signal quality. Tarsono et al. (2017) improved the research on the extender by developing a universal extender for GPON, a Time Wavelength-Division Multiplexing PON (TWDM-PON) and a 10G Symmetrical PON (XGSPON). The extender or Active Coexistence (ACEX) unit was demonstrated experimentally over a coexistence PON. The ACEX unit was then extended to the field in a Telekom Malaysia network that served HSBB subscribers.

Our paper considers the field-trial demonstration of an active coexistence unit with one PON port of a GPON system in a brownfield area. The existing GPON PON port served broadband users at a distance of 16 km from the central office. The existing users subscribe to triple play services with different packages of high-speed internet (from 10 Mbps to 100 Mbps).

2 SYSTEM DESIGN

The field trial was conducted in Bangi, Selangor, Malaysia on TM's existing GPON system and physical infrastructure under the brand name UNIFI. The total number of UNIFI users in Bangi served by TM was 17,329 (as at 2017). The ACEX unit acting as a reach extender was connected to a PON port on an OLT currently serving 29 active users (maximum users up to 32) in Bangi Node (CO), thus preserving the passive optical network and simplifying future maintenance, as shown in Figure 1. The ACEX unit is also equipped with a remote monitoring link to enable our personnel to monitor its performance from the TM Research & Development office in Cyberjaya.

The total distance from the OLT to the customer premises is 15.7 km and the E-side cabling is supported with a 1+1 protection scheme to the Fiber Distribution Cabinet (FDC).

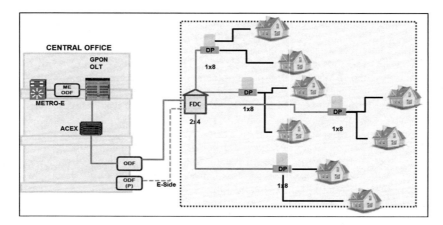

Figure 1.   Network architecture of ACEX field trial in Bangi, Malaysia (ODF – optical distribution frame; FDC – fiber distribution cabinet; DP – distribution point).

Table 1.   ACEX field trial link information.

	GPON	
ITU standard	G.984 (2003)	
Optical path loss class	B+ (MAX loss 28 dB)	
	GPON OLT	GPON ONU
Mean launched power MIN (Tx) (dBm)	1.5	0.5
Mean launched power MIN (Tx) (dBm)	5.0	5
MIN sensivity (Rx) (dBm)	−28	−27
MIN overload (dBm)	−8	−8
MIN field trial link loss (dB)	24.7	
MAX field trial link loss (dB)	34.7	

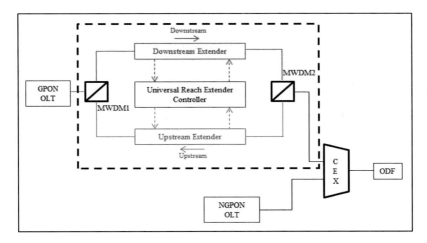

Figure 2. Bidirectional active coexistence unit.

TM commonly practices two-stage splitting in the PON: a 2×4 splitter is placed at the FDC with a 1×8 splitter in each Fiber Distribution Panel (FDP). The link loss recorded among the 29 active users involved in the field trial is between 24.7 and 34.7 dB, as shown in Table 1. The variation in link loss is due to aging and other external factors such as weather, wild animals, and other unspecified causes.

In this field trial, the target for the ACEX unit was to improve the current link losses to at least 28 dB, as specified in the TM guidelines, but at the same time not overload customers within the acceptable link loss range. The ACEX unit is based on a bidirectional optical extender, shown in Figure 2, which enables a 1490 nm Downstream (DS) signal and a 1310 nm Upstream (US) signal to be amplified simultaneously. Both DS and US extenders utilize SOA, and a pair of Multi-Wavelength-Division Multiplexers (MWDM1 and MWDM2) is used to isolate the signals of both directions. The extender is also equipped with monitoring and a controller to ease monitoring and configuration of the extender.

3 RESULT AND DISCUSSION

Figure 3 shows the gain as a function of SOA current for both directions. From the figure, the maximum gain obtained for DS and US signals is 6 dB and 17 dB, respectively. The configuration of the ACEX unit is based on physical study, data collection and a survey of the area to avoid exceeding of power received by all affected users.

The performance in terms of ONU received power for a sampled Distribution Point (DP) is shown in Figures 4 and 5. The DP consists of eight live customers with different broadband packages. The performance of ONU received power was obtained through the TM Network Monitoring System (NMS). At the sampled DP, one user (User 1) demonstrated low received power of −33 dBm, which is in the unacceptable range of ONU received power for a GPON system. As illustrated in Figure 4, all users at the sampled DP experienced an improvement in ONU received power of 5 dBm after installation of the ACEX unit.

Figure 5 represents the performance of ONU received power for all users of the sampled DP during four months of monitoring, and shows that all the users experienced consistency in ONU received power. This demonstrated the steady performance of the ACEX unit in the field. Furthermore, no trouble tickets were raised by the customers during the monitoring period.

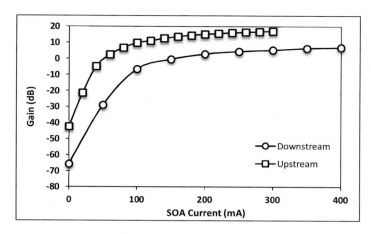

Figure 3.   Gain vs SOA current for DS and US signals.

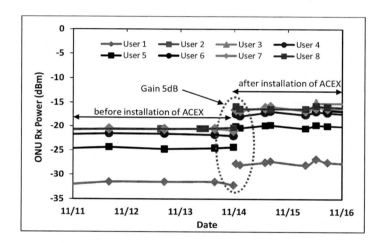

Figure 4.   ONU received power vs time before and after installation of ACEX.

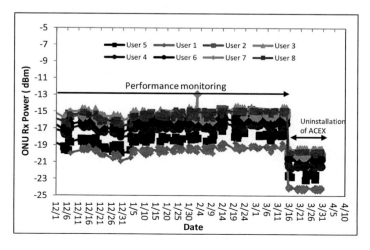

Figure 5.   Performance of ONU received power for eight users of sampled DP.

4 CONCLUSION

The field trial of the ACEX unit with a GPON system was successfully executed in the selected FTTH network area. The existing network could also accommodate higher splitting and more distance with the implementation of the ACEX unit. This is supported by the verification of the network performance as well as the uninterrupted services provided to the GPON customers. Ultimately, this trial validated the reach extender implemented at the CO and the feasibility for future reference and planning of long-reach connectivity of PON technologies.

ACKNOWLEDGMENTS

This activity was supported by Telekom Malaysia (TM) and TM Research & Development. Highest gratitude to our colleagues in the AVENGERS team, NATP, ST, AD, ANM, ISPNM and TM NOC/NMO HSBB who have provided the insights and expertise for this project.

REFERENCES

Dalla Santa, M., Antony, C., Talli, G., Krestnikov, I. & Townsend, P.D. (2016). Burst-mode analysis of XGPON Raman reach extender employing quantum-dot lasers. *Electronics Letters, 52*(1), 1157–1158.

Le Guyader, B., Saliou, F., Guillo, L., Le Roux, M., Charbonnier, B., Chanclou, P. & Pascal, J.-M. (2012). Dual reach extender based on optical amplification for GPON and symmetrical 10G-PON systems. In *Proceedings National Fiber Optic Engineers Conference, 4–8 March, Los Angeles, CA, United States* (paper NTu1 J.2). Washington, DC: Optical Society of America. doi:10.1364/NFOEC.2012.NTu1 J.2.

MyIX. (2017, August 23). *MyIX records over 1200% growth in internet traffic within five years.* Kuala Lumpur, Malaysia: Malaysia Internet Exchange. Retrieved from http://myix.my/news/myix-records-over-1200-growth-in-internet-traffic-within-five-years.

Ok, K.K. & Seok, C.H. (2016). Real-time demonstration of extended 10G-EPON capable of 128-way split on a 100 km distance using OEO-based PON extender. In *International Conference on Information and Communication Technology Convergence, 19–21 October, Jeju, South Korea* (pp. 930–932). Piscataway, NJ: Institute of Electrical and Electronics Engineers. doi:10.1109/ICTC.2016.7763333.

Tarsono, D., Ahmad, A., Sharif, K.A., Othman, M.H., Khairi, K., Ngah, N.A. & Manaf, Z.A. (2017). Optical and network performance analysis of XGS-PON system over active co-existence PON systems. *Optics and Photonics Journal, 7*(8), 40–48.

Frontier Research and Innovation in Optoelectronics Technology and Industry – Habib & Lewis (Eds)
© 2019 Taylor & Francis Group, London, ISBN 978-1-138-33178-5

Diffractive optical elements with chiral-focusing properties for optical-trapping applications

A. Vijayakumar, M.R. Rai & J. Rosen
Department of Electrical and Computer Engineering, Ben-Gurion University of the Negev,
Beer-Sheva, Israel

B. Vinoth & C.-J. Cheng
Institute of Electro-Optical Science and Technology, National Taiwan Normal University, Taipei, Taiwan

I.V. Minin
Tomsk State University, Tomsk, Russia
FGUP "SNIIM", Novosibirsk, Russia

O.V. Minin
Tomsk State University, Tomsk, Russia

ABSTRACT: Two Diffractive Optical Elements (DOEs), namely chiral Fresnel zone plate and chiral axicon, were designed according to the principle of zone rotation in order to generate chiral beams with rotating intensity patterns. The DOEs were introduced in an optical-trapping experiment to trap and rotate yeast cells.

1 INTRODUCTION

Diffractive Optical Elements (DOEs) have the ability to modulate the characteristics of an optical field and produce any desired phase and intensity profile (Vijayakumar & Bhattacharya, 2017). One of the many areas of research that have been revolutionized by DOEs is optical trapping (Padgett & Bowman, 2011). Optical trapping can be broadly classified into two categories according to the type of trap forces used, namely gradient force with Gaussian beams, and Orbital Angular Momentum (OAM) with vortex-like beams (Padgett & Bowman, 2011). The type of force that is used in most optical-trapping setups is the gradient force of the Gaussian intensity profile (Ashkin, 1970). Vortex beams are used to avoid absorptive heating and optical damage to the trapped particles.

The reason that the design and fabrication of DOEs for the generation of vortex beams is so complicated is because of the phase profiles generated. For instance, a spiral phase plate, which is one of the basic vortex-generating DOEs, has a staircase-type phase profile that is difficult to manufacture (Cheong et al., 2004). The same problem exists for other vortex-generating DOEs.

The concept of the microwave chiral zone plate lens antenna, introduced by Minin et al. (2005) for a square configuration, was later investigated in order to improve the hexagonal zone plate antenna sidelobe performance (Minin & Minin, 2006; Stout-Grandy et al., 2008). In this paper, we introduce a family of binary DOEs based on the zone-rotation principle, with chiral-focusing characteristics to generate beams with rotating intensity patterns in optical bands. Two different DOEs, namely Chiral Square Fresnel Zone Plate (CSFZP) and Chiral Square Axicon (CSA), are designed to generate hybrid beams with a gradient force at the center, with a rotating intensity pattern and only rotating intensity pattern, respectively, for optical-trapping applications.

Figure 1. Square Fresnel zone plate with classical and rotating zones.

2 BASIC PRINCIPLES OF ZONE ROTATION

Let us consider a plane waves diffraction problem in a system of embedded one-by-one alternate (transparent/non-transparent) rectangular screens. As viewed from above, in this case a field intensity distribution at the distant zone in the first approximation is characterized by the function SINC.

The goal of the optimization is to eliminate the pronounced regular (cross-shaped) structure of the side maximums and lower the level of the side maximums. To solve this problem, Fresnel zones have been offered for rotation with respect to each other by some angle that depends on the zone's number – see Figure 1 (Minin et al., 2005).

The diffraction of the zone plates is characterized by superposition of the diffracted waves on each embedded zone. Thus the field intensity distribution at the distant zone can be described by the following function:

$$I(x,y) = \left| \sum_{n=1,3,5...}^{N} \left[\begin{array}{l} \sin c(\pi A a_n (x\cos\varphi_n + y\sin\varphi_n)) \bullet \\ \sin c(\pi A a_n (-x\sin\varphi_n + y\cos\varphi_n)) 2 A a_n^2 - \\ \sin c(\pi A a_{n-1} (x\cos\varphi_n + y\sin\varphi_n)) \bullet \\ \sin c(\pi A a_{n-1} (-x\sin\varphi_n + y\cos\varphi_n)) 2 A a_{n-1}^2 \end{array} \right] \right|^2$$

where φ_n is constrained by the condition: square bounds a_{2i} and a_{2i-1} must not intersect each other, that is:

$$-c(a_n) \le \varphi_n \le c(a_n), \text{ where } C(a_n) = \begin{cases} \dfrac{\pi}{4} : a_n > \sqrt{2} \cdot a_{n-1} \\ \dfrac{\pi}{4} - \arccos\left(\dfrac{a_n}{\sqrt{2} \cdot a_{n-1}} \right) \end{cases},$$

$$(a_0 = a_{-1} = 0), \varphi_0 = 0, A = const = \frac{2}{\lambda z}; a_i = \sqrt{i\lambda F + i^2 \lambda^2 \cdot 0.25}, i = 1,2,3...; F, \lambda, z = const$$

3 METHODOLOGY

The DOEs, namely CSFZP and CSA, are designed by the rotation of the half-period zones with respect to one another following a linear trend $\theta = Km$, where $m = 2n$ and K is the angular step size. The size of every zone is estimated for CSFZP and CSA using the equations $W_n = (n^2\lambda^2 + 2nf\lambda)^{1/2}$ and $W_n = n\Lambda/2$, where f is the focal length of CBSA, λ is the wavelength, Λ is the period of the CBSA, and $n = 0, \frac{1}{2}, 1, \dots$ (Vijayakumar et al., 2017). The DOEs were

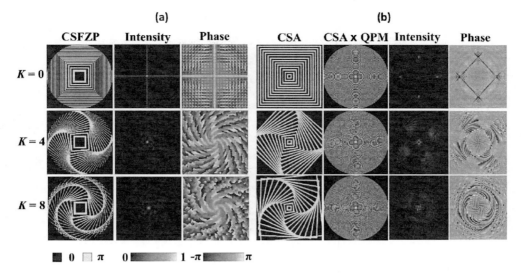

(a) CSFZP | Intensity | Phase

(b) CSA | CSA × QPM | Intensity | Phase

$K = 0$

$K = 4$

$K = 8$

■ 0 □ π   0 ▬ 1   -π ▬ π

Figure 2.   Simulation of intensity and phase for $K = 0°$, 4° and 8° for: (a) CSFZP; (b) CSA.

designed with binary phase values [0, π] in order to obtain the maximum binary efficiency (40%) (Vijayakumar & Bhattacharya, 2017). In the case of CSA, in order to bring the far-field intensity pattern within a finite distance, a quadratic phase function (QPM) is multiplied with it.

The intensity and phase profiles generated by the two DOEs are simulated using a scalar diffraction formulation, as shown in Figure 2. From the simulation, it can be seen that with an increase in the value of K, the phase pattern twists about the optical axis. Second, it was noticed that there was a redistribution of light from the focal plane of CSFZP to other axial planes, resulting in an increase in the focal depth as K increased. On the other hand, in the case of CSA, there was a redistribution of light to the center, creating a ring pattern, with an increase in K. In both cases, the intensity pattern was found to rotate about the optical axis.

3 EXPERIMENTS AND RESULTS

The intensity patterns generated by the DOEs were experimentally studied using an optical setup, as shown in Figure 3, by displaying the DOE masks on a phase-only Spatial Light Modulator (SLM).

The intensity patterns recorded for CSFZP and CSA for $K = 4°$ are shown in Figure 3 as insets. In the case of CSA, an additional QPM with a focal length equal to the distance between the SLM and the image sensor is multiplied with the DOE.

The optical-trapping experiment was carried out using the setup shown in Figure 4. The optical tweezer setup consists of two laser sources emitting at wavelengths $\lambda_1 = 532$ nm and $\lambda_2 = 632.8$ nm for trapping and imaging, respectively. The light from the green laser is spatially filtered and collimated and is incident on a reflective phase-only SLM (Jasper Display Corp., Hsinchu City, Taiwan; pixel numbers 1920 × 1024; pixel pitch 6.4 μm) at an angle of 12°. After the SLM, a 4F system is used, followed by a 100x objective lens with a Numerical Aperture (NA) = 1.3. We have used the yeast *Candida rugosa* (ATCC® 200555™) as a specimen for trapping and this is recorded by an imaging system, as shown in Figure 4. The sample was not only trapped but also rotated about the optical axis using the DOEs. The image of the trapping of the *Candida rugosa* (ATCC® 200555™) sample with the beam pattern by chiral Fresnel zone plate for $K = 1°$ is shown in Figure 5.

The images of *Candida rugosa* (ATCC® 200555™) samples before and after trapping using the beam pattern for $K = 4°$ by chiral axicon are shown in Figure 6. When the yeast specimen

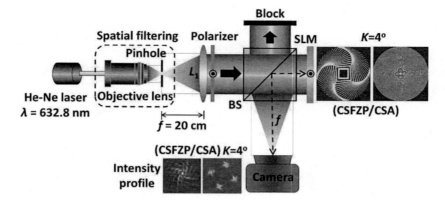

Figure 3. Experimental setup for evaluating the DOEs.

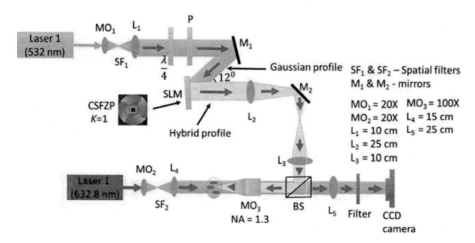

Figure 4. Schematic of the optical-trapping experiment using the DOEs.

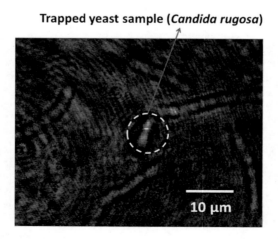

Figure 5. Optical trapping of *Candida rugosa* with the beam pattern generated by CSFZP for $K = 1°$.

(a) Before trapping

(b) After trapping

(c) After rotation

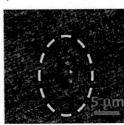

Figure 6. Microscopic images of *Candida rugosa*: (a) before trapping; (b) after trapping; (c) after rotation for $K = 4°$ by chiral axicon.

was overlapped with the trap beam, the yeast sample was rotated until it reached a final stable position.

4 CONCLUSION

In conclusion, we have introduced a family of DOEs based on the zone-rotation principle with chiral-focusing properties for optical-trapping applications. The optical-trapping results showed a rotation of the *Candida rugosa* sample, indicating the optical-trapping and manipulation capabilities of the chiral DOEs. In the current study, only DOEs with a chirality of four have been studied. It is possible to design DOEs with a smaller or larger order of chirality by either decreasing or increasing the number of vertices of the DOEs.

REFERENCES

Ashkin, A. (1970). Acceleration and trapping of particles by radiation pressure. *Physical Review Letters*, *24*(4), 156–159.

Cheong, W.C., Lee, W.M., Yuan, X.C., Zhang, L.S., Dholakia, K. & Wang, H. (2004). Direct electron-beam writing of continuous spiral phase plates in negative resist with high power efficiency for optical manipulation. *Applied Physics Letters*, *85*(23), 5784–5786.

Minin, I.V. & Minin, O.V. (2006). Array of Fresnel zone plate lens antennas: Circular, hexagonal with chiral symmetry and hexagonal boundary. In *Digest of the Joint 31st International Conference on Infrared and Millimeter Waves and 14th International Conference on Terahertz Electronics, September 18–22, Shanghai, China* (p. 270). doi:10.1109/ICIMW.2006.368478.

Minin, I.V., Minin, O.V., Danilov, E.G. & Lbov, G.V. (2005). Parameters optimization algorithm of a new type of diffraction optics elements. In *Proceedings of 5th IEEE-Russia Conference MEMIA, December 13–15, Novosibirsk, Russia* (pp. 177–185).

Padgett, M. & Bowman, R. (2011). Tweezers with a twist. *Nature Photonics*, *5*(6), 343–348.

Stout-Grandy, S.M., Petosa, A., Minin, I.V., Minin, O.V. & Wight, J. (2008). Fresnel zone plate antenna with hexagonal-cut zones. *Microwave and Optical Technology Letters*, *50*(3), 672–676.

Vijayakumar, A. & Bhattacharya, S. (2017). *Design and fabrication of diffractive optical elements with MATLAB*. Bellingham, WA: Society of Photo-optical Instrumentation Engineers (SPIE).

Vijayakumar, A., Vinoth, B., Minin, I.V., Rosen, J., Minin, O.V. & Cheng, C.J. (2017). Experimental demonstration of square Fresnel zone plate with chiral side lobes. *Applied Optics*, *56*(13), F128–F133.

Frontier Research and Innovation in Optoelectronics Technology and Industry – Habib & Lewis (Eds)
© 2019 Taylor & Francis Group, London, ISBN 978-1-138-33178-5

The research on a dual-wavelength gain competition mechanism for a sensing application

Qiang Wang & Shun Wang
Laboratory of Optical Information Technology, Wuhan Institute of Technology, Wuhan, Hubei Province, China

ABSTRACT: We propose and demonstrate a gain competition mechanism based on a Dual-Wavelength Erbium-Doped Fiber Laser (DWEDFL), for a sensing application. The two wavelengths in the DWEDFL share the same gain medium (i.e. EDF), forming gain competing with each other, which is very sensitive to the cavity loss caused by external parameters. Utilizing this, we implemented parametric measurements, including Refractive Index (RI) and temperature. Experimental results demonstrate high sensitivities of −231.1 dB/RIU for RI, and −1.83 dB/°C for temperature, respectively. Thanks to the high sensitivity and easy demodulation, our scheme offers an option for increasing the measuring sensitivity in optical fiber sensing applications.

1 INTRODUCTION

In the past decades, some research effort has been devoted toward Novel Optical Fiber Sensing (OFS) development to provide alternative novel approaches. Due to their unique structure and performance, Photonic Crystal Fiber (PCF) based fiber sensors (Tian et al., 2016; S. Liu et al., 2016) are especially popular and widely used, although they involve complicated production processes. A variety of techniques have been developed to further improve the fiber sensors' working performance, including coating (Zhao et al., 2013), micro-machining (Sun et al., 2016), the use of micro/nano structures (Martinez-Rios et al., 2012), multiple interference (Luo et al., 2014), and the Vernier effect (Xu et al., 2015). Besides, Fiber Laser Sensors (FLS) are shaping up as being an important area of research and development, for their excellent performance regarding high optical power and narrow bandwidth (Guan et al., 2012). Among them, the Dual-Wavelength Erbium-Doped Fiber Laser (DWEDFL) opens up new possibilities for fiber-based applications in the fields of optical fiber communication systems (Sun et al., 2012), but is rarely used in optical fiber sensing systems (Liu et al., 2007).

In this paper, we demonstrate a series of sensors based on the gain competition mechanism in a DWEDFL. The DWEDFL is in a simple ring-cavity configuration, which includes filtering devices: two FBGs with similar bandwidth but slightly different central wavelengths. The two wavelengths share a common gain medium, which makes it extremely sensitive to external perturbation. High sensitivities for Refractive Index (RI) and temperature sensing are experimentally realized, proving them to be good candidates for enhancing the measuring sensitivity in optical fiber sensing applications.

2 LASER CONFIGURATION AND SETUP PRINCIPLE

DWEDFL is in a simple ring-cavity configuration, as shown in Figure 1, which includes two ring cavities: C1 (980 LD-WDM-ISO-OC$_2$-Circulator1-Sensing element-FBG$_1$-OC1-EDF) and C2 (980 LD-WDM-ISO-OC$_2$-Circulator2-VOA-FBG$_2$-OC1-EDF). The 980 LD pump is a 10 m EDF (gain factor 6.5 ± 1.0 dB/m, OFS Fitel, LLC) through a Wavelength Division

Multiplexing (WDM). The fiber isolator (ISO) in the cavity keeps the laser working in a counterclockwise direction. Two optical couplers, OC1 and OC2 (splitting ratio 50:50), are used to divide the laser cavity into two parts.

A Variable Optical Attenuator (VOA) is embedded in C2 to adjust the power balance of the two wavelengths during the gain competing process. The sensing element is embedded in C1 to measure the external measurands. The filtering devices, FBG1 and FBG2, have similar reflection and 3-dB bandwidth (both ~85% and < 0.15 nm, respectively) but slightly different center wavelengths ($\lambda 1$ = 1,551.02 nm, $\lambda 2$ = 1,550.12 nm). An optical spectrum analyzer (OSA, YOKOGAVA AQ6370C) with a resolution of 20 pm is used to observe the output spectra of our DWEDFL. Its output is shown in the dashed box in Figure 1, and is almost consistent with the FBGs.

In addition, the stability of our DWEDFL will have an impact on the afterwards measurements in terms of operating point shift and detecting accuracy. Thus, the intensity fluctuation over 50 minutes of the dual wavelengths is illustrated in Figure 2; a maximum power fluctuation of 0.75 dB at $\lambda 1$ and 0.96 dB at $\lambda 2$ can be generalized. This can be further optimized by controlling the ambient temperature more precisely, keeping the cavity loss stable, and averaging the repeated data measurements.

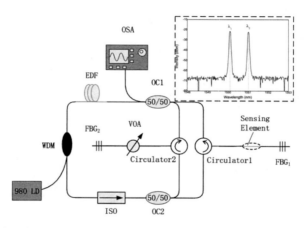

Figure 1. Schematics of gain competition mechanism based on DWEDFL.

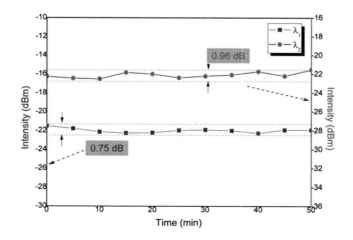

Figure 2. Dual-wavelength power fluctuation over 50 minutes.

3 EXPERIMENTAL RESULTS AND DISCUSSION

Based on the DWEDFL experimental platform, experiments on refractive index and temperature are implemented to test the sensing performance of our scheme. The detailed experiments are described as follows.

3.1 *Refractive index sensor*

It should be noted that the RI sensing element is embedded in C2 while the VOA is in C1 in this part of the experiment. The RI sensing element is a section of microfiber, which is stretched from a standard Single-Mode-Fiber (SMF) by using flame-heating and taper-drawing technology. The microfiber has a diameter of ~14 μm and length of ~1.5 cm, fuse-spliced between two SMF fibers, as shown in Figure 3. The fiber line is fixed between two stages to keep the fiber straight and stable. RI matching solutions are dropped in the MgF_2 substrate plane to keep the microfiber immersed in the solution.

Figure 4(a) illustrates the output spectra variation with the RI ranging from 1.300 to 1.355, and Figure 4(b) the linear fit of our experimental data. The intensity at 1,551 nm decreases, along with the RI increasing, due to the increasing optical loss of our sensing element, and power at 1,550 nm changes inversely owing to the gain competition between the two wavelengths. The overall results show a sensitivity of 42.6 dB/RIU at λ_1 and −231.1 dB/RIU at λ_2, with linearity of 0.984 and 0.943, respectively. What implies to be a novel and highly sensitive RI sensor based on our DWEDFL is realized by applying a gain competition mechanism in the range from 1.300 to 1.335 within a step of 0.005.

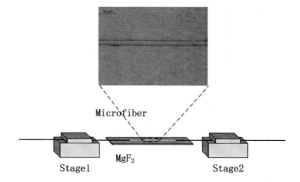

Figure 3. Schematic of microfiber for refractive index sensing.

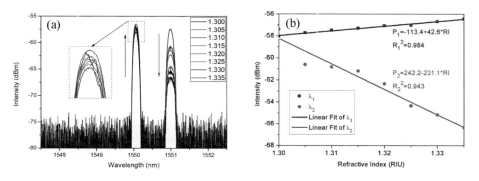

Figure 4. (a) Output spectrum of our sensor in RI matching solutions with RI 1.300–1.355; (b) Intensity of the peak power at dual wavelengths as a function of the surrounding RI.

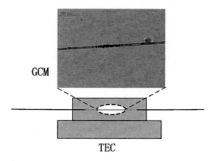

GCM

TEC

Figure 5. Schematic of graphene-coated microfiber for temperature sensing.

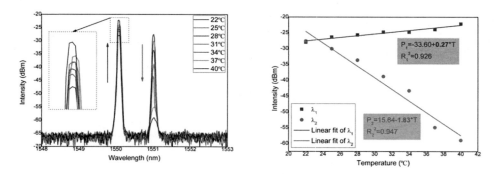

Figure 6. Output optical spectral of the DWEDFL in different temperature fields.

3.2 Temperature sensor

As shown in Figure 1, the temperature-sensing element is embedded in C1. The temperature-sensing element is shown in Figure 5, in which a microscope-magnified map shows the Graphene Coated Microfiber (GCM) structure. The GCM structure here is formed from a microfiber with 6.7 μm diameter and 30 mm length by using a 'thermophoresis effect' (Sun et al., 2014). A Thermoelectric Controller (TEC) is used in our experiment for controlling the temperature fields around the GCM.

Similarly, with the RI sensing, Figure 6 shows the output spectral variation in different temperature fields. High sensitivities of 0.27 dB/°C and −1.83 dB/°C within the range 22–40°C are realized, which are several times higher than the reference (Q. Sun et al., 2016). Due to the resolution of the commercial optical power meter being as much as 0.001 dB, the corresponding temperature resolutions are 0.0037°C and 0.0005°C, respectively.

In addition, it is worth mentioning that our gain competition mechanism based on DWEDFL has the potential for sensing in applications not only in the fields investigated in this article, but also in other fields such as strain, stress and acoustics. It is a good candidate for high-sensitivity sensing.

4 CONCLUSION

This paper presents and discusses a gain competition mechanism based on the DWEDFL, in which the two output wavelengths share the same gain medium EDF. Thus, the two wavelengths will scramble the only gain medium, making it extremely sensitive to external measurands. To test the sensing performance, experiments in terms of RI and temperature are carried out. High sensitivities of −231.1 dB/RIU for RI and −1.83 dB/°C for temperature are experimentally demonstrated respectively. The merits of high sensitivity and ease to demodulate makes our scheme a good candidate for sensing applications.

REFERENCES

Guan, B.O., Jin, L., Zhang, Y. & Tam, H.Y. (2012). Polarimetric heterodyning fiber grating laser sensors. *Journal of Lightwave Technology*, *30*(8), 1097–1112.

Liu, D., Ngo, N.Q., Tjin, S.C. & Dong, X. (2007). A dual-wavelength fiber laser sensor system for measurement of temperature and strain. *IEEE Photonics Technology Letters*, *19*(15), 1148–1150.

Liu, S., Wang, Z., Hou, M., Tian, J. & Xia, J. (2016). Asymmetrically infiltrated twin core photonic crystal fiber for dual-parameter sensing. *Optics & Laser Technology*, *82*, 53–56.

Luo, H., Sun, Q., Xu, Z., Liu, D. & Zhang, L. (2014). Simultaneous measurement of refractive index and temperature using multimode microfiber-based dual Mach-Zehnder interferometer. *Optics Letters*, *39*(13), 4049.

Martinez-Rios, A., Monzon-Hernandez, D., Salceda-Delgado, G., Cardenas-Sevilla, G.A. & Villatoro, J. (2012). Optical microfiber mode interferometer for temperature-independent refractometric sensing. *Optics Letters*, *37*(11), 1974.

Sun, X.Y., Chu, D.K., Dong, X.R., Zhou-Chu & Li, H.T. (2016). Highly sensitive refractive index fiber inline Mach-Zehnder interferometer fabricated by femtosecond laser micromachining and chemical etching. *Optics & Laser Technology*, *77*, 11–15.

Sun, Q., Sun, X., Jia, W., Xu, Z., Luo, H., Liu, D. & Zhang, L. (2016). Graphene-assisted microfiber for optical-power-based temperature sensor. *IEEE Photonics Technology Letters*, *28*(4), 383–386.

Sun, X., Sun, Q., Jia, W., Xu, Z., Wo, J., Liu, D. & Zhang, L. (2014). Graphene coated microfiber for temperature sensor. In *Fiber-Based Technologies and Applications* (pp. FF4B-3).

Sun, Q., Wang, J., Tong, W., Luo, J. & Liu, D. (2012). Channel-switchable single-/dual-wavelength single-longitudinal-mode laser and THz beat frequency generation up to 3.6 THz. *Applied Physics B*, *106*(2), 373–377.

Tian, J., Lu, Z., Quan, M., Jiao, M. & Yao, Y. (2016). Fast response Fabry-Perot interferometer microfluidic refractive index fiber sensor based on concave-core photonic crystal fiber. *Optics Express*, *24*(18), 20132.

Xu, Z., Sun, Q., Li, B., Luo, Y., Lu, W., Liu, D., … Zhang, L. (2015). Highly sensitive refractive index sensor based on cascaded microfiber knots with Vernier effect. *Optics Express*, *23*(5), 6662–6672.

Zhao, Y., Pang, F., Dong, Y., Wen, J. & Chen, Z. (2013). Refractive index sensitivity enhancement of optical fiber cladding mode by depositing nanofilm via ALD technology. *Optics Express*, *21*(22), 26136.

Frontier Research and Innovation in Optoelectronics Technology and Industry – Habib & Lewis (Eds)
© 2019 Taylor & Francis Group, London, ISBN 978-1-138-33178-5

Optical absorption in an InAs/GaSb based type II quantum well system

X.F. Wei
Research Center of Atoms Molecules and Optical Applications, West Anhui University, Luan, China

W.Y. Wang
School of Physics and Electronics Information, Shangrao Normal University, Shangrao Jiangxi, China

ABSTRACT: The dependence of optical absorption on AlSb barrier widths was investigated in InAs/AlSb/GaSb based Quantum Wells (QWs). The optical absorption coefficients were calculated in InAs/AlSb/GaSb based QWs by employing the balance equation method to solve the Boltzmann equation. Two peaks in optical absorption coefficient were observed due to the intraband transitions. The AlSb cap layer was inserted between the InAs layer and the GaSb layer to reduce the Generation–Recombination (G–R) noises induced by the electron and hole transition in different material layers. The optical absorption coefficients induced by interband transition were significantly reduced when the width of the AlSb cap layer reached up to 1 nm. The positions of the absorption peaks lie in the mid-infrared region and perform well at high temperatures. The results suggest that InAs/GaSb based type II and broken-gap QWs can be employed as two-color photodetectors working at mid-infrared bandwidth at relatively high temperatures up to room temperature.

Keywords: Optical absorption, InAs/GaSb based type II quantum well, photodetectors

1 INTRODUCTION

Infrared (IR) photodetectors have been widely used in medical, industrial, military, fire-fighting and environmental monitoring. The major photon detection technologies include mercury cadmium telluride (Rogalski, 1999), Quantum Well IR Photodetectors (QWIPs) (Levine, 1993) and Quantum Dot IR Photodetectors (QDIPs) (Pan et al., 2000). Commercially available infrared detectors can be categorized as interband, which are HgCdTe and InAsSb, or intersub-band quantum well infrared detectors. QWIPs are an excellent choice for state-of-the-art photodetection because of their easy growth, fast response time, high homogeneity and spatial resolution. However, in the process of photoelectric conversion, not only the effective signals are presented, but also the noise signals, such as thermal noise, dark current noise, scattering noise and so on. These noise signals greatly reduce the detection performance of the detector and the signal-to-noise ratio of the system. Therefore, it is of great significance to analyze the nature and influence factors of noise and take effective solutions. The InAs/GaSb based quantum wells and supper lattice systems have been widely used in IR photodetectors. Unfortunately, there are some fundamental problems, namely the fast auger recombination in interband detectors and the high thermal generation rate in intersub-band detectors, which drastically decreases their ability for near room temperature operation in the long-wavelength infrared range.

In this study, we focus on the influence of the width of the AlSb cap layer on the optical absorption coefficients in GaSb/AlSb/InAs Quantum Well (QW) systems. We analyze the influencing factor of detector noise signal, and design an effective optimization structure of QWs to reduce the noise and improve the performance of the detector. There are many internal noises in photoelectric detectors. The main noise sources are thermal noise and shot noise. The thermal noise has the greatest influence on the detection ability. Thermal noise exists in any conductor and semiconductor. It comes from the irregular thermal motion of the internal free electron or the charge carrier of the resistor. When there is no external field, the electronic conductor in the random thermal motion, non-directional migration, there is no current, but due to the number of fluctuations, two electrons moving in the opposite direction are not completely equal, resulting in potential fluctuations in conductor and semi-conductor noise voltage, causing the fluctuations of current.

2   THEORETICAL CONSIDERATIONS AND APPROACHES

When a linearly polarized light field is present, the Hamiltonian Equation to describe such a two-body system can be written as:

$$H = H_e + H_h + H'_{e-o} + H'_{h-o} \tag{1}$$

Here, $H_i = P_i^2/2m_i^* + U_i(z_i)$ (i = e, h) is the single-particle Hamiltonian Mechanism for an electron and a hole, respectively, with m_i^* being the effective mass of an electron or a hole. The energy is measured from the bottom of the conduction band in the electron layer. P_i is the momentum operator. $U_i(z_i)$ is the confining potential energy for an electron or a hole along the growth direction. The electron and hole wave functions along with the corresponding energy spectra can be written respectively as:

$$| e >= e^{ik \cdot r} \psi_n^e(z_e), | h >= e^{ik \cdot r} \psi_n^h(z_h), E_n^e(k) = \frac{\hbar^2 k^2}{2m_e^*} + \varepsilon_n^e, E_n^h(k) = -\frac{\hbar^2 k^2}{2m_h^*} + \varepsilon_n^h,$$

Here, **k** is the electron or hole wave vector along the 2D-plane. $\psi_n^i(z_i)$ and the sub-band energy ε_n^i are the solutions of the Schrödinger Equation along the growth direction for an electron or a hole.

In the present study, the transfer matrix approach is employed to solve the Schrödinger Equation numerically, which has been proven simple and accurate (Ying et al., 2010). As the effect of the self-consistent potential only shifts the band edges of the QWs, which is reasonably neglected in the present calculations? The semi-classic Boltzmann Equation is employed to study the response of the carriers to the applied radiation fields. For an electron (i = e) or a hole (i = h), we have:

$$\frac{\partial f_n^i(k,t)}{\partial t} = g_s \sum_{j,k',n'} [F_{n'n}^{ji}(k',k,t) - F_{nn'}^{ij}(k,k',t)], \tag{2}$$

Here, $F_{nn'}^{ij}(k,k',t) = f_n^i(k,t)[1 - f_{n'}^j(k',t)]W_{nn'}^{ij}(k,k')$, is the electronic sub-band index, $f_n^i(k,t)$ is the momentum-distribution function for an electron or a hole at a state |k, n>, $g_s = 2$ counts for spin-degeneracy, and $W_{nn'}^{ij}(k,k')$ is the steady-state electronic transition rate for scattering of an electron or a hole from a state |k, n> in layer i to a state |k', n'> in layer j. For cases where the electronic transition is induced by electron or hole interactions with the radiation field, the transition rate can be obtained by using Fermi's golden rule, which reads:

$$W_{nn'}^{ij}(k) = \frac{2\pi}{\hbar} \left(\frac{e\hbar F_0}{m_i^* \omega} \right)^2 |X_{nn'}^{ij}|^2 \delta[E_n^i(k) - E_{n'}^j(k') + \hbar\omega]$$

where F_0 and ω are the electric field strength and frequency of the EM field, respectively: $X_{nn'}^{ij} = \int dz \psi_{n'}^{j*}(z) d\psi_n^i(z)/dz$.

In this work, we apply the usual balance equation approach to solve the problem. For the first moment, the energy-balance equation (Lei & Horing, 1987) can be derived by multiplying $g_s \Sigma_k E_n^i(k)$ to both sides of Equation 2. In doing so, we obtain two energy-balance equations for an electron and for a hole, respectively. Then, the total electronic energy transfer rate due to electron/hole interactions with photons is obtained as:

$$P = P_e + P_h = \sum_{ij} P_{ij}$$

where $P_i = g_s \partial [\Sigma_n \Sigma_k E_n^i(k) f_n^i(k,t)]/\partial t$ is the electronic energy transfer rate for electrons or holes and:

$$P_{ij} = 4\hbar\omega \sum_{n,n',k} f(E_n^i(k))[1 - f(E_n^i(k))]W_{nn'}^{ij}(k). \tag{3}$$

Here, we have used a statistical energy distribution such as the Fermi–Dirac function as the electron/hole distribution function at a steady-state. Namely, we have taken $f_n^i(k,t) \approx f(E_n^i(k)) = [1 + e^{(E-E_F)/k_BT}]^{-1}$ and E_F being the Fermi energy. The optical absorption coefficient induced by electron and hole interactions with the EM field can be calculated through (Lei & Liu, 2000):

$$\alpha = \alpha_0(2\hbar P/e^2 F_0^2) = \sum_{ij} \alpha_{ij} \tag{4}$$

where $\alpha_0 = e^2/(\hbar\sqrt{\kappa}\varepsilon_0 C)$, κ and ε_0 are the dielectric constants of the material and the free space, respectively and C is the velocity of the light in vacuum. Furthermore, for intra-layer transition:

$$\alpha_{ii} = \beta_i \frac{k_BT}{\pi} \sum_{n',n} \frac{|X_{nn'}^{ii}|^2 B_{n'n}^i \Gamma_i}{(\varepsilon_{n'}^i - \varepsilon_n^i - \hbar\omega)^2 + \Gamma} \tag{5}$$

and for inter-layer transition:

$$\alpha_{eh} = \beta_e M^*/m_e^* \sum_{n',n} \Theta(\varepsilon_{n'}^h - \varepsilon_n^e - \hbar\omega)|X_{n'n}^{he}|^2 F_{nn'}^{eh}$$

$$\alpha_{he} = \beta_h M^*/m_h^* \sum_{n',n} \Theta(\varepsilon_{n'}^h - \varepsilon_n^e + \hbar\omega)|X_{n'n}^{he}|^2 F_{nn'}^{eh} \tag{6}$$

Here $\beta_i = \alpha_0(8\hbar/m_i^*\omega)$, $1/M^* = 1/m_e^* + 1/m_h^*$, $B_{nn}^i = [A_{n'}^i/(A_n^i - A_{n'}^i)]\ln[A_n^i(1+A_{n'}^i)/A_{n'}^i(1+A_n^i)]$, $A_n^i = \exp[(\varepsilon_n^i - E_F)/k_BT]$, $F_{nn'}^{ij} = f(x_j^-)[1 - f(x_i^+)]$, $x_i^\pm = (m_e^*\varepsilon_n + m_h^*\varepsilon_{n'} \pm m_i^*\hbar\omega)/(m_e^* + m_h^*)$.

Furthermore, for the case of intra-layer scattering, in Equation 5 we have taken $\delta(x) \to (\Gamma_i/\pi)/(x^2 + \Gamma_i^2)$, with Γ_i being the broadening of the scattering state.

3 NUMERICAL RESULTS AND DISCUSSION

In this study, we examined the optical absorption coefficients in InAs/AlSb/GaSb based quantum wells where the AlSb cap layer is inserted between the InAs layer and the GaSb layer to reduce the noise of the optical absorption. The transfer matrix approach is employed to solve the Schrödinger Equation to get the sub-band energies as well as the wave functions both for the electron and hole. A set of typical QW widths LInAs = 15 nm and LGaSb = 10 nm are taken throughout the calculations. The static dielectric constants in the InAs layer and the GaSb layer are taken as $\kappa_e = 15.15$ and $\kappa_h = 15.69$, respectively. The reference point of the energy is put at the bottom of the conduction band in the InAs layer.

In the present study we take the widths of the InAs layer and the GaSb layer as 15 nm and 10 nm, respectively. The wave functions for electron and hole are shown in Figure 1 with the widths of the AlSb cap layer as 0 nm and 2 nm, respectively. As can be seen, the wave

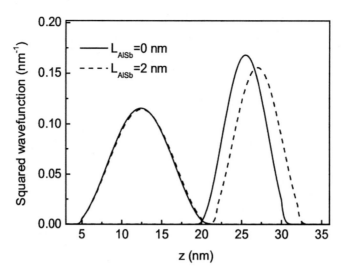

Figure 1. The ground state wave functions for electron and hole at different widths of the AlSb cap layer. The solid line and dashed line represent wave functions at $L_{AlSb} = 0$ nm and $L_{AlSb} = 2$ nm, respectively.

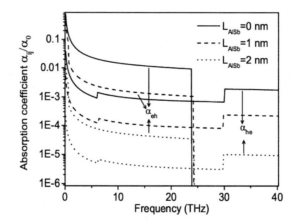

Figure 2. The interband optical absorptions induced by the carrier transitions between different material layers at T = 300 K. The solid line, dash line and dot line represent to the interband transitions with $L_{AlSb} = 0$, $L_{AlSb} = 1$ nm and $L_{AlSb} = 2$ nm, respectively.

functions for electron and hole overlap at the interface of the quantum well which leads to the penetration of the electron and hole into different materials. When the width of the AlSb cap layer reaches up to 2 nm, the overlap of the electron and hole can be neglected. As a result, the interband optical absorption can be significantly affected by the AlSb cap layer. The optical absorption coefficients induced by the interband transitions are shown in Figure 2 with different widths of the AlSb cap layer. As can be seen, there are several platforms in optical absorption coefficients α_{ij} induced by interband transition due to the fact that several transition channels open in the presence of the radiation field. The optical coefficient α_{eh} is significantly higher than α_{he} which reflects the fact that the optical transition of the electron from the InAs layer to the GaSb layer dominates in the interband transition due to the lesser mass of the electron. It should be noted that, the optical coefficient α_{ij} can be significantly affected by the AlSb cap layer. Additionally, the optical coefficient α_{ij} is reduced by four orders of magnitude when the width of the AlSb cap layer reaches up to 2 nm. The optical

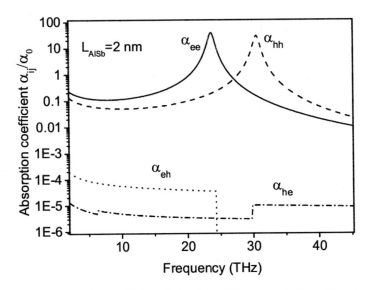

Figure 3. The optical absorption coefficients induced by different carrier transition channels with the width of the AlSb cap layer L_{AlSb} = 2 nm at T = 300 K. The solid line and dashed line refer to the optical absorption coefficients induced by electron and hole transitions in the intraband transition. The dotted line and dashed line refer to the optical absorption coefficients induced by the interband transitions.

transitions of electron and hole between different materials can lead to Generation–Recombination (G–R) noise in the application of photoelectric detectors (Rumyantsev et al., 2001).

The temperature-dependent Fermi energy (or chemical potential) is determined by the condition of charge number conservation. In this study, we set the Fermi energy at the intercross points of the electron and hole energy dispersions. We get the same electron and hole densities correspondingly. In such a structure, we take the two lowest electron sub-bands and two highest hole sub-bands which are occupied respectively by electrons and holes. Therefore, the electron and hole can transit in the same material layer and different material layer and several transition channels open in the presence of the radiation field. The optical absorption coefficients α_{ij} are shown in Figure 3 with the width of the AlSb cap layer = 2 nm. The optical absorption coefficients $\alpha_{ee}(\alpha_{hh})$ represents the electron (hole) transition in the same material layer. The optical absorption coefficients α_{eh} (α_{he}) represents the carrier transition in the different material layers. The optical transitions in the same material layer dominate in InAs/AlSb/GaSb based quantum wells which can be used in photo detectors. The optical transitions in different material layers arise from the electron–hole combination in the quantum well structure which can cause G–R noise. In the present study, we focus on the design of the quantum well to reduce the G–R noise. The G–R noise can be significantly reduced when the width of the AlSb cap layer reaches up to 2 nm. Two optical absorption peaks are observed in the mid-infrared regime due to the optical transitions in the same material layers indicating that InAs/AlSb/GaSb quantum wells can be used at room temperature in an infrared photodetector with a low G–R noise in accordance with the experiments (Yuping et al., 2013).

4   CONCLUSIONS

We investigated the optical absorption properties in InAs/AlSb/GaSb quantum wells to reduce the G–R noise in the applications of the photodetectors. Two peaks in optical absorption coefficient were observed in the mid-infrared bandwidth. The strength of the peaks depend slightly on the temperature, indicating that the photodetector based on InAs/AlSb/GaSb quantum wells can work at room temperature. The AlSb cap layer was inserted to reduce the

G–R noise arising from the electron–hole transitions in different material layers. The optical transitions of the carriers in different material layers were reduced significantly when the width of the AlSb cap layer reaches up to 2 nm. Our theoretical results show that the InAs/AlSb/GaSb quantum wells can be used as low noise semiconductor photodetectors.

ACKNOWLEDGMENTS

This work was supported by the National Natural Science Foundation of China No. 11474310, Key projects of Anhui Provincial Department of Education (No. KJ2017A406, No. KJ2017A401, No. KJ2016A749, No. KJ2015A150), Quality Engineering Projects of Anhui Province (No. 2015gkk015, No. 2015jyxm284, No. 2016gxx153) and Program of West Anhui University.

REFERENCES

He Ying, Zhang Fanming, Yang Yanfang & Li Chunfang, (2010). Energy eigenvalues from an analytical transfer matrix method. *Chinese Physics B*, *19*(4), 040306.
Levine, B.F. (1993). Quantum-well infrared photodetectors. *Journal of Applied Physics, 74*, R1–R81.
Lei X.L. & Horing N.J., (1987). Nonlinear balance equations for hot-electron transport with finite phonon-relaxation time. *Physical Review B Condensed Matter*, *35*(12), 6281.
Lei X.L. & Liu S.Y., (2000). Nonlinear free-carrier absorption of intense THz radiation in semiconductors. *Journal of Physics: Condensed Matter*, *12*, 4655.
Pan, D., E. Towe, & S. Kennerly, (2000). Photovoltaic quantum-dot infrared detectors. *Applied Physics Letters, 76*, 3301–3303.
Rogalski, A. (1999). Assessment of HgCdTe photodiodes and quantum well infrared photoconductors for long wavelength focal plane arrays. *Infrared Physics and Technology*, *40*, 279–294.
Rumyantsev, S.L., Pala, N., Shur, M.S. & Borovitskaya. E. (2001). Generation-recombination noise in gan/algan heterostructure field effect transistors. *IEEE Transactions on Electron Devices*, *48*(3), 530–534.
Yuping Zeng, Chien-I Kuo, Rehan Kapadia et al. (2013). Two-dimensional to three-dimensional tunneling in InAs/AlSb/GaSb quantum well heterojunctions. *Journal of Applied Physics, 114*, 024502.

Porthole structural stray light analysis and suppression in star sensor

Kun Yu, Mingyu Cong & Biao Guo

Research Center for Space Optical Engineering, School of Astronautics, Harbin Institute of Technology, Harbin, Heilongjiang, China

ABSTRACT: Considering the needs of aerodynamic layout and high temperature protection, the star sensor is usually installed inside the near space hypersonic vehicle, and implements attitude measurement through porthole attached to the surface of the vehicle. Ground simulation experiments have shown that the detection capability of the star sensor is reduced, due to the increased stray light received at the detector, after the porthole structure is introduced into the optical path of the star sensor. In this paper, the phenomenon of structural stray light caused by the porthole is simulated. Stray light sources, system geometry and surface optical properties are modeled respectively in the simulation scenario. Point source transmittance is used as evaluation indices for the stray light suppression effect. The simulation calculations found that the surface scattering of the porthole glasses is the main reason that the porthole structure causes the stray light suppression effect to decrease. This structural stray light is mainly composed of scattered light formed by multiple scattering between porthole glasses and scattered light formed by scattering between the inner wall of the system and the porthole glasses. Compared with the case without porthole structure, the point source transmittance of the star sensor can reach a maximum of 10^{-4}. The star sensor has been unable to meet the detection limit requirements. From the viewpoint of improving the surface optical properties of the system and optimizing the system structure, the suppression method for porthole structural stray light is studied. The results show that reducing the total integrated scatter of the system structure surface (including the system inner wall surface and the porthole glass surface) and using a single-layer porthole can decrease the influence of structural stray light caused by the porthole. The two methods are used to determine the final optimization scheme. The optimized system satisfies the detection requirements of the star sensor. Outside the stray light protection angle, the point source transmittance of the optimized system is below 3×10^{-5}.

1 INTRODUCTION

The star sensor is a high precision attitude measuring device that uses the celestial coordinate system as a reference system and targets stars in space. Compared with other attitude measuring devices, the star sensor has higher measurement accuracy, strong anti-interference ability and autonomous navigation capability, and has been widely used in the attitude measurement of spacecraft in various countries of the world (Liang, 2016).

The near space hypersonic vehicle generally flies at speeds above Mach 3, and under the influence of the aerodynamic heating effect, thousands of degrees of high temperature are formed on its surface (McClinton, 2005). In order to maintain the aerodynamic layout and high temperature protection of the vehicle, the star sensor is usually installed inside the near space hypersonic vehicle, and implements attitude measurement through porthole attached to the surface of the vehicle. The porthole structure is generally composed of porthole glass and metal shell, and the star sensor realizes optical imaging and star map recognition through the porthole glass (Zhang, 2005).

The star sensor is weak light detection device, which is highly susceptible to stray light during detection. The direct radiation of the sun is generally the main source of stray light of the star sensor (Kawano, 2005). Stray light is usually suppressed by adding a lens hood in front of the star sensor light path (Boggess, 1978). By designing vanes structure and adding extinction coatings to the surface, stray light can be suppressed before it is incident on the detector (Breault, 1995). In the design process of the lens hood, it is generally considered that there is no other influencing factor before the light path of the star sensor outside the lens hood, and the design of the vanes structure usually only considers the suppression on the direct incident stray light. However, for the star sensor installed in near space hypersonic vehicle, the porthole structure will change the light path of the star sensor. The ground simulation experiments show that due to the introduction of the porthole structure, the stray light that could be suppressed can enter the detection optical path, causing the star sensor to fail. The stray light additionally introduced by the porthole structure is defined as the structural stray light of the system. At present, there are few studies on the mechanism and suppression method of structural stray light.

In this paper, the star sensor and its protective porthole structure on a near space hypersonic vehicle are analyzed. The stray light suppression effect of the system is studied by simulation, and the mechanism of the stray light caused by the porthole structure on the star sensor is determined. By improving the surface optical properties of the system and optimizing the system structure, a structural stray light suppression optimization scheme that satisfies the star sensor detection requirements is obtained.

2 STAR SENSOR STRUCTURAL STRAY LIGHT ANALYSIS METHOD

2.1 Stray light evaluation index

The stray light suppression capability of the optical system is usually evaluated using Point Source Transmittance (PST) (Zhao, 2016). PST is defined:

$$PST(\theta) = \frac{E_d(\theta)}{E_i(\theta)} \tag{1}$$

where θ is off-axis angle, $E_d(\theta)$ is the irradiance of the light source (pointolite or parallel light source) received by the detector, $E_i(\theta)$ is the irradiance of the light source at the optical system inlet. PST embodies the stray light suppression capability of the optical system itself, independent of the radiation intensity of the stray source. In the case of the same off-axis angle, the smaller the PST value, the stronger the stray light suppression capability of the system. In general, if the PST of an optical system is below the order of 10^{-5}, it is considered to have better stray light suppression capability.

2.2 Stray light source

Considering the space environment and working mode of the near space hypersonic vehicle, the direct sunlight is used as the source of the structural stray light of the star sensor system with the porthole structure. The visible illuminance of the sun is about 1.27×10^5 Lux, simulated by a parallel light source. The off-axis angle is used to describe the transmission direction of stray light, which is defined as the angle between the direction of the light and the optical axis of the optical system, ranging from 0 to 90°. In the analysis process, by changing the off-axis angle of the light source, the ability of the system to suppress stray light in different directions can be calculated.

2.3 Structural surface optical property model

The star sensor structure including the porthole installed on the near space hypersonic vehicle analyzed in this paper is shown in Figure 1. The porthole structure is composed of two parts: the porthole glass and the porthole shell.

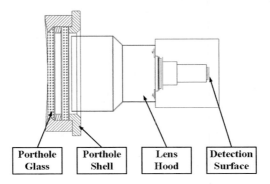

| Porthole Glass | Porthole Shell | Lens Hood | Detection Surface |

Figure 1.   Star sensor detection system with porthole composition chart.

The geometry of the system can be directly imported into the stray light analysis software through its 3D modeling file. Before the stray light simulation calculation and analysis, the surface optical properties of each part of the system need to be set. The optical scattering properties of the material surface are typically described using a Bidirectional Reflectance Distribution Function (BRDF) (Dittman, 2002). BRDF is defined:

$$f_r(\theta_i,\varphi_i,\theta_r,\varphi_r)=\frac{dL_r(\theta_i,\varphi_i,\theta_r,\varphi_r)}{dE_i(\theta_i,\varphi_i)} \tag{2}$$

where θ_i, φ_i are the zenith and azimuth of the incident light, θ_r, φ_r are the zenith and azimuth of the reflected light, dL_r is the reflection radiance of dA in the (θ_r, φ_r) direction, dE_i is the incident irradiance in the (θ_i, φ_i) direction. Total Integrated Scatter (TIS) of the material can be obtained by integrating the BRDF in the hemisphere space. TIS is defined:

$$\mathrm{TIS}=\int_0^{2\pi}\int_0^{\pi/2}BRDF(\theta_r,\varphi_r)\cos\theta_r\sin\varphi_r\,\mathrm{d}\theta_r\,\mathrm{d}\varphi_r \tag{3}$$

In the porthole structure, the structures that affect the incident light are the porthole shell inner wall and the porthole glass. The inner wall of the porthole shell and the lens hood are stray light suppressed by spraying darken paint. The darken paint has substantially the same scattering properties in all directions, and its BRDF can be described by the Lambertian model. Lambertian model is defined:

$$BRDF_L=\frac{\mathrm{TIS}}{\pi} \tag{4}$$

without scattering light, the remaining incident light will be absorbed.

The porthole glass surface can be thought of as a smooth optical mirror, and its BRDF can be selected from the Harvey-Shack model(Harvey, 1995). Harvey-Shack model is defined:

$$BRDF_H(\theta,\theta_0)=b_0\left(1+\left(\frac{\sin\theta-\sin\theta_0}{L}\right)^2\right)^{\frac{s}{2}} \tag{5}$$

where θ is the scattering angle of the scattering light, θ_0 is the specular reflection angle, b_0, s, L are the control parameters of the model.

S is the skewness factor of the model, and its value range is generally between -2.5 and -1. L is the approximate specular flip angle of the Harvey-Shack model. For planes with higher smoothness, its value is usually less than or equal the order of 10^{-2}. b_0 can be calculated by TIS of the material, and it is defined:

$$b_0 = L^s \cdot \text{TIS} \cdot \frac{s+2}{2\pi} \qquad (6)$$

In addition to considering the surface scattering properties, porthole glass as a transparent material should also include its transmittance and refractivity during the optical property modeling process.

3 MECHANISM OF PORTHOLE STRUCTURAL STRAY LIGHT

3.1 *Simulation calculation parameters*

In this paper, the FRED software is used to analyze the stray light suppression capability of the star sensor system with a porthole. In order to ensure the reliability of the calculation results, the number of trace rays of the light source is set to 10^6, the relative energy cutoff threshold of the ray tracing is set to 10^{-14}, and the maximum number of splits of the scattered/reflected rays is 5 times.

The parameters of the star sensor system are obtained through actual measurement. The detection maximum PST of the star sensor is 3×10^{-5}, the semi-field angle is $10°$, and the stray light protection angle of the hood is $30°$. The geometry parameters of the system include: the porthole inlet diameter is 135 mm, the star sensor hood inlet diameter is 85 mm, and the star sensor detector diameter is 20 mm. The optical properties parameters of each part of the system are shown in Table 1.

3.2 *Porthole structural stray light analysis*

After installing the porthole, the star sensor optical path is added with the porthole glass and the inner wall of the porthole shell. The two parts are added in sequence to calculate the stray light suppression ability of the star sensor, thereby determining the mechanism of the porthole structural stray light. The simulation results are shown in Figure 2.

When the star sensor is not equipped with the porthole structure, the PST of the system is less than the detection limit in the case where the stray light off-axis angle is greater than the stray light protection angle. When the porthole shell is installed in front of the star sensor but the porthole glass is not installed, the system's PST is further reduced and still meets the detection limit. After the installation of the complete porthole structure (porthole glass and porthole shell), the system's suppression effect on stray light is significantly reduced, and the PST increases with the decrease of the off-axis angle. For the stray light with off-axis angles of less than 75°, the detection limit cannot be met. At this time, the system has a PST of 1.50×10^{-4} for the off-axis angle 40° stray light and a PST of 8.65×10^{-5} for the off-axis angle 70° stray light. It can be seen that the addition of the porthole glass is the main reason for the reduction of the stray light suppression ability of the system.

Table 1. Surface optical properties table.

Structure name	Optical parameters	Value
System inner wall surface	Scattering model	Lambert
	TIS_w	5.0%
Porthole glass surface	Transmission(τ_g)	95.0%
	Refractivety(n_g)	1.46
	Scattering model	Harvey-Shack
	TIS_g	3.0%
	s	−1.5
	L	0.005
	b_0	6.75

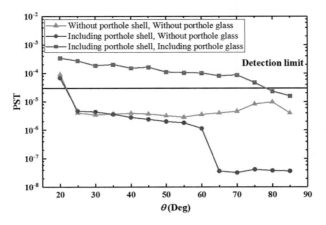

Figure 2.   Stray light suppression effect of different structural systems.

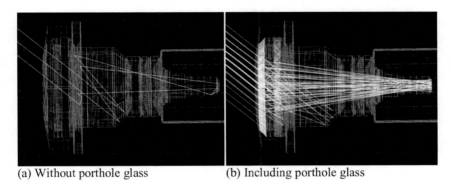

(a) Without porthole glass   (b) Including porthole glass

Figure 3.   Stray light transmission path.

The transmission path of the porthole structural stray light is further analyzed by ray tracing. The ray tracing result of the porthole glass at the off-axis angle of 40° is shown in Figure 3. In the absence of porthole glass, the stray light received by the detector is mainly due to the scattering of incident light by the inner wall of the system. This part of the stray light can be effectively suppressed, and the actual energy reaching the detector is not enough to affect the star sensor. After adding the porthole glass, the incident light scatters between the two layers of porthole glass, causing scattering light to reach the detector directly. This part of the stray light is not suppressed by the star sensor hood, which is the main source of the stray light energy enhancement of the detector. In addition, the scattering of scattering light from the inner wall of the system through the surface of the porthole glass to the detector is also a secondary source of increased stray light energy.

4   PORTHOLE STRUCTURAL STRAY LIGHT SUPPRESSION METHOD

4.1   *Surface optical properties optimization*

Without changing the structure of the system, first consider the method of reducing the TIS of the porthole glass to improve the stray light suppression ability of the system. The TIS of the porthole glass was reduced to 1.0%, and the surface optical properties of others remained unchanged. The simulation results are shown in Figure 4. After the TIS of the porthole glass

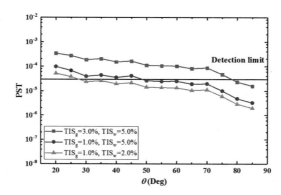

Figure 4. Stray light suppression effect of different optical properties system of double glass porthole.

is reduced, the PST of the system is also reduced, but for the stray light with an off-axis angle of less than 50°, the detection limit cannot be met.

Since only reducing the TIS of the porthole glass can't meet the detection limit of the system, the method of reducing the total reflectance of the inner wall of the system by the fault is further considered to further improve the stray light suppression capability of the system. While the TIS of the porthole glass is reduced to 1.0%, the TIS of the inner wall of the system is reduced to 2.0%, and other parameters remain unchanged. The simulation results are shown in Fig. 4. The system's stray light suppression ability has been further improved, and the system can suppress the stray light outside the protection angle to meet the detection limit. At this time, the system has a PST of 1.95×10^{-5} for the off-axis angle 40° stray light and a PST of 1.10×10^{-5} for the off-axis angle 70° stray light.

4.2 Structural layout optimization

The optimization scheme without changing the structure of the system has high requirements on the optical properties of the system surface, and there are great difficulties in engineering implementation. The existing materials and processing precision are difficult to meet the requirements. Therefore, it is a practical method to improve the system structure layout to improve the stray light suppression effect. Consider optimizing the porthole structure to eliminate multiple scattering of stray light between the porthole glass by reducing the number of porthole glass.

The porthole structure was changed to a single-layer glass, and the inner porthole glass closer to the detector was removed. The optical properties of the system remained unchanged as shown in Table 1. The simulation results of the stray light suppression effect at this time are shown in Figure 5. The single-layer glass porthole structure can improve the stray light suppression capability of the system, but at this time, the PST still cannot meet the detection limit of the system, and there is still a re-inward scattering of the outward scattering light between the inner wall and the porthole glass that has an effect on the detector. This part of the structural stray light can be suppressed by improving the structure of the hood, for example, the vanes can be changed from vertical to the optical axis to have a camber angle.

Based on the conclusions in Section 3.1, this structural stray light can also be suppressed by improving the optical properties parameters of the system surface. In the case of single-layer porthole glass, the TIS of the inner wall is reduced to 3.0%, the TIS of the porthole glass is reduced to 1.5%, and other parameters of the system remain unchanged. The simulation results are shown in Figure 5. At this time, the system's stray light suppression ability can meet the detection limit requirements, and the system has a PST of 1.48×10^{-5} for the off-axis angle 40° stray light and a PST of 1.27×10^{-5} for the off-axis angle 70° stray light. Compared with the optimization scheme without changing the system structure, this optimization scheme has lower requirements on the optical properties of the system surface and has higher engineering feasibility.

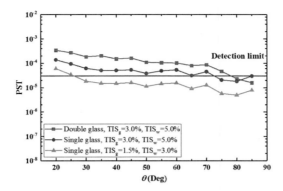

Figure 5.   Stray light suppression effect of different optical properties system of single glass.

5   CONCLUSION

Aiming at the problem of structural stray light caused by the star sensor system with porthole structure, this paper studies the stray light analysis modeling method, structural stray light mechanism and suppression method. The simulation results show that the surface scattering of the porthole glass is the reason for the reduced stray light suppression ability of the system. This structural stray light is mainly composed of scattered light formed by multiple scattering between porthole glasses and scattered light formed by scattering between the inner wall of the system and the porthole glasses. This part of the structural stray light can be suppressed by improving the optical properties of the system surface and optimizing the porthole structure. The final optimization scheme uses a single-layer porthole glass structure, reduces the TIS of the system inner wall to 3.0% and the TIS of the porthole glass to 1.5%. This scheme enables the star sensor to meet the detection limit requirements, and the system's PST is below 1.50×10^{-5}. This paper provides a theoretical and simulation reference for such stray light optimization design work including the porthole structure optical system. The optimization scheme also needs to carry out feasibility analysis in engineering practice.

REFERENCES

Boggess, A., Carr, F.A. & Evans, D.C. 1978. The IUE spacecraft and instrumentation. *Nature*: 275(5679): 372.

Breault, R.P. 1995. Control of stray light. *Handbook of Optics*: 1: 38.1–38.35.

Dittman, M.G. 2002. Contamination scatter functions for stray-light analysis. *Optical System Contamination: Effects, Measurements, and Control VII. International Society for Optics and Photonics*: 4774: 99–111.

Harvey, J.E. & Thompson, A.K. 1995. Scattering effects from residual optical fabrication errors. *International Conference on Optical Fabrication and Testing. International Society for Optics and Photonics*: 2576: 155–175.

Kawano, H., Shimoji, H. & Yoshikawa, S. 2005. Suppression of sun interference in the star sensor baffling stray light by total internal reflection. *Optical Design and Engineering II. International Society for Optics and Photonics*: 5962: 59621R.

Liang, B., Zhu, H.L. & Zhang, T. 2016. Research status and development tendency of star tracker technique. *Chinese Optics*: 9(1): 16–29.

McClinton, C.R., Rausch, V.L. & Nguyen, L.T. 2005. Preliminary X-43 flight test results. *Acta Astronautica*: 57(2–8): 266–276.

Zhang, J.X. & Liu, Q.Z. 2005. The Application and Simplified Thermal Analysis of the Porthole Thermal Cover Technique for Earth Observation Spacecraft. *Journal of Astronautics (China)*: 26(4): 395–399.

Zhao, H.L., Jian, K.Z. & Liang, X. 2016. Analysis and calibration of precision for point source transmittance system. *Acta Physica Sinica*: 65(11): 150–156.

Frontier Research and Innovation in Optoelectronics Technology and Industry – Habib & Lewis (Eds)
© *2019 Taylor & Francis Group, London, ISBN 978-1-138-33178-5*

Performance prediction for coherent optical transponder using accurate model

Yang Yue, Qiang Wang & Jon Anderson
Juniper Networks, Sunnyvale, CA, USA

ABSTRACT: One of the key transmission performance indicators for coherent optical transponder is the Bit Error Rate (BER) versus Optical Signal to Noise Ratio (OSNR) characteristics. We have developed a model to predict BER versus OSNR at various Receiver Optical Power (ROP). The model has three parameters, which are related to BER noise floor, filter mismatching, and OSNR value without noise loading. The model is applied to transponders with high baud rate and Quadrature Amplitude Modulation (QAM). By considering the influence of baud rate on the fitting parameters, accurate prediction of performance for coherent transponder can be achieved over various baud rates. Novel applications enabled by this model include in-field measurement of BER versus OSNR, simple abstraction of coherent transponder, accurate OSNR monitor and coherent optical channel monitor.

1 INTRODUCTION

To meet the ever-growing requirement of Internet traffic, latest commercial optical communication systems is moving towards 400 and beyond Gb/s per wavelength (Roberts, K. et al. 2015, Wei, J. et al. 2015). Besides increasing the modulation baud rate, higher data rate per wavelength is achieved through coherent detection and high-spectral-efficiency modulation formats. e.g., polarization division multiplexed (PDM) and quadrature amplitude modulation (QAM) (Peng, W. et al. 2013, Yue, Y. et al. 2018). Multiplexing two orthogonal polarizations is proved to be a very effective method to increase the capacity and spectral efficiency of an optical communications system by a factor of 2. Furthermore, by using the complex in-phase (I) and quadrature (Q) domain, record high 4096-QAM optical signals have recently been demonstrated, which gives $>10 \times$ increase for the capacity and spectral efficiency simultaneously (Terayama, M. et al. 2018, Olsson, S. et al. 2018).

Figure 1 shows the architecture of packet optical integration which closely integrates the packet forwarding engine (PFE) application specific integrated circuit (ASIC), the digital signal processing (DSP) ASIC, and the coherent optical transponders. For example, the coherent optical transponder can be a CFP2 form factor analog coherent optics (CFP2-ACO) module. The IP traffic is converted into optical signal through coherent transmitter and sent through the long-haul optical communication system which can include multiple reconfigurable optical add drop modules (ROADM). During the transmission, multiple optical impairments can accumulate. After coherent detection at the receiver and analog-to-digital (ADC) conversion, most impairments are compensated by the DSP ASIC in the digital domain.

In the long-haul optical communication systems, optical amplifiers are used to compensate the attenuation introduced by optical fiber. While optical signal is boosted, additional noise is also added, which causes the decrease of optical signal noise ratio (OSNR) and the increase of bit error ratio (BER). Eventually, transmission distance can be limited, when BER is too high for forward error correction (FEC) to correct. Thus, BER vs. OSNR curve is one of the most important characteristic of coherent transponder for long-reach application. However a complicated setup is required to measure the BER vs. OSNR curve. This makes it difficult and complicated to measure this curve in the manufacturing environment and during the

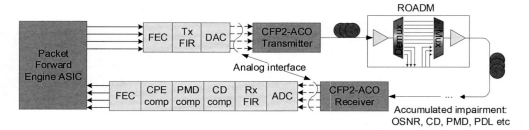

ROADM

Figure 1. Architecture for packet-optical integration. FEC: forward error correction; FIR: finite impulse response; DAC: digital to analog conversion; OSNR: optical signal to noise ratio; CD: chromatic dispersion; PMD: polarization modal dispersion; PDL: polarization dependent loss; ADC: analog to digital conversion; CPE: carrier phase estimation.

in-field operation. Thus, a laudable goal would be to have a simple way to measure this performance for coherent transponder.

In this paper, we develop an accurate analytical model to predict the BER vs. OSNR performance (Wang, Q. et al. 2018). Experimental studies show that the model can function well for high baud rate, up to 86 Giga baud per second (GBd/s), and advanced modulation format (PDM-16QAM). By considering the influence of baud rate on model parameters, we accurately predict the BER vs. OSNR curve for various baud rates.

2 ANALYTICAL MODEL AND EXPERIMENTAL RESULTS

2.1 *Analytical model*

In (Torrengo, E. et al. 2011), an analytical model has been developed to predict the performance of BER vs. OSNR curve as shown in Eq. (1). There are two fitting parameters, η (Eta, related to filter mismatching) and κ (Kappa, related to the noise floor),

$$OSNR_{cal} = \frac{10^{\wedge}(OSNR^{dB}/10)*bw}{2*B}, \frac{1}{SNR} = \frac{1}{\kappa} + \frac{1}{OSNR_{cal}},$$

$$BER_{fit} = erfc\left(\sqrt{\eta * SNR}\right)$$

(1)

$$\text{Minimize } Err_{rms} = \frac{1}{N}\sum_{i=1}^{N}\left(\frac{BER_{fit} - BER_{measure}}{BER_{fit}}\right)^2 \text{ to solve } \eta \text{ and } \kappa$$

where $OSNR^{dB}$ is the OSNR value in dB, SNR is the signal to noise ratio, B is the baud rate of electrical signal including FEC overhead, bw is 12.5GHz (0.1 nm) where OSNR is usually measured against, $OSNR_{cal}$ is the OSNR value after normalization against B and bw, $erfc()$ is the complementary error function. κ and η are solved by minimizing the relative error (Err_{rms}) between measurement result ($BER_{measure}$) and curve-fitting result (BER_{fit}). The model works very well over different receiver optical power (ROP). However for each ROP value, a different set of η and κ are needed. So there is a strong need to expand the model so that the influence of ROP is included.

The model assumes additive white Gaussian noise (AWGN). With coherent detection, electrical field containing both the signal and amplified spontaneous emission (ASE) noise is linearly recovered. Thus, the noise source from OSNR can be added together with other noise source by AWGN assumption.

2.2 *Experimental setup*

Figure 2(a) shows the experimental setup to measure the BER vs. OSNR curve. As seen, the setup of BER vs. OSNR measurement is complicated. So there is a strong need to reduce the

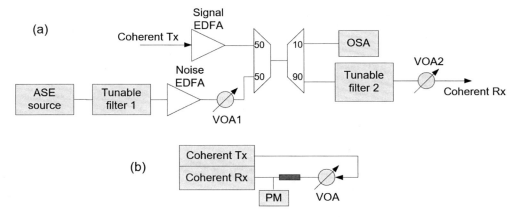

Figure 2. Experimental setup for (a) BER vs. OSNR, (b) BER vs. ROP measurement.

complexity of this measurement, so that the BER vs. OSNR measurement can be performed in the field operation.

Another type of optical communication system is unamplified link. The sensitivity (BER vs. ROP) of coherent transceiver determines the transmission distance. The measurement setup only requires a VOA and a power meter (PM), which is very simple as shown in Fig. 2(b). Many coherent transponders, like CFP2-ACO module, have integrated VOA and PM. So a simple loopback between the transmitter and the receiver is sufficient. The BER vs. OSNR curve and BER vs. ROP curve are intrinsically correlated. In BER vs. OSNR measurement, the noise increases while the signal remains constant when OSNR decreases; in BER vs. ROP measurement, the signal decreases while the noise remains constant when ROP decreases. Both scenarios lead to the decrease of SNR and the increase of BER.

2.3 Experimental result

Figure 3 shows the overlay between the curves of BER vs. OSNR at 0dBm ROP with the curves of BER vs. ROP at the different baud rate. For BER vs. ROP curve, the x axis value is $ROP+10*log_{10}(\rho)$, where ρ (Rho) corresponds to the OSNR value at the transmitter output. ρ can be determined from the required shift for BER vs. ROP curve to overlay with the BER vs. OSNR curve. ρ remains relatively unchanged over different transponders. So it is possible to measure these parameters during the design verification testing (DVT) and apply the same parameters over different transponders.

From those observations, we can expand the BER vs. OSNR model to take consideration of ROP influence. The influence of ROP is treated as Gaussian noise, which can be added together with other noise source. We can also extract $[\eta^{SEN}, \kappa^{SEN}]$ parameters from BER vs. ROP curves, then we can use those parameters to predict BER vs. OSNR curve.

Equation (2) shows the expanded model to predict BER vs. OSNR curve. The extracted fitting parameter $[\eta^{SEN}, \kappa^{SEN}]$ from BER vs. ROP curve using the setup shown in Fig. 2(b), and the ρ parameters determined during DVT, are used to predict BER vs. OSNR curve. Here ROP^{dB} is the ROP value in dBm, ROP_{cal} is the ROP value after normalization against B and bw.

$$
\begin{aligned}
OSNR_{cal} &= \frac{10^{\wedge}(OSNR^{dB}/10)*bw}{2*B}, \\
ROP_{cal} &= \frac{10^{\wedge}(ROP^{dB}/10)*bw}{2*B}, \\
\frac{1}{SNR} &= \frac{1}{\kappa^{SEN}} + \frac{1}{OSNR_{cal}} + \frac{1}{\rho ROP_{cal}}, \\
BER_{pred} &= erfc\left(\sqrt{\eta^{SEN}*SNR}\right)
\end{aligned}
\tag{2}
$$

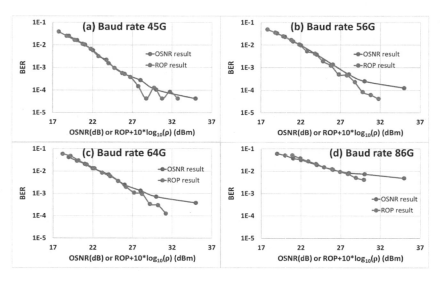

Figure 3. Correlation between BER vs. OSNR curves and BER vs. ROP curves for different baud rates. Rho: ρ. The amount of the shift applied to ROP curve.

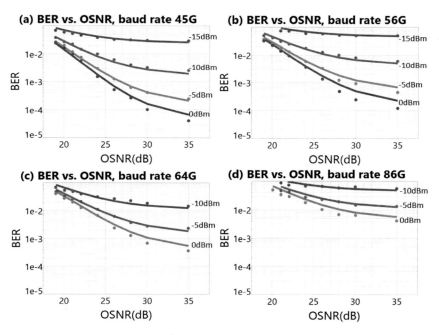

Figure 4. BER vs. OSNR measurement and prediction. Symbol: measurement; line: prediction.

Figure 4 shows BER vs. OSNR curves at various baud rates (45, 56, 64 and 86 GBd/s) over different ROP value. We measured 0, −5, −10, and −15 dBm for 45 GBd/s and 56 GBd/s; we measured 0, −5, and −10 dBm for 64 GBd/s and 86 GBd/s due to the fact that BER at −15 dBm was above the FEC threshold. The symbols in the plots correspond to the measurement results, and the curves in the plots correspond to the analytical model. As seen, two results are very close with each other, particularly at the low OSNR region where most long-haul optical communication systems operate at.

Next, we study how the fitting parameters change over different baud rate. In our experiment, the receiver bandwidth is set at $0.8*B$. Since white noise is proportional to the receiver bandwidth and the signal remains unchanged, ρ, which is OSNR at 0dBm ROP, will be inversely proportional to baud rate B.

$$\rho(B) = \rho(B_0)*B_0/B \qquad (3)$$

where $\rho(B)$ is the value of ρ at baud rate B, and $\rho(B_0)$ is the value of ρ at baud rate B_0. We choose 45 GHz as B_0. And the value of $\rho(B_0)$ is ~2370. This indicates that the OSNR of coherent Tx is 33.75 dB at 0 dBm output power.

Ideally, the filter mismatching between coherent receiver and signal should remain nearly constant since the receiver bandwidth is $0.8*B$. However, we have noticed that the filter mismatching parameter η changes linearly with the baud rate at a small slope as shown in Fig. 5(a). This deviation is likely due to non-perfect implementation of coherent transponder within the experimental setup. For example, the arbitrary waveform generator works at 92G sample/s. At 86G baud rate, the sample per bit is 1.07, well below the 2 samples per bit required by Nyquist theorem. This will introduce distortion to the coherent signal and cause η to change at various baud rates. The relationship between η and B can be expressed as a linear curve fitting:

$$\eta^{SEN}(B) = \alpha_0*B + \beta_0 \qquad (4)$$

where α_0 and β_0 are the curve fitting parameters for the η values derived from BER vs. ROP curves. One can also notice that the η value calculated from BER vs. OSNR curves agree well with the η value calculated from BER vs. ROP curves. Figure 5(b) shows the κ values calculated from BER vs. OSNR curves and the κ values calculated from BER vs. ROP curves. As expected, two groups of values are very close to each other. Two linear curve-fittings are plotted against the measurement data. However the experimental results do not match with linear curve-fitting very well, which prompts us to develop a better model for κ vs B.

We noticed the noise floors of BER in BER vs. OSNR curves increase when the baud rates increase. This is similar for BER vs. ROP curves. From BER vs. ROP curves, we can extract model parameters $[\eta^{SEN}, \kappa^{SEN}]$, and we can calculate noise floor of BER from Eq. (1).

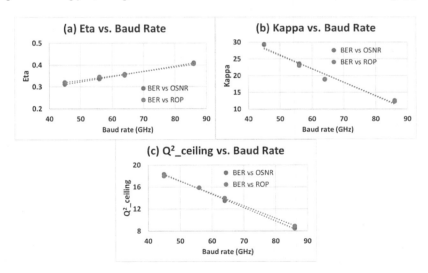

Figure 5. Analytical model parameters vs. baud rate. (a) η vs. baud rate, (b) κ vs. baud rate, (c) $Q^2\_$ ceiling vs. baud rate. Two sets of parameters and their linear fittings are shown. One is extracted from the BER vs. OSNR curves, the other one is extracted from the BER vs. ROP curves. As seen, two sets of parameters agree with each other.

$$BER_{floor}(B) = erfc\left(\sqrt{\eta\kappa}\right) \tag{5}$$

Furthermore, we converted the noise floor of BER to the $Q^2\_ceiling$ factor. By plotting $Q^2\_ceiling$ vs. B, we noticed that $Q^2\_ceiling$ changes linearly against the baud rate, which is shown in Fig. 5(c). This can be easily understood as following: higher baud rate requires larger receiver bandwidth, and leads to larger amount of white Gaussian noise; with the signal power remained constant, SNR or Q^2 factor degrades linearly vs. B. Thus, we get an accurate result for κ vs B as shown below:

$$Q^2_{ceiling}{}^{SEN}(B) \approx \alpha_1 * B + \beta_1, \ \eta^{SEN}(B) = \alpha_0 * B + \beta_0$$
$$BER_{floor}{}^{SEN}(B) = erfc\left(\sqrt{Q^2_{ceiling}{}^{SEN}(B)/2}\right) = erfc\left(\sqrt{\eta^{SEN}\kappa^{SEN}}\right) \tag{6}$$
$$\kappa^{SEN}(B) = Q^2_{ceiling}{}^{SEN}(B)/2/\eta^{SEN}(B) = (\alpha_1 * B + \beta_1)/2/(\alpha_0 * B + \beta_0)$$

Having determined how the fitting parameters change over different baud rates, we can use those values like $[\rho(B_0), \alpha_0, \beta_0, \alpha_0, \beta_0]$ to determine the corresponding $[\eta^{SEN}, \kappa^{SEN}]$ at any baud rate. Thus we can predict the BER vs. OSNR performance at the different baud rate without even measuring the BER vs. ROP curve. This offers significant benefit.

3 CONCLUSION

In this paper, we have demonstrated an accurate model allowing the prediction of BER vs. OSNR performance of coherent transponders over various baud rates and modulation formats. The model's parameters can be extracted from the BER vs. ROP curves measured at a few baud rates. The prediction results from the model agree well with the measurement results.

By adopting this model, one can easily predict the most important performance metrics for coherent transponder, which is BER vs. OSNR performance, without using a cumbersome and complicated setup. It allows the abstraction of coherent transponder into a set of modeling parameters, well suited for application in software defined network. This model also allows the accurate monitoring of OSNR from BER values without requiring extra hardware or DSP building block. A coherent optical channel monitor based on this model will be a powerful monitoring tool for whole optical network.

REFERENCES

Olsson, S.L.I., Cho, J., Chandrasekhar, S., Chen, X., Burrows, E.C. & Winzer, P.J. 2018. Record-High 17.3-bit/s/Hz Spectral Efficiency Transmission over 50 km Using Probabilistically Shaped PDM 4096-QAM. In *Optical Fiber Communication Conference'2018*, paper Th4C.5.

Peng, W., Tsuritani, T. & Morita, I. 2013. Transmission of High-Baud PDM-64QAM Signals. *J. Lightwave Technol.* 31(13): 2146–2162.

Roberts, K., Foo, S., Moyer, M., Hubbard, M., Sinclair, A., Gaudette, J. & Laperle, C. 2015. High Capacity Transport 100G and Beyond. *J. Lightwave Technol.* 33(3): 563–578.

Terayama, M., Okamoto, S., Kasai, K., Yoshida, M. & Nakazawa, M. 2018. 4096 QAM (72 Gbit/s) Single-Carrier Coherent Optical Transmission with a Potential SE of 15.8 bit/s/Hz in All-Raman Amplified 160 km Fiber Link. In *Optical Fiber Communication Conference'2018*, paper Th1F.2.

Torrengo, E., Cigliutti, R., Bosco, G., Carena, A., Curri, V., Poggiolini, P., Nespola, A., Zeolla, D. & Forghieri, F. 2011. Experimental Validation of an Analytical Model for Nonlinear Propagation in Uncompensated Optical Links. *Optics Express* 19(26): B790-B798.

Wang, Q., Yue, Y., He, X., Vovan, A. & Anderson, J. 2018. Accurate model to predict performance of coherent optical transponder for high baud rate and advanced modulation format. *Optics Express* 26(10): 12970–12984.

Wei, J., Cheng, Q., Penty, R., White, I. & Cunningham, D. 2015. 400 Gigabit Ethernet using advanced modulation formats: performance, complexity, and power dissipation. *IEEE Commun. Mag.* 53(2): 182–189.

Yue, Y., Wang, Q. & Anderson, J. 2018. Transmitter skew tolerance and spectral efficiency tradeoff in high baud-rate QAM optical communication systems. *Optics Express* 26(11): 15045–15058.

Frontier Research and Innovation in Optoelectronics Technology and Industry – Habib & Lewis (Eds)
© 2019 Taylor & Francis Group, London, ISBN 978-1-138-33178-5

Frequency-quadrupled millimeter-wave photonic mixer with a tunable phase shift

Conghui Zhang, Yu Qiao, Yikun Zhao, Caili Gong, Kai Sun & Yongfeng Wei
Institute of Electronic Information Engineering, Inner Mongolia University, Hohhot, Inner Mongolia, China

Tianwei Jiang & Xinlu Gao
State Key Laboratory of Information Photonics and Optical Communications, Beijing University of Posts and Telecommunications, Beijing, China

ABSTRACT: A microwave photonic mixer that can realize the generation, up/down conversion, amplitude/phase control of a millimeter-wave (mm-wave) signal is proposed and proven. In the scheme, a Dual-Parallel Mach–Zehnder Modulator (DPMZM) that is modulated by Radio Frequency (RF) signals is utilized to generate two second-order sidebands. The two sidebands are then separated by a Mach–Zehnder Interferometer (MZI). The –2nd-order sideband is injected into a Phase Modulator (PM) to get a tunable phase shift. The +2nd-order sideband is sent to the second DPMZM which is modulated by Local Oscillator (LO) signals in order to implement frequency up/down conversion or amplitude control. After an Optical Coupler (OC) and a Photodiode (PD), an mm-wave signal with tunable phase shift is produced. The results show that up conversion signals from 41 to 45 GHz, down conversion signals from 35 to 39 GHz and an amplitude-controlled signal at 40GHz are produced, meanwhile, tunable phase shifts are accompanied.

Keywords: frequency-quadrupled, mm-wave signal generation, mm-wave up conversion, tunable phase shift, amplitude control, mm-wave down conversion

1 INTRODUCTION

Microwave photonic technologies have attracted a lot of research interest, such as photonic generation of mm-wave signal, photonic mixer, mm-wave amplitude control and photonic phase shifter, which can be applied in phased array radar and Radio Over Fiber (ROF) systems (Capmany & Novak, 2007; Minasian et al., 2013; Yao, 2009). In recent years, microwave photonic technologies have been developing toward multi-function technologies to meet the needs of the high integration of modern communication systems.

Schemes combining a microwave photonic mixer with a photonic phase shifter have been proposed (Li et al., 2018; Wang et al., 2017; Zhang et al., 2017; Zhai et al., 2018). In these schemes, the interconversion of the Intermediate Frequency (IF) signal and the RF signal can be realized, and the phase of the converted signals can be tuned. However, it is difficult to realize the frequency conversion and tunable phase shift in the mm-wave band. A number of mm-wave signal generators that can produce an mm-wave signal with a tunable phase shift have been presented (Li et al., 2014; Li et al., 2016; Zhang & Pan, 2016). Nevertheless, in order to produce mm-wave signals of different frequencies, RF signals of different frequencies are required, that means in a multichannel mm-wave communication system, a great deal of microwave signal sources or microwave mixers are needed.

In this paper, we propose a photonic mixer that can generate the phase-tunable mm-wave signal with amplitude control, up conversion and down conversion, respectively. In the proposed system, the first DPMZM modulated by the RF signals is employed to produce two second-order RF sidebands. In the DPMZM, two Dual-Drive Mach–Zehnder Modulators (DDMZM) and a main-MZM are included. The two DDMZMs work at Maximum Transmission Point (MATP), and the DC voltage on the main-MZM is equal to the half-wave voltage of the first DPMZM. A MZI is then utilized to separate the two sidebands. The −2nd-order RF sideband is introduced to a PM and the phase is controlled by the DC voltage on the PM. The +2nd-order RF sideband is injected into the second DPMZM which is driven by the Local Oscillator (LO) signals, and the two DDMZMs of the second DPMZM work at Minimum Transmission Point (MITP). By changing the DC voltage on the main-MZM of the second DPMZM, the single sideband modulation with carrier suppression (CS-SSB) is realized. In the meantime, by setting the amplitude of the LO signals to zero and adjusting the DC voltages on the two DDMZMs, the amplitude control of the +2nd-order RF sideband can be implemented. After an OC and a PD, an mm-wave up conversion signal or down conversion signal or a frequency-quadrupled mm-wave signal with amplitude control is generated, and the phase of the generated signal is tunable. This has the advantage of employing a RF source with fixed frequency and producing the mm-wave signals with different frequencies by using the LO signals with different frequencies, therefore, it needs only one RF source and multiple IF local oscillators instead of multiple RF sources or microwave mixers to generate mm-wave signals with different frequencies in a multichannel mm-wave system.

2 PRINCIPLES

A schematic diagram of the proposed photonic mixer is shown in Figure 1. A light wave from a laser diode is introduced to the first DPMZM which is driven by the RF signals and then two second-order RF sidebands are generated. A MZI is used to separate the two sidebands. The −2nd-order RF sideband is sent to a PM and the phase is controlled by the DC voltage on the PM, and the +2nd-order RF sideband is modulated by the LO signals in the second DPMZM. The frequency up-shift or down-shift of the +2nd-order RF sideband is controlled by changing the DC voltage on the main-MZM of the second DPMZM, and the amplitude control is realized by adjusting the DC voltages on the two sub-MZMs of the second DPMZM. After a PD, an mm-wave up conversion signal or down conversion signal or a frequency-quadrupled mm-wave signal with amplitude control is generated, and the phase of the generated signal is tunable.

The first DPMZM is driven by the RF signals, and the phase shift between the RF signal on the top DDMZM and the RF signal on the bottom DDMZM is 90°, in addition, the two

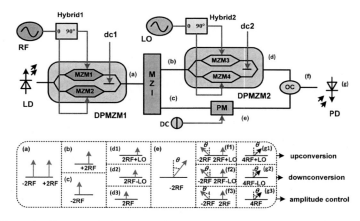

Figure 1. Schematic diagram of the proposed device. LD: laser diode; RF: radio frequency signal; DPMZM: dual-parallel Mach–Zehnder modulator; MZI: Mach–Zehnder interferometer; LO: local oscillator; PM: phase modulator; dc: direct voltage; DC: direct voltage source; OC: optical coupler; PD: photodiode.

494

DDMZMs work at MATP and the DC voltage on the main-MZM is equal to the half-wave voltage of the first DPMZM. After the DPMZM, the output can be written as:

$$E_a(t) = E_0 J_2(\beta) \exp(j\omega_0 t - j2\omega_m t) + E_0 J_2(\beta) \exp(j\omega_0 t + j2\omega_m t) \tag{1}$$

where E_0 and ω_0 are the amplitude and angular frequency of the optical carrier, respectively, ω_m denotes the angular frequency of the RF signals, $J_n(\beta)$ is the Bessel function of the first kind and n is an integer, $\beta = \pi V_m / V_\pi$, V_m denotes the amplitude of the RF signals and V_π is the half-wave voltage of the first DPMZM. The two RF sidebands are separated after a MZI, and the expression can be written as:

$$E_b(t) = \frac{\sqrt{2}}{2} E_0 J_2(\beta) \exp\left(j\omega_0 t + j2\omega_m t + j\frac{\pi}{4}\right) \tag{2}$$

$$E_c(t) = \frac{\sqrt{2}}{2} E_0 J_2(\beta) \exp\left(j\omega_0 t - j2\omega_m t + j\frac{3}{4}\pi\right) \tag{3}$$

The +2nd-order RF sideband is modulated in the second DPMZM driven by LO signals. The two sub-MZMs work at MITP, and the phase difference between the LO signal on the top DDMZM and the LO signal on the bottom DDMZM is 90°. The half-wave voltage of the second DPMZM is equal to the half-wave voltage of the first DPMZM. The frequency up-shift of the +2nd-order RF sideband is expressed as Equation 4 when the DC voltage on the main-MZM of the second DPMZM is $-V_\pi/2$. The frequency down-shift is expressed as Equation 5 when the DC voltage on the main-MZM is $V_\pi/2$, and the amplitude control is expressed as Equation 6 when the DC voltages on the two sub-MZMs are adjusted and the amplitude of the LO signals is 0.

$$E_{d1}(t) = -\frac{\sqrt{2}}{2} E_0 J_2(\beta) J_1(\gamma) \exp\left[j\left(\omega_0 t + 2\omega_m t + \omega_L t + \frac{\pi}{4}\right)\right] \tag{4}$$

$$E_{d2}(t) = \frac{\sqrt{2}}{2} E_0 J_2(\beta) J_1(\gamma) \exp\left[j\left(\omega_0 t + 2\omega_m t - \omega_L t + \frac{\pi}{4}\right)\right] \tag{5}$$

$$E_{d3}(t) = \frac{\sqrt{2}}{4} j E_0 J_2(\beta) \sin(\pi V_M / V_\pi) \exp\left[j\left(\omega_0 t + 2\omega_m t + \frac{\pi}{4}\right)\right] \tag{6}$$

where $\gamma = \pi V_L / V_\pi$, V_L denotes the amplitude of the LO signals, ω_L represents the angular frequency of the LO signals, V_M is the value of the DC voltage on the upper arm of MZM3 (Figure 1), and the value of the DC voltage on the upper arm of MZM4 is $-V_M$. When the amplitude control is realized, the amplitude of the LO signals is 0, and there are no DC voltages on the bottom arm of MZM3 and the bottom arm of MZM4. The phase of the −2nd-order RF sideband is controlled after a PM, and the equation can be expressed as:

$$E_e(t) = \frac{\sqrt{2}}{2} E_0 J_2(\beta) \exp\left[j\left(\omega_0 t - 2\omega_m t + \frac{3}{4}\pi + \pi\frac{V_D}{V_P}\right)\right] \tag{7}$$

where V_D denotes the DC voltage on the PM and V_P is the half-wave voltage of the PM. After an OC and a PD, the generated electrical mm-wave signals can be written as:

$$I_{f1}(t) = -E_0^2 J_2^2(\beta) J_1(\gamma) \sin\left(4\omega_m t + \omega_L t - \pi\frac{V_D}{V_P}\right) \tag{8}$$

$$I_{f2}(t) = E_0^2 J_2^2(\beta) J_1(\gamma) \sin\left(4\omega_m t - \omega_L t - \pi\frac{V_D}{V_P}\right) \tag{9}$$

$$I_{f3}(t) = \frac{1}{2} E_0^2 J_2^2(\beta) \sin(\pi V_M / V_\pi) \cos\left(4\omega_m t - \pi\frac{V_D}{V_P}\right) \tag{10}$$

The DC components are ignored. Equation 8 indicates the signal of frequency up conversion of frequency-quadrupled mm-wave signal, Equation 9 denotes the signal of frequency down conversion of frequency-quadrupled mm-wave signal and Equation 10 indicates the frequency-quadrupled mm-wave signal with amplitude control. In addition, the phases of the three signals can be tuned by adjusting the DC voltage on the PM.

3 RESULTS AND ANALYSIS

The experiments were carried out on a simulation platform called a VPI transmission maker to verify the proposed system. An optical carrier with a frequency of 193.1 THz and power of 10 dBm was emitted from a LD, and then introduced to the first DPMZM with a half-wave voltage of 5 V. The RF signals at a frequency of 10 GHz, with an amplitude of 3 V were employed to drive the first DPMZM. In the DPMZM, two sub-MZMs work at MATP, and the two RF signals that drive the two sub-MZMs have a phase difference of 90°. In addition, the DC voltage on the main-MZM is 5 V. Two second-order RF sidebands were produced after the DPMZM, as shown in Figure 2(a). A MZI was then utilized to separate the two sidebands, and the +2nd-order RF sideband is shown in Figure 2(b), at the same time, the −2nd-order RF sideband is shown in Figure 2(c).

The −2nd-order RF sideband was injected into a PM, and its phase controlled by the DC voltage on the PM. The +2nd-order RF sideband turns into the second DPMZM driven by the LO signals.

The amplitude of the LO signals was 1.5 V and the frequency was 5 GHz. In the second DPMZM, two sub-MZMs work at MITP, and the two LO signals that drive the two sub-MZMs have a phase difference of 90°. In the meantime, the half-wave voltage of the DPMZM was 5 V. When the DC voltage on the main-MZM of the second DPMZM was −2.5 V, the frequency up-shift of the +2nd-order RF sideband was realized. When the DC voltage on the main-MZM was 2.5 V, the frequency down-shift of the +2nd-order RF sideband was realized. When the amplitude of the LO signal was zero and the DC voltage on the main-MZM was 5 V, the +2nd-order RF sideband of amplitude control was obtained. After an OC, the three output signals of the DPMZM were respectively combined with the −2nd-order RF sideband and are shown in Figure 3(a), Figure 3(b) and Figure 3(c), respectively. An optical amplifier with 20 dB gain was placed between OC and PD. After the PD, a frequency up conversion mm-wave signal with a frequency of 45 GHz was produced, as shown in Figure 3(d). A frequency down conversion signal with a frequency of 35 GHz was generated, as shown in Figure 3(e), and only the DC voltage on the main-MZM of the second DPMZM was changed. A 40 GHz mm-wave signal with amplitude control was obtained as displayed in Figure 3(f).

To verify the frequency tunability of the generated mm-wave up conversion signal and down conversion signal, the LO signals with a frequency from 1 to 5 GHz were used, and the produced up conversion mm-wave signals with frequency variation from 41 to 45 GHz are shown in Figure 4(a). Meanwhile, the generated down conversion mm-wave signals with a frequency variation from 35 to 39 GHz are displayed in Figure 4(b). Setting the amplitude value of the LO signals to 0, and adjusting the value of V_M, the power variation of the

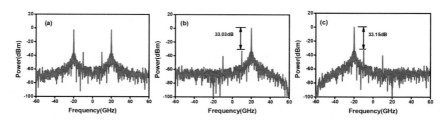

Figure 2.   Optical spectrum of two second-order RF sidebands (a), separated +2nd-order RF sideband (b), and −2nd-order RF sideband (c). 0 corresponds to 193.1 THz.

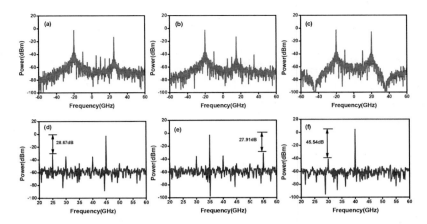

Figure 3. Optical spectrum of the −2nd-order RF sideband and up-shift sideband of +2nd-order RF sideband (a), down-shift sideband (b), amplitude control sideband (c). Electrical spectrum of mm-wave up conversion signal (d), down conversion signal (e), amplitude control signal of frequency-quadrupled mm-wave (f). 0 corresponds to 193.1 THz in the optical spectrum.

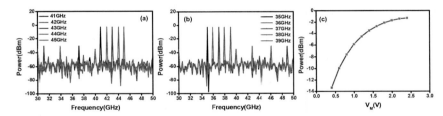

Figure 4. Electrical spectrum of mm-wave up conversion signals with a frequency from 41 to 45 GHz (a), and down conversion signal with a frequency from 35 to 39 GHz (b). The power variation of the generated frequency-quadrupled signal with a frequency of 40 GHz versus the variation of V_M (c).

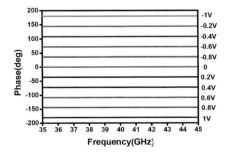

Figure 5. Phase response of the generated mm-wave signal at different DC voltages on the PM versus the frequency variation of the produced mm-wave signal.

generated frequency-quadrupled mm-wave signal with a frequency of 40GHz is shown in Figure 4(c), and indicates that the amplitude control of the produced frequency-quadrupled mm-wave signal was implemented.

According to Equations 8, 9 and 10, a phase compensation of 180° is added to the mm-wave up conversion signal, and a phase compensation of −90° is added to the mm-wave amplitude control signal. The phase of the generated mm-wave signals can be tuned by changing the DC voltage on the PM as shown in Figure 5. The half-wave voltage of the PM was 1 V. A full 360° phase shift of the produced mm-wave signals was realized when the DC voltage on the PM varied from −1 V to 1 V, and the phase deviation was less than 2°.

4 CONCLUSIONS

A frequency-quadrupled mm-wave photonic mixer with tunable phase shift was proposed and demonstrated. It can produce a frequency-quadrupled mm-wave signal with amplitude control and tunable phase shift. By controlling the frequency of the LO signals, up conversion and down conversion of the produced frequency-quadrupled mm-wave signal can be achieved. In the meantime, the phase of all the generated mm-wave signals can be controlled by adjusting the DC voltage on the PM. In multichannel mm-wave communication systems, only one high-frequency microwave signal source with a single frequency and multiple intermediate-frequency LO signal sources with different frequencies were used to produce multiple mm-wave signals with different frequencies by the scheme, and that reduces the cost of the system. Hence, it has the potential to be applied in future multichannel mm-wave communication systems. In addition, it can be applied in future phased array radar systems due to the tunable phase shift of the produced mm-wave signal.

ACKNOWLEDGMENTS

The work was supported by the National Natural Science Foundation of China (No.61561037, No. 61605015, No. 61601414, No. 61461035, No. 61362027, No. 61563038), the Natural Science Foundation of Inner Mongolia Autonomous Region (No. 2017MS0609, 2015MS0602) and Program for Innovative Research Team in Universities of Inner Mongolia Autonomous Region (No. NMGIRT-A1609).

REFERENCES

Capmany, J. & Novak, D. (2007). Microwave photonics combines two worlds. *Nature Photonics, 1*(6), 319–330.

Li, H. et al. (2014). Photonic generation of frequency-quadrupled microwave signal with tunable phase shift. *IEEE Photonics Technology Letters, 26*(3), 220–223.

Li, T. et al. (2018). Broadband photonic microwave signal processor with frequency up/down conversion and phase shifting capability. *IEEE Photonics Journal, 10*(1), 1–12.

Li, Y. et al. 2016. Filter-less frequency-doubling microwave signal generator with tunable phase shift. *Optics Communications, 370*, 91–97.

Minasian, R.A., Chan, E.H.W. & Yi, X. (2013). Microwave photonic signal processing. *Optics Express, 21*(19), 22918–22936.

Wang, Y. et al. (2017). All-optical microwave photonic downconverter with tunable phase shift. *IEEE Photonics Journal, 9*(6), 1–8.

Yao, J. (2009). Microwave photonics. *Journal of Lightwave Technology, 27*(3), 314–335.

Zhai, W. et al. (2018). A multi-channel phase tunable microwave photonic mixer with high conversion gain and elimination of dispersion-induced power fading. *IEEE Photonics Journal, 10*(1), 1–10.

Zhang, J. et al. (2017). Broadband microwave photonic sub-harmonic downconverter with phase shifting ability. *IEEE Photonics Journal, 9*(3), 1–10.

Zhang, Y. & Pan, S. (2016). Frequency-multiplying microwave photonic phase shifter for independent multichannel phase shifting. *Optics Letters, 41*(6), 1261–1264.

Frontier Research and Innovation in Optoelectronics Technology and Industry – Habib & Lewis (Eds)
© *2019 Taylor & Francis Group, London, ISBN 978-1-138-33178-5*

A filterless frequency-sextupling microwave signal processor with tunable phase shift

Xiaoyu Zhang, Conghui Zhang, Ruiying He, Xiaoqiang Song, Caili Gong & Yongfeng Wei
Institute of Electronic Information Engineering, Inner Mongolia University, Hohhot, Inner Mongolia, China

ABSTRACT: A new filterless frequency-sextupling microwave signal processor with full-range phase shift is presented and demonstrated. A single Dual-Polarization Quadrature Phase-Shift Keying (DP-QPSK) modulator is used to generate two ±3rd-order sidebands along orthogonally polarized directions with other sidebands suppressed. After passing through a polarization controller, the polarization directions of the two sidebands are aligned with the two principal axes of a Polarization Modulator (POLM). By adjusting the Direct Current (DC) voltage applied to the POLM, full-range phase shift is implemented. A 45° polarizer then provides the output signals with the same polarization state. By beating ±3rd-order orthogonal sideband signals at a photodetector, a frequency-sextupled microwave signal is generated. Verified by simulation, the results show that for RF signals from 1 to 10 GHz, output signals from 6 to 60 GHz can be achieved with phase deviation in the generated signals of less than 3°.

Keywords: analog system, frequency-sextupled, microwave signal processor, tunable phase shift

1 INTRODUCTION

As one of the most promising technologies in the next generation of broadband wireless access, Radio-Over-Fiber (ROF) has attracted much attention and research. As an essential component of ROF systems, microwave signal processors used to generate millimeter-wave (mm-wave) signals are also being investigated by researchers. Compared with conventional electronic microwave signal processors, microwave photonics signal processors have come to the forefront of research because of their special advantages, such as wide bandwidth, light weight, and immunity to electromagnetic interference (Yao, 2009). Much attention is paid to optical generation of mm-wave signals as an aspect of microwave photonics technology. The ability to generate a mm-wave signal in the optical domain rather than the electric domain would greatly simplify the equipment required. At present, because of the limited bandwidth of filters and LiNbO$_3$ Mach–Zehnder Modulators (MZMs), mm-wave signal processors are limited in their ability to generate a high-frequency signal. To solve this problem and generate signals with wide bandwidth and good tunability, researchers have proposed schemes without optical filters. In research by Lin et al. (2008), a filterless single Dual-Drive MZM (DDMZM) is used to generate quadrupled frequency, with the generated signal achieving up to 60 GHz. Zhu et al. (2015) used an integrated nested MZM to realize 12-tupled frequency, generating a high-quality 120 GHz mm-wave signal, with an Optical Sideband Suppression Ratio (OSSR) of higher than 37 dB.

In addition, more practical applications require phase-shift ability to be provided to equipment such as phased-array radars. Conventional electronic phase shifters face many problems, such as limited response time and poor linearity. These issues can be solved by photonics methods and can be better applied to mm-wave phased-array beamforming networks and microwave filters. There are many optical methods to realize 0–360° phase shift. W. Li et al. (2014) used a conventional DDMZM cascade as an optical bandpass filter to achieve full-range phase shift, while using a filter to limit the system's bandwidth. Wang et al. (2017) proposed and demonstrated a technique without a broadband phase-shifting operation. Recently, some schemes have employed a new type of optical component to control the phase of specific sidebands, for example, a single Dual-Polarization Dual-Parallel Mach–Zehnder Modulator (DP-DPMZM) with a new Optical Phase Shifter (OPS) component to realize phase shift; the polarization-dependent OPS controls the optical carrier phase and RF signal simultaneously (Niu et al., 2016).

In addition, studies have combined the two functions mentioned above within a system that is more suitable for practical applications. A scheme using a frequency-doubling signal generator with the function of full-range phase shift is proposed (Y. Li et al., 2016). A Dual-Parallel Polarization Modulator (DP-POLM) and a cascading Electro-Optical Phase Modulator (EOPM) are the core components in this system. The Polarization Controller (PC) can change the direction of two orthogonally polarized sidebands simultaneously. By changing the Direct Current (DC) of the EOPM, the phase of the specific sideband can be controlled. In Yang et al. (2017), a microwave photonic phase shifter with the capability of frequency quadrupling is proposed. A DP-POLM is the key component in generating ±2nd-order sidebands. By adjusting the angle of the polarizer, the phase of two orthogonally polarized sidebands can be controlled flexibly.

In this paper, we propose an architecture for a frequency-sextupling microwave signal processor with tunable phase shift. By properly setting the inherent static phase difference of the Dual-Polarization Quadrature Phase-Shift Keying (DP-QPSK) modulator and the phase deviation of the RF signal, only ±3rd-order sidebands are obtained. The output signal is then put through a PC, in which the internal parameters are set accurately. The polarization states of the two ±3rd-order sidebands are aligned with the principal axes of the Polarization Modulator (POLM). It is known that there is a mathematical relationship between DC voltage and the phase variation of the POLM. Thus, by changing the DC voltage applied to the POLM, a phase difference θ between two orthogonally polarized sidebands is introduced, allowing the phase of one polarization state to be controlled. Then a 45° polarizer matches the polarization states of the two sidebands. After passing through a photodiode, a frequency-sextupled microwave signal can be generated.

2 PRINCIPLES

The schematic diagram of the frequency-sextupling microwave signal processor with full-range phase shift is shown in Figure 1. The laser beam is linearly polarized at 45° via a PC. Then the light wave is modulated by a RF signal by means of a DP-QPSK modulator, which consists of two Dual-Parallel Mach–Zehnder Modulators (DPMZMs), a Polarization Beam Splitter (PBS) and a Polarization Beam Combiner (PBC). Then the output signal is applied to a PC that aligns the polarization directions of the two sidebands with the principal axes of the POLM. By adjusting the DC voltage of the POLM, an angle θ is achieved with respect to one of its polarization axes. The 45° polarizer gives the polarization states of the two sidebands the same polarization direction. Finally, when the output signal is sent to the Photodetector (PD), a frequency-sextupled microwave signal with full-range phase shift can be generated.

A DP-QPSK modulator is the key component to realizing two orthogonally polarized ±3rd-order sidebands and suppressing other sidebands. The sub-MZMs and parent MZMs of the two DPMZMs are set to the Minimum Transmission Point (MITP) and the Maximum Transmission Point (MATP), respectively. The DP-QPSK modulator is driven by two quadrature RF signals that can be expressed as $\sin(\omega t + \varphi)$ and $\cos(\omega t + \varphi)$, respectively, as shown in Equation 1:

$$
\begin{bmatrix} E_X \\ E_Y \end{bmatrix} = -\frac{1}{8} e^{j\omega_0 t} E_0 \begin{bmatrix} e^{j\beta\sin(\omega_{RF}t+\varphi)} - e^{-j\beta\sin(\omega_{RF}t+\varphi)} + e^{j\beta\sin(\omega_{RF}t+\varphi+\Delta\phi)} - e^{-j\beta\sin(\omega_{RF}t+\varphi+\Delta\phi)} \\ e^{j\beta\cos(\omega_{RF}t+\varphi)} - e^{-j\cos(\omega_{RF}t+\varphi)} + e^{j\beta\cos(\omega_{RF}t+\varphi+\Delta\phi)} - e^{-j\beta\cos(\omega_{RF}t+\varphi+\Delta\phi)} \end{bmatrix}
$$

$$
= -\frac{1}{8} e^{j\omega_0 t} E_0 \sum_{-\infty}^{+\infty} J_{2n+1}(\beta) e^{j(2n+1)\omega_{RF}t} \begin{bmatrix} e^{j(2n+1)\varphi} + e^{j(2n+1)(\varphi+\Delta\phi)} \\ e^{j(2n+1)\varphi} + e^{j(2n+1)(\varphi+\Delta\phi)} \end{bmatrix} \quad (1)
$$

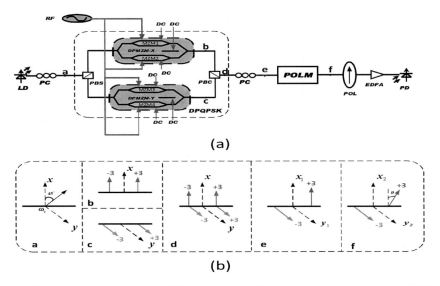

(a)

(b)

Figure 1. (a) Schematic of the proposed frequency-sextupling microwave signal processor (LD – laser diode; PC – polarization controller; PBS – polarization beam splitter; DP-QPSK – dual-polarization quadrature phase-shift keying; PBC – polarization beam combiner; POL – polarizer; POLM – polarization modulator; EDFA – erbium-doped fiber amplifier; PD – photodetector); (b) evolution of the polarization directions.

where $\beta = \pi V_{RF}/V_{\pi}$ is the modulation depth of the MZM; ω_0 and ω_{RF} are the angular frequencies of the input carrier signal and RF signal, respectively; V_{π} is the half-wave voltage; V_{RF} and E_0 are the amplitude of the RF signal and optical carrier, respectively; $J_x()$ is the Bessel function of the x-order sideband; φ and $\Delta\varphi$ are the primary phases of the RF signal and the phase difference between the two RF signals, respectively. To suppress even-order sidebands, all sub-MZMs of the DP-QPSK modulator work at the MITP and DC voltages are set to V_{π}. The ±7th-order sidebands and above can be ignored, because they have little impact on the system. From Equation 1 it can be seen that when:

$$J_1(\beta) = 0 \tag{2a}$$

$$1 + e^{\pm j5\Delta\varphi} = 0 \tag{2b}$$

the ±1st-order and ±5th-order sidebands can be suppressed. Using Equations 2a and 2b, the result can be derived as $\beta = 3.868$ and $\Delta\varphi = 36°$. Although the modulation depth of the MZM is beyond the commercial modulator, the smaller half-wave voltage of the MZM can realize the required goal (Kondo et al., 2005). The output signal can be expressed mathematically as:

$$\begin{bmatrix} E_X \\ E_Y \end{bmatrix} = -\frac{1}{4}E_0 e^{j\omega_0 t}J_3(\beta)\begin{bmatrix} e^{-j3\omega_{RF}t} - e^{j3\omega_{RF}t} \\ -je^{-j3\omega_{RF}t} - je^{j3\omega_{RF}t} \end{bmatrix} \tag{3}$$

The output signal is then put though the PC which contains three wave plates to align the polarization directions of the two sidebands with the principal axes of the POLM. Based on Jones' theory, the transfer function of a wave plate can be expressed as:

$$J_{WP}(\delta,\phi) = \begin{bmatrix} \cos\dfrac{\delta}{2} + j\sin\dfrac{\delta}{2}\cos2\theta & -j\sin\dfrac{\delta}{2}\sin2\theta \\[3mm] -j\sin\dfrac{\delta}{2}\sin2\theta & \cos\dfrac{\delta}{2} - j\sin\dfrac{\delta}{2}\cos2\theta \end{bmatrix} \tag{4}$$

where θ is the rotation angle of the wave plate and δ is the phase difference between the signal on the two orthogonal axes after the wave plate. The PC comprises three wave plates, in which the first and the third are quarter-wave plates, and the second is a half-wave plate. The optical vector phase differences on the orthogonal axis processed by the three wave plates are $\delta_1 = \pi/2$, $\delta_2 = \pi$ and $\delta_3 = \pi/2$. The rotation angles of the three wave plates are set to $\theta_1 = 0$, $\theta_2 = -\pi/8$ and $\theta_3 = -\pi/2$. The output signal through the PC can be expressed as:

$$\begin{bmatrix} E_{x1} \\ E_{y1} \end{bmatrix} = \begin{bmatrix} \cos\dfrac{\delta}{2} + j\sin\dfrac{\delta}{2}\cos 2\theta & -j\sin\dfrac{\delta}{2}\sin 2\theta \\ -j\sin\dfrac{\delta}{2}\sin 2\theta & \cos\dfrac{\delta}{2} - j\sin\dfrac{\delta}{2}\cos 2\theta \end{bmatrix} \begin{bmatrix} E_x \\ E_y \end{bmatrix}$$

$$= -\frac{1}{2}E_0 J_3(\beta)e^{j\omega_0 t}\begin{bmatrix} -je^{j3\omega_{RF}t} \\ -e^{j3\omega_{RF}t} \end{bmatrix} \tag{5}$$

Thus, two orthogonally polarized ±3rd-order sidebands can be obtained after the PC. Then two orthogonally polarized sidebands are introduced to the POLM. By adjusting the DC voltage of the POLM, an angle θ is achieved with respect to one of its polarization axes. The equation can be written as:

$$\begin{bmatrix} E_{x2} \\ E_{y2} \end{bmatrix} = -\frac{1}{2}E_0 J_3(\beta)e^{j\omega_0 t}\begin{bmatrix} -je^{j3\omega_{RF}t} \\ -e^{-j3\omega_{RF}t}e^{j\theta} \end{bmatrix} \tag{6}$$

where $\theta = \pi V_P / V_{P\pi}$ is the phase difference of the two orthogonally polarized sidebands, V_p is the DC voltage applied to the POLM, and $V_{P\pi}$ is the half-voltage of the POLM. The output signal from the polarizer to the PD for square-law detection is:

$$I(t) = \mu E_{out}E_{out}^* \propto \mu E_0^2 J_3^2(\beta)\sin(6\omega_{RF}t - \theta) \tag{7}$$

From Equation 7 it can be seen that the microwave signal is frequency-sextupled and the phase shift θ is determined by the DC voltage applied to the POLM. When the DC voltage is between −1 V and +1 V, full-range phase shift is realized.

3 RESULTS AND ANALYSIS

To verify the quality of the frequency-sextupling microwave signal processor with tunable full-range continuous phase shift, a simulation based on the configuration shown in Figure 1 is carried out using the commercially available simulation software VPItransmissionMaker (VPIphotonics GmbH, Berlin, Germany). The center frequency of the Continuous Wave (CW) laser is 193.1 GHz, and the frequency of RF is 6 GHz. The light wave is linearly polarized with an angle of 45° by the PC before it is sent into the DP-QPSK modulator. By adjusting the DC voltage applied to sub-MZMs and parent MZMs in DPMZM-x and DPMZM-y to 5 V, 0 V and 0 V, respectively, even-order sideband-suppressed modulation is implemented. By setting the modulation depth of the sub-MZMs to $\beta = 3.868$, the ±1st-order sidebands can be suppressed. The phase difference of RF signals applied to DPMZM-x and DPMZM-y is set to 36° respectively so as to suppress the ±5th-order sidebands. The phase difference between the two DPMZMs is 90°. Two orthogonally polarized ±3rd-order sidebands, with other sidebands suppressed, are realized. The output spectrum of the DP-QPSK modulator is monitored with the VPIphotonics Analyzer, as shown in Figure 2a. It can be seen that the OSSR is 26 dB. Then the output signal is put through a PC. There are two inter-parameters for the PC in the simulation software. The two arguments are set to 0° and −45°, respectively. To observe the polarization states of the two sidebands, a PBC was placed after the PC. In the output spectrum shown in Figure 2b, two orthogonally polarized ±3rd-order sidebands can be seen.

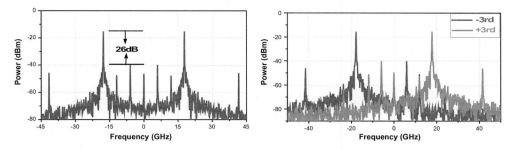

Figure 2.   Optical spectrum of: (a) the output of the DP-QPSK modulator; (b) the two orthogonally polarized ±3rd-order sidebands.

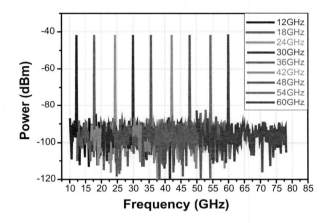

Figure 3.   Electrical spectrums of frequency-sextupled signals from different input frequencies.

To verify the capability of frequency sextupling, different frequencies of input signals are applied to the microwave signal generator. When sweeping the RF signal from 2 to 10 GHz in 1-GHz steps, the electrical spectrums in Figure 3 show the corresponding output signal frequencies as 12, 18, 24, 30, 36, 42, 48, 54 and 60 GHz. The results indicate that the processor is equipped with frequency-sextupling function over a relatively wide range of frequencies. Amplitude variation is nearly flat when microwave signals of different frequencies are produced.

We then set the frequency of the RF signal to 6 GHz to indicate the phase and amplitude variation of produced signals versus the variation of DC voltage. By changing the DC voltage applied to the POLM, a phase difference is introduced between two ±3rd-order orthogonally polarized sidebands. When the DC is varied from −1 V to +1 V, a continuous phase shift from −180° to +180° can be achieved. As is clearly indicated in Figure 4, with ten points marked, the phase shift is linear to the DC voltage. During the transduction processes, the amplitude is stable.

To further test the phase shifting of the generated signal, the relationship between input signal frequency, DC voltage applied to the POLM, and phase shift is simulated. When the simulations are conducted over a frequency range of 2–10 GHz, the frequencies of the output signals range from 12 to 60 GHz. It can be seen in Figure 5 that flat phase responses for all the phase shifts over a relatively wide frequency range are obtained, and the phase deviation is less than 3°. Theoretically, such a wide bandwidth can satisfy the requirements of practical applications.

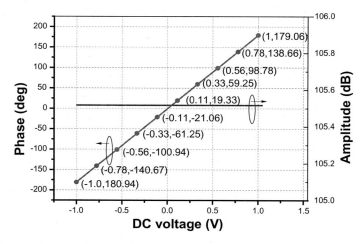

Figure 4.   Phase and amplitude of produced signal vs DC voltage.

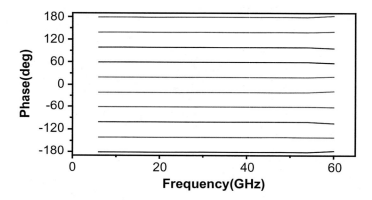

Figure 5.   Phase response of the proposed microwave photonic link.

4   CONCLUSIONS

A novel frequency-sextupling microwave signal processor with full-range phase shift based on a filterless architecture is presented and analyzed. In this scheme, a DP-QPSK modulator cascades to a PC to generate two orthogonally polarized ±3rd-order sidebands. By setting the DC voltage applied to the POLM, full-range shift is realized. When the frequency of the RF signal varies from 2 to 10 GHz in 1-GHz steps, the frequency of the output signal is six times as high as that of the input signal. By adjusting the DC voltage of the POLM from −1 V to +1 V over a bandwidth of 12–60 GHz, the phase deviation is less than 3°. Theoretically, this dual-function photonic microwave signal processor may be more suitable for practical applications than those currently available.

ACKNOWLEDGMENTS

The work was supported by the National Natural Science Foundation of China (nos. 61561037, 61605015, No.61601414, 61461035, 61362027 and 61563038), and the Natural Science Foundation of Inner Mongolia Autonomous Region (nos. 2017MS0609 and 2015MS0602).

REFERENCES

Kondo, J., Aoki, K., Iwata, Y., Hamajima, A., Ejiri, T., Mitomi, O. & Minakata, M. (2005). 76-GHz millimeter-wave generation using MZ LiNb03 modulator with drive voltage of 7Vp-p and 19 GHz signal input. In *International Topical Meeting on Microwave Photonics, 14 October 2005* (pp. 363–366). Piscataway, NJ: Institute of Electrical and Electronics Engineers. doi:10.1109/MWP.2005.203613

Li, W., Sun, W.H., Wang, W.T., Wang, L.X., Liu, J.G. & Zhu, H.Z. (2014). Photonic-assisted microwave phase shifter using a DMZM and an optical bandpass filter. *Optics Express*, *22*(5), 1–6.

Li, Y., Pei, L., Li, J., Wang, Y. & Yuan, J. (2016). Filter-less frequency-doubling microwave signal generator with tunable phase shift. *Optics Communications*, *370*, 91–97.

Lin, C.-T., Shih, P.T., Chen, J., Peng, P.-C., Dai, S.-P., Xue, W.-Q. & Chi, S. (2008). Generation of carrier suppressed optical mm-wave signals using frequency quadrupling and no optical filtering. In *Optical Fiber Communication Conference and Exposition and The National Fiber Optic Engineers Conference, 24–28 February, San Diego, CA* (paper JThA73). Piscataway, NJ: Institute of Electrical and Electronics Engineers. doi:10.1109/OFC.2008.4528106

Niu, T., Wang, X., Chan, E.H.W., Feng, X. & Guan, B.O. (2016). Dual-polarization dual-parallel MZM and optical phase shifter based microwave photonic phase controller. *IEEE Photonics Journal*, *8*(4), 1–14.

Wang, X., Zhang, J., Ghan, E.H.W., Feng, X. & Guan, B.O. (2017). Ultra-wide bandwidth photonic microwave phase shifter with amplitude control function. *Optics Express*, *25*(3), 1–12.

Yang, P., Wei, Y. & Bai, F. (2017). A frequency-quadrupling microwave photonic phase shifter based on dual-polarization MZM. In *Asia Communications and Photonics Conference, 10–13 November, Guangzhou, Guangdong, China* (paper Su4E.3). Washington, DC: Optical Society of America. doi:10.1364/ACPC.2017.Su4E.3

Yao, J. (2009). Microwave photonics. *Journal of Lightwave Technology*, *27*(3), 314–335.

Zhu, Z., Zhao, S., Li, Y., Chen, X. & Li, X. (2015). A novel scheme for high-quality 120 GHz optical millimeter-wave generation without optical filter. *Optics & Laser Technology*, *65*, 29–35.

Frontier Research and Innovation in Optoelectronics Technology and Industry – Habib & Lewis (Eds)
© *2019 Taylor & Francis Group, London, ISBN 978-1-138-33178-5*

Rainfall interference reduction for a distributed fiber optical intrusion sensor system

Hui Zhu & Yang Zhou
The 28th Research Institute of China Electronic Technology Group Corporation, Nanjing, Jiangsu, China

ABSTRACT: In this paper, a method of rainfall interference reduction for a Distributed Fiber Optical Intrusion Sensor System (DFOISS) is proposed and verified by experiment. The DFOISS consists of two subsystems, one of which multiplexes Continuous Wave (CW) light with pulsed light of different wavelengths of 1310 and 1550 nm to detect intrusion along the sensing fiber, and the other launches CW light to obtain a rainfall signal to reduce rainfall interference. An on-site experimental system was established to test the intrusion detection and its location in a rainy environment. The results show that DFOISS can effectively reduce rainfall interference, detecting and locating intrusion when the average number of periodic backscatter signals reaches 1,000.

Keyword: rainfall interference reduction, distributed fiber optical intrusion sensor, intrusion location

1   INTRODUCTION

Distributed fiber optic sensing technology has been extensively studied and a variety of fiber optic sensing systems have been implemented to locate breakpoints and detect intrusions, such as perimeter security and pipeline monitoring (Juarez et al., 2005; Wan & Leung, 2007). However, when the sensing cable is exposed to the atmosphere, rainfall directly acts on it. Because phase modulation due to raindrop collision is mixed in with the sensor signal, it is difficult to distinguish between normal interference and rainfall effects (Mahmoud & Katsifolis, 2009). Rainfall can cause a lot of false positives, making the sensor system unreliable during the rainy season. So far, little attention has been paid to the reduction of rainfall interference in distributed fiber optic sensing systems. In the past, we have proposed rain compensation techniques, but the rate of false positives due to optical inequality is too high and does not work at the location of the invasion (Zhu et al., 2014).

In this paper, we demonstrate a rainfall interference reduction method using a Distributed Fiber Optical Intrusion Sensor System (DFOISS). This DFOISS consists of two subsystems, one of which multiplexes Continuous Wave (CW) light with pulsed light of different wavelengths of 1310 and 1550 nm to detect intrusion along the sensing fiber, and the other launches CW light that is transmitted to the rain reference fiber to obtain a rain signal. We established a DFOISS response model for rainfall and intrusion and describe the principle of rainfall disturbance reduction. The intrusion can be detected by comparing the reflected signals of the CW lamp, and the intrusion can be located by averaging a plurality of backscattering trajectory periods caused by the pulsed light under rainy conditions. An experimental system was established to conduct on-site testing of intrusion detection and location under rainy conditions.

2 SYSTEM SCHEME AND MODELING OF SIGNAL

The DFOISS with rainfall interference reduction is shown in Figure 1. It consists of two subsystems, one of which is the sensing subsystem and the other is the rainfall reference subsystem. Pulsed light of 1550 nm was injected into the sensing subsystem, and 1310 nm of CW light was injected into the two subsystems through the fiber splitter. By using two Wavelength Division Multiplexers (WDM), the pulsed light and the CW light are multiplexed. The pulsed light is attenuated, and the CW light is reflected at the end of the sensing subsystem. The CW light injected into the rainfall reference subsystem is also reflected at the end. In order to keep the CW optical paths of the two subsystems consistent, two WDMs are added to the reference subsystem and the lengths of the fibers are equal. A light receiving and processing module is used to receive light from the subsystem. The sensing fiber is placed as a sensing element at the periphery to detect intrusion, and the rain reference fiber is fixed near the perimeter and should only produce a signal when disturbed by rainfall. Intrusion can be detected by comparing the reflected signal of the CW lamp in a rainy environment.

The interferometer consists of two 3 dB fiber couplers and one long delay fiber L_d. There are two possibilities by which CW light may enter the sensing fiber, through two paths (direct or delayed) and the signal returned is exactly the same. Ultimately, three contributions appear in the output: two non-interfering signals (direct-direct and delayed-delayed) and one interference signal (delayed-direct or direct-delayed). The two non-interfering signals can be considered as background signals because they are not sensitive to intrusion. Ignoring the power of the two non-interfering signals, when rainfall and intrusion act on the sensing fiber, the reflected power of the CW light in the sensing subsystem can be described (Zhu et al., 2014) as:

$$P_s(t) = \frac{1}{16} Pe^{-\alpha(L_d + 2L_s)}[1 - \cos(\Delta\varphi_r(t) + \Delta\varphi_v(t))] \qquad (1)$$

where P is the power of the light source at 1310 nm, α is the fiber loss coefficient at 1310 nm, $\Delta\varphi_r(t) = \varphi_r(t) - \varphi_r(t + \frac{L_d}{v})$ is the phase difference between coherent paths introduced by the rainfall, and $\Delta\varphi_v(t) = \varphi_v(t) - \varphi_v(t + \frac{L_d}{v})$ is the phase difference introduced by the intrusion in the sensing subsystem when $\varphi_r(t)$ and $\varphi_v(t)$ are phase-modulated by the rainfall and intrusion.

Simultaneously, the reflected power in the rainfall reference subsystem can be described as:

$$P_r(t) = \frac{1}{16} P_1 e^{-\alpha(L_d + 2L_s)}[1 - \cos(\Delta\varphi_r(t + \tau))] \qquad (2)$$

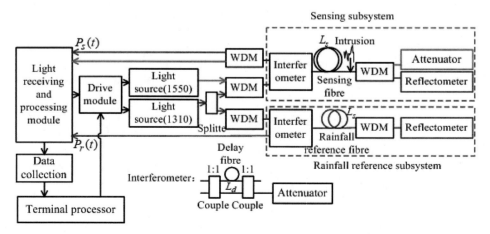

Figure 1. DFOISS with rainfall interference reduction (L_s – length of the sensing fiber and the rainfall reference fiber; L_d – length of the delay fiber).

where $\Delta\varphi_r(t+\tau)=\varphi_r(t+\tau)-\varphi_r(t+\tau-\frac{L_d}{v})$ is the phase difference between coherent paths introduced by the rainfall in the rainfall reference subsystem.

The phase differences are infinitesimal values because $\frac{L_d}{v}$ is infinitesimal, and the ratio between light powers can be derived as:

$$\frac{P_s}{P_r}=\left[\frac{\Delta\varphi_r(t)+\Delta\varphi_v(t)}{\Delta\varphi_r(t+\tau)}\right]^2 \tag{3}$$

$\Delta\varphi_r(t)$ and $\Delta\varphi_r(t+\tau)$ have the same range of values, so we can obtain:

$$\frac{\max(P_s)}{\max(P_r)}>1 \tag{4}$$

When the intrusion is applied to the sensing fiber, the maximum power of the sensing subsystem is greater than the maximum power of the rain reference subsystem. By comparing the maximum value of the reflected light power in the two subsystems, the intrusion can be detected under rainy conditions.

When an intrusion is detected, the light source at 1550 nm sets a light pulse to locate the intrusion. According to Optical Time-Domain Reflectometer (OTDR) theory, we can locate intrusions and raindrop collisions (Aoyama et al., 1981; Pan et al., 2012), but need to distinguish between the locations associated with the invasion and the rainfall. The most common location will be that of the intrusion after many pulse periods because such intrusion is continuous at one location whereas rainfall is random. By averaging multiple backscatter trajectory periods, random signals can be filtered out and intrusions can be located.

3 FIELD EXPERIMENTAL SETUP AND RESULTS

In a high-tech park in Nanjing, the capital of Jiangsu Province, China, a DFOISS to reduce rainfall interference was established for perimeter security. As shown in Figure 2, the sensing fiber was fixed on the perimeter fence, and the rain reference fiber was located below the perimeter fence and was not subject to intrusion. The maximum power of the light sources at 1310 and 1550 nm are 10 mW and 40 mW, respectively. The 1550 nm source produces a light pulse with a repetition period of 0.5 ms and a width of 1 μs. The length of the delay fiber is approximately 5 km and the length of the sensing fiber is approximately 10 km.

The output signals of the continuously reflected light in the two subsystems when it rains are shown in Figure 3. To avoid interference, the maximum voltage is the average of the three maximum values in the output signal. When the value of the sensing subsystem is 0.2

Sensing fibre

Rainfall reference fibre

Figure 2. DFOISS with rainfall interference reduction.

509

V higher than the value of the rainfall reference subsystem, we consider that the sensing subsystem has an intrusion. In Figure 3a, the maximum voltage in the sensing subsystem is 2.47 V, while in Figure 3b, the maximum voltage of the rain reference subsystem is 2.51 V. Comparing these values indicates that there is no intrusion.

Using a vibrating rod to simulate an invasion under rainfall conditions, Figure 4 shows the output signal of the reflected light. In Figure 4a, the maximum voltage of the sensing subsystem is 3.36 V. In Figure 4b, the maximum voltage of the rain reference subsystem is 2.48 V. We can conclude that the sensing fiber has been invaded.

When an intrusion is detected, the 1550 nm wavelength source starts working to locate the intrusion. Figure 5 shows the output signal of the backscattered light, averaging 1,000 pulse periods. Random noise caused by rainfall is filtered out. In Figure 5, the black line is the backscattering trajectory when there is no intrusion, and the red line is the backscattering

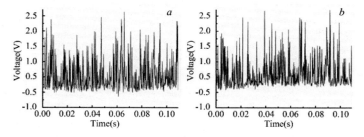

Figure 3. Output signals of the reflected light under rainy conditions: (a) in the sensing subsystem; (b) in the rainfall reference subsystem.

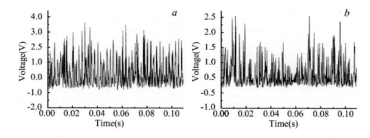

Figure 4. Output signals of the reflected light under rainy conditions and intrusion conditions: (a) in the sensing subsystem; (b) in the rainfall reference subsystem.

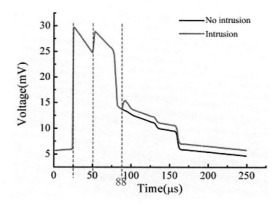

Figure 5. Output signals of the backscattered light with an average of 1,000 pulse periods.

510

trajectory with intrusion. The rise and fall of the black trace is due to the two non-interfering paths brought by the interferometer as background signals. The first and second rises correspond to the illumination position and exit of the interferometer. The third rise in the red line is the location of the invasion. In Figure 5, the time interval between the second and third rises is 38 µs and the calculated intrusion position on the sensing fiber is at 3.8 km distance.

The results show that comparing the maximum value of the reflected light power in the two subsystems is an effective way to obtain the location of the intrusion in the sensing fiber under rainy conditions. At the same time, the system's false positive rate is reduced. When an intrusion is confirmed, the use of pulsed light and an intentional increase in the average number of periodic backscatter signals can detect the intrusion position effectively, enabling a DFOISS to work in harsh weather conditions.

4 CONCLUSION

In this paper, we demonstrated a method to reduce rainfall interference in a DFOISS. Criteria for intrusion detection under rainy conditions are proposed, and the random noise caused by rainfall is filtered out by the averaging method for backscattered signals. We established an experimental system and conducted field tests under rainy conditions. The results show that a DFOISS can detect intrusion under different rainfall conditions. When the average number of periodic backscatter signals reaches 1,000, the system can effectively locate the intrusion.

REFERENCES

Aoyama, K.I., Nakagawa, K. & Itoh, T. (1981). Optical time domain reflectometry in a single-mode fiber. *IEEE Journal of Quantum Electronics*, *17*(6), 862–868.

Juarez, J.C., Maier, E.W., Choi, K.N. & Taylor, H.F. (2005). Distributed fiber-optic intrusion sensor system. *Journal of Lightwave Technology*, *23*(6), 2081–2087.

Mahmoud, S.S. & Katsifolis, J. (2009). Elimination of rain-induced nuisance alarms in distributed fiber optic perimeter intrusion detection systems. *SPIE Conference Proceedings*, *7316*, 731604. doi:10.1117/12.818096

Pan, C., Zhu, H., Yu, B., Zhu, Z. & Sun, X. (2012). Distributed optical-fiber vibration sensing system based on differential detection of differential coherent-OTDR. In *Proceedings of IEEE Sensors 2012, 28–31 October, Taipei, Taiwan* (pp. 870–872). Institute of Electrical and Electronics Engineers: Piscataway, NJ. doi:10.1109/ICSENS.2012.6411215

Wan, K.T. & Leung, C.K. (2007). Applications of a distributed fiber optic crack sensor for concrete structures. *Sensors and Actuators A: Physical*, *135*(2), 458–464.

Zhu, H., Pan, C. & Sun, X. (2014). Rainfall compensation scheme in distributed optical-fiber vibration sensor engineering system. *SPIE Conference Proceedings*, *9113*, 91130F. doi:10.1117/12.2050393.

Author index